T0155786

IFIP Advances in Information and Communication Technology 576

IFIP – The International Federation for Information Processing

IFIP was founded in 1960 under the auspices of UNESCO, following the first World Computer Congress held in Paris the previous year. A federation for societies working in information processing, IFIP's aim is two-fold: to support information processing in the countries of its members and to encourage technology transfer to developing nations. As its mission statement clearly states:

> *IFIP is the global non-profit federation of societies of ICT professionals that aims at achieving a worldwide professional and socially responsible development and application of information and communication technologies.*

IFIP is a non-profit-making organization, run almost solely by 2500 volunteers. It operates through a number of technical committees and working groups, which organize events and publications. IFIP's events range from large international open conferences to working conferences and local seminars.

The flagship event is the IFIP World Computer Congress, at which both invited and contributed papers are presented. Contributed papers are rigorously refereed and the rejection rate is high.

As with the Congress, participation in the open conferences is open to all and papers may be invited or submitted. Again, submitted papers are stringently refereed.

The working conferences are structured differently. They are usually run by a working group and attendance is generally smaller and occasionally by invitation only. Their purpose is to create an atmosphere conducive to innovation and development. Refereeing is also rigorous and papers are subjected to extensive group discussion.

Publications arising from IFIP events vary. The papers presented at the IFIP World Computer Congress and at open conferences are published as conference proceedings, while the results of the working conferences are often published as collections of selected and edited papers.

IFIP distinguishes three types of institutional membership: Country Representative Members, Members at Large, and Associate Members. The type of organization that can apply for membership is a wide variety and includes national or international societies of individual computer scientists/ICT professionals, associations or federations of such societies, government institutions/government related organizations, national or international research institutes or consortia, universities, academies of sciences, companies, national or international associations or federations of companies.

More information about this series at http://www.springer.com/series/6102

Michael Friedewald · Melek Önen ·
Eva Lievens · Stephan Krenn ·
Samuel Fricker (Eds.)

Privacy and Identity Management

Data for Better Living: AI and Privacy

14th IFIP WG 9.2, 9.6/11.7, 11.6/SIG 9.2.2
International Summer School
Windisch, Switzerland, August 19–23, 2019
Revised Selected Papers

Editors
Michael Friedewald
Fraunhofer ISI
Karlsruhe, Germany

Eva Lievens
Ghent University
Ghent, Belgium

Samuel Fricker
Fachhochschule Nordwestschweiz
Windisch, Switzerland

Melek Önen
EURECOM
Biot, France

Stephan Krenn
Austrian Institute of Technology
Vienna, Austria

ISSN 1868-4238 ISSN 1868-422X (electronic)
IFIP Advances in Information and Communication Technology
ISBN 978-3-030-42506-7 ISBN 978-3-030-42504-3 (eBook)
https://doi.org/10.1007/978-3-030-42504-3

This Springer imprint is published by the registered company Springer Nature Switzerland AG
The registered company address is: Gewerbestrasse 11, 6330 Cham, Switzerland

Preface

This volume contains the proceedings of the 14th IFIP Summer School on Privacy and Identity Management – "Data for Better Living: Artificial Intelligence and Privacy" – which took place during August 19–23, 2019, in Brugg/Windisch, Switzerland.

The 14th IFIP Summer School was a joint effort among IFIP Working Groups 9.2, 9.6/11.7, 11.6, Special Interest Group 9.2.2 in co-operation with the International Association for Cryptologic Research (IACR), and several European and national projects: the EU H2020 projects PAPAYA, CyberSec4Europe, e-SIDES, SPARTA, and the German Privacy Forum (Forum Privatheit). It was hosted and also supported by the University of Applied Sciences Northwestern Switzerland (FHNW).

This IFIP Summer School brought together more than 60 junior and senior researchers and practitioners from different parts of the world from many disciplines, including many young entrants to the field. They came to share their ideas, build a network, gain experience in presenting their research, and have the opportunity to publish a paper through these proceedings.

One of the goals of the IFIP Summer School is to encourage the publication of thorough research papers by students and emerging scholars. To this end, it had a three-phase review process for submitted papers. In the first phase, authors were invited to submit short abstracts of their work. Abstracts within the scope of the call were selected for presentation at the school. After the school, authors were encouraged to submit full papers of their work and received two to three reviews by members of the Program Committee. They were then given time to revise and resubmit their papers for inclusion in these proceedings. In total, 31 abstracts were submitted, out of which 22 were finally accepted, including the papers by Shukun Tokas, Olaf Owe, and Toktam Ramezanifarkhani on "Language-Based Mechanisms for Privacy by Design" and Teresa Anna Steiner on "Differential Privacy in Online Dating Recommendation Systems," which were judged to be the summer school's best student papers.

In addition to the submitted papers, this volume also includes reviewed papers summarizing the results of workshops and tutorials that were held at the summer school as well as papers contributed by several of the invited speakers.

We are grateful to all contributors of the summer school and especially to the Program Committee for reviewing the abstracts and papers as well as advising the authors on their revisions. Our thanks go to all supporting projects and especially to FNWH for their support in hosting the school.

January 2019

Michael Friedewald
Melek Önen
Eva Lievens
Stephan Krenn
Samuel Fricker

Organization

Program Chairs

Michael Friedewald	Fraunhofer ISI, Germany
Eva Lievens	Ghent University, Belgium
Melek Önen	EURECOM, France

General Chairs

Samuel Fricker	University of Applied Sciences Northwestern Switzerland, Switzerland
Stephan Krenn	AIT Austrian Institute of Technology GmbH, Austria

Program Committee

Petra Asprion	University of Applied Sciences Northwestern Switzerland, Switzerland
Rocco Bellanova	University of Amsterdam, The Netherlands
Felix Bieker	Unabhängiges Landeszentrum für Datenschutz Schleswig-Holstein, Germany
Michael Birnhack	Tel Aviv University, Israel
Franziska Boehm	Karlsruhe Institute of Technology, Germany
Sébastien Canard	Orange Labs, France
José M. Del Álamo	Universidad Politécnica de Madrid, Spain
Roberto Di Pietro	Hamad Bin Khalifa University, Qatar
Josep Domingo-Ferrer	Universitat Rovira i Virgili, Spain
Gerard Draper Gil	European Commission - Joint Research Centre (JRC), Italy
Kaoutar Elkhiyaoui	EURECOM, France
Orhan Ermis	EURECOM, France
Simone Fischer-Hübner	Karlstad University, Sweden
Pedro Freitas	Universidade Católica Portuguesa, Portugal
Lothar Fritsch	Karlstad University, Sweden
Gloria Gonzalez Fuster	Vrije Universiteit Brussel, Belgium
Thomas Gross	Newcastle University, UK
Marit Hansen	Unabhängiges Landeszentrum für Datenschutz Schleswig-Holstein, Germany
Meiko Jensen	Kiel University of Applied Sciences, Germany
Stefan Katzenbeisser	University of Passau, Germany
Reto Koenig	Bern University of Applied Sciences, Switzerland
Eleni Kosta	Tilburg University, The Netherlands
Joachim Meyer	Tel Aviv University, Israel

Evgeni Moyakine	University of Groningen, The Netherlands
Andreas Nautsch	EURECOM, France
Sebastian Pape	Goethe University Frankfurt, Germany
Norberto Patrignani	Politecnico of Torino, Italy
Robin Pierce	Tilburg University, The Netherlands
Jo Pierson	Vrije Universiteit Brussel, Belgium
Tobias Pulls	Karlstad University, Sweden
Charles Raab	The University of Edinburgh, UK
Kai Rannenberg	Goethe University Frankfurt, Germany
Kjetil Rommetveit	University of Bergen, Norway
Arnold Roosendaal	Privacy Company, The Netherlands
Ina Schiering	Ostfalia University of Applied Sciences, Germany
Stefan Schiffner	University of Luxembourg, Luxembourg
Daniel Slamanig	AIT Austrian Institute of Technology GmbH, Austria
Christoph Striecks	AIT Austrian Institute of Technology GmbH, Austria
Bibi Van Den Berg	Leiden University, The Netherlands
Simone Van Der Hof	Leiden University, The Netherlands
Marc van Lieshout	TNO Information and Communication Technology, The Netherlands
Frederik Zuiderveen Borgesius	Radboud University, The Netherlands
Harald Zwingelberg	Unabhängiges Landeszentrum für Datenschutz Schleswig-Holstein, Germany
Rose-Mharie Åhlfeldt	University of Skövde, Sweden

Additional Reviewers

Rasmus Dahlberg
Dara Hallinan
Eelco Herder
Dirk Kuhlmann
Thomas Lorünser
Sebastian Ramacher
Kai Samelin
Valerie Verdoodt

Contents

Invited Papers

Workshop and Tutorial Papers

Language and Privacy

Invited Papers

Privacy as Enabler of Innovation

Daniel Bachlechner[1,2(✉)], Marc van Lieshout[3], and Tjerk Timan[4]

[1] Fraunhofer Institute for Systems and Innovation Research ISI, 76139 Karlsruhe, Germany
daniel.bachlechner@isi.fraunhofer.de,
daniel.bachlechner@fraunhofer.at

[2] Fraunhofer Austria Research GmbH, 6112 Wattens, Austria

[3] iHub – Interdisciplinary Research Centre on Security, Privacy and Data Governance, Radboud University, 6525 HT Nijmegen, The Netherlands

[4] Research Department on Strategy and Policy, TNO, 6525 GA Den Haag, The Netherlands

Abstract. Privacy has long been perceived as a hindrance to innovation. It has been considered to raise costs for data governance without providing real benefits. However, the attitude of various stakeholders towards the relationship between privacy and innovation has started to change. Privacy is increasingly embraced as an enabler of innovation, given that consumer trust is central for realising businesses with data-driven products and services. In addition to building trust by demonstrating accountability in the processing of personal data, companies are increasingly using tools to protect privacy, for example in the context of data storage and archiving. More and more companies are realising that they can benefit from a proactive approach to data protection. A growing number of tools for privacy protection, and the emergence of products and services that are inherently privacy friendly indicate that the market is about to change. In this paper, we first outline what "privacy as enabler of innovation" means and then present evidence for this position. Key challenges that need to be overcome on the way towards successful privacy markets include the lack of profitability of privacy-friendly offerings, conflicts with new and existing business models, low value attached to privacy by individuals, latent cultural specificities, skill gaps and regulatory loopholes.

Keywords: Privacy · Innovation · Data protection · GDPR · Data protection by design · Fundamental rights

1 The Role of Privacy in Innovation Processes

Both in popular business and scholarly literature, the relationship between privacy and innovation is described in antagonistic terms. On the one side, evidence is presented for the detrimental consequences of having to deal with privacy in business processes, given the multitude of data sources, the various origins of these data sources and the problem of keeping track of the origins of data, let alone whether data has been collected on a legitimate basis in the first place. Hemerly, for instance, recalls the problems of having specific data be personal data [1]. Boats have an owner and as such identification

Published by Springer Nature Switzerland AG 2020
M. Friedewald et al. (Eds.): Privacy and Identity 2019, IFIP AICT 576, pp. 3–16, 2020.
https://doi.org/10.1007/978-3-030-42504-3_1

numbers of boats belong to the category of personal data. This linkage prevents innovative research in identifying the impact of specific boats and vessels on fisheries. A study by London Economics arrived at the conclusion that the introduction of the General Data Protection Regulation (GDPR) could amount to losses of 58 billon UK pounds due to failing businesses because of overtly mingling of privacy regulations with ordinary businesses [2]. Zarsky provides anecdotal evidence underscoring that a privacy-minded Europe has not been able to compete with a lesser privacy-minded USA in technology-related business innovations [3]. After having introduced five hypotheses that deal with the relationship between privacy and innovation, Zarsky uses the well-known innovation paradox to argue that the absence of clear leadership in Europe with respect to innovations based on information and communication technology (ICT) at least lends support to the statement that privacy has not led to a better position of Europe in this market. He pleads however for a nuanced view since it is very hard to convincingly demonstrate causality between different events (such as the stricter regime on privacy in Europe versus the higher rate of innovation in the USA) that are linked by a variety of factors.

On the other side, we can find utterances of business leaders who proclaim that adherence to privacy regulation results in a positive business case. One argument that can be heard is that the need to keep a register of data sources leads to greater transparency with respect to what data is kept for what purpose, amongst others resulting in less doubling of data and a stricter management of data sources within organisations [4]. A scarcely mentioned side-effect of cleaning up data is increased overall data quality, which makes data much more useful for the implementation of artificial intelligence (AI) applications [5]. Additionally, the increased need to encrypt data also contributes to greater security of data and fewer data breaches with high impact [6]. Finally, trust of consumers is enhanced when data policies are transparent, and consumers are kept informed and are asked for their consent [7].

Using different arguments and focusing on different points of view is nothing new. In reality, several of the arguments presented may be right at the same moment. So, while Europe has stricter privacy laws it still cannot be argued that this by itself will diminish the innovative capacity of Europe. We might argue, for instance, that these companies could rely upon a relatively friendly investment climate, with investors considering that investments in activities that promote current regulations is a promising investment. As we will see, however, the counter argument for this position has some truth in it as well: investors in the USA show more risk appetite and are willing to invest in specifically these kind of companies, i.e. in companies that within the USA might face a harder future than they might face in Europe but that still are able to blossom in the USA rather than in Europe just because of them being more risk prone.

Probably the most relevant driver for privacy as enabler of innovation is the institutional rearrangement that has taken place in the EU through the introduction of the GDPR. This regulation can be considered a turning point in the protection of persons with respect to the processing of their data [8]. For one, it introduces a strict accountability regime for controllers and processors. For another, its applicability overcomes the geographical limitation of "only" being relevant for the EU. Data subjects from places outside the EU still may rely on the same protective measures as inhabitants of an EU

Member State, as long as these data subjects do business with companies that are situated within the EU or are themselves in one of the EU Member States.[1] Data intensive companies that want to do business within the EU need to comply with the GDPR. This has already resulted in other countries and regions copying the approach of the GDPR in their own legislation.[2] The GDPR thus sets a number of clear indications concerning responsibilities of controllers and processors that – as we will demonstrate – paves the way for innovative approaches of dealing with personal data.

In this paper, we present an in-depth exploration of what precisely is at stake and outline ways forward. Section 2 starts by outlining the fundamental rights perspective that has become a common perspective on the relevance of privacy and data protection.[3] The section underscores the GDPR as the – temporary – outcome of a process that shows how the domain of privacy has been invaded by ICT. This invasion has led to a blurred distinction between privacy and data protection. When it comes to innovation, it is relevant to keep the focus on the interrelationship between privacy, data protection, and measures and tools that promote or hinder the development of new products and services. Section 3 deals with this topic and presents two takes on the rise and emergence of privacy markets. To that end, it outlines what should be understood by a privacy market. Section 4 continues with presenting additional empirical evidence for how privacy and innovation are interrelated focussing on key challenges and possible ways forward. Section 5, in the end, presents the conclusions that can be drawn on the basis of the presented insights and that form the present perspective on the emergence of a market for privacy.

2 Privacy and Data Protection – Two Sides of the Coin

Privacy being a fundamental right was first acknowledged in the United Nations Declaration of Human Rights (1948). The Declaration was a response to the atrocities of the Second World War [10]. It referred to a famous speech of President Roosevelt in 1941 in which he phrased four freedoms that should be safeguarded, "the freedom of speech and expression, the freedom of worship, the freedom from want, and the freedom from fear" (quoted in Morsink [10]). Article 12 of the Declaration explicitly states the freedom from arbitrary interference with privacy, family, home or correspondence. Isaiah Berlin elaborated the concept of the freedoms in two directions: a negative freedom or the absence of coercion that should safeguard civilians from arbitrary interference by public authorities in their activities and a positive freedom or self-mastery that should

---

[1] Article 3 deals with the territorial scope of the GDPR. Contrary to its predecessor, Directive 95/46/EC, the GDPR is enforced as is it is in all EU Member States, thus enabling the indication of a territorial scope (which was absent in the previous Directive).

[2] Examples are the Data Protection Bill of Kenya (http://www.ict.go.ke/wp-content/uploads/2016/04/Kenya-Data-Protection-Bill-2018-14-08-2018.pdf; last accessed: 31/10/2019) and the Californian Consumer Privacy Act. The Kenyan Data Protection Bill is not yet accepted. The Californian Consumer Act has been passed in June 2018 [9].

[3] The phrase "data protection" may cause confusion, since it is not so much the data that needs protection but rather the person to whom these data refer. Because the term has become rather established, we will use it as well, keeping in mind that it should be read differently (namely: the protection of persons with respect to the processing of their data).

enable citizens to develop themselves as autonomous persons in self-chosen directions [11]. While this perspective focuses specifically on the individual dimension of privacy, privacy has a collective, a group and a public dimension as well [12, 13]. The freedom of communication refers to the interaction between two or more persons, leveraging privacy above a merely individual level. Some have also referred to the need for safeguarding the freedom of association, which refers to the group dimension of privacy [14]. The impact of ICT on all these dimensions of privacy can be demonstrated in a rather straightforward manner. Interestingly, these days both public and private actors have considerable possibilities to infringe upon the privacy of individuals. Reference to the data crunchers that collect personal data in a large variety of different ways is sufficient to demonstrate that individual freedoms may be at stake in light of the business practices that these companies are deploying.[4] Official surveillance programmes go alongside with secret snooping and snuffing in personal data, and new ways of getting a foot in the private homes of families by voice recognition techniques such as Amazon's Alexa, Apple's Siri and Google Home, or collecting DNA samples [17, 18].

Still we can safely conclude that privacy is far from dead, notwithstanding the remark by Scott McNealy, then CEO of Sun Microsystems, made twenty years ago, of the opposite point of view[5]. Despite popular utterings of privacy being dead (meaning, privacy as a social value), privacy as a right, at least in Europe, is firmly embedded in constitutions [19]. Taking the Netherlands as an example, privacy is safeguarded in four consecutive articles in the Dutch constitution: article 10, 11, 12 and 13. Article 10 offers a safeguard for private life, including the protection of personal data. Article 11 deals with the integrity of the human body. Article 12 presents safeguards for the home, being the physical place that may not be trespassed by outsiders without permission of the landlord. Article 13 protects the secrecy of communications. Not going into depth regarding constitutional privacy protection, it suffices to point out that privacy spans many aspects of a person's life and is not only concerned with personal data. However, due to vast and rapid datafication of many if not all aspects of daily life, the role of personal data protection as a way to safeguard privacy is growing in importance.

2.1 From Privacy to Data Protection

Turning the perspective to the protection of persons with regard of the processing of their data, the OECD privacy principles play a crucial role in setting the stage. The OECD guidelines, which include the privacy principles, were formulated in 1980, and the most recent revision dates from 2013 [20]. The privacy principles are still very relevant. They cover principles concerning the limitation of the collection of data, data quality, purpose

[4] The Cambridge Analytica case is an illustrative example. Having published original work that demonstrates the potential impact of analysing "Likes" at Facebook, Cambridge Analytica became an enemy of its own business approach when using its knowledge to nudge voters in the USA during the last presidential elections in a specific direction [15]. Cadwalladr also investigated Cambridge Analytica [16].

[5] The argument that "privacy is dead" is recurring throughout the last decades, often followed by the introduction of a novel ICT paradigm. In that sense, public perception of privacy follows a wave-pattern, not dissimilar (and perhaps the inverse of) the Gartner innovation hype cycle.

specification, use limitation, security safeguards, openness, individual participation and accountability. The Council of Europe Convention 108 saw the light in 1981, and was the first European institutional framework that set out guidelines for dealing with personal data. The Convention has been revised in 2018, so as to align with the GDPR that is enforced from 25 May 2018 onwards [21]. The original Convention has been ratified by 55 countries [22]. The Convention was the first legally binding instrument in the field of data processing. Interestingly, the principles of the GDPR are still very much alike the principles that are present in the original Convention; both refer to the rights of data subjects and the obligations of controllers and processors, the tasks and responsibilities of supervisory authorities and the rules to abide in case of transborder data flows.

An obvious distinction that can be made between the legal instruments dealing with privacy and legal instruments dealing with data protection is the focus of the instruments. With privacy, the focus is on the substantive core of what is at stake. Privacy concerns an infringement of personal or relational space either in physical or in virtual terms. This infringement results in a form of damage, which can be physical, financial, reputational or psychological such as fear of being targeted, of being suspected, of not feeling safe in one's own surroundings anymore. With data protection, the focus is on procedures followed and taken into account, such as having performed a risk assessment, or having informed data subjects over the data that is collected. Contrary to the previous Data Protection Directive (and its implementations in national laws), the GDPR is not solely focused on procedural elements but has some substantive aspects as well. This is for instance eminent in the risk approach that is part of the GDPR. Though one might stipulate the risk assessment as described in the GDPR is a procedural one – requesting the need to identify risks and to take necessary precautionary measures – part of the GDPR is the explanation of what should be considered high risks and thus in need of a full-fledged data protection impact assessment. This clearly is more than just a procedural obligation. The same goes for a novel aspect within the GDPR concerning data protection by design and by default. Again, one might argue that the GPDR itself does not offer much support to understanding how data protection by design (or by default) should be interpreted. Along the present indications in the GDPR concerning data minimisation and pseudonymisation, the European Data Protection Board (EDPB) has released a draft version of its guidelines on data protection by design and by default.[6]

The GDPR is large, far-reaching and includes substantive and procedural elements, which may give rise to innovative practices. A drawback is however the present uncertainty on how to understand specific elements from the GDPR. The uncertainty about how

[6] The stipulations of the EDPB concerning data protection by design and by default (https://edpb.europa.eu/our-work-tools/public-consultations-art-704/2019/guidelines-42019-article-25-data-protection-design_en; last accessed: 29/11/2019) are in line with international developments taking place in standardisation organisations such as ISO. Work on privacy management issues is performed by ISO/IEC JTC 1/SC 27/WG 5. Other organisations, such as the International Privacy Engineering Network, which has been established by the European Data Protection Supervisor, are also working on templates and guidelines to help organisations in practical tooling for integrating data protection by design and by default (https://edps.europa.eu/data-protection/our-work/subjects/ipen_en; last accessed: 23/10/2019). They all contribute to the substantive kernel of what should be considered legitimate implementations of data protection by design and by default.

for instance data protection by default should be interpreted limits interest in developing products or services that may help in implementing data protection by default.

2.2 Companies Active in Data Protection

For sure, the GDPR has given an impetus to companies offering products and services to organisations that need to comply with the GDPR. The offers relate to practical issues such as keeping a registry of data processing activities, keeping track of reception of requests of data subjects, organising procedures in case of identified data breaches, etc. Some companies offer more advanced tooling that enables organisations to track the maturity of their approach, to install data governance procedures, to have advanced access management and logging protocols, etc.

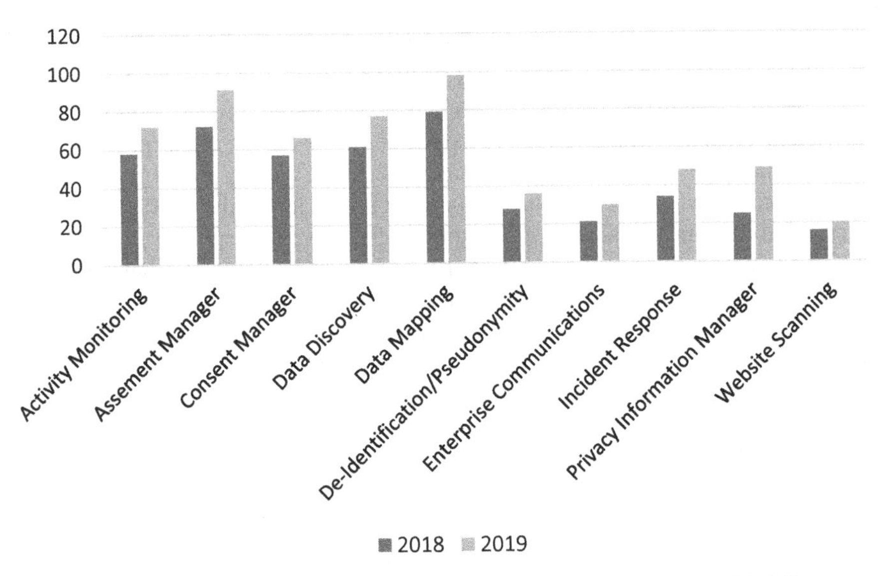

Fig. 1. Number of vendors active per product category 2018–2019 [23, 24].

The International Association of Privacy Professionals (IAPP) produces annual reports (starting in 2018) in which it presents an overview of vendors active on the market of data protection products and services. The reports are based on self-selection, i.e. vendors were offered the opportunity to provide information about their activities in the report.

The overall growth of companies offering data protection products and services was around 20% between 2018 and 2019. Figures presented in older reports show that the rise was rather linear over the past few years [24]. Figure 1 illustrates the absolute increase in vendors per category over the last few months. Although the figures in these reports should be interpreted with caution (given that they are based on self-selection), they clearly indicate a flourishing and maturing market for data protection products and services. We will turn to these arguments in the following section.

3 Two Takes on Privacy Markets

When it comes to innovation, it is relevant to keep the focus on the interrelationship between privacy, data protection, and measures and tools that promote or hinder the development of new data-driven products and services. With advanced technologies like AI taking off, the need for data is greater than ever and much of it comes from consumers.

The GDPR and other modern laws dealing with privacy and data protection such as the California Consumer Privacy Act become an operational reality and catch the attention of corporate leadership. Countries and jurisdictions around the world are increasingly adopting their own privacy-focused regulations. China and Russia are already installing local data residency requirements for citizens. At the same time, awareness for privacy is growing on the individual and the organisational level. This is not only the result of laws in general and new data breach notification obligations in particular but also of growing public interest in and increased sensitivity for the topic. People are beginning to understand that their data is vulnerable to disclosure without their consent. They demand organisations to take the accountability for securing their data and to comply with the laws. Privacy and security breaches pose an augmenting threat not only to users but also to system operators and designers.

The recent developments affect the market in two fundamental ways. First, the number of companies that offer privacy tools is on the rise. This is an emerging subsector of the industry that is related to the data storage sector and the security sector. Second, companies in all sectors, especially those where personal data plays a key role, not only comply with the law but go beyond it to distinguish themselves positively from their competitors.

3.1 Tools Focusing on Privacy Protection

The rising demand for accountability forces organisations to adopt privacy technology. As stated in the previous section, the privacy technology vendor market continues to mature as more and more organisations adopt tools that help automate and streamline privacy functions [24]. More specifically, the global market for privacy technology (the study uses the term "privacy management software") is accounted to 521.3 million US dollars (107.9 million in Europe) in 2018 and is expected to grow at a compound annual growth rate (CAGR) of 13.7% during the forecast period 2019–2027, to account to 1,585.8 million US dollars by 2027 [25]. Another study even expects a CAGR of 33.1% for the period 2019–2025. This study accounts the 2018 global market to 450 million US dollars [26]. By 2025, the study authors expect the market to account to 3,289 million US dollars.

Although the expected growth rates differ considerably, the studies agree that the global market will grow strongly in the coming years. Europe is anticipated to be the fastest growing market and North America to be the highest revenue contributor. Companies are emerging to capitalise on the growing demand for data privacy tools, both for regulatory compliance and consumer peace of mind [27]. Several such companies including, for instance, OneTrust, TrustArc, Privitar and BigID have raised sizable sums of cash for various privacy, data protection and compliance offers.[7]

The market is fragmented with the presence of several industry sectors and the lack of international coordination regarding regulations. Competition is expected to intensify in the coming years. One factor that will affect the dynamics is how international companies – that go beyond paying lip service to the laws – approach privacy. Generally, they pursue one of two approaches. They apply the tightest standards on a global basis or assess the risks and act by region. Apple stated that it is modifying its products to comply with the GDPR, and the modification will be worldwide for everyone; Facebook, in contrast, stated that they might implement extra protections for Europeans to comply with the GDPR, which will not be rolled out to people in other jurisdictions [29]. In the medium run, it is possible that the EU will become an exporter of norms that have the potential to lead to technological changes globally. However, it is also possible that certain cutting-edge technologies may not access, or have delayed access to the EU market due to regulations [30].

3.2 Privacy-Friendly Products and Services

In a survey conducted in 2018, 75% of adults stated that they will not buy a product from a company – no matter how good its products are – if they do not trust the company to protect their data [31]. Companies compete for information and derive revenues both from purchases as well as, in a secondary market, from disclosing consumer information. Casadesus-Masanell and Hervas-Drane expect consumer awareness of disclosure practices and familiarity with the implications of the disclosure of personal data to increase [32]. They found competition to drive the provision of services with a low level of consumer information disclosure (i.e., a high level of privacy).

Protecting privacy met with considerable public interest with the refusal of Apple to grant US law enforcement backdoor access to the iPhone of a known terrorist. Indeed, headlines about the matter directly addressed the notion of using privacy as a strategy by referring to the government prosecutors' quote that Apple's refusal "appears to be based on its concern for its business model and public brand marketing strategy" [33]. Beyond Apple, which seems to be willing to sacrifice some profit for the sake of privacy to bolster its image as a company that protects consumers [34], companies increasingly compete on the basis of strong privacy protections. The term "privacy as a strategy" is used to refer to the phenomenon of using data protection approaches for competitive differentiation [35].

Using privacy as a strategy remains viable as long as companies compete in markets where measures for privacy protection can be differentiated and are valued by consumers. Higher competition intensity in the marketplace does not necessarily improve privacy when consumers exhibit low willingness to pay. Therefore, privacy-friendly products and services remain niche offers. Products and services such as search engines, e-mail clients, web browsers or messenger services (e.g., DuckDuckGo, ProtonMail, Tor, Signal) that

[7] End of October 2019, another announcement was made by Very Good Security, a US-based company, that was able to raise 35 million US dollars for developing tools and services that help protecting consumer privacy [28].

protect the privacy of users usually do not strive for economic success, but are willing to sacrifice business profits in turn of remaining close to privacy idealism.

4 Challenges and Ways Forward

Privacy and ICT innovation are interrelated. While the market for privacy protection tools is expected to continue to grow steadily over the next decade, the market for privacy-friendly products and services remains a niche market for the time being. Although awareness for privacy is growing, most decision makers are not willing to invest more than necessary into making products and services more privacy friendly or to use privacy-friendly products and services. Privacy and data protection have the potential to enable innovation but several challenges need to be addressed in order to make privacy markets a success across the spectrum. Key challenges are lack of profitability of privacy-friendly offerings, conflicts with new and existing business models, low value attached to privacy by individuals, latent cultural specificities, skill gaps and regulatory loopholes. These challenges, which make matching privacy harms with adequate technical tools extremely difficult, hamper the scaling and adoption of standard privacy-preserving technologies, and are reinforced by the multitude of sectoral and regulatory requirements, are outlined in the following paragraphs.

Making products and services privacy friendly leads to additional costs for both developers and users. Acquisti et al. clearly state that there are costs associated with the act of protecting privacy [36]. These costs must be offset by the expected benefits. Only then will a higher level of privacy protection make sense from an economic point of view. Examples for costs that may be incurred include costs for hardware and software as well as costs caused by user inconvenience [37]. In addition to regulatory compliance, the benefits may include reduced employee need to deal with privacy breaches (e.g., for dispute resolution) and improved reputation. There is no evidence that privacy-friendly products and services will lead to increased sales for developers or justify significantly higher prices.

Data-driven innovation may involve methods and usage patterns that neither the entity collecting the data nor the data subject considered or even imagined at the time of data collection. Putting privacy principles such as purpose limitation or data minimisation into practice may thus be in conflict with current or desired business models [38]. Moreover, conflicts may arise from different treatments of special categories of data and legal rules governing decision-making processes. Closely related to business model conflicts is the trade-off between privacy protection and the utility of data. It was found that increased data protection limits flexibility and innovation in contexts such as health care or smart cities [39, 40]. One should however be cautious in taking this argument strictly at face value. In the Netherlands, health data scientists objected to the strict rules they had to follow, complaining that this would curtail innovation. However, this approach met with opposition, accusing the data scientists of behaving irresponsibly and having insufficient eye for the potential to use the data within the limitations of the regulatory framework.[8] Moreover, carelessness, or recklessness in dealing with personal data increasingly adds to reputational damage and lack of trust, indirectly also leading to a slowing down of innovation, especially if seemingly obscure personal data processing practises are being exposed [41].

Individuals have the potential to exert significant pressure on actors in the data value chain. However, it was found that where there is a privacy difference between companies, the slightly cheaper but less privacy-friendly company typically obtains a greater market share [42]. Moreover, individuals do not always act in a fully rational way in situations where their privacy is affected. Therefore, privacy valuations cannot be precisely estimated [43]. Privacy concerns and expectations are remarkably context-dependent and very difficult to predict. For instance, people perceive losses differently than gains and are more willing to prevent a loss than achieve a similar gain, are risk averse, tend to overvalue immediate rewards and undervalue long-term rewards, tend to mimic behaviour shown by predecessors and behave differently in the absence of real choices [44].

Acknowledging that privacy preferences and practices vary among nations and regions is important. A universal regulatory approach to privacy would ignore cultural and societal differences. Millberg et al. state clearly that "one size does not fit all" with respect to regulatory implementation [45]. Cultural values can influence people's privacy perceptions such that countries with tighter privacy regulations tend to experience fewer privacy problems [46]. Bellman et al. found that cultural values have an impact on the extent to which errors in databases and unauthorised secondary use raise privacy concerns [47]. Opinions differ as to whether data-driven innovation leads rather to de-individualisation and discrimination, or to personalisation. Van Wel and Royakkers, for instance, state that anonymous profiling could be harmful, and lead to possible discrimination and de-individualisation [48].

Adapting to a new mindset seems to be necessary as data has become a strategic business asset and privacy a threatened value. Failures in data security and governance regularly create public embarrassments for companies [49]. Today, according to Miller, those in charge of data must have skills ranging from math and statistics, machine learning, decision management and computer science to data ethics, law and information security [50]. Several of these skills are essential for developers to make sure privacy-preserving features are properly integrated into products and services as well as daily business practices. According to Kshetri, data-driven products and services are likely to affect the welfare of unsophisticated, vulnerable and technologically inexperienced users more negatively than others [51]. Digital literacy is a key skill in the age of big data [52].

The flexible interpretation of privacy and data protection is both a blessing and a curse for practitioners. Specific rules for the protection of special categories of data (i.e., sensitive data), which are included in the GDPR and other regulations, are embraced to a different extent by professionals. Some healthcare professionals, for instance, have seen

[8] Financieel Dagblad, "Medisch onderzoek in het gedrang door strenge privacyregels", 23/09/2019. In response, the Minister of Health released a letter indicating the opportunities that researchers may refer to in having health data processed for research purposes. Minister VWS. "Reactie Artikel FD over secundair gebruik data." 1587082-195476-DICIO. 04/10/2019.

strict privacy protection as an impediment for epidemiological research [53]. Regulations are also considered to have gaps and loopholes with respect to inferred data. The vast amount of data sources and their linkability points towards a direction when everybody would be identifiable through various data relations and as Purtova states data protection law would apply to everything [54].

Within the scope of the concluding session of the 14th IFIP Summer School on Privacy and Identity Management, an expert panel discussed ways forward towards privacy-friendly big data. Much attention was paid to the relationship between privacy and innovation. There was consensus among the experts that an increase in the funding available for related research and innovation is essential. While significant progress was made over the last couple of years with respect to privacy-preserving technologies, societal and economic aspects as well as the interplay of technologies and regulations on the one side and people and organisations on the other side is not yet sufficiently understood. Intensified research and innovation activities, however, are not enough. The experts also agreed that measures must be taken to support the transfer of results from the laboratories into practice. Relevant measures, mentioned by the experts, include involving a wide variety of stakeholders in research and innovation activities and promoting standardisation. It was considered as important that success stories and good practices are shared. However, it must be made sure that the specifics of different application contexts are adequately taken into account.

5 Conclusions

Flourishing start-ups offering tools for privacy protection and an increasing availability of products and services that are inherently privacy friendly show that privacy and data protection have the potential to be enablers of innovation. However, to make privacy markets a success across the spectrum, not only challenges related to profitability and business models need to be overcome, but also challenges related to the individual valuation of privacy, cultural differences, skills and regulations. Consumer trust is central for realising businesses with data-driven products and services. Therefore, being able to demonstrate accountability in processing personal data as well as security and transparency in data archiving and storage are essential for businesses. A proactive privacy approach, which means that the legal framework is not played down but taken seriously, makes this much easier.

We are convinced that responsible use of data and being innovative fit together well; it is not necessary to choose one or the other. In the short term, taking not only laws fully into account but also additional measures to protect privacy may be more of an effort, but it will pay off in the medium and long term. Increased research and innovation and the exchange of good practices will encourage a change of mindset. This perspective was shared by the panel of experts at the IFIP Summer School. Concerning privacy as an innovation opportunity, Hasselbalch and Tranberg compared data ethics with being eco-friendly [55], "Being eco-friendly has become an investor demand, a legal requirement, a thriving market and a clear competitive advantage. Data ethics will develop similarly – just much faster".

Acknowledgements. The research leading to the presented results has received funding from the European Union's Horizon 2020 research and innovation programme under grant agreements No. 731873 and No. 732630.

References

1. Hemerly, J.: Public policy considerations for data-driven innovation. Computer **46**(6), 25–31 (2013)
2. London Economics: Study on the Economic Benefits of Privacy-Enhancing Technologies (PETs) (2010). https://londoneconomics.co.uk/wp-content/uploads/2011/09/17-Study-on-the-economic-benefits-of-privacy-enhancing-technologies-PETs.pdf
3. Zarsky, T.Z.: The privacy-innovation conundrum. Lewis Clark Law Rev. **19**(1), 115–168 (2015)
4. CISCO: Maximizing the Value of your Data Privacy Investments. Data Privacy Benchmark Study (2019). https://www.cisco.com/c/dam/en_us/about/doing_business/trust-center/docs/dpbs-2019.pdf
5. Dhanda, R.: Data Privacy Regulations' Implications on AI. https://securityboulevard.com/2019/09/the-implications-of-data-privacy-regulations-on-ai/. Accessed 29 Nov 2019
6. Ponemon Institute: Cost of a Data Breach Report 2019 (2019). https://www.ibm.com/downloads/cas/ZBZLY7KL
7. The Harris Poll: IBM Survey Reveals Consumers Want Businesses to Do More to Actively Protect Their Data. https://theharrispoll.com/ibm-survey-reveals-consumers-want-businesses-to-do-more-to-actively-protect-their-data/. Accessed 31 Oct 2019
8. Fleck, M.: How GDPR is Unintentionally Driving the Next Decade of Technology. https://www.securityweek.com/how-gdpr-unintentionally-driving-next-decade-technology. Accessed 27 Oct 2019
9. Kelly, H.: California passes strictest online privacy law in the country. https://money.cnn.com/2018/06/28/technology/california-consumer-privacy-act/index.html. Accessed 31 Oct 2019
10. Morsink, J.: The Universal Declaration of Human Rights. Origins, Drafting, and Intent. University of Pennsylvania Press, Philadelphia (2009)
11. Berlin, I.: Two concepts of liberty. In: Berlin, I. (ed.) Four Essays on Liberty, pp. 118–172. Oxford University Press, Oxford (1969)
12. Westin, A.F.: Privacy and Freedom. Simon & Schuster, New York (1967)
13. Bennett, C.J., Raab, C.D.: The Governance of Privacy. Policy Instruments in Global Perspective. MIT Press, Cambridge (2006)
14. Finn, R.L., Wright, D., Friedewald, M.: Seven types of privacy. In: Gutwirth, S., Leenes, R., de Hert, P., Poullet, Y. (eds.) European Data Protection. Coming of Age, pp. 3–32. Springer, New York (2013). https://doi.org/10.1007/978-94-007-5170-5_1
15. Laterza, V.: Cambridge Analytica, independent research and the national interest. Anthropol. Today **34**(3), 1–2 (2018)
16. Cadwalladr, C.: The Cambridge Analytica files. 'I made Steve Bannon's psychological warfare tool': meet the data war whistleblower. https://www.theguardian.com/news/2018/mar/17/data-war-whistleblower-christopher-wylie-faceook-nix-bannon-trump. Accessed 31 Oct 2019
17. Nagenborg, M.H.: Hidden in plain sight. In: Timan, T., Newell, B.C., Koops, B.-J. (eds.) Privacy in Public Space. Conceptual and Regulatory Challenges, pp. 47–63. Edward Elgar, Northampton (2017)

18. Scherr, A.E.: Privacy in public spaces. The problem of out-of-body DNA. In: Timan, T., Newell, B.C., Koops, B.-J. (eds.) Privacy in Public Space. Conceptual and Regulatory Challenges, pp. 211–241. Edward Elgar, Northampton (2017)
19. Koops, B.-J., Newell, B.C., Timan, T., Škorvánek, I., Chokrevski, T., Galič, M.: A typology of privacy. Univ. Pennsylvania J. Int. Law **38**(4), 483–575 (2017)
20. OECD: The OECD Privacy Framework (2013). https://www.oecd.org/sti/ieconomy/oecd_privacy_framework.pdf
21. Council of Europe: Convention 108+. Convention for the Protection of Individuals with regard to the Processing of Personal Data (2018). https://rm.coe.int/convention-108-convention-for-the-protection-of-individuals-with-regar/16808b36f1
22. Council of Europe: Chart of signatures and ratifications of Treaty 108. https://www.coe.int/en/web/conventions/full-list/-/conventions/treaty/108/signatures?p_auth=jHKJqtBd. Accessed 30 Nov 2019
23. IAPP: 2018 Privacy Tech Vendor Report (2018). https://iapp.org/media/pdf/resource_center/2018-Privacy-Tech-Vendor-Report.pdf
24. IAPP: 2019 Privacy Tech Vendor Report (2019). https://iapp.org/media/pdf/resource_center/2019TechVendorReport.pdf
25. The Insight Partners: Privacy Management Software Market to 2027. https://www.theinsightpartners.com/reports/privacy-management-software-market/?HK+MD. Accessed 30 Nov 2019
26. Market Study Report: Global Privacy Management Software Market Size, Status and Forecast 2019-2025. https://www.marketstudyreport.com/reports/global-privacy-management-software-market-size-status-and-forecast-2019-2025. Accessed 09 Dec 2019
27. Sawers, P.: 5 data privacy startups cashing in on GDPR. https://venturebeat.com/2019/07/23/5-data-privacy-startups-cashing-in-on-gdpr/. Accessed 27 Oct 2019
28. Sawers, P.: Very Good Security raises $35 million to protect companies' private customer data. https://venturebeat.com/2019/10/24/very-good-security-raises-35-million-to-protect-companies-private-customer-data/. Accessed 31 Oct 2019
29. Hern, A.: What is GDPR and how will it affect you? https://www.theguardian.com/technology/2018/may/21/what-is-gdpr-and-how-will-it-affect-you. Accessed 27 Oct 2019
30. Bachlechner, D., La Fors, K., Sears, A.M.: The role of privacy-preserving technologies in the age of big data. In: Proceedings of the 13th Pre-ICIS Workshop on Information Security and Privacy (WISP 2018). AIS, Atlanta (2018)
31. IBM News Room: New Survey Finds Deep Consumer Anxiety over Data Privacy and Security. https://newsroom.ibm.com/2018-04-15-New-Survey-Finds-Deep-Consumer-Anxiety-over-Data-Privacy-and-Security. Accessed 30 Nov 2019
32. Casadesus-Masanell, R., Hervas-Drane, A.: Competing with privacy. Manage. Sci. **61**(1), 229–246 (2015)
33. Lichtblau, E., Apuzzo, M.: Justice Department Calls Apple's Refusal to Unlock iPhone a 'Marketing Strategy'. https://www.nytimes.com/2016/02/20/business/justice-department-calls-apples-refusal-to-unlock-iphone-a-marketing-strategy.html. Accessed 27 Oct 2019
34. Love, J.: Apple 'privacy czars' grapple with internal conflicts over user data. https://www.reuters.com/article/us-apple-encryption-privacy-insight-idUSKCN0WN0BO. Accessed 27 Oct 2019
35. Martin, K.D., Murphy, P.E.: The role of data privacy in marketing. J. Acad. Mark. Sci. **45**(2), 135–155 (2017)
36. Acquisti, A., Taylor, C., Wagman, L.: The economics of privacy. J. Econ. Lit. **54**(2), 442–492 (2016)
37. Khokhar, R.H., Chen, R., Fung, B.C.M., Lui, S.M.: Quantifying the costs and benefits of privacy-preserving health data publishing. J. Biomed. Inform. **50**, 107–121 (2014)

38. Zarsky, T.Z.: Incompatible: the GDPR in the age of big data. Seton Hall Law Rev. **47**(4), 995–1020 (2017)
39. Iyengar, A., Kundu, A., Pallis, G.: Healthcare informatics and privacy. IEEE Internet Comput. **22**(2), 29–31 (2018)
40. Mazhelis, O., Hamalainen, A., Asp, T., Tyrvainen, P.: Towards enabling privacy preserving smart city apps. In: Proceedings of the 2016 IEEE International Smart Cities Conference (ISC2), pp. 1–7. IEEE, Piscataway (2016)
41. Pilkington, E.: Google's secret cache of medical data includes names and full details of millions – whistleblower. https://www.theguardian.com/technology/2019/nov/12/google-medical-data-project-nightingale-secret-transfer-us-health-information. Accessed 29 Nov 2019
42. Jentzsch, N., Preibusch, S., Harasser, A.: Study on Monetising Privacy. An Economic Model for Pricing Personal Information (2012). https://www.enisa.europa.eu/publications/monetising-privacy/at_download/fullReport
43. Acquisti, A., John, L.K., Loewenstein, G.: What is privacy worth? J. Legal Stud. **42**(2), 249–274 (2013)
44. van Lieshout, M.: The value of personal data. In: Camenisch, J., Fischer-Hübner, S., Hansen, M. (eds.) Privacy and Identity 2014. IAICT, vol. 457, pp. 26–38. Springer, Cham (2015). https://doi.org/10.1007/978-3-319-18621-4_3
45. Milberg, S.J., Smith, H.J., Burke, S.J.: Information privacy. Corporate management and national regulation. Organ. Sci. **11**(1), 35–57 (2000)
46. Dolnicar, S., Jordaan, Y.: A market-oriented approach to responsibly managing information privacy concerns in direct marketing. J. Advert. **36**(2), 123–149 (2007)
47. Bellman, S., Johnson, E.J., Kobrin, S.J., Lohse, G.L.: International differences in information privacy concerns. A global survey of consumers. Inf. Soc. **20**(5), 313–324 (2004)
48. van Wel, L., Royakkers, L.: Ethical issues in web data mining. Ethics Inf. Technol. **6**(2), 129–140 (2004)
49. Duhigg, C.: How Companies Learn Your Secrets. https://www.nytimes.com/2012/02/19/magazine/shopping-habits.html. Accessed 27 Oct 2019
50. Miller, S.: Collaborative approaches needed to close the big data skills gap. JOD **3**(1), 26 (2014)
51. Kshetri, N.: Big data's impact on privacy, security and consumer welfare. Telecommun. Policy **38**(11), 1134–1145 (2014)
52. Segura Anaya, L.H., Alsadoon, A., Costadopoulos, N., Prasad, P.W.C.: Ethical implications of user perceptions of wearable devices. Sci. Eng. Ethics **24**(1), 1–28 (2018)
53. Nyrén, O., Stenbeck, M., Grönberg, H.: The European Parliament proposal for the new EU General Data Protection Regulation may severely restrict European epidemiological research. Eur. J. Epidemiol. **29**(4), 227–230 (2014)
54. Purtova, N.: The law of everything. Broad concept of personal data and future of EU data protection law. Law Innov. Technol. **10**(1), 40–81 (2018)
55. Hasselbalch, G., Tranberg, P.: Data Ethics. The New Competitive Advantage. Valby PubliShare, Copenhagen (2016)

Fair Enough? On (Avoiding) Bias in Data, Algorithms and Decisions

Francien Dechesne(✉)

eLaw Center for Law and Digital Technologies, Leiden University Law School, Steenschuur 25, 2311 ES Leiden, The Netherlands
f.dechesne@law.leidenuniv.nl

Abstract. This contribution explores bias in automated decision systems from a conceptual, (socio-)technical and normative perspective. In particular, it discusses the role of computational methods and mathematical models when striving for "fairness" of decisions involving such systems.

Keywords: Bias · Data analytics · Algorithmic decision systems · Fairness

1 Introduction

This contribution reflects on the discussion of fairness in so-called algorithmic decisions from a foundational and interdisciplinary perspective, based on a few of my favourite readings. It is a write-up of the keynote with the same title, delivered on the first day of the IFIP Summerschool. With the title, I have made a deliberate choice to highlight the following notions: bias, data, algorithms, and decisions. Data refers to the input as a possible point for intervention, preprocessing. Algorithms to the computational processing of the data. Decisions to the outcome - but as I hope to make clear: not just the outcome of the computational process.

If you are a scholar with a budding interest in the debate on fairness of algorithmic decision systems, you may search the internet for "What is bias?", and be surprised to find a tutorial titled: "All about Fabric Bias - Grain vs Bias and Cutting on the Bias."[1] Indeed, the third of four meanings on Dictionary.com explains this:

1. A particular tendency, trend, inclination, feeling, or opinion, especially one that is preconceived or unreasoned: illegal bias against older job applicants; the magazine's bias toward art rather than photography; our strong bias in favour of the idea.

[1] Cf. https://www.cucicucicoo.com/2017/02/fabric-bias-vs-grain/ - last accessed 3 Nov. 2019.

Published by Springer Nature Switzerland AG 2020
M. Friedewald et al. (Eds.): Privacy and Identity 2019, IFIP AICT 576, pp. 17–26, 2020.
https://doi.org/10.1007/978-3-030-42504-3_2

2. Unreasonably hostile feelings or opinions about a social group; prejudice: accusations of racial bias.
3. An oblique or diagonal line of direction, especially across a woven fabric.
4. Statistics. A systematic as opposed to a random distortion of a statistic as a result of sampling procedure.

What we can see in this definition is that essentially, bias is a deviation or inclination away from a certain standard, that can be either objective (as in meanings 3 and 4) or a matter of judgment (meanings 1 and 2). The relevant meanings of bias in the debate about so-called *algorithmic decisions* have to do both with systematic inclinations as in the fourth meaning, and with the judgments of meanings 1 and 2. Let us try to understand how they are connected.

2 Abstract Objectivity and Human Construction

There is a paradox in the way we talk about Artificial Intelligence, as something separate and somehow opposite from us. Doing so is attractive as it gives room for hope that it can deal with problems we cannot solve without us having to think about it. But at the same time, the autonomy that we attribute to it, feeds the fear that AI will soon figure out that human kind is the problem - so needs to be controlled. The fear, in short, that by granting AI too much autonomy, it will deprive us from ours.

We may recognise this fear in the discussion about so-called "algorithmic decision systems". Over the past half century we have both figured out that humans operate with bounded rationality (cf. the seminal work of [1]) and developed *universal machines* that operate strictly according to computational logics. So while we are bound to act according to biases and prejudices, which leads to suboptimal and/or unfair decisions, the machines are our hope to realise the modern ideal of rational actors. We hope the *artificial* part of AI can free us from our harmful irrationalities and prejudices.

"It is all human construction" said philosopher Maxim Februari in the 2017 Godwin lecture about the influence of digital technologies (in particular Big Data) on society [2]. Not only is AI a human creation, it is created to work in and for our human society, another human construction. The data we gather about behaviour, emotions, political views, cultural or ethnical groups, etc. are not merely things that are *given* to us (which is what *'data'* means in Latin); we constructed that social reality. Natural language is a great example of this, and we will get back to it later.

When algorithmic systems seem to be external to us, and objective, it is easy for us to believe that they -unlike us- have access to some external objective reality. This would make them suitable to be objective in their "decisions", without bias, so "objectively fair". The point here is that we have to be realistic however: algorithmic systems, the data we use in them, the interpretation of the outcome into a decision, are all in some way or another human/societal constructions, as is a normative notion like *fairness*. We should therefore look at questions of fairness through a socio-technical lens, acknowledging that the formal and computational gets meaning, including normative load, from the social context of application. But as we will see: formal and computational methods can also help in better understanding the problems around fairness, and contribute significantly to possible ways of dealing with them.

3 How Is Bias Relevant in Computational Technologies?

Although not exactly the start, the Propublica discussion of the COMPAS recidivism scoring system [3] can be seen as a high profile catalyst for the debate on possible discriminatory effects of the application of data-analytics in all kinds of societal decision making. This case is interesting because the academic discussion of it has shown that mathematical tools may not *automatically* remove bias from our judgments. But it is particularly interesting because the discussion has demonstrated mathematically that different mathematical measures for fairness give different assessments [4] - and that some combinations are actually inconsistent, meaning that we cannot have them all.

Why would we even look to mathematics to help achieve fairness? Well, simply said, one could say that fair is treating equal cases equally, or similar cases similarly. This requires a theory that says when two cases can be considered the same, and mathematics works with abstract concepts like equivalence classes, or metric spaces which allow to define measures that indicate how close or far apart two different cases are in a certain sense. Such metric spaces play a central role in statistics and machine learning as well.

But still, while mathematics offers such tools, it will not tell you which measure to choose, which parameters are most important in determining the distance between cases, what characteristics are most salient to give a good representation of the issue in the abstract model you build of it. Neither does it tell you what "fair treatment" substantively means. Fairness is a normative concept. We can abstract it into maths, but only after deciding what type of "equal treatment" we find most appropriate. For example (see Fig. 1), when dividing limited resources, do give each individual the same amount (equality)? Or do we take into account what people actually need to achieve the same outcome (equity)? But we could also look for a way to reshape the situation so that the problem disappears. In the picture, for example: what if we removed the fence?

What I think the picture in Fig. 1 nicely illustrates, is Melvin Kranzberg's famous quote in the presidential address to the Society for the History of Technology [6]:

"Technology is neither good nor bad, nor is it neutral."

It points to the inherent interaction between societal constructions, and technical ones. How is technology not neutral? To understand this, I find Don Ihde's notion of technological *mediation* a useful one: we perceive and construct our understanding of the world partially through the tools that we have at hand. It changes how we perceive the world as it *is*, as it *can be* (what we consider

Fig. 1. A visualisation of two arguably fair, but incompatible ways of distributions of limited resources: equality (left) and equity (right). This visualisation went viral: [5].

to be possible) and also: as it *should be* (what we consider to be desirable or necessary) [7].

If we go back and think about AI in this way, it influences how we think about what is natural, what is real, what is human, what is fair... etc. This in turn influences how we design solutions for what we see as problems. For the particular case of "algorithmic decisions" I would like to recommend two books that elaborate how this transformation unfolds of how we perceive the world, what we see as possible and what we think should be the case.

The first one is legal scholar Harcourt's *Against Prediction* [8]. Rather than speaking in terms of AI or algorithms, it discusses the effect of relying on statistical methods, correlations, in the judicial practice and the idea these methods are effective in preventing crime. In the book, Harcourt develops three main critiques. Firstly, the over-reliance on rational action theory in claims about the effectivity of these methods in reducing crime. Secondly, the so-called *ratchet effects*: disproportionate policing on situations that come out of the predictions as *high-risk* will increase statistical bias, and may effectively increase actual crime by leaving other areas under-policed (and crime there under-registered). It is important to also acknowledge that such practices may have an impact on the community through the transformations of the social meaning of police conduct, which affect social norms on priorities in law enforcement. In that way, thirdly, technical advances implicitly reshape the justice system.

Another book comes from a mathematician, who, working in the commercial practice of data-analytics, saw how mathematical tools can be applied with fundamental societal impact: O'Neil's *Weapons of Math Destruction* [9]. In the book she identifies three main issues with data-driven (risk) scoring mechanisms: the feedback loops they establish in whatever they measure (cf. Harcourt's *ratchet effects*), their scale of application across contexts, and their opacity. The opacity not only comes from the fact that mathematical methods feel impenetrable (and unobjectionable) to non-mathematicians, but also because applications of machine learning are *inherently* opaque. And on top of that, there is the opacity that comes with the fact that many systems that are used, also in public institutions, are secret under trade secrets protection laws.

O'Neil makes an important observation regarding the source of bias in the use of algorithmic systems, and that is that the definition of what constitutes success for the system is where biases originate. This steers the ship in a certain direction (such as efficiency, profitability, cost minimisation etc.) and may severely compromise other goals or (non-functional) requirements of the system.

4 Biases in Natural Language

As indicated before, natural language and natural language processing demonstrate interestingly how bias, machine learning and the remark that "it's all human construction" come together. As we know from Google search suggestions and Google Translate, machine learning techniques are very successful in predicting what we might search for, or translating words and phrases between languages. These systems learn from huge amounts of textual data that reflect how we actually use language.

But such techniques then also absorb tendencies of usage that are not predefined in the dictionary meanings of words. In Google search suggestions, the type of suggestions that popped up if you would type in the search bar: "men are...", would be surprisingly (and sometimes offensively) different from the suggestions for "women are..." [10]. Similar effects could be noticed in translations, for example when translating back and forth from Turkish, which has gender neutral pronouns, to English [11].

The phenomenon of biases in our use of natural language has been studied using machine learning techniques such as Word2Vec. Tolga Bolukbasi et al.'s 2016 paper with the telling title: "Man is to Computer Programmer as Woman is to Home-maker" [12] uses this technique to show implicit gender biases in our use of words that are gender neutral in principle. The paper also proposes ways of removing such biases by identifying such gender neutral words and performing an intervention on their vector representations as induced from the texts (giving such words the same distance to "he" as to "she"). While this debiases the representation of our natural language use, it of course does not debias our use of natural language itself, nor the social, cultural and historical structures and meanings that it encodes (as [13] in fact demonstrates).

One could question how fundamentally effective and meaningful this is, and the authors of the paper initiate this discussion themselves. But in any case, for the Google Search and Translate examples mentioned above, Google has in the meantime indeed intervened. You will notice that certain queries of the form "[group of people] are..." no longer generate any suggestions, and also the identified issue with translating back and forth to gender neutral languages has been acted upon [15].

5 Should We Do Something About Bias?

At this point it is maybe good to take a step back and revisit the concept of *bias*. While the word bias in everyday language (and most writing about algorithmic decision systems) is primarily used to indicate (unjustified and/or harmful) prejudices, remember that it in essence refers to an inclination away from a certain standard: e.g. bias in statistics is about structural misrepresentation, and bias in behavioural decision theory is about tendencies that deviate from the *rational* choice. In principle bias does not have to have a negative normative load, but calling something a bias does imply an assumed norm or standard (e.g. rational choice).

Deviation from a norm depends on the norm and the measure, and whether bias is a bad thing depends on the normative context. Because many of the examples about bias and algorithmic systems touch upon discrimination, it is also important to highlight that discrimination (in those discourses) is legally codified. Discrimination as a legal concept is introduced against harmful distinctions between groups of people, defined in law in terms of certain sensitive characteristics (such as race, gender, religion etc.). Not only intended, explicit differential treatment (in US law: *disparate treatment*) counts as discrimination, but also implicit harmful differences with effect on certain groups of people (*disparate impact*). An excellent overview of how algorithmic systems may fit legal definitions of discrimination in the latter sense can be found in the paper "Big Data's Disparate Impact" by Barocas and Selbst [14].

Bias in algorithmic systems enters at several layers, and it is impossible to completely disentangle the role of different sources of bias. At the level of the **data**: they may be incorrect, incomplete, or non-representative in a certain systematic way. Bias in data is introduced in what we can efficiently and accurately measure, how good our proxies work and what parameters we think are significant - for the definition of success, which in turn is determined by one stakeholder, and interpreted and validated by others. When training our systems of machine learning **algorithms**, bias is introduced by the availability and our selection of the training data, the choice of the algorithm, the choice what to optimize for in the learning process and what error margins are considered to be acceptable. This in turn is closely related to how we interpret and apply the outcome of such a system. While we may refer to the outcome of the computation as the (algorithmic) **decision**, in essence, the number is just that: the outcome of a (symbolic) computation. It is our (human) interpretation of that outcome, and the translation of that interpretation into a judgment or action (or our ascription of an intention to that outcome), that elevates the outcome of a computation into a decision. Here I would like to refer back to Kranzberg's quote that technology - including the *application* of computational techniques - is not neutral. It mediates and co-shapes our ideas of what a decision is - and also what a *good* decision is.

Through algorithmic systems, that are very powerful in highlighting patterns in data, we may attribute to those systems the powers to overcome human biases. But we have to realise that our biases seep into these systems at all layers, as trade-offs need to be made and will be made by people who can not oversee all eventual uses of the system - so they have to act on certain preconceptions. For example, underlying the claims of fairness of the recidivism scoring algorithm Compas ([3], see above), there was an "assumption that the algorithm's predictions were inherently better than human ones" [16]. It can actually be shown that in this case the system's predictions do not outperform human judgments - at least not in accuracy.

6 Formal Methods as Tools for Understanding

What do these reflections mean for what we can do to deal with harmful effects of biases in algorithmic systems? By now, different papers have appeared presenting different (technical) interventions to address bias at the level of data and algorithms [17–19]. Also, the first comparative studies appear, finding that a large number of measures are essentially similar but just make different trade-offs. It turns out there is a very high dependency on the way the data are preprocessed [20].

The outcomes of the systems only have meaning in their (historical, societal, cultural, domain-specific) context, and so does what counts as fair. This points at the fact that there can be no mathematical solution bringing us universal and objective fairness. Trade offs need to be made at all points, and the models that are trained to represent some physical or conceptual reality, can only be representations of a part of that reality - and it can at its best be made to be accurate for another part of that.

While mathematical methods cannot provide universal fairness solutions, they *can* be used to clearly point us at their limitations - which is useful knowledge in the light of underlying assumptions about some kind of technological superiority. It will help us understand in which ways the systems can improve human decision making, but also how this improvement depends on an implicit mapping between "reality" and its symbolic representation within the data and the algorithmic system. In the paper "The (im)possibility of fairness" by Friedler, Venkatasubramanian and Scheidegger [21], the authors use formal methods to pull apart the different *spaces* involved: the (intractable) construct space (the aspect you would like to measure), the observed space (the proxy for the construct, something measurable, the inputs for the decision system) and the decision space (the outputs of the system). They use this formal model of algorithmic decisions to mathematically prove that certain different notions of fairness can only be realised together under extreme, implausible metaphysical assumptions. This shows that it is really necessary to make normative choices of what is the most appropriate measure of fairness for a given situation.

Another paper, by Zliobaite and Custers [22], demonstrates mathematically how bias mitigating measures taken at different layers may interact - or rather: counteract. Data protection, as for example codified in the European Union's General Data Protection Regulation (GDPR [23]), has data minimisation as one of its principles in order to protect data subjects. In particular, Art. 9 of the GDPR *in principle* prohibits the processing of special categories of personal data ('sensitive data') such as data revealing racial or ethnic origin or religious beliefs. But if we leave out those attributes to prevent disparate treatment, this also means that we make it impossible to apply certain automated methods to control for harmful biases inherited in the data.

While law is often accused of being too slow to adapt to changing circumstances, one could say that mathematical laws in fact do not move at all. The application of any mathematical formalism comes with underlying assumptions on the structure of the problem it models and addresses. It thereby has a tendency to abstract away at least some forms of change or feedback loops of the use of the model, that may over time lead to violation of those underlying assumptions. There is also mathematical work demonstrating this for fairness measures [24].

7 Concluding Remarks

In this contribution, I have reflected on a few years of attention from the mathematical and computational sciences to bias in data, algorithms and decisions. Whereas one might expect neutrality and objectivity from computational systems, I have discussed how taking actual decisions involves human interpretations and judgments - in the data sets the system works with, the design of the algorithmic system, and turning the outcome into a decision. Assumptions, selections and priority judgments are made, on the basis of availability, efficiency, or other reasons, and through all of these, biases enter the picture.

Bias in its most neutral description is some systematic deviation from a standard. To which extent bias can be harmful or unfair, depends on the circumstances. Fairness in decisions based on machine learning is inherently sociotechnical and context sensitive. Formalisation of the problem will not provide us with universal solutions for fairness in real life, but the good news is: mathematical methods do bring us some insights in where the limitations lie both in formal approaches and in real life solutions to unfairness. It can point us to the necessity of looking at the temporal dimension as well, to capture feedback loops and historical factors.

Different fairness measures may be provably inconsistent in most realistic cases. Working with different measures and formally comparing them is very useful, in particular for analysing the problems and deciding which measure is most fitting where. Ultimately, whether a system is fair enough depends on whether the trade-offs we embed in the system, implicitly or explicitly, are right for the societal context of application. And what is right is subject to constant societal re-evaluation. This may also lead to the conclusion that for certain types of decisions, in the interest of fairness and justice, algorithmic systems are not the right tools.

Acknowledgment. This work is part of the SCALES project funded by the Dutch Research Council NWO MVI-program under project number 313-99-315.

References

1. Kahneman, D., Paul, P.S., Tversky, A.: Judgment under Uncertainty: Heuristics and Biases. Cambridge University Press, Cambridge (1982)
2. Februari, M.: "Het is allemaal mensenwerk" (in Dutch). Godwin lecture. De Correspondent, 5 May 2017. https://decorrespondent.nl/6692/de-datahonger-van-staten-en-bedrijven-zet-veel-meer-op-het-spel-dan-uw-privacy-alleen/1514846110316-d1a5748d
3. Angwin, J., Larson, J., Mattu, S., Kirchner, L.: Machine Bias - There's software used across the country to predict future criminals. And it's biased against blacks. ProPublica, 23 May 2016. https://www.propublica.org/article/machine-bias-risk-assessments-in-criminal-sentencing
4. Narayanan, A.: 21 definitions of fairness. In: Tutorial at the FAT* Conference, New York, February 2018. https://www.youtube.com/watch?v=jIXIuYdnyyk
5. Froehle, C.: The evolution of an accidental meme. How one little graphic became shared and adapted by millions. Medium, 14 April 2016. https://medium.com/@CRA1G/the-evolution-of-an-accidental-meme-ddc4e139e0e4
6. Kranzberg, M.: Technology and history: "Kranzberg's laws". Technol. Cult. **27**(3), 544–560 (1986). https://doi.org/10.2307/3105385
7. Verbeek, P.-P.: Animation: Explaining Technological Mediation, June 2017. https://vimeo.com/221545135
8. Harcourt, B.: Against Prediction. University of Chicago Press (2006)
9. O'Neil, C.: Weapons of Math Destruction - How Big Data Increases Inequality and Threatens Democracy. Crown (2016)
10. Lapowsky, I.: Google Autocomplete Still Makes Vile Suggestions. Wired, 2 December 2018. https://www.wired.com/story/google-autocomplete-vile-suggestions/
11. Sonnad, N.: Google Translate's gender bias pairs "he" with "hardworking" and "she" with lazy, and other examples. Quartz, 29 November 2017. https://qz.com/1141122/google-translates-gender-bias-pairs-he-with-hardworking-and-she-with-lazy-and-other-examples/
12. Bolukbasi, T., Chang, K.-W., Zou, J.Y., Saligrama, V., Kalai, A.T.: Man is to computer programmer as woman is to homemaker? Debiasing word embeddings. In: Advances in Neural Information Processing Systems, pp. 4349–4357 (2016)
13. Caliskan, A., Bryson, J.J., Narayanan, A.: Semantics derived automatically from language corpora contain human-like biases. Science **356**(6334), 183–186 (2017)
14. Barocas, S., Selbst, A.D.: Big data's disparate impact. Calif. Law Rev. **104**, 671 (2016)
15. Kuczmarski, J.: Reducing gender bias in Google Translate. Google Blog, 2 December 2018. https://www.blog.google/products/translate/reducing-gender-bias-google-translate/
16. Yong, E.: A popular algorithm is no better at predicting crimes than random people. The Atlantic, 17 January 2018. https://www.theatlantic.com/technology/archive/2018/01/equivant-compas-algorithm/550646/

17. Zafar, M.B., Valera, I., Gomez Rodriguez, M., Gummadi, K.P.: Fairness beyond disparate treatment & disparate impact: learning classification without disparate mistreatment. In: Proceedings of the 26th International Conference on World Wide Web (WWW 2017), pp. 1171–1180 (2017). https://doi.org/10.1145/3038912.3052660
18. Verma, S., Rubin, J.: Fairness definitions explained. In: Fair-Ware 2018: IEEE/ACM International Workshop on Software Fairness, 29 May 2018, Gothenburg, Sweden. ACM, New York, 7 p. (2018). https://doi.org/10.1145/3194770.3194776
19. Zliobaite, I.: A survey on measuring indirect discrimination in machine learning (2015). arXiv:1511.00148 [cs.CY]
20. Friedler, S.A., Scheidegger, C., Venkatasubramanian, S., Choudhary, S., Hamilton, E.P., Roth, D.: A comparative study of fairness-enhancing interventions in machine learning. In: Proceedings of the Conference on Fairness, Accountability, and Transparency (FAT* 2019). ACM, New York, pp. 329–338 (2019). https://doi.org/10.1145/3287560.3287589
21. Friedler, S.A., Scheidegger, C., Venkatasubramanian, S.: On the (im)possibility of fairness (2016). arXiv:1609.07236 [cs.CY]
22. Zliobaite, I., Custers, B.: Using sensitive personal data may be necessary for avoiding discrimination in data-driven decision models. Artif. Intell. Law **24**, 183–201 (2016). Available at SSRN: https://ssrn.com/abstract=3047233
23. European Council: General Data Protection Regulation. Regulation (EU) 2016/679
24. Liu, L.T., Dean, S., Rolf, E., Simchowitz, M., Hardt, M.: Delayed impact of fair machine learning. In: Proceedings of the 35th International Conference on Machine Learning, PMLR, vol. 80, pp. 3150–3158 (2018). See also the blog-post at https://bair.berkeley.edu/blog/2018/05/17/delayed-impact/

Workshop and Tutorial Papers

Opportunities and Challenges of Dynamic Consent in Commercial Big Data Analytics

Eva Schlehahn[1], Patrick Murmann[2], Farzaneh Karegar[2], and Simone Fischer-Hübner[2](✉)

[1] Unabhängiges Landeszentrum für Datenschutz Schleswig-Holstein, Kiel, Germany
[2] Karlstad University, Karlstad, Sweden
simofihu@kau.se

Abstract. In the context of big data analytics, the possibilities and demands of online data services may change rapidly, and with it change scenarios related to the processing of personal data. Such changes may pose challenges with respect to legal requirements such as a transparency and consent, and therefore call for novel methods to address the legal and conceptual issues that arise in its course. We define the concept of 'dynamic consent' as a means to meet the challenge of acquiring consent in a commercial use case that faces change with respect to re-purposing the processing of personal data with the goal to implement new data services. We present a prototypical implementation that facilitates incremental consent forms based on dynamic consent. We report the results gained via two focus groups which we used to evaluate our design, and derive from our findings implications for future directions.

Keywords: Dynamic consent · EU General Data Protection Regulation (GDPR) · Human-computer interaction (HCI) · Notification · Re-purposing

1 Introduction

Big data analytics can provide organisations with valuable insights. In particular, it may enable companies to understand their customers' preferences and provide these customers with the right information, experiences or services that are of value for them. With big data analytics, new types of data are usually derived that could be utilised for new purposes, which were possibly initially even unforeseen. For protecting privacy and ensuring compliance with the EU General Data Protection Regulation (GDPR), the use of the newly derived data for new data processing purposes could be legitimised by the consent of the individuals concerned (i.e., the data subjects).

Instead of confronting data subjects with long and barely comprehensible consent statements covering all possible future cases of derived data usages at

Published by Springer Nature Switzerland AG 2020
M. Friedewald et al. (Eds.): Privacy and Identity 2019, IFIP AICT 576, pp. 29–44, 2020.
https://doi.org/10.1007/978-3-030-42504-3_3

the time when they subscribe to a data service, a more specific consent for the use of newly derived data and/or the processing of data beyond the initially stated data processing purposes could be requested dynamically in the context and at the time when it becomes relevant for the data subject.

Within the scope of the SPECIAL EU H2020 project[1] in cooperation with the Privacy&Us[2] and PAPAYA[3] projects, the concept and management of dynamic consent, which has been initially suggested for the medical domain, has been further refined for commercial use cases.

This paper first discusses the concept and legal motivation for dynamic consent and then presents results of our research in regard to the following research questions:

1. How can user interfaces (UIs) for a commercial use case be implemented to facilitate repurposing via dynamic consent?
2. How is dynamic consent perceived by domain experts?

For addressing these research questions, we have first been developing mock-ups of UIs for managing dynamic consent requests for the re-purposing of TV-viewing profiles for the purpose of targeting customers with event notifications in several iteration cycles. These event notifications concern offline events like concerts, sports, or theatre events that may match the individual interests of the above mentioned customers. These mockups have then been discussed and evaluated in two focus groups with privacy researchers that were held at the IFIP Summer School in August 2019 in Brugg/Switzerland.

The rest of this paper is structured as follows: Sect. 2 reflects on the requirements pertaining to transparency and consent from the perspective of law and human-computer interaction (HCI), from which we derive a definition of dynamic consent as a means to address the challenges described in a use case related to big data analytics. Section 3 describes the methodology we applied to investigate our research questions by means of designing and evaluating a prototype that implements dynamic consent. Section 4 reports the results we obtained by designing and evaluating our prototype to address our two research questions. In Sect. 5, we discuss these results and their implications for the design of usable means that facilitate dynamic consent. Section 6 demarcates our contribution from related work, and Sect. 7 concludes the paper.

2 Background and Motivation

For personal data processing operations based on the legal basis of consent, Art. 6 para. 1 (a) and Art. 4 (11) GDPR, the latter demands a freely-given, informed, unambiguous, and specific indication of a data subject's wish to accept the processing of her personal data by a statement or a clear affirmative action.

---

[1] https://www.specialprivacy.eu/.
[2] https://privacyus.eu/.
[3] https://www.papaya-project.eu/.

This individual should be able to clearly understand what data are being collected by what party for what purposes. Moreover, the nature and location of the processing need to be communicated by the controller, as well as with whom which data will be shared [3].

Recital 32 of the GDPR indicates that electronic means can be used for the request and provision of consent, whereas choosing technical settings for information society services is mentioned as one example. However, especially for complex processing operations, as well as for situations where the use of small mobile screens is involved, this poses a real challenge for data controllers. We address this challenge by a new approach called 'dynamic consent'.

2.1 The Concept of Dynamic Consent

'Broad consent' [9,14] refers to consent from data subjects for the complete scope anticipated for a particular data processing scenario, including but not limited to, dimensions such as data types, purposes of processing, and additional data. However, such often 'global', long, and jargon-filled consent requests with attached privacy policies usually do not contribute well to a data subject's comprehension about the processing. Rather, they lead to difficulties in comprehension and/or are ignored by users [4,11]. However, for situations of changes to the processing, such as further processing of data beyond the original purpose (re-purposing), controllers lack a communication channel to ask for additional consent. This is all the more complex once special categories of personal data (Art. 9 GDPR) are involved, for which the consent needs to be explicit.

Our dynamic consent approach involves the controller asking the data subject for permission to communicate with them at a later stage after setting up an initial processing operation and requesting the correlating consent for it. Such a permanent communication may be used to facilitate incremental, context-specific consent requests (or at least notifications) addressing changes or intended changes to the original processing operation. These new consent requests are triggered 'dynamically' in the context, whenever the data subject uses other services of the controller.

We define *dynamic consent* as incremental, context-specific consent that will be asked any time after an initial consent was collected. Dynamic consent facilitates the characteristic of being freely-given since it provides granular, context-driven control and a real choice for the data subject instead of having a 'take it or leave it' situation for any complete scope of a data processing scenario. It helps to achieve a specific and unambiguous consent, as the data processing scenarios, especially with respect to data processing purposes, can be defined when they are clear without the demand of generalisation. Dynamic consent facilitates informed consent, as instead of confronting data subjects with all information regarding the data processing scenarios at once, shorter and more specific privacy notices are presented over time, which can help them to afford the time to read and understand what they are requested to agree to.

Service providers, i.e. controllers, can benefit from dynamic consent, as it can legitimise new forms of data processing which go beyond to what the data subjects initially consented to. Moreover, it allows service providers to gradually learn in detail about their users' privacy preferences. Thus, they can adapt their service to meet user expectations, leading to higher satisfaction.

Despite several possible advantages of dynamic consent compared to the traditional consent, dynamic consent may also suffer from a couple of problems. This includes the consent fatigue and habituation, as the number of consent forms users are supposed to handle increases. Hence, we cannot benefit from the ultimate potential of dynamic consent without considering its HCI implications and how users understand and perceive the concept of dynamic consent. Therefore, in this paper, one of our research objectives is to investigate the users' understanding of novel dynamic consent forms designed for a commercial use case.

2.2 Imaginary Scenario

The above mentioned dynamic consent approach has been developed within a specific imaginary scenario suitable for mobile devices. The prototype developed has two phases in its UI design, namely the installation of an app, and later notifications handled by this app.

This scenario involves a data subject who already is in a contractual relationship with a data controller for a specific digital service. In this case, this is a digital TV viewing service, for which the data subject at some earlier point in time already has (per assumption for this basic setting) validly consented to the processing of his or her TV viewing behaviour.

The data controller (in our scenario called 'Apricot Ltd.') offers many other digital services, one of them a service which provides recommendations for various events, such as concerts, parties, theatre plays or the like. This service offers the possibility to build a user profile over time in order to give better event recommendations (processing purpose). Depending on what the data subject agrees to, this user profile may consist of data categories such as location of the mobile device the app is installed on, or additional data categories to fine-tune the service to the interests of the user.

These additional data categories could also be information about the data subject's TV-viewing behaviour collected previously. To enable this in a GDPR-compliant way, the data controller needs to ask the data subject for consent to re-purpose this exact information to add it to the event recommendation interest profile.

This imaginary scenario provides a use case that showcases our dynamic consent approach to issues like changes to the processing operation, user comprehension and control, as well as establishing a steady communication channel between controller and data subject.

3 Methodology

3.1 Designing the Prototype

Designing the prototype was carried out in multiple iterations. Based on the use case described in Sect. 2.2, we started by assessing the legal requirements pertaining to the scenario. We complemented them with the informational contents needed to satisfy these requirements and used both to create initial mock-ups. The early generations of the designs mainly disregarded aspects related to HCI and principles of usability as regards the future implementation, which necessitated rethinking the design process to accommodate these dimensions.

We therefore started redesigning the prototype in consideration of HCI requirements and discussed each iteration of the design with respect to whether they fulfilled the legal requirements that had been specified earlier. However, the resulting prototypical design, as described in Sect. 4.1, reflected our own interpretation of how the concept of dynamic consent could be implemented. We therefore decided to seek out independent parties to reflect on the design, which we hoped to accomplish by means of an extended user study.

3.2 Evaluating the Prototype

At this early stage of our research, we were mainly interested in receiving feedback related to the conceptual aspects of dynamic consent, which is why we decided to rely on domain experts rather than laypersons to help us evaluate our prototype. To get access to a large variety of privacy researchers and practitioners, we conducted a workshop during the IFIP Summer School 2019, and designed it as a hybrid between a focus group and a cognitive walkthrough. We call it a focus group because its primary purpose was to gauge opinions, feelings and attitudes about the high-level concepts of dynamic consent in the context of re-purposing the processing of personal data in an everyday scenario. Our goal was to obtain as much feedback as possible and therefore encouraged lively discussion among the participants. Additionally, the study exhibited the characteristics of a cognitive walkthrough in that we used a paper prototype to gauge whether the order in which we had designed the operation steps of the prototype helped our test subjects understand the concept we tried to convey. By receiving immediate feedback on individual interaction steps, we tried to ascertain how well the conceptual approach of our design was received.

Participating in our workshop was voluntary. Before beginning the workshop, all participants were informed that no personally identifying information would be collected and that only non-personalised notes would be taken during the group discussions.

We opened the workshop by giving our participants a brief presentation of the use case, including a description of the existing relationship between a data subject and data controller (Sect. 2.2), and pointed out that during the focus group they would be taking the role of said data subject. We split up the cohort of a total of ten participants into two separate groups. All participants were PhD students at various stages of their studies, and all supplied the domain knowledge of their respective discipline to their group. We distributed the participants evenly among the two groups according to their backgrounds, which covered disciplines such as computer science, information systems, law and psychology, and of which we took note in non-personalised form.

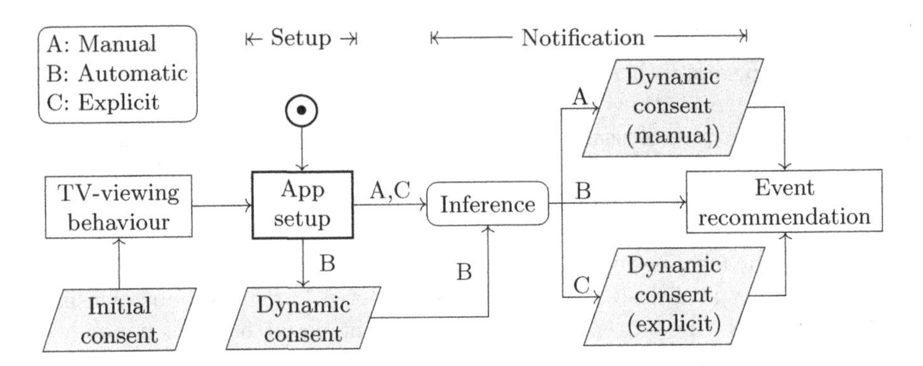

Fig. 1. Schema of improving event recommendations by analysing TV-viewing behaviour. UIs used to collect consent are shaded. Alternative decision paths: A: Dynamic consent based on manual approval, B: dynamic consent based on automatic approval, and C: dynamic consent based on explicit consent.

During the first part of the group discussions, each group acted independently, headed by pairs of the four authors who served as moderator and minute-taker, respectively. At specific stages while going through the prototype, the moderators asked questions from a list we had prepared earlier. Hence, the discussions followed a semi-structured form. The questions helped us to both direct the flow of the discussion and gauge the participants' perception with respect to specific aspects of the scenario. Discussing a list of predefined questions in both groups also helped us achieve consistency for collecting feedback related to our research questions. Each group was equipped with a copy of A3-sheets that represented the screen of the prototype, and stacks of coloured post-it notes they could stick onto the sheets. Additionally, each participant received her own copy of A4-prints that she could refer to at her leisure.

At the end of the workshop, all participants reconvened to compare and discuss the results each group had gathered. We contributed to the discussion by giving our own interpretation of how we thought dynamic consent could help facilitate personal data processing in the context of rapidly developing big data scenarios.

4 Results

4.1 Implementing Dynamic Consent

Our design is based on the premise of breaking down bulky traditional privacy policies into multiple smaller parts that require less cognitive effort to read and understand. The modular character of dynamic consent would potentially allow for better transparency in that our approach sought to bridge the causal relationship between the change detected in a user's TV-viewing behaviour and the change she would experience thereupon with respect to receiving event recommendations. The approach would enable data subjects to exercise more granular control as regards individual aspects of how their personal data will be processed compared to the extensive scope of traditional consent (Fig. 1).

The use case at hand (Sect. 2.2) targeted users of mobile devices, which called for a technological artefact in the form of an app running on a mobile phone. We split the prospective run-time behaviour of our app into two distinctive operational phases: a preparatory setup phase and a subsequent notification phase. During the *setup phase*, users of the app, i.e. data subjects, configure the app to reflect their privacy preferences. They do this by specifying to what extent the service provider, i.e. data controller, is permitted to draw from and link multiple sources of the user's personal data to infer from them appropriate candidates for recommending events. We considered three sources of personal data that can potentially be employed to customize event recommendations:

Pre-selected event categories help narrow down the user's interest prior to using the recommendation service. We included this source for the sake of making the prototype appear more authentic. It did, however, not touch upon the concept of re-purposing the processing of personal data.

Location data narrow down the geographical area of events that serve as possible candidates for a recommendation. The use of this data source was tied to a dedicated consent form.

TV-viewing behaviour provides behavioural input as to what kind of events a user might potentially be interested in. This is the data source that the user study focused on.

It was up to users to decide how many data sources they permitted the app to tap into, knowing that a combination of more sources would potentially lead to recommendations that would more adequately reflect their actual interests regarding events.

To provide users with the transparency necessary to understand the causal relationship between their TV-viewing behaviour and the change they would experience with respect to the recommendations they will receive, we relied on a *notification phase* that complemented the setup phase. Delivered via the notification center of the mobile device, notifications were issued in response to changes detected in the user's TV-viewing behaviour, provided these changes impacted the recommendations users would receive at a later time, or in cases when their consent was required to lawfully process their personal data as an effect of the change. These notifications were sent irrespective of any event recommendations a user would receive based on her preferences regarding events.

If users chose to have their TV-viewing behaviour being considered for receiving customised recommendations for events, they were able to select between two modes of operation that determined how they would receive notifications related re-purposing the processing of their personal data (Fig. 2):

Manual (option A). New categories must be acknowledged *manually* as soon as they were detected as a result of a change of the user's TV-viewing behaviour (Figs. 1, 3 (top) and 4a).

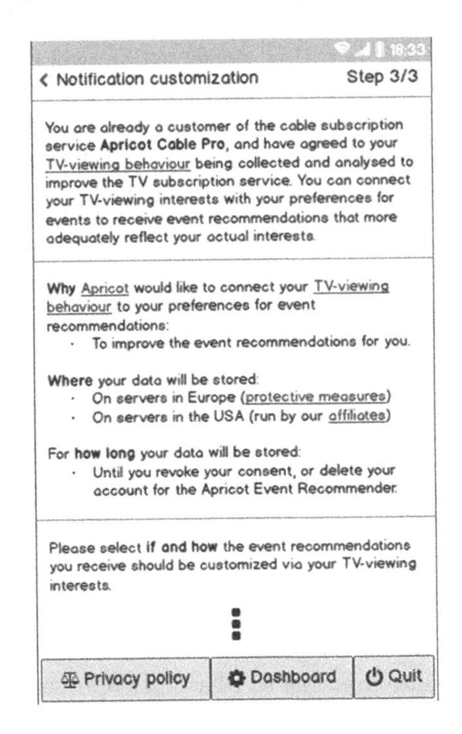

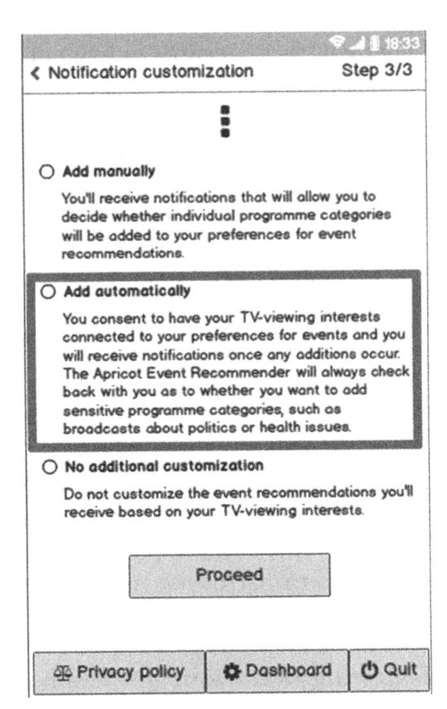

Fig. 2. Setup phase: Choose how TV-viewing behaviour will affect the event recommendations one will receive. Automatic adaptation of the recommendations requires consent *a priori* (framed, highlighting was not part of the original prototype). The screen on the left-hand side continues on the figure to the right.

Automatic (option B). Once detected, new categories are added *automatically* without requiring acknowledgement (Figs. 1, 3 (middle) and 4b).

We conceptualised a third type of notification that would always be issued irrespective of the mode of operation selected previously. It covered cases in which 'special categories of personal data' (GDPR Art. 9) were about to be processed, in which case the data subject's explicit consent would be required (Fig. 1 option C and Fig. 3b). Such cases include, but are not limited to, TV programmes that might allow a data processor to infer the data subject's political opinion or sex life.

We designed the mockup of the app as a click-through prototype. Each of the eleven screens (19 including screens related to stylised secondary information, such as the dashboard or the privacy policy used for collecting initial consent) represented an interaction phase navigable by the user. However, since the purpose of the workshop was to discuss high-level concepts rather than detecting usability issues, we decided to print each screen on paper. That way, we allowed for high accessibility and readability, and ensured that each participant had access to her own copy of the prototype.

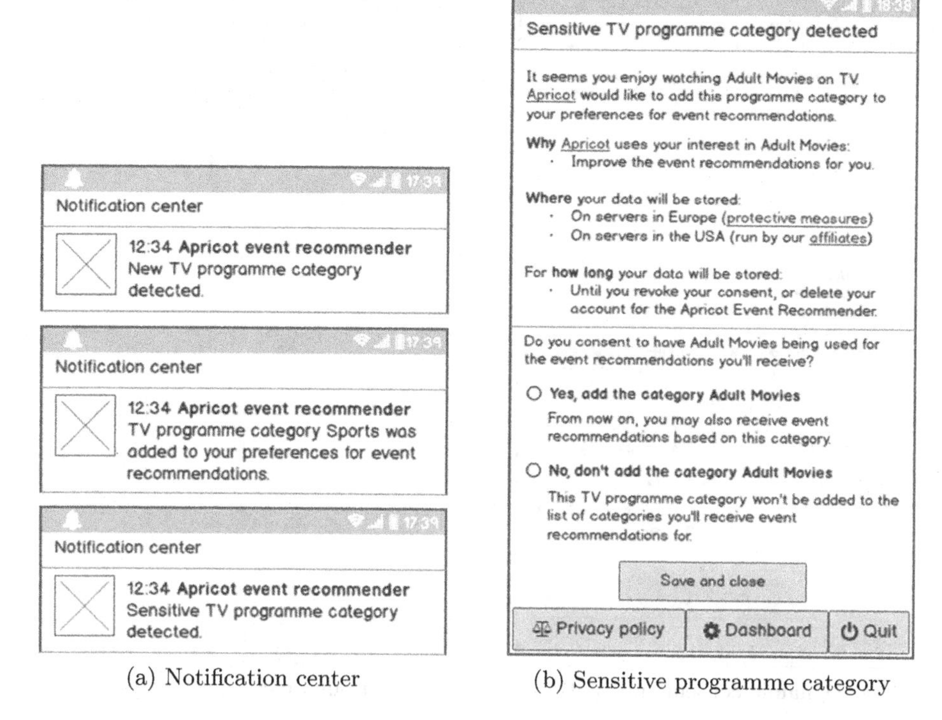

(a) Notification center

(b) Sensitive programme category

Fig. 3. (a) Messages delivered via the notification center of the mobile device (A: manual (top), B: automatic (middle), C: sensitive (bottom)), and (b) request of dynamic consent due to the processing of special categories of data (C).

4.2 Perception of Dynamic Consent

The members of both focus groups were well accustomed to traditional privacy policies, knew how to read them, and were capable of weighing up the pros and cons of individual components of consent forms. When they discussed the consent form related to the processing of location data, e.g., the opinions of individual members varied as regards the completeness and necessity of various aspects, but they all backed up their arguments with a wealth of previous knowledge due to first-hand experience with the subject matter. Most participants appreciated the brief and concise presentation of facts provided throughout the consent forms they discussed. However, some of them requested additional information as regards individual statements made as part of the forms, such as the identity of the data processor outside of Europe.

The concept and usefulness of dynamic consent was perceived differently by the two groups. The members of group 1 seemed capable of conceptually following the scenario as they went through the various stages of the prototype. They understood well that they had earlier consented to the profiling of their TV-viewing behaviour when first they had subscribed to the TV-cable service.

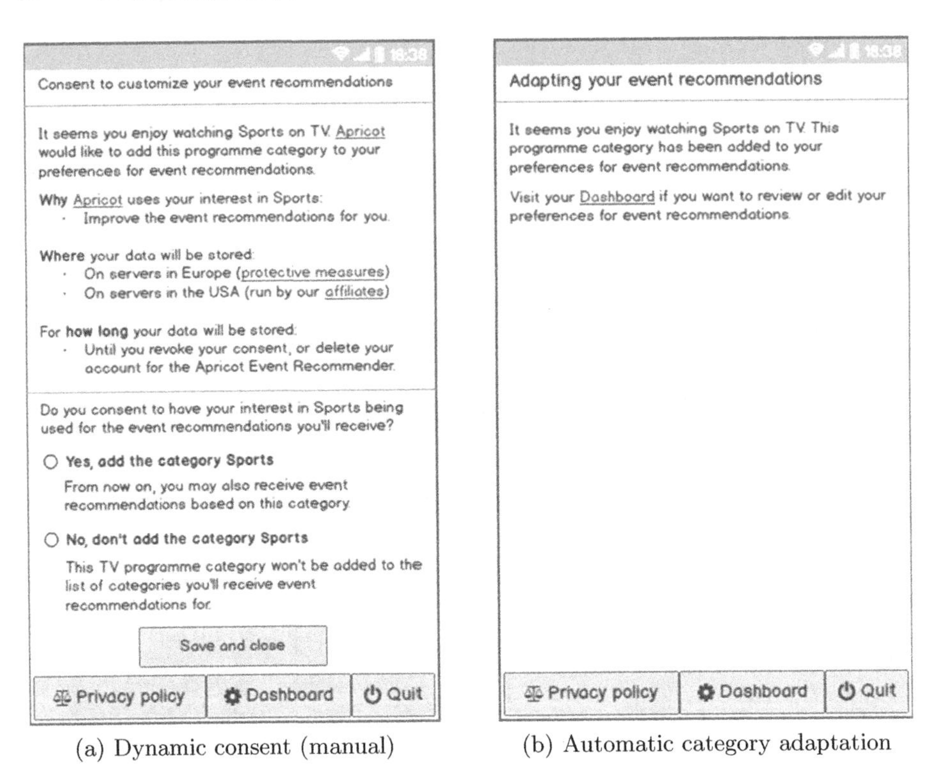

(a) Dynamic consent (manual)

(b) Automatic category adaptation

Fig. 4. Notifications received in response to a change of TV-viewing behaviour depending on whether users consented to have their event recommendations adapted automatically.

They realised that the purpose of dynamic consent was to permit the data controller to link that profile to their profile used to customize the event recommendations they would receive. Conversely, group 2 seemed largely unable to follow the train of thought the way in which we had expected them to. They were seemingly unable to see the big picture of the data processing scenario in that they were unable to make a connection between (1) the existing relationship between them, in their role as data subjects, with the data controller as explained in the introductory presentation, (2) their choice to use another online service of the same data controller, and (3) the data controller asking them for additional consent for the purpose of connecting their personal data collected for the first service to their profile that facilitated the second service.

Our participants wanted to stay abreast of any changes detected in their TV-viewing behaviour. When asked to make a choice during the setup phase (Fig. 2), both groups decided to go for manual consent (Figs. 4a and 1 option A) rather than to rely on automatic consent (Figs. 4b and 1 option B). Group 1 appreciated the increased level of control they were able to exercise in terms of choosing how they were notified about changes. The members of group 1 noted that in general, dynamic consent is more adapted to the context, as it appears in the context when it is relevant and that it provides good opportunities for mobile phones to provide "step by step policies". By this, users may get "educated" on privacy. However, the members of group 1 also noted that the consequences of the consent could be better clarified in terms of what it implies to be categorised beyond receiving event notifications in the future, e.g. whether further profile details would be added if the user later actually attended a recommended event. They noted that now the question for consent (Fig. 4) is phrased as: "Do you want us to use our profile that we already have for your benefits?" This means that they criticised that consequences were rather stated in terms of utility of receiving customised recommendations than in terms of privacy.

The members of group 1 remarked that they would have liked even more control in terms of exercising intervenability, such as having access to shortcuts in the form of buttons that would allow them to revoke their consent by means of the notification itself. This would save them time and effort compared to invoking the dashboard, searching for the required option in question, and then making the change. Both groups noted that the dashboard should be more visible, and that further instructions would be required on how transparency and intervenability rights can be exercised.

Group 1 members discussed the dynamic explicit consent requests for sensitive TV-programme categories (Fig. 3). It was noted that these explicit consent requests can be embarrassing and compromising, e.g. if they appear when someone is together with her partner or a colleague, or at any other unexpected occasion when others can watch this consent request. Hence, dynamic consent requests for sensitive categories should not appear directly, which emphasises the need for a 'notification center', as we have implemented in our mockups (Fig. 3a (bottom)).

Conversely, group 2 seemed confused by the repetitive nature of the follow-up consent forms, which they had to go through as part of receiving notifications about changes related to their TV-viewing behaviour. They were unable to relate the consent forms to choices they had made earlier during the setup phase and perceived them as redundant and tedious.

5 Discussion

5.1 Reflecting on Dynamic Consent

Sections 2 and 6 motivate dynamic consent as regards its characteristics of being specific, freely given and informed, compared to traditional consent. Despite the fact that dynamic consent can help break down a comprehensive, anticipatory consent form into the parts that are actually applicable to the data processing scenario in question, dynamic consent will not, in itself, solve the inherent issue of readability and intelligibility. Privacy notifications will still have to be designed such that they cater towards the specific needs of the individual reading the contents, that they clearly communicate risks and consequences in terms of privacy that will or may arise due to acting and not acting in response to the notification, and that they provide support and guidance in case recipients are unable to make an informed decision solely based on the information currently at her disposal [12]. In this respect, dynamic consent can not mitigate the effect of a consent form that was designed poorly in relation to its content or structure.

Clarity will be indispensable in cases when the circumstances under which a notification has been issued are delicate and when particular care is required. This will, e.g., be the case when notifications report about or ask consent for the processing of sensitive personal data (GDPR Art. 9). Multilayered policies could be a key enabler as regards providing a customisable depth of information for dynamic consent [3,12]. By doing so, the cognitive load imposed upon readers of information provided on the highest layer would be comparatively low, while lower layers would still allow for more detailed information upon request. As was requested by some of our participants, such information could, e.g., elaborate on specific statements made in the consent form, or on the exact consequences that will arise for a data subject should she choose to link a particular TV-programme category to the profile used for sending her event recommendations.

In our study, we did not attempt to cover the longitudinal aspect introduced by the asynchronous nature of notifications as such. Notifications may not only arrive at inopportune moments, as discussed by focus group 1, but also at times when a data subject may not immediately be able to comprehend the causal relationship between a message and the cause that has triggered it [16]. Despite the fact that informational content conveyed as part of the notification may in itself be more comprehensible compared to traditional consent forms, assessing the causal relationship of temporally asynchronous events may impose additional cognitive load on recipients of notifications. Privacy notifications will therefore have to implement appropriate contextual cues to support users in understanding the contextual relationship reported by the notification [12].

Some of our participants requested additional means of intervenability from within the notifications themselves. This is in line with previous findings in the literature in that privacy notifications should provide actionable choices in addition to the informational content provided in the message [6]. It corresponds to the legal requirement stipulated in GDPR Art. 7 (3) and the recommendations of the Art. 29 Working Party [1] in that consent should be as easily revocable by data subjects, as it was given. Hence, designers of future prototypes should consider adding actionable choices to notifications, which would help data subjects exercise specific data subject rights according to the situation at hand [12].

More research will be required to investigate the longitudinal aspect of dynamic consent as regards user perception. Authentic results may be obtained via studies in which users of a data service make decisions that affect the processing of their actual personal data, and in which they provide situational feedback on the perceived utility of dynamic consent as part of the process.

5.2 Limits

As is pointed out in Sect. 3.2, we specifically targeted domain experts for the initial evaluation of our prototype. The participants of our focus groups were not only highly educated, but also comparatively young. Our findings can therefore not be generalised for the general public. Older, less knowledgeable users or users who are less familiar with employing mobile devices in the context of IT might not exhibit the same level of expertise as regards the socio-technological ecosystem reflected in our study. For such an audience it might be even more difficult to fully comprehend the longitudinal process accompanying the concept of dynamic consent, and might therefore be unable to make informed decisions as regards the processing of their personal data.

The contents reflected in the focus groups were relatively broad and required considerable time to process by our participants. In particular, many of the screens and UIs discussed during the focus groups were not strictly related to the concept of dynamic consent. We had included some of the screens primarily for the sake of authenticity, i.e. to make our prototype and the underlying use case more tangible. Future studies would be well advised to focus on the aspects that are essential for conveying the concept of dynamic consent, namely the consent form by which users choose how they will be informed about future changes (Fig. 2), and the notifications used to facilitate the subsequent consent management (Figs. 3 and 4).

Similarly, some of the dimensions considered during the study did not contribute to convey the concept of dynamic consent as such. Location data, e.g., were not relevant for discussing repurposing TV-viewing behaviour to facilitate event recommendations. If such marginalia were removed, the study would focus even more on the actual core concept of the subject matter.

6 Related Work

Privacy nutrition labels [10], multi-layered short policies summarising key data practices [2,3], privacy icons and images [5,7], and comic-based interfaces to convey policy information [17] are examples of proposed solutions to solve the issues of traditional consent forms discussed in Sect. 2. Nonetheless, none of the methods proposed could adequately address the consent problems in various contexts. This motivated researchers to propose other solutions for achieving consent, such as Just-In-Time Click-Through Agreements (JITCTAs) [13] and dynamic consent.

The concept of JITCTA proposed by Patrick and Kenny [13] is based on subdividing a “large, complete list of service terms” [13] into smaller ‘agreements’ that collect consent from data subjects as needed, i.e. once conditions apply that require personal data to be processed in a specific way. Kay and Terry [8] interpret JITCTAs as segmenting a larger legal agreement into smaller parts that are presented at situationally appropriate times. Both groups of authors argue that by confronting users with small pieces of context-sensitive information compared to a large amount of anticipatory information, the cognitive load necessary to process such information can be reduced, while the level of specificity increases.

Conversely, dynamic consent as we consider it in this paper does not rely on the largest possible superset of a privacy policy defined *a priori*, which is dealt out in smaller portions once the need arises. We instead assume that the necessity to process personal data changes over time, potentially in a way that could not have been foreseen when the preceding version of the policy was issued. Due to the causal relationship between observed user behaviour and change of how data are processed, however, we share with these authors the opinion as regards an increase in specificity and, potentially, an increased level of understanding on the part of users of such data services.

Dealing with multiple researchers and projects in biobank research makes it difficult to obtain informed consent from participants for all future uses of data at the time of recruitment into the biobank [9]. Kaye et al. [9] argue that re-consenting, to overcome the issues of informed consent in the biobank research is costly and time-consuming in practice and might lead to high drop-out rates due to the difficulty in locating people. A practical solution, as discussed in [9,14], to open-ended projects in biobank research is broad consent. However, using the broad consent model may not help to achieve informed consent [9,14].

An alternative solution for consent in biobanks in some form of a dynamic consent approach was initiated for the first time in the EnCoRe project[4] (June 2008 to April 2012). In this solution, collecting consent for three biobanks was supposed to be replaced with broad consent [9]. Researchers elaborated on the pros and cons of dynamic consent in biobanks [9,15] and discussed the implementation issues of the dynamic consent model in practice [9]. Although dynamic consent is participant-centred, allows interactions over time, and protects participants' autonomy over their data [9,14], it requires, above all of the physical resources, a commitment to such a vision by stakeholders involved in the biobank research such as clinicians and researchers, health-care services, and governments for its deployment in real scenarios [9].

Kaye et al. [9] argue that while dynamic consent is proposed as a concept related to a biobank project, it is an approach that can be applied more broadly in other fields beyond healthcare. To apply dynamic consent in other contexts, we require more research regarding its applicability to solve the issues of current traditional consent forms. Hence, we redefined the concept of dynamic consent for online service providers, specifically for commercial data services using big data analytics based on their users' consent. Nonetheless, there are some differences in practice between the dynamic consent in biobank research and the dynamic consent we defined in this paper. For the former, the data collected, processed, and re-purposed are always special categories of data; thus requiring explicit consent by law. For the latter, on the contrary, there are different levels of sensitivity for the data being processed which relieves the data controller of the need to always obtain explicit consent. It introduces more opportunities for implementing dynamic consent, which can potentially fulfil different users' requirements. However, it comes with its own challenges as how to facilitate consent in cases when dynamic consents is not requested explicitly (see Fig. 1 option A and option B). This calls for more investigation on the effects

[4] http://www.encore-project.info/.

of dynamic consent on users' experience, their understanding of their data flow and conditions of consent, and their sense of oversight and control.

7 Conclusion

This paper presents a novel approach for obtaining dynamic consent for a commercial scenario, which was evaluated by expert focus groups at the IFIP Summer School 2019. Our expert evaluations showed that our approach involving alternative paths for obtaining dynamic consent was not easily understood by all experts. Nevertheless, those that understood how the concept of dynamic consent was used in our scenario also appreciated the approach of incremental consent requests. These can more specifically describe the current context and provide increased user control and transparency. However, at the same time, emphasis must also be put on informing users well about privacy consequences when choices need to be made. This dynamic way of collecting or altering the data subject's permissions over time should come along with meaningful ways to exercise intervenability, including the data subject's rights entailed in the GDPR. This in particular needs to provide the user with direct access to functions for easily revoking a previously given consent at the moment when a request to dynamically extend this consent appears. Future directions for the design of dynamic consent should address these results of our expert evaluations.

Acknowledgements. The research presented in this paper was jointly conduced by the SPECIAL, Privacy&Us and PAPAYA EU projects. The project SPECIAL (Scalable Policy-awarE linked data arChitecture for prIvacy, trAnsparency and compLiance) has received funding from the EU's Horizon 2020 research and innovation programme under grant agreement No. 731601. The Privacy&Us project has been supported by the EU's Horizon 2020 Research and Innovation Programme under the Marie Skłodowska-Curie Grant 675730 and the project PAPAYA (A Platform for Privacy Preserving Data Analytics) is funded by the H2020 Framework of the European Commission under grant agreement No. 786767.

We thank Harald Zwingelberg (ULD) and Rigo Wenning (ERCIM/W3C) for their valuable insight, ideas and contributions to the concept of dynamic consent, and also the participants of the two focus groups for their valuable feedback.

References

1. Article 29 Data Protection Working Party: Guidelines on consent under regulation 2019/679 (2018)
2. Article 29 Data Protection Working Party: Opinion 10/2014 on more harmonised information provisions. Accessed 25 Nov 2004
3. Article 29 Data Protection Working Party: Guidelines on transparency under Regulation 2016/679. Accessed 11 Apr 2018
4. Cate, F.H.: The limits of notice and choice. IEEE Secur. Priv. **8**(2), 59–62 (2010)
5. Cranor, L.F., Guduru, P., Arjula, M.: User interfaces for privacy agents. ACM TOCHI **13**(2), 135–178 (2006)

6. Egelman, S., Cranor, L.F., Hong, J.: You've been warned: an empirical study of the effectiveness of web browser phishing warnings. In: Proceedings of SIGCHI Conference on Human Factors in Computing Systems, pp. 1065–1074. ACM (2008)
7. Holtz, L.E., Zwingelberg, H., Hansen, M.: Privacy policy icons. In: Camenisch, J., Fischer-Hübner, S., Rannenberg, K. (eds.) Privacy and Identity Management for Life, pp. 279–285. Springer, Heidelberg (2011). https://doi.org/10.1007/978-3-642-20317-6_15
8. Kay, M., Terry, M.: Textured agreements: re-envisioning electronic consent. In: Proceedings of the Sixth Symposium on Usable Privacy and Security, p. 13. ACM (2010)
9. Kaye, J., Whitley, E.A., Lund, D., Morrison, M., Teare, H., Melham, K.: Dynamic consent: a patient interface for twenty-first century research networks. Eur. J. Hum. Genet. **23**(2), 141 (2015)
10. Kelley, P.G., Cesca, L., Bresee, J., Cranor, L.F.: Standardizing privacy notices: an online study of the nutrition label approach. In: Proceedings of the CHI, pp. 1573–1582. ACM (2010)
11. Luger, E., Moran, S., Rodden, T.: Consent for all: revealing the hidden complexity of terms and conditions. In: Proceedings of the CHI, pp. 2687–2696. ACM (2013)
12. Murmann, P.: Eliciting design guidelines for privacy notifications in mhealth environments. Int. J. Mob. HCI **11**(4), 66–83 (2019)
13. Patrick, A.S., Kenny, S.: From privacy legislation to interface design: implementing information privacy in human-computer interactions. In: Dingledine, R. (ed.) PET 2003. LNCS, vol. 2760, pp. 107–124. Springer, Heidelberg (2003). https://doi.org/10.1007/978-3-540-40956-4_8
14. Ploug, T., Holm, S.: Meta consent: a flexible and autonomous way of obtaining informed consent for secondary research. BMJ **350**, h2146 (2015)
15. Prictor, M., Teare, H.J., Kaye, J.: Equitable participation in biobanks: the risks and benefits of a "dynamic consent" approach. Front. Public Health **6**, 253 (2018)
16. Schaub, F., Balebako, R., Cranor, L.F.: Designing effective privacy notices and controls. IEEE Internet Comput. **21**(3), 70–77 (2017)
17. Tabassum, M., Alqhatani, A., Aldossari, M., Richter Lipford, H.: Increasing user attention with a comic-based policy. In: Proceedings of the CHI, pp. 200:1–200:6. ACM (2018)

Identity Management: State of the Art, Challenges and Perspectives

Tore Kasper Frederiksen[1(✉)], Julia Hesse[2(✉)], Anja Lehmann[4(✉)], and Rafael Torres Moreno[3(✉)]

[1] Alexandra Instituttet, Aarhus, Denmark
tore.frederiksen@alexandra.dk
[2] IBM Research – Zurich, Zurich, Switzerland
jhs@zurich.ibm.com
[3] University of Murcia, Murcia, Spain
rtorres@um.es
[4] Hasso-Plattner-Institute, Potsdam, Germany
anja.lehmann@hpi.de

Abstract. Passwords are still the primary means for achieving user authentication online. However, using a username-password combination at every service provider someone wants to connect to introduces several possibilities for vulnerabilities. A combination of password reuse and a compromise of an iffy provider can quickly lead to financial and identity theft. Further, the username-password paradigm also makes it hard to distribute authorized and up-to-date attributes about users; like residency or age. Being able to share such authorized information is becoming increasingly more relevant as more real-world services become connected online. A number of alternative approaches such as individual user certificates, Single Sign-On (SSO), and Privacy-Enhancing Attribute-Based Credentials (P-ABCs) exist. We will discuss these different strategies and highlight their individual benefits and shortcomings. In short, their strengths are highly complementary: P-ABC based solutions are strongly secure and privacy-friendly but cumbersome to use; whereas SSO provides a convenient and user-friendly solution, but requires a fully trusted identity provider, as it learns all users' online activities and could impersonate users towards other providers.

The vision of the OLYMPUS project is to combine the advantages of these approaches into a secure and user-friendly identity management system using distributed and advanced cryptography. The distributed aspect will avoid the need of a single trusted party that is inherent in SSO, yet maintain its usability advantages for the end users. We will sketch our vision and outline the design of OLYMPUS' distributed identity management system.

Keywords: Privacy · Security · Identity management · Digital identities · Distributed cryptography

This is a workshop paper for the OLYMPUS project. OLYMPUS has received funding from the European Union's Horizon 2020 research and innovation program under grant agreement No 786725.

Published by Springer Nature Switzerland AG 2020
M. Friedewald et al. (Eds.): Privacy and Identity 2019, IFIP AICT 576, pp. 45–62, 2020.
https://doi.org/10.1007/978-3-030-42504-3_4

1 Introduction

Data has become the new oil: the advent of advanced AI algorithms for data analytics and cheap storage has allowed analysis based on millions of individuals' personal information. This includes health, location and perhaps most importantly; any kind of online activity. The latter is of particular importance due to the sheer magnitude of data in that category. An important example in this setting is Facebook, who with its user base of over one billion, holds a treasure trove of personal information, ready to be used for social analysis [24] and manipulation. In this setting the Cambridge-Analytica scandal provides interesting insight into how high a value personal data, and in particular subjective attributes, such as likes, can have [5].

A crucial lesson learned from this scandal is that we must become more aware of how much personal information is available about us online and how it is possible for us to limit this. One approach is of course to simply not share anything with sites we do not fully trust. However, even trusted sites suffer from breaches that leak personal information [19,22] and sometimes this information is even collected without our knowledge [26]. If we still wish to be part of the connected world the Internet facilitates then we cannot be completely safe and protect against all non-consensual collection of information. However, we can try to prevent theft of personal information by limiting the locations at which our data is stored and by tuning how easy it is to gain access to this information without a complete take-over of the service provider.

The Need for User Authentication. Personal data at different service providers are usually protected through some sort of authentication mechanism which allows users to prove that they are who they say they are. Such authentication is done by proving ownership of a given username, meaning that a user must prove that it holds some secret information that was used during account creation.

This secret information is most often a password that was picked by the user. Unfortunately, despite advised otherwise, people tend to either use low-entropy passwords or reuse their passwords (or slight variations thereof) at different service providers [31]. This introduces a large attack surface for an adversary wishing to take-over a user's account; he/she could try to guess low entropy passwords or try to find users who reused a password between a fully compromised service provider and an un-compromised service provider.

In fact, the latter not only leverages the users' tendency of password re-use, but also the availability of massive password leaks through compromised service providers. More than one billion personal data records including email address and password combination have been reported stolen to date [27]. Even when properly salted and hashed, the low entropy in human-memorizable passwords makes it trivial to brute-force the plaintext passwords using modern hardware. Increasing entropy through measures such as two-factor authentication can be seen as a trade-off between security and usability. Such methods are often not adequately efficient to handle for users quickly wanting to access a service.

Using a cryptographic key for a digital signature scheme, as is the case when authentication is based on X.509 certificates, or an access token authorized to the user through a different service provider, known as an Identity Provider (IdP), alleviates some of the worst security issues with passwords. In particular these approaches do not require users to trust every single service provider they wish to access, to keep our password safe.

However these solutions have other drawbacks: Certificates, for example, are not very user-friendly and require users to manage large, non-memorizable keys between all the devices they use. When the certificates contain user-specific attributes they also require disclosure of all these attributes with every usage, which violates the privacy paradigm of data minimization. Online IdPs and Single Sign-On (SSO) are quite complementary: they are easy and convenient to use, as users only need to authenticate to the IdP which then issues short-term authentication tokens towards service providers. The IdP can thereby certify only the minimal amount of attributes needed for each access request. On the negative side, the IdP must be a fully trusted entity, as it learns about all the services its users accesses, and—if compromised—could impersonate the users towards all service providers.

Modern cryptography has afforded another approach to identity management, handling some of the issues with passwords, certificates and IdPs. This approach is known as Privacy-Enhancing Attribute-Based Credential (P-ABC) [3,9,10,28]. A P-ABC scheme allows a user to hold a special token, known as a credential, which contains an assertion on the user's identity and features from a trusted provider. The user can use the credential to generate single-use tokens for authentication towards different service providers. Like for certificates and IdPs this approach also avoids having to trust all the service providers a user wishes to access *and* it handles the privacy issues that are otherwise associated with IdPs and conventional certificates. Unfortunately, P-ABCs struggle when it comes to user-friendliness: they have the same limitations as certificates, and on top require complex cryptography and policy management that needs to be handled by users and service providers.

The OLYMPUS project is aiming to combine the best of each to create a more secure approach to online identity management, while avoiding the drawbacks of the current solutions. The idea behind it is very simple: It follows the SSO paradigm, but instead of relying on a single trusted IdP, the trust is distributed among a set of servers, which might be run by different institutions. This removes the single point of failure in a system where compromise of the single trusted server has devastating consequences for the user [19]. Further, using techniques from P-ABC systems, one can realize the distributed IdP in an *oblivious* manner, i.e., the IdPs won't be able to learn or track users' online activities.

Roadmap. The document starts with an introduction to identity management and deployed technologies such as X.509 certificates and Single Sign-On (SSO) in Sect. 2. Afterwards, we discuss the idea of P-ABCs in Sect. 3, which constitutes a more privacy-friendly but also costly way for the user to demonstrate her identity herself. Lastly, we sketch the project's vision of creating privacy-friendly *and* user-friendly identity management in Sect. 4, including a summary of the discussion at the workshop.

2 Identity Management and Existing Solutions

As things are now, most people have several different online accounts with several different providers, each offering a specific service which requires knowledge of some specific attributes about its user. For example, Tinder wants pictures, age, location, residency, occupation and sexual preferences, an airline wants knowledge of your residency, citizenship and passport number and Facebook wants whatever you are willing to give. These are just examples of a few of the potential online accounts we can have. Most people have many such accounts and thus must repeatedly supply the same information/personal attributes to each of these different providers.

2.1 The Problem with Online Identity

Besides the annoyance of having to supply the same information several times to different parties, the main issue with this is how our information is protected at these sites. The trivial approach to account management is to pick a username and password for each account and then upload attributes to the provider. However, this yields significant issues in relation to breaches and linkability as discussed below.

Breach Issues. Despite the risk of heavy fines through legislation such as the GDPR, some providers do not ensure the safety of personal data and might, either through negligence or financial motivation, leak users' personal information [5]. This might in particular be true for smaller and more sleazy providers. However, sharing of personal information to a provider is not the only issue. A leakage of an account database, even without personal information, can have tremendous consequences. A database leakage of username and password pairs (regardless of whether hashing was used) can provide an adversary amble opportunity to learn a user's password for another provider. If the user has reused its password, this can then be used to compromise the user at other sites who might otherwise employ good security measures. However, once an adversary has the password, even strong security measures might not prevent illegitimate access.

In the recent years, companies have started to add measures to prevent illegitimate take-over of accounts, *even* if the adversary is in possession of a user's password. These include things such as two-factor authentication and machine learning. However, there are not employed globally and may not be sufficient to quench all compromises. Especially not targeted attacks, where combinations of security issues might also cause a compromise, even without a single compromised password [21].

Linkability Issues. The fact that the same personal attributes are stored at distinct providers allows collaborating providers to link users. The most simple way of linking a user is directly through the same username/mail address. However, the linking can also be done by simply comparing attributes. For example, gender, age and address is often enough to uniquely identify an individual.

2.2 Current Identity Management Solutions

In the following we will discuss some of the currently used ways of doing online identity management in a more integrated fashion than simply constructing new username/password-based accounts at every provider.

2.2.1 X.509 Certificates

It is possible to use a standard digital signature scheme in a Public Key Infrastructure (PKI). This can either be based on a global PKI (as is for example done on the web through root Certificate Authorities (CAs)) or based on a self-constructed PKI within a network or organization. In this setting all service providers will trust root certificates which are used to sign one or more intermediary CA certificates. Each of these intermediary CA certificates are used to sign certificates for each user on the network. Such a user certificate contains a handle on a specific user and could also include specific attributes asserted by the intermediary CA. When a user wishes to use a certain service provider it sends it an access request. The service provider then returns a message which it wishes to get signed by the user, in order to verify its identity. The user can then use the private signing key associated with its certificate, and return a signature on the message requested, along with its public certificate to the service provider. The service provider can then verify the *chain of trust* of the certificate. That is, if it trusts the root CA and the certificate of the intermediary CA. If so, it can accept the identity of the user.

Advantages. The main advantage of this approach over the username/password approach is that this offers a minimal need for trust. It is only necessary for the user to trust the CAs, in particular a compromise of a service provider will not yield any adversarial advantage in user impersonation. That said, if the signed info the user provides to the service provider includes personal attributes of the user and the service provider stores these offline, then we still have the issue that a compromise of the service provider would leak the user's attributes.

A great advantage of this approach however is that the user is only dependent on the CA being online when it constructs its certificate. That is, the user and the service provider are the only parties required to be online when a user wishes to sign in to the service provider.

Problems. Despite being simple and easy to implement on top of the PKI we already have in place in all web browsers, this solution still has several significant issues that make it unsuitable to be used in a non-enterprise setting.

First of all, all different service providers must agree to use this system (although any approach besides the individual username/password approach will require this) and to trust any *credible* intermediary CA that a user could use. However, a more pressing issue is that an individual user now has responsibility for its own private keys. That is, if they are lost the user will not be able to sign in anywhere and would need to create a new account. Even creating a new account would still pose a problem as an adversary would now be able to claim that they lost the private key of a legitimate user's account. Thus the request to create a new account (and close the old one) must be verified. Still, in case it is a sincere request then the legitimate user will not have its private key anymore and thus will not be able to verify the request. Even if a user-friendly solution is constructed to this problem, a perpetual issue of any PKI system still remains; distribution of revocation information. Specifically the service providers must have a method to check whether a user's certificate has been revoked. This could be done using an online service. However, that would remove the main advantage of this approach; that pervasive availability is not otherwise necessary.

Perhaps the biggest issue with using this approach with regular end-users is that real-world evidence has shown [14] users are terrible at protecting and remembering secrets. Furthermore, as most users have more than a single device, it would require that the private key must be distributed across the user's different devices. Thus the usability of this solution leave a lot to be desired. Finally we note that unless the user has a pseudonym and certificate for each service provider, the issue of linkability remains since the public certificate of the user will allow for unique identification across service providers.

2.2.2 Federated Identity

Instead of ensuring security using the weakest link among all the service providers, we might instead choose to centralize trust to a single, strongly trusted party. This is the idea behind using a *Federated Identity Provider* (Fig. 1). Basically we store all the attributes associated with a user at a trusted identity provider (IdP), and service providers rely on identity assertions of the IdP when granting users access to their services.

Whenever a user wishes to sign in to a service provider (perhaps including some attributes) it establishes contact with the service provider, who returns a request of the user's identity (and perhaps certain attributes). The user then checks the request and then signs in to the IdP (using a username/password approach). After signing in, it passes on the request to the IdP. The IdP then constructs a token based on the request and a timestamp (and potentially relevant attributes), and returns it to the user. The user can then pass it on to the service provider. The service provider can then verify the token either based on a signature of the IdP (known as a *bearer token*) or through a round of communication with the IdP. If the verification goes through and the attributes that have been asserted fit the information the service provider requires, then the service provider accepts the user and signs it in. This process is sketched in Fig. 2.

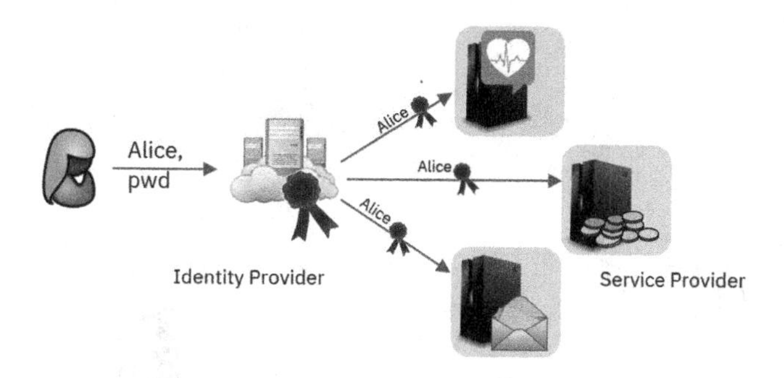

Fig. 1. Federated identity management

Federated Identity can, under a liberal definition, be considered *Single Sign-On* (SSO) as we only require signing on to the IdP in order to provide authentication for other servers (the service providers). However, under the "conservative" definition of SSO, it is required that a user only signs in once to the IdP, no matter how many different service providers it wishes to access. However, since tokens are often going to be short-lived and specific to a single service provider it is generally required of a user to contact the IdP every time it wishes to sign on to an new service provider (although it might not require singing in to the IdP again if it already has an open session).

Advantages. Similar to the X.509 solution, the main advantage of this approach is that we remove the requirement of trust of every service provider we wish to use and instead centralize the trust at an IdP. But also similar to the X.509 solution is that user attributes might still be stored at a service provider and could thus be leaked in case of a service provider compromise.

A great advantage of this approach, handling one of the main shortcomings of X.509, is that there is no need to store anything confidential user-side. Everything is based on a *single* username and password. This removes the burden of the user to manage secrets, or even multiple passwords, and allows it to sign in on any, possibly new, device.

Furthermore, since log-ins are now performed by signed single-use tokens that have a short validity only, this solution removes most of the complication of providing a reliable revocation mechanism across all service providers. Only the account with the IdP must provide an appropriate revocation or closing mechanism, which will automatically disable all other accesses. Since all relevant user attributes are now stored at a central IdP, it also removes the burden of the user to repeatedly upload and authenticate these with every service provider.

However, even more advantageous is the fact that since a trusted party (IdP) is now in possession of authenticated attributes, it can facilitate a great level of granularity on these in requests. Say for example that a user is born November 26, 1978. This might be asserted by a government authority and shared (and verified) by an IdP. Now when a service wishes to verify that a user is above the age of 18, it is not necessary to leak the birthday and year, even though it

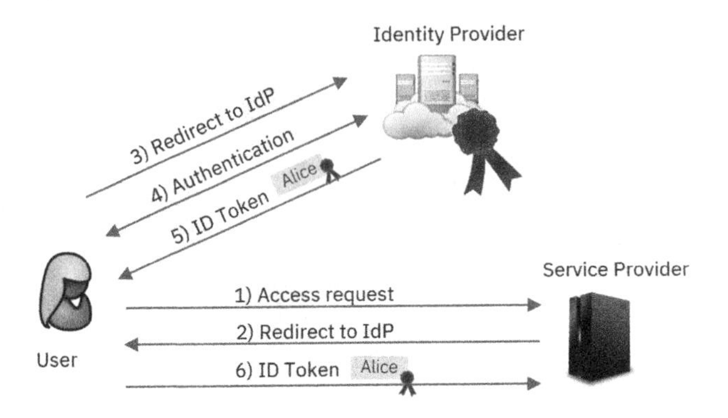

Fig. 2. Single Sign-On process in federated identity management

might have been asserted and signed in a monolithic manner, but instead the IdP can now simply assert that the user is above 18 based on the info it has. This of course applies to other attributes as well, for example location. Instead of sharing the exact address of a user, it might be enough to simply share which city, or even country, the user is residing in. This is a great way to ensure *granular and selective disclosure.*

Problems. Having all attributes at an IdP is not only an advantage, but also a liability. In case the IdP does get compromised, all the private information of the users is leaked. Like for the X.509 solution, the problem of linkability still remains. The user's account name is supplied to every service provider. Thus, any two service providers that compare logs will be able to identify if the same user has been using both services. Besides linkability, we now also have the issue of *traceability*, meaning that we now have a single party, the IdP, which will learn about *every* service provider that the user wishes to access.

SAML. There are different possibilities for realizing a federated identity solution. One of the oldest standardized approaches to this is SAML [29]. This approach is mainly focused on the enterprise setting rather than catering for end-users on the Internet. SAML has been specified by the Internet Engineering Task Force (IETF) in a Request For Comments (RFC) and also in an OASIS standard. The first version was standardized in 2002 and the second, major, revision was standardized in 2005. Messages in SAML are XML and communication between the different parties take part over HTTP, using the SOAP protocol.

The focus of this standard is not really efficiency, as can be seen from the high reliance on XML. That said, the protocol is relatively straightforward and specifies the token issued by the IdP to be signed. This means that the service provider must hold the public certificate of the IdP and checks the validity of the token by verifying the signature of the IdP, along with its timestamp. In particular this means that anyone holding the token will be able to impersonate the user towards the specific service provider. Hence it is crucial that such tokens have a sufficiently time-constrained validity.

OAuth. OAuth is a new standard used to achieve a federated identity solution. The standard leaves open a lot of aspects for deployment and is thus not as straightforward to implement without first making certain decision about which of the plethora of options to implement and support. The main spec of OAuth stems from 2012 [20] and is also specified by the IETF through an RFC. Unlike SAML, messages are sent in OAuth are based on JSON and is working directly on top of TLS.

The specification allows different flows and types of tokens. In one flow, which is for example used with OpenID Connect, bearer tokens are used, such that they are time-constrained and authenticated based on a signature of the IdP (similar to SAML). However, the OAuth specification also allows for verification of token validity through the service provider contacting the IdP. The OpenID Connect flow of OAuth is of particular interest as this has been implemented by several social federated identity providers such as Facebook, Google and Twitter.

3 Privacy-Enhancing Attribute-Based Credentials

In the previous section we have shown the benefits and drawbacks of deployed technologies such as X.509 certificates and federated identity solutions. A shared drawback is their little to no guarantees when it comes to the users' *privacy*: they either expose more user information than necessary or require a trusted party that can track all the users' online behaviour.

In response to this problem, Privacy-Enhancing Attribute-Based Credentials (P-ABCs) [3,9] have been proposed as a possible solution (Fig. 3). P-ABCs follow the same "offline" approach as X.509 certificates, i.e., users receive a *credential* from a trusted issuer. The credential contains a set of certified user attributes that the issuer vouches for, e.g., the user's name, age or address, and that can be used to convince a service provider of the validity of such claimed attributes. Unlike classical X.509 certificates, which have to be disclosed entirely and expose a static value whenever used, these P-ABC credentials allow the user to derive dedicated one-time use tokens that reveal only the information that is minimally necessary.

P-ABCs in a Nutshell. More precisely, when the user wants to access an account or resources at a service provider (SP), the provider first responds with a *presentation policy*, stating the requirements the user has to fulfill. These requirements can range from simple re-authentication requests w.r.t. a certain account, to requests for proofs of certain attributes, e.g., that the user is older than 18 years and lives in a European country. If the user's credential can satisfy the policy, it derives a so-called *presentation token* from the credential. The presentation token contains only the minimal amount of attributes requested by the policy. P-ABCs thereby support not only minimal disclosure techniques, but also predicate proofs. That is, instead of revealing the full date of birth for proving that she is over 18, the user can compute a dedicated proof revealing nothing beyond that fact. Importantly, the user can derive only presentation tokens that are consistent with the information certified in the credential, and the service provider can verify the token against the policy and be convinced about its correctness.

A crucial concept in P-ABCs is that of *user-controlled linkability*. Per default, presentation tokens are unlinkable, which means that a receiver cannot tell whether two presentation tokens are stemming from the same or two different users (unless this is revealed by the disclosed attributes). As full unlinkability would render P-ABCs useless in many applications, the user can also choose to create tokens w.r.t. a certain *pseudonym*. The pseudonym can be seen as a privacy-enhancing version of standard public keys: they are derived from a user secret key that is implicitly embedded in the credential, but the user can derive arbitrarily many and unlinkable pseudonyms from the same secret key. The user can choose to re-use an established pseudonym, which makes all tokens released for the same pseudonym linkable, but unlinkability across different pseudonyms is still guaranteed.

P-ABCs also support a number of additional features and concepts, such as privacy-friendly revocation or conditional disclosure and inspection. We refer to [3] for a detailed overview of these concepts and possible realizations.

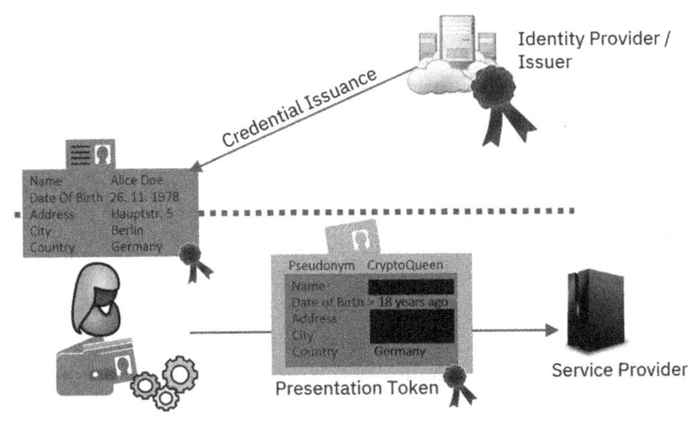

Fig. 3. Privacy-preserving authentication via P-ABCs

Advantages. The clear advantage of P-ABCs is their user centric and privacy-preserving behaviour. As P-ABCs follow the offline approach, the Identity Provider is only involved once, at credential issuance, but does not need to be contacted at every authentication request. This immediately solves the privacy problem posed by SSO solutions, but at the same time offers the same (or even better) minimal disclosure capabilities as an online IdP. The user has full control over the information she discloses and can authenticate in an unlinkable yet certified manner. The latter refers to the use of pseudonyms, which allow users to generate certified partial identities when desired, which can be backed-up by authenticated attributes. Thus, in terms of security and privacy, P-ABCs are clearly superior to conventional certificates and SSO-based solutions.

Problems. Despite being around for almost two decades now, and being available through mature realizations such as IBM's Identity Mixer [6,7] or Microsoft's U-Prove [28], P-ABCs have not seen noticeable adoption yet. The main reason is their bad usability. P-ABCs inherit the same struggle that conventional certificate-based solutions such as X.509 have, as it requires users to securely manage their credentials and key material. Any compromise of these credentials and keys will allow the attacker to impersonate the user, hence they must enjoy strong protection, e.g., be kept on trusted hardware tokens, such as smart cards. While being sufficiently secure, this is not a user-friendly approach as users need to have appropriate smart-card readers and apps for all their devices.

Furthermore, the handling and verification of P-ABCs tokens is more complex than of conventional signatures and credentials. The multitude of features and privacy-enhancing capabilities require an expressive policy language, credential matching and ontologies to verify whether the revealed information satisfies the required policy. The realizations of P-ABCs also require a careful combination of advanced cryptographic techniques, such as zero-knowledge proofs, which is not available through simple API calls of widespread cryptographic libraries. Instead, users and service provides must use tailored software packages in order to parse, create or verify such P-ABCs.

Another and more subtle problem of existing P-ABC scheme is their reliance on a single Identity Provider that issues the attribute-based credentials. While, for most schemes, the issuer does not have to be trusted for the privacy-related properties to hold, it still is the main entity and root of trust for the provided identity assertions. Thus, the issuer also presents a single point of failure, although is by far not as critical as in SSO solutions.

4 Single Sign-On with Distributed Trust

The OLYMPUS project is aiming to combine the best of the discussed approaches to create a more secure approach to online identity management, while avoiding the drawbacks of the current solutions. Roughly, it follows the Single Sign-On identity management approach, and uses tools from distributed and privacy-enhancing cryptography to obliterate a single point of failure. The resulting system's goals are as follows.

- Feature strongly secure and privacy-preserving authentication (e.g., minimal disclosure of attributes).
- No single point of failure or fully trusted third party.
- No secure hardware and credential management on the side of the user.
- Simple integration for service providers (integration into simple authentication standards such as, e.g., SAML and OAuth)

Let us first recap why existing solutions described in the preceding sections fail to achieve all these goals simultaneously.

X.509 certificates require the user to manage her secret keys and use a trusted device. The method is also not integratable into authentication standards such as SAML or OAUTH. While all this is not an issue for the federated identity approach, here the problem lies in the single point of failure that the fully trusted IdP constitutes. The Privacy-ABCs described in the previous section enable strong user privacy and minimal attribute disclosure, but require heavy machinery on the side of the user as well as the SP when verifying the credential – and there is no hope to make these systems work with simple authentication token standards. As we see, all systems have their advantages and excel in fulfilling some of the above listed goals – but none of them is able to fulfill all of them simultaneously.

4.1 The OLYMPUS Approach

Within the OLYMPUS project, an SSO scheme is designed that fulfills all of the above mentioned goals. The main idea is to modify the federated identity solution described in Sect. 2 and remove the single point of failure from it. To remind the reader, this single point of failure is the Identity Provider (IdP) which is in charge of storing the user's attributes, verifying its identity and issuing access tokens for the authenticated user. Within OLYMPUS, advanced cryptographic protocols are used to *distribute* the task of the single IdP to a set of n *different* IdPs (Fig. 4). None of these IdPs needs to be fully trusted. The system's privacy and security guarantees hold as long as *not all of the IdPs are corrupted.* The set of all n IdPs is called a *virtual IdP* (vIdP).

Essentially, the vIdP needs to perform all activities jointly that the former IdP of the federated identity scheme took care of. This boils down to two main tasks: (1) Verifying the password of the user and (2) issuing an access token. Within OLYMPUS, cryptographic solutions for distributing these two tasks among a set of servers are selected and carefully combined. We will give more details on the cryptographic part next.

Cryptography Part I: Distributed Password Verification. Password-based authentication is the most prominent form of user authentication towards an IdP. Normally, the IdP stores the password in form of, e.g., a hash $H(pw)$ – this is often called a *password file.* If the password is chosen only from a small dictionary, a malicious or compromised IdP can easily find out the user's password and potentially impersonate the user towards other IdPs. Server compromise, where attackers can steal billions of user passwords, is nowadays the main threat to password security.

Remembering long passwords and moreover typing them error-free into mobile devices puts a lot of burden on the user. However, password size can be dramatically reduced if (a) servers throttle login attempts and (b) the password file is protected from getting leaked through a server compromise attack. Within the OLYMPUS project, a system is designed where the role of the single

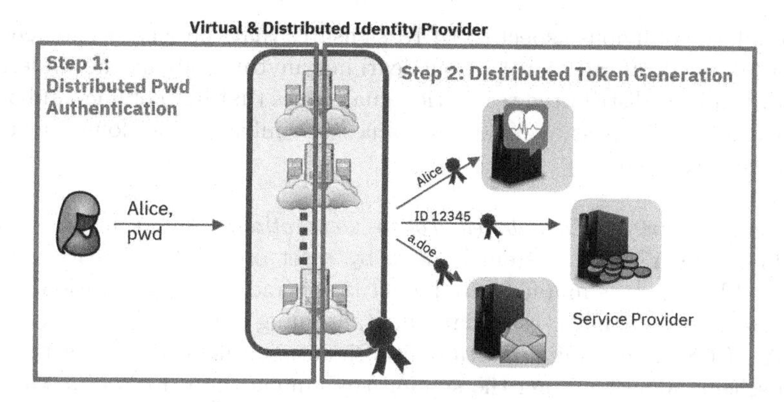

Fig. 4. The OLYMPUS privacy-preserving SSO system: the core concept is the virtual Identity Provider that consists of several servers of which none has to be fully trusted. The single servers need to jointly perform tasks such as password verification and token generation of the single IdP in a federated identity system.

and fully trusted IdP is distributed among n different servers. Intuitively, one can think of sharing the password file $H(pw)$ among all n IdPs, thus keeping an attacker from offline attacking pw as long as not all n of them are compromised. Realizations for such distributed password-verification protocols have already been proposed in [1,8,12] and make use of a *(distributed) oblivious pseudorandom function.*

Let us look at this primitive in a bit more detail, assuming the reader is already familiar with the concept of a pseudorandom function (PRF). An **oblivious** PRF (OPRF) is simply a PRF interpreted as an interactive 2-party protocol, where a user provides the input x to the function, the server contributes with a PRF key K and the user obtains $y \leftarrow \mathsf{PRF}_K(x)$. The server learns neither x nor y, and the user does not learn anything about K besides what he can derive from y. Oblivious PRFs have been introduced in [16] and have since then been extensively used in password-based protocols. Such OPRFs already allow the server to generate password files of the form $\mathsf{PRF}_K(pw)$ without the user ever having to reveal the password towards the server.

To protect against a single server running offline attacks by continuously evaluating the PRF and comparing it against $\mathsf{PRF}_K(pw)$, a **distributed** OPRF [23] (also implicitly used in [8]) can be used. Here, the PRF key K is *shared* among many servers, who then need to participate in the interactive evaluation protocol, contributing with their key share. As long as not all of them are compromised, offline attacks remain infeasible. Further, as every evaluation of the OPRF requires participation of all n servers, the remaining honest servers can refuse or pause the evaluation when they detect suspicious behaviour such as unusually high amounts of evaluation requests, which might indicate online guessing attacks.

Overall, the oblivious aspect of OPRFs ensures that the servers can generate password-derived information without learning anything about the underlying password; and the distributed realization guarantees that both offline- and online attacks against the password-derived values are infeasible as long as not all servers are corrupted.

Cryptography Part II: Distributed Token Generation. Generating the access token in the OLYMPUS system is done by a set of n IdPs instead of a single trusted IdP such as in a federated identity scheme. Token generation, from a cryptographic viewpoint, is essentially digital signing done by the IdP. Thus, for the OLYMPUS system a *distributed* digital signature scheme (DSIG) is required. Such a scheme allows sharing the signing key among the n IdPs in a way such that only if all of the IdPs participate in the distributed signing protocol the user will obtain a valid signature on his access request.

Distributed signature schemes can be obtained from the RSA assumption [2,4,13,15]. The main benefit is that verification of distributed RSA signatures does not introduce an overhead over verification in the non-distributed case. It is also possible to construct a distributed signature scheme based on (EC)DSA [11,25], where distributed key generation is more straightforward, but distributed signing less natural than with RSA-based signatures. Regardless of whether the underlying cryptographic assumption is RSA or a discrete-log type assumption, any DSIG scheme should guarantee *unforgeability* of signatures, even though the attacker might hold up to all but one signing key share. This property yields a system where access token cannot be forged unless all servers are corrupted.

Recently, also distributed versions for P-ABCs, or rather their main signature building block, have been proposed. More precisely, Sonnino et al. [30] and Gennaro et al. [18] have shown how distributed issuance can be realized for pairing-based credentials such as CL- or BBS-schemes. Despite the distributed issuance, the format of the resulting credential is preserved, which means that the user's derivation of presentation tokens is not impacted by this change. We sketch below how distributed P-ABCs could be used to instantiate the distributed token generation in a more privacy-preserving manner.

4.2 Open Questions

While the virtual IdP at the core of the OLYMPUS system resolves many security concerns, the system does not satisfy all our goals yet. The most crucial drawback is that each IdP can track the online behavior of the user, even amplifying the privacy problem of standard SSO. We will therefore investigate how the distributed system needs to be extended to prevent such tracking and ensure unlinkability of the users' requests.

P-ABCs and OLYMPUS*?* A straightforward approach to achieve the aforementioned privacy guarantees is to integrate P-ABCs into our Olympus solution and replace the standard distributed signatures (as described above) with a distributed issuance protocol of a P-ABC credential (see Fig. 5). That is, the user no longer receives signature shares of the final SSO token from the IdPs, but rather shares of a credential containing all her user attributes. We then leverage the power of P-ABCs to let the user derive the final SSO token as a P-ABC presentation token from the freshly received credential, inheriting the unlinkability and minimal disclosure features from P-ABCs.

A crucial design goal of Olympus is to avoid any trust assumptions for the clients' devices. Thus, we must ensure that a corruption of the user's device does not result in a permanent compromise of all her accounts. This can be realized by making the user credential *short-lived* only. More precisely, an additional *epoch* attribute gets introduced into every credential that must always be revealed in a presentation token and the service provider checks the epoch's validity.

However, avoiding the need for trusted hardware on the user side is pretty much the only advantage compared to standalone P-ABCs: users and service providers still need the full-blown P-ABC stack to generate and verify the presentation tokens. Thus, we will also investigate more lightweight alternatives to this approach.

Proactive Security. Last but not least, finding an adequate threat model is not straightforward for a distributed system such as OLYMPUS. While it is clear that not all servers can be under full control of the adversary, the system could be strengthened by allowing the adversary to *transiently* compromise all servers, as long as not all of them are compromised at the same time. For this, the OLYMPUS system would have to specify cryptographic primitives that allow *recovering* from adversarial corruption.

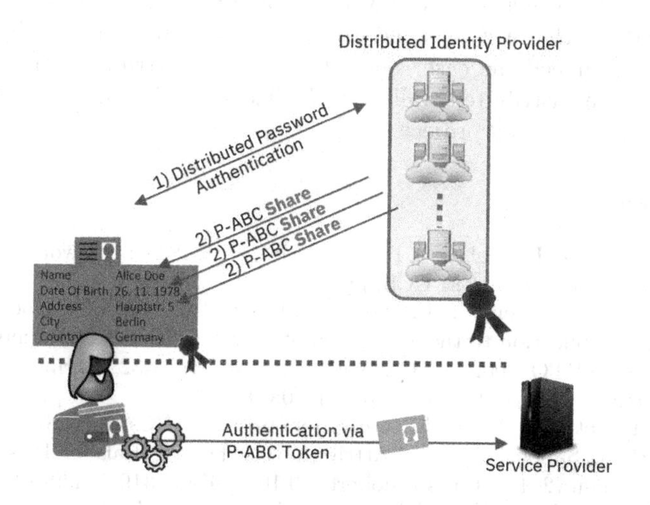

Fig. 5. A P-ABC based OLYMPUS system

4.3 Discussions at the Workshop

One question raised at the workshop targeted the P-ABC-based Olympus system sketched above. While guaranteeing strong privacy properties, this solution would inherit almost all disadvantages of a standalone P-ABC system. In particular, service providers need to run dedicated P-ABC libraries to parse and verify the complex presentation tokens, which contradicts the initial intention of seamless integration of SSO. It was discussed whether the verification of such tokens can be outsourced to an oblivious, and possibly distributed, party as well. Intuitively, outsourcing the entire process will be hard to realize, as a considerable part of the verification deals with parsing (xml-based) policies and matching them against the statements that are actually proven in the presentation token. Thus, this part does not seem very amendable to a blinded operation. The verification of the cryptographic evidence might be a more promising target, possibly using techniques from commuting signatures which maintain their verifiability even when being encrypted [17].

Another question was whether usage of *digital online wallets* could be helpful with online identity management. A digital wallet is essentially a storage for cryptographic objects such as digital coins, secret keys or passwords. Surely, using a cryptographic key from a wallet to authenticate to an IdP provides better security against a compromised IdP than authenticating with a password. Instead, the user protects access to her wallet with a password. Nonetheless, the wallet has to be stored on all devices the user wants to use a service from, or alternatively at a cloud or wallet provider. Essentially, using this approach, we mainly introduce more points of failure, since we now also have to trust the entity hosting the wallet, e.g., the wallet provider. Workshop participants further discussed whether a *distributed wallet solution* could remedy the situation. And this seems indeed the case, since then trust is distributed among a set of providers. Such a distributed wallet would likely require similar techniques as used within the OLYMPUS project.

References

1. Agrawal, S., Miao, P., Mohassel, P., Mukherjee, P.: PASTA: password-based threshold authentication. In: ACM CCS, pp. 2042–2059 (2018)
2. Algesheimer, J., Camenisch, J., Shoup, V.: Efficient computation modulo a shared secret with application to the generation of shared safe-prime products. In: Yung, M. (ed.) CRYPTO 2002. LNCS, vol. 2442, pp. 417–432. Springer, Heidelberg (2002). https://doi.org/10.1007/3-540-45708-9_27
3. Bichsel, P., et al.: An architecture for privacy-ABCs. In: Rannenberg, K., Camenisch, J., Sabouri, A. (eds.) Attribute-Based Credentials for Trust, pp. 11–78. Springer, Cham (2015). https://doi.org/10.1007/978-3-319-14439-9_2
4. Boneh, D., Franklin, M.K.: Efficient generation of shared RSA keys. J. ACM **48**(4), 702–722 (2001)
5. Cadwalladr, C., Graham-Harrison, E.: Revealed: 50 million Facebook profiles harvested for Cambridge analytica in major data breach. The Guardian. https://www.theguardian.com/news/2018/mar/17/cambridge-analytica-facebook-influence-us-election

6. Camenisch, J., Herreweghen, E.V.: Design and implementation of the idemix anonymous credential system. In: ACM CCS, pp. 21–30 (2002)
7. Camenisch, J., Krenn, S., Lehmann, A., Mikkelsen, G.L., Neven, G., Pedersen, M.Ø.: Formal treatment of privacy-enhancing credential systems. IACR Cryptology ePrint Archive 2014/708 (2014)
8. Camenisch, J., Lehmann, A., Neven, G.: Optimal distributed password verification. In: ACM CCS, pp. 182–194 (2015)
9. Camenisch, J., Lysyanskaya, A.: An efficient system for non-transferable anonymous credentials with optional anonymity revocation. In: Pfitzmann, B. (ed.) EUROCRYPT 2001. LNCS, vol. 2045, pp. 93–118. Springer, Heidelberg (2001). https://doi.org/10.1007/3-540-44987-6_7
10. Chaum, D.: Untraceable electronic mail, return addresses, and digital pseudonyms. Commun. ACM **24**(2), 84–88 (1981)
11. Doerner, J., Kondi, Y., Lee, E., Shelat, A.: Threshold ECDSA from ECDSA assumptions: the multiparty case. In: IEEE Symposium on Security and Privacy, SP, pp. 1051–1066 (2019)
12. Everspaugh, A., Chatterjee, R., Scott, S., Juels, A., Ristenpart, T.: The Pythia PRF service. In: 24th USENIX Security Symposium pp. 547–562 (2015)
13. Frankel, Y., MacKenzie, P.D., Yung, M.: Robust efficient distributed RSA-Key generation. In: Proceedings of the Thirtieth Annual ACM Symposium on the Theory of Computing, pp. 663–672 (1998)
14. Frauenfelder, M.: 'I forgot my PIN': an epic tale of losing $30,000 in bitcoin. Wired. https://www.wired.com/story/i-forgot-my-pin-an-epic-tale-of-losing-dollar30000-in-bitcoin/
15. Frederiksen, T.K., Lindell, Y., Osheter, V., Pinkas, B.: Fast distributed RSA key generation for semi-honest and malicious adversaries. In: Shacham, H., Boldyreva, A. (eds.) CRYPTO 2018. LNCS, vol. 10992, pp. 331–361. Springer, Cham (2018). https://doi.org/10.1007/978-3-319-96881-0_12
16. Freedman, M.J., Ishai, Y., Pinkas, B., Reingold, O.: Keyword search and oblivious pseudorandom functions. In: Kilian, J. (ed.) TCC 2005. LNCS, vol. 3378, pp. 303–324. Springer, Heidelberg (2005). https://doi.org/10.1007/978-3-540-30576-7_17
17. Fuchsbauer, G.: Commuting signatures and verifiable encryption. In: Paterson, K.G. (ed.) EUROCRYPT 2011. LNCS, vol. 6632, pp. 224–245. Springer, Heidelberg (2011). https://doi.org/10.1007/978-3-642-20465-4_14
18. Gennaro, R., Goldfeder, S., Ithurburn, B.: Fully distributed group signatures (2019). https://www.orbs.com/wp-content/uploads/2019/04/Crypto_Group_signatures-2.pdf
19. Goel, V., Perlroth, N.: Yahoo says 1 billion user accounts were hacked. The New York Times. https://www.nytimes.com/2016/12/14/technology/yahoo-hack.html
20. Hardt, D.: The OAuth 2.0 Authorization Framework. RFC 6749, October 2012. https://tools.ietf.org/html/rfc6749
21. Honan, M.: How Apple and Amazon security flaws led to my epic hacking. Wired. https://www.wired.com/2012/08/apple-amazon-mat-honan-hacking/
22. Hong, N., Hoffman, L., Andriotis, A.: Capital one reports data breach affecting 100 million customers, applicants. Wall Street J. https://www.wsj.com/articles/capital-one-reports-data-breach-11564443355
23. Jarecki, S., Kiayias, A., Krawczyk, H.: Round-optimal password-protected secret sharing and T-PAKE in the password-only model. In: Sarkar, P., Iwata, T. (eds.) ASIACRYPT 2014. LNCS, vol. 8874, pp. 233–253. Springer, Heidelberg (2014). https://doi.org/10.1007/978-3-662-45608-8_13

24. Kosinski, M., Matz, S., Gosling, S., Popov, V., Stillwell, D.: Facebook as a research tool for the social sciences. Am. Psychol. **70**, 543–556 (2015). https://doi.org/10.1037/a0039210
25. Lindell, Y., Nof, A.: Fast secure multiparty ECDSA with practical distributed key generation and applications to cryptocurrency custody. In: ACM CCS, pp. 1837–1854 (2018)
26. Newman, L.H.: Equifax officially has no excuse. Wired. https://www.wired.com/story/equifax-breach-no-excuse/
27. O'Flaherty, K.: Collection 1 breach - how to find out if your password has been stolen. Forbes (2019). https://www.forbes.com/sites/kateoflahertyuk/2019/01/17/collection-1-breach-how-to-find-out-if-your-password-has-been-stolen/#37aa36382a2e
28. Paquin, C., Zaverucha, G.: U-prove cryptographic specification v1. 1. Technical report, Microsoft Corporation (2011)
29. Rekhter, Y., Li, T.: Security Assertion Markup Language (SAML) V2.0. OASIS, March 2015. http://saml.xml.org/saml-specifications
30. Sonnino, A., Al-Bassam, M., Bano, S., Meiklejohn, S., Danezis, G.: Coconut: threshold issuance selective disclosure credentials with applications to distributed ledgers. In: Network and Distributed System Security Symposium, NDSS (2019)
31. Wang, C., Jan, S.T.K., Hu, H., Bossart, D., Wang, G.: The next domino to fall: empirical analysis of user passwords across online services. In: Proceedings of CODASPY (2018)

SoK: Cryptography for Neural Networks

Monir Azraoui[1], Muhammad Bahram[2], Beyza Bozdemir[3], Sébastien Canard[1], Eleonora Ciceri[4], Orhan Ermis[3(✉)], Ramy Masalha[2], Marco Mosconi[4], Melek Önen[3], Marie Paindavoine[5], Boris Rozenberg[2], Bastien Vialla[1], and Sauro Vicini[4]

[1] Applied Crypto Group, Orange Labs, Caen, France
{monir.azraoui,sebastien.canard,bastien.vialla}@orange.com
[2] IBM Haifa, Haifa, Israel
{muhammad,Ramy.Masalha,borisr}@il.ibm.com
[3] EURECOM, Sophia Antipolis, France
{beyza.bozdemir,orhan.ermis,melek.onen}@eurecom.fr
[4] MediaClinics, Lissone, Italy
{e.ciceri,m.mosconi,s.vicini}@mediaclinics.it
[5] Cybersecurity Research, Renault, Paris, France
marie.paindavoine@renault.com

Abstract. With the advent of big data technologies which bring better scalability and performance results, machine learning (ML) algorithms become affordable in several different applications and areas. The use of large volumes of data to obtain accurate predictions unfortunately come with a high cost in terms of privacy exposures. The underlying data are often personal or confidential and, therefore, need to be appropriately safeguarded. Given the cost of machine learning algorithms, these would need to be outsourced to third-party servers, and hence protection of the data becomes mandatory. While traditional data encryption solutions would not allow accessing the content of the data, these would, nevertheless, prevent third-party servers from executing the ML algorithms properly. The goal is, therefore, to come up with customized ML algorithms that would, by design, preserve the privacy of the processed data. Advanced cryptographic techniques such as fully homomorphic encryption or secure multi-party computation enable the execution of some operations over protected data and, therefore, can be considered as potential candidates for these algorithms. However, these techniques incur high computational and/or communication costs for some operations. In this paper, we propose a Systematization of Knowledge (SoK) whereby we analyze the tension between a particular ML technique, namely, neural networks (NN), and the characteristics of relevant cryptographic techniques.

Keywords: Privacy · Neural networks · Homomorphic encryption · Secure multi-party computation

Published by Springer Nature Switzerland AG 2020
M. Friedewald et al. (Eds.): Privacy and Identity 2019, IFIP AICT 576, pp. 63–81, 2020.
https://doi.org/10.1007/978-3-030-42504-3_5

1 Introduction

Artificial Intelligence (AI) is a generic term used to designate any system that is capable of learning and solving problems based on the perception of its environment. AI is today divided into several sub-fields, depending on the technical considerations, and, several tools related to AI are now capable of solving a lot of difficult problems related to computer science. These tools such as neural networks (including deep learning), Bayesian networks, or classifiers are well-known today. In this paper, we focus on neural networks (NN) that are inspired by the architecture of neurons in the human brain. NN have two modes of operations: a *training phase* (also called learning phase) in which the network learns a new capability from a training dataset and a *querying phase* (also known as prediction or classification phase), where this capability is tested over new data. The first phase takes as input the training dataset that permits the "neurons" to create their data model used in the second phase. While goal of NN is to learn from data, the European General Data Protection Regulation (GDPR) [2] aims to protect the data often considered as personal and hence, privacy sensitive. NN and GDPR cannot a priori live together, and several advanced cryptographic techniques are used to reconcile them.

Among these advanced cryptographic techniques, Secure Multi-Party Computation (MPC) [27] allows several parties to put in common their own input to obtain a unique output. While the latter can be made publicly available or kept private at the end of the protocol, each input should remain confidential at any time. Nowadays, several practical constructions exist [4,27,38,58,59], based on garbled circuits, secret sharing, or oblivious transfer, and some implementations are available. Another important cryptographic technique is homomorphic encryption. When only additions or multiplications are performed in the "encrypted world" using RSA [50], ElGamal [21] or Paillier [48], which is quite limited in practice, the possibility to have a "fully" homomorphic encryption (FHE), capable to perform both additions and multiplications in the encrypted world in an arbitrary manner, is quite recent. And since the first construction of FHE in 2009, several papers [9,17,18,22] have appeared on this subject or the way to use such encryption schemes with practical algorithms [7,11,26,53]. As for MPC, several implementations for FHE exist. In this case, the entity encrypting the data will necessarily be the one who will be able to decrypt the result.

The possibility to use advanced cryptographic techniques for AI has first been suggested in 2000 by Lindell and Pinkas [37], in the case of data mining and decision trees. Regarding neural networks, to the best of our knowledge, the first paper on the subject is the work by [47] in 2007. However, by the second half of 2010s, things accelerated a lot and many papers have been published, either focusing on the training phase [34], or working on the classification phase [7,26,39,45], which is the most frequent case as it is the easiest, but also the most useful one.

In this work, we propose the first Systematization of Knowledge (SoK) paper that compares the different approaches and results of privacy-preserving NNs based on advanced cryptographic techniques. As state-of-the-art non-linear NN layers (pooling and activation) are too complex to be directly executed in the encrypted world, there is a strong need for approximating them. We first compare the different strategies for these approximations, giving the ones we have found in the literature and the resulting accuracy. Then, this allows us to exhibit the NN operations that need to be performed on protected data (such as polynomial evaluation or comparison). Finally, we present a performance evaluation to compare the advanced cryptographic techniques on particular NN models designed for arrhythmia classification and image classification.

The rest of the paper is organized as follows. In the next section, we overview the Neural Networks operations and identify the main privacy and security requirements in our context. In Sect. 3, we compare MPC and FHE in their use to provide a privacy-preserving NN. We detail the solutions in Sect. 4. Section 5 presents a performance evaluation for two NN classifier based on underlying advanced cryptographic techniques. Finally, our conclusive remarks and some potential future works are given in Sect. 6.

2 Neural Networks

In this section, we briefly define neural networks and describe their underlying operations. We further discuss how these operations can be approximated so that cryptographic tools can support them.

2.1 Definition

A Neural Network (NN) is a particular case of machine learning techniques. It consists of several interconnected nodes called *neurons* that are structured in layers. Each neuron performs one operation depending on which layer it belongs to. The NN layers are described are as follows:

- **Convolution layer (optional):** The basic idea behind a convolution layer is to slide a filter, or kernel, over the original input to obtain information about the similarity between the chunk of the original image covered by the filter and the filter itself. On input a (ℓ_1, ℓ_2)-matrix $\mathbf{X}$ representing the data, and a smaller (ℓ_1', ℓ_2')-matrix K representing the kernel, the convolution function outputs a matrix $\mathbf{Y}$ as follows:

$$\mathbf{Y}[n,m] = \sum_{i=1}^{\ell_1'} \sum_{j=1}^{\ell_2'} \Big(K[i,j] \cdot \mathbf{X}[n+i-1, m+j-1] \Big).$$

  $\mathbf{Y}$ is a map corresponding to the filter K, which was slid over the image with a stride of 1. This function is executed for all filters considered in the layer to obtain multiple maps.

- **Activation layer:** The goal of the activation layer is to determine whether the pattern of a filter is actually present at a given position in the data. There are different kinds of activation functions. We consider the three following ones:
  - sigmoïd σ as $\mathbf{y} = \frac{1}{1+e^{\mathbf{x}}}$;
  - hyperbolic tangent tanh as $\mathbf{y} = \frac{e^{\mathbf{x}}-e^{-\mathbf{x}}}{e^{\mathbf{x}}+e^{-\mathbf{x}}}$;
  - the rectified linear unit (ReLU)[1] as $\mathbf{y} = \mathsf{max}(0, \mathbf{x})$. When using the ReLU activation functions, it is highly recommended to use normalization by adding a batch normalization layer prior to each activation layer to obtain a stable and normalized distribution before the execution of the activation function [35].
- **Pooling:** The max pooling operations consists in reducing the spatial size of the input in order to make it more manageable.
- **Fully Connected layer:** The fully connected layer correlates the output of the previous layer with the features of each class.

2.2 Architecture

In a scenario, NN are outsourced to a third party server, we basically consider two actors:

- a client $\mathcal{C}$ having the input i.e. some data $\mathbf{X}$;
- a server $\mathcal{S}$ having already received a trained neural network model (M).

The main goal is to delegate the *querying* phase to $\mathcal{S}$. Hence, at the end of the process, $\mathcal{C}$ wants to obtain $\mathbf{R} = \mathsf{M}(\mathbf{X})$, where M consists of a set of the predefined functions. In order to ensure data privacy, namely the privacy of $\mathbf{X}$ against $\mathcal{S}$, some NN operations unfortunately cannot easily/efficiently be supported by cryptographic techniques. These operations should therefore be approximated into polynomial operations without having a significant impact on the accuracy of the overall NN. In the next section, we discuss the different approximation methods applied for each particular layer.

2.3 Approximation of NN Layers

As previously mentioned, some NN layers contain nonlinear operations so that it may be hard to directly execute them on the encrypted input. The best idea is to approximate them to simplify, and then improve the efficiency of such execution without sacrificing from the NN accuracy.

- **Approximation of pooling functions:** As the max pooling function is not linear, the literature [26] suggests to approximate it by either summing up all values or computing their average.

[1] The ReLU is currently the activation function that is mostly used. There are also some variants, such as the parametric version PReLU and the Exponential Linear Units ELU.

- **Approximation of the activation function:** There are several ways to perform the activation function but, regarding the literature, the most suitable one in terms of efficiency and prediction accuracy is to use square function, which directly computes x^2 for any given input x.

Operations: Regarding the description of the different main functions (directly or using an approximation) that should be executed during an NN evaluation (activation and pooling), we have extracted the main basic functions that should be operated on encrypted data. We have obtained the following results.

- **Addition:** activation, pooling, fully connected, and convolutional layer.
- **Multiplication:** activation, fully connected, and convolutional layer.
- **Polynomial evaluation:** activation.
- **Comparison:** activation, pooling.

2.4 Security Requirements

In order to identify the main privacy requirements, we first define the overall context where neural networks are used with privacy sensitive data. We therefore consider a scenario whereby an entity, such as a hospital (or an SME), is collecting or has already collected some data from some *data subjects* such as patients (or customers). The entity wishes to infer some information about some clinical diagnosis (or customer habits), or predict the diagnosis of the next patient (or the behavior of the next customer), using this collected data and the NN tool. This entity that is usually considered as the *data controller* will outsource both the data and the relevant computations for analyzing the data and performing the prediction to a powerful cloud server, which is defined as the *data processor*. The data owner or any authorized party can further query the model. In the sequel of this paper, this party is called the *querier*.

With GDPR and given the sensitiveness of the collected and processed data, there is a strong need for the data subjects to be protected. The first and foremost requirement to satisfy is to protect their data against unauthorized access by third parties during the entire lifetime of the data, i.e., from their collection until their analysis or even their deletion. Entities who can be considered as unauthorized to access are of three types: First, the *external parties* do not play any role in the collection, storage and analysis of the data. Second, the *cloud server* acts as the data processor and is considered as a third-party server that only provides storage and computational resources to the data controller. The cloud server is considered as a honest-but-curious adversary. Finally, the *data collector* collects the data (such as the hospital or the SME in the previous example) and can also be sometimes prevented from accessing the cleartext content.

In addition to the collected data, the query of the NN prediction/classification should remain private against unauthorized parties. The query (and sometimes the corresponding result) should not reveal any information to some potentially malicious adversaries. Those can be external parties who basically should not learn any information (neither the data nor the queries and results). Even if the cloud server can be a malicious party, it should be able to process the query without discovering any information about the data being processed. In some cases, the cloud server should even not learn the classification result.

Finally, in the context of neural networks, even the model can reveal some privacy sensitive information and therefore needs some protection. Similarly to the previous two information, the model should not be revealed to external parties. The model should also be protected against the querier or the data subjects as it is obtained based on all collected data. In some cases, the model should even remain private against the cloud server.

3 Cryptographic Techniques

This paper investigates the suitability of two advanced cryptographic techniques to neural networks, namely: fully homomorphic encryption (FHE) and secure multi-party computation (MPC), and overviews the state-of-the-art solutions that succeed in obtaining privacy preserving neural networks by applying some approximations on the underlying operations.

3.1 Multi-party Computation

Secure multi-party computation is introduced in early 1980s by Yao [58,59] who focused on the two-party computation (2PC) case by defining Yao's Millionnaire problem. Then, by Goldreich et al. in [28], the problem was generalised to multiple parties.

Definition. Secure multi-party computation (MPC) is defined as a system in which a group of *data owners* can jointly compute a function of their private inputs without disclosing the underlying inputs, but the output of the function. Formally, let $P_1, \ldots, P_n$ be n parties and each of them having input $x_1, \ldots, x_n$, respectively. The parties $P_1, \ldots, P_n$ want to jointly compute the function f over all inputs $\{x_1, \ldots, x_n\}$ and learn the output without revealing their input.

MPC should ensure the following two properties, at least: (i) *input privacy*, i.e., parties' inputs should remain private and only the output of the function is learned; (ii) *correctness*, i.e., even if some parties misbehave, the correct output should be obtained.

Building Blocks. Existing MPCs leverage Yao's Garbled circuits [58,59] and secret sharing (additive or Boolean) [4]. We briefly explain each method in the following paragraphs. Before going into details, we first introduce Oblivious Transfer (OT) method.

Oblivious Transfer. Oblivious transfer (OT) [49] is a fundamental cryptographic primitive that is used as a building block in MPC. OT allows a party to choose k out of n secrets from another party without disclosing which secrets have been chosen. Usually, the 1-out-of-2 OT is used, ensuring that one secret out of two of them is retrieved: Let Alice have two inputs x_0 and x_1, and Bob selects a bit b and wants to obtain x_b. OT ensures that Bob does not learn x_{1-b} and does not reveal b to Alice.

Yao's Protocol. Yao's protocol (*a.k.a.* Garbled Circuits (GC)) is a secure two-party computation that allows the two parties to evaluate a function $f(x_1, x_2)$ in the presence of semi-honest adversary (i.e., this adversary has to truly follow the protocol yet s/he can try to extract information during the execution of protocol), where inputs x_1 and x_2 are provided by two parties, namely Alice and Bob. Let Alice be the *garbler* and Bob be the *evaluator*. Alice builds a garbled version of a circuit for the function f by obfuscating all possible outputs in the truth table. The garbled circuit and Alice's garbled input $GI(x_1)$ are sent to Bob. Alice also provides a map from the garbled-circuit outputs to the actual bit values. After receiving the circuit, Bob uses 1-out-of-2 OT [49] with Alice to obliviously obtain his garbled circuit values $GI(x_2)$ without revealing it to Alice. Bob further evaluates the function $f(x_1, x_2)$ using $GI(x_1)$ and $GI(x_2)$.

The function f is evaluated through a Boolean circuit. The *garbler* assigns two keys that correspond to bit values 0 or 1 for each wire of the circuit. Then, Alice, the garbler, computes four ciphertexts for each binary gate with the input wires and the output wire. After obtaining ciphertexts, Alice randomly orders these four outcoming values. The evaluator, Bob, can decrypt the correct row from the table if he successfully obtains the pair of keys from Alice via OT.

Secret Sharing. Alternatively to Yao's Garbled Circuits, MPC solutions based on secret sharing consist of distributing secrets among parties involved in the system and further evaluate the function defined as a circuit accordingly. The GMW protocol [28] relies on Boolean shares and mainly support XOR operations over single bits. The function to be evaluated is encoded as a Boolean circuit and OT is used during the circuit evaluation. The Boolean circuit takes as inputs bit u from Alice and bit v from Bob. These bits are first secret-shared between the parties as $u = u_1 \oplus u_2$ and $v = v_1 \oplus v_2$, where share 1 belongs to Alice and share 2 to Bob. Then both parties evaluate the circuit gate by gate. For example, given shared values, an XOR gate with input bits u and v and output bit w is evaluated locally (i.e. without communication) by each party by computing $w_i = u_i \oplus v_i$. Value w can be retrieved by exchanging and XORing the shares. Some other solutions use arithmetic circuit whereby inputs are additively shared and addition gates (respectively multiplication gates) correspond to XOR gates (resp. AND gates).

Available Implementations. Several practical open-source implementations for 2PC/MPC systems have been proposed in recent years. Some consist of high-level description languages and corresponding compilers used to specify the function to be securely evaluated and to translate it into a Boolean or arithmetic circuit; for example, Fairplay [41] and its extension to multiple parties, FairplayMP [5]. Other implementations offer libraries for MPC such as SCAPI [20]. Finally, some other solutions propose more comprehensive frameworks consisting of libraries, languages and their compilers, runtime environments and OT tools such as TASTY [33], ABY [19] or EMPtoolkit [57].

3.2 Fully Homomorphic Encryption

Homomorphic encryption allows to process encrypted data without learning neither the input data nor the computation result. The data owner, Alice, can thus delegate some of her computation over sensitive data to a non-trusted party, Bob. Alice encrypts her data under her own public key, and sends the encryption to Bob. Once received, Bob can evaluate a circuit over Alice's inputs, obtaining a result still encrypted under Alice's public key. Bob sends the result back to Alice, who is the only party able to decrypt it. Formally, a homomorphic encryption scheme is composed of four procedures, defined as follows:

- KeyGen: the key generation procedure takes as input the security parameter λ and outputs the public-secret key pair $(\mathsf{pk}, \mathsf{sk})$.
- Enc: the encryption procedure uses the public key pk to transform a message M into a ciphertext c.
- Eval: the evaluation procedure takes as inputs a circuit $\mathcal{C}$, and ciphertexts $c_1, \ldots c_\ell$ such that $c_i = \mathsf{Enc}(M_i, \mathsf{pk})$. It outputs another ciphertext $c_{\text{res}} = \mathsf{Eval}(\mathcal{C}, (c_1, \ldots c_\ell))$.
- Dec: the decryption procedure uses the secret key sk to transform back a ciphertext c to the message M.

Homomorphic encryption has been known since many decades. The concept, initially called privacy homomorphism, has been introduced in 1978 by Rivest, Adleman and Dertouzos [50] and several "basic" schemes verifying this property followed: such as the well-known RSA [50] (multiplicatively), ElGamal [21] (multiplicatively, or additively, depending on the variant) and Paillier [48] (additively). In 2005, the Boneh-Goh-Nissim encryption [6] scheme was able to perform an arbitrary number of additions, and one single multiplication (hence becoming one of the first somewhat homomorphic encryption schemes). The first FHE scheme was proposed in 2009 by Craig Gentry [24] with an ingenious idea called bootstrapping. Further, since research on FHE has been prolific and new schemes have been proposed based on Gentry's first idea to improve the efficiency of the original but impractical schemes. Both somewhat and leveled homomorphic schemes include some noise in the ciphertexts. This noise grows throughout computations. A refreshing procedure, *bootstrapping*, can be added to manage the noise growth. The bootstrapping operation remains a bottleneck when evaluating circuits in the encrypted domain.

While Gentry's work is today considered as the first FHE generation, the second generation has been marked by Brakerski and two important schemes were published in 2012: Brakerski-Gentry-Vaikuntanathan (BGV) [9] and Fan-Vercauteren (FV) [22], for which many optimizations have later been proposed. This second generation of FHE enjoys a huge efficiency increase in comparison to the first one [29,30]. The third generation has started in 2013 with the

Gentry-Sahai-Waters (GSW) scheme [25] and a different way to represent keys. However, this generation is less used, because existing optimizations are not compatible with such a kind of representation. Recently, the fourth generation has been introduced by the scheme named TFHE [17,18], which is based on a mathematical object called a torus, allowing to have the advantage of both second and third generations. Today, existing implementations are mostly based on the second and fourth generations.

An entire line of work is dedicated to the FHE over the integers. These solutions rely on the Approximate Greatest Common Divisor (AGCD) problem. They achieve mostly the same properties as LWE-based FHE. In [16], the authors proved that both AGCD and LWE problems are equivalent. Another line of work has designed FHE schemes based on the NTRU problem, such as LTV [40]. Finally, a scheme allowing to do computation directly on floating point numbers was introduced [15]. This scheme had major implications in applied homomorphic encryption by facilitating the implementation of algorithms having a lot of numerical computation such as neural networks.

Available Implementations. There are multiple implementations of (fully) homomorphic encryption. Most of them provide a leveled homomorphic scheme (without bootstrapping). The Simple Encrypted Arithmetic Library (SEAL)[2] is edited by Microsoft and provides an implementation of the FV scheme in C++. A Python version also exists, Pyfhel[3]. Unlike FV-NFLlib[4], which is based on the NFLlib library dedicated to lattice cryptography, SEAL does not require any dependencies. PALISADE[5] is a standalone library written in C++ that lets the user choose between four schemes: FV, BGV, LTV and Stehlé Steinfield. Finally $\Lambda \circ \lambda$[6] is a Haskell library that offers a refinement of BGV scheme. HElib[7] implements BGV in C++ and provides bootstrapping. FHEW[8] implements a fast bootstrapping procedure. Finally, TFHE[9] implements one of third generation of FHE schemes that features boostrapping under 0.1 s. A performance comparison between the most used libraries published in [43].

4 Existing Solutions

4.1 MPC-Based Privacy Preserving NN Solutions

We analyze the method of secure MPC-based privacy preserving neural network solutions. Most of the early solutions use the 2PC-based approach between two

[2] https://www.microsoft.com/en-us/research/project/simple-encrypted-arithmetic-library/.
[3] https://github.com/ibarrond/Pyfhel.
[4] https://github.com/CryptoExperts/FV-NFLlib.
[5] https://git.njit.edu/palisade/PALISADE.
[6] https://hackage.haskell.org/package/lol.
[7] https://github.com/shaih/HElib.
[8] https://github.com/lducas/FHEW.
[9] https://github.com/tfhe/tfhe.

entities, namely the client and the server. One such example for 2PC-based solutions is SecureML [45] which aims at building a privacy preserving training and classification solution for neural networks using secure multiparty computation. SecureML uses the stochastic gradient descent method to build the model and supports secure arithmetic operations on shared decimal numbers. In SecureML, evaluation of ReLU using Garbled circuits and they further a secure computation of polynomial approximation for activation functions (i.e., *sigmoïd* and *softmax* functions) are provided. Another secure 2PC-based study is MiniONN proposed by Liu et al. [38] for providing privacy preserving convolutional neural networks (CNN). To ensure data privacy, MiniONN defines oblivious transformations for each CNN operation. Moreover, DeepSecure [51] relies on Yao's Garbled circuits to securely compute deep learning models. Rouhani et al. [51] use sigmoïd and *tanh* as activation functions due to the optimization of Garbled circuits. Ball et al. [3] propose an extension to DeepSecure that is a secure evaluation of the NN classifier based on garbled circuits. Unlike DeepSecure, authors in [3] make use of arithmetic circuits instead of Boolean circuits and further use some improved techniques for the computation of matrix multiplication, activation function, and max-pooling. Chameleon proposed by Riazi et al. [52] is a hybrid protocol to securely compute function evaluation where two parties jointly perform a function without disclosing their inputs. Chameleon is called a hybrid framework since two parties can use Garbled circuits, the Goldreich-Micali-Wigderson (GMW) protocol, and arithmetic sharing. Similar to Chameleon, another solution EzPC [13] is a cryptographic cost-aware secure 2PC protocol generator and makes use of arithmetic and Boolean circuits for a secure NN classification. This scheme is useful for the one who does not have sufficient knowledge on the cryptographic techniques to compute a secure NN classification since EzPC takes source code as an input and outputs 2PC protocols.

Recently, MPC-based solutions also proposed for NN classification and training. ABY^3 proposed by Mohassel et al. [44] is an extension to SecureML. In SecureML, there are two non-colluding servers and clients share their private inputs among them whereas in ABY^3 there are three non-colluding servers; hence, arithmetic, Boolean and Yao's sharing are redefined across these three servers. SecureNN [55] presents secure a 3-party computation protocols for NN operations. This scheme supports both NN training and NN classification on convolutional neural networks based on MNIST dataset. SecureNN does not use garbled circuits and oblivious transfer to obtain performance gain in terms of the communication cost. The proposed protocols are secure against not only semi-honest adversary but also malicious adversary. Furthermore, a protocol proposed by Ohrimenko et al. [46] use trusted SGX processors for NN training.

4.2 FHE-Based Privacy Preserving NN Solutions

The first work leveraging homomorphic encryption for NN is CryptoNets [26]. The selected homomorphic encryption scheme is FV [22], without its bootstrapping procedure. In order to minimize the computation depth, they approximate the activation function by the square function, which only consumes one level of homomorphic evaluation. While effectively transforming a neural network into a FHE-friendly circuit, the square function is only a good approximation of the RELU on restricted distributions.

In [12], the authors introduce a batch normalization layer before each activation layer in order to stabilize the distribution and achieve better accuracy. Where CryptoNets gets an accuracy of 98.95 %, Chabanne et al.'s work [12] permits to obtain 99.30 %. Moreover, in both protocols, the number of neurons that can be handled is bound by the initially chosen parameters.

Encoding real numbers in a way that preserves the operations is a core problem that might affect performances. CryptoNets first convert the real number to a fixed precision number, and then embed it into a polynomial whose coefficients are its binary decomposition. The inverse mapping consists in evaluating the polynomial at 2. While neural networks operate on real numbers, with natural arithmetic, plaintext spaces for homomorphic encryption are finite field of polynomials (modular arithmetics). Therefore, different encoding methods have been proposed to operate with integers, as in the case of BGV [8] or FV [22].

In order to avoid this encoding step, the authors in [7] only consider discretized neural networks that directly operate on integers. They use a particularly restrictive form of neural network, Binarized Neural Networks, where weights are set in $\{1, -1\}$. In order to preserve this property, they select the `sign` function as approximation function, where negative integers are mapped to -1 and positive ones to 1. The underlying homomorphic scheme is TFHE [17,18], with bootstrapping. They modified this scheme to compute both bootstrapping and the `sign` function at the same time. Due to the use of bootstrapping, their scheme can evaluate networks with arbitrary numbers of neurons. Although the use of bootstrapping allows a huge increase in the number of layers in the network, the discretization of the network incurs a non negligible loss of accuracy (obtaining "only" 96%).

A recent work of [34] addresses the problem of both training and classification over encrypted data. Their solution follows the approach of approximating activation functions with low-degree polynomials by introducing a new method based on Chebyshev-like orthogonal polynomials. Besides the authors resort to homomorphic encryption (HElib) to encrypt both the inputs and (unlike previous work [7,26]) the models. To handle the noise in the ciphertexts during the computations, they do not consider bootstrapping since it imposes a high computational overhead, but instead propose to have the server (i) check the noise level and (ii) ask the client to recrypt the ciphertext when the noise reaches a predefined threshold.

4.3 Hybrid Solution

The classification protocol of GAZELLE [36] combines FHE and MPC (via Garbled Circuits) to compute neural network classifications privately. Fully connected and convolutional layers are computed via FHE. Activation functions and max pooling layers are computed via MPC. Transitions between FHE and MPC are performed by each participant having an additive secret sharing of the intermediate result. This allows to take FHE with very low noise capacity (which results in efficient computation). Transitions between FHE and MPC act effectively as bootstrapping as they reset the noise. Another feature of these transitions is that computational and communication costs grow only linearly with network depth. The transitions between FHE and MPC and the specialized algorithms for linear layers in FHE could be applied regardless of the given cryptographic primitives which realize FHE and MPC. However, another avenue of innovation for GAZELLE lies in the FHE implementation and parameter choice. In GAZELLE, the Brakerski-Fan-Vercauteren (BFV) scheme [23] is used. GAZELLE is secure for semi-honest adversaries, that is, neither the server nor the client recovers any information if they follow the protocol. The protocol does not reveal the weights of each layer or their exact size. The performance of GAZELLE was evaluated by using the MNIST dataset. The neural network used in this evaluations consists of one convolution layer and two consecutive fully connected layers. The offline runtime is 0.15 ms, and online runtime is 0.05 ms (overall 0.2 ms), which is better than other approaches [38,45].

5 Performance Study

In this section, we present a performance study that motivates the usage of neural networks in the context of health data and image processing. For this respect, we compare the performance of FHE-based, 2PC-based and Hybrid (GAZELLE-based) solutions for arrhythmia and image classification.

5.1 Arrhythmia Case Study

Heart arrhythmia is a set of conditions in which the heartbeat is not regular. Most types of arrhythmia a patient can be subjected to, are not causes for concern, as they neither cause damages to the heart nor make the patient experience symptoms. Unfortunately, several arrhythmia types cause symptoms that range from tolerable ones (e.g., lightheadedness) to more serious ones (e.g., short breath), and some others predispose patients to heart failure and stroke, resulting in grave consequences such as cardiac arrests. For this reason, it becomes vital to monitor chronic patients' ECG signals to identify arrhythmias at their onset, and prevent the aggravation of patients' conditions.

Nowadays, several commercial services that perform arrhythmia detection on ECG signals can be found in the market. These services collect patients' ECG data via dedicated wearable devices, analyse them to detect arrhythmias and report the results to a healthcare professional, who creates a report. As the identification of arrhythmia in this case is done by a machine, there is a need for building reliable and accurate algorithms for the identification of critical ECG sections. In this context, the algorithmic paradigm of deep learning represents a valid tool for improving the performance of automated ECG analysis [31].

Unfortunately, there are limitations to this approach. Indeed, the burden of analyzing long streams of ECG data for a large number of patients may be difficult to be handled on premises, where the potential lack of computational resources would limit the performance. To overcome this issue, one could acquire ECG data on premises and outsource them to an external environment (with more resources), where the arrhythmia detection would be performed. Nevertheless, moving from a trusted environment to an untrusted one would endanger the protection of personal data. This aspect becomes particularly critical when analyzing health-related data, as the most recent regulations on data protection (such as the GDPR) impose strict analysis constraints for the so-called *special categories of data* (as per Article 9). Hence, it becomes essential, in this case, to protect data before outsourcing them to the untrusted environment, e.g., via the use of advanced cryptographic techniques.

5.2 Image Classification Case Study

Image classification [32] is the study of processing an image and extract valuable information from its content. Image classification has various application areas that spans from face [14] or finger print [56] recognition for biometrics to video surveillance systems [54], hand gestures recognition for sign language [10], etc.

Classifying an image involves computationally intensive operations. With the recent developments in information systems, particularly with the rise of Graphical Processing Units (GPUs), the popularity of image classification has increased again for researchers in machine learning. Thus, many small and medium organizations started developing new applications based on the purpose of image classification for either providing better services for their customers such as face recognition for access control rather than using password based access control, or surveilling an area, building, room, etc. Although GPUs provide extra computation power for the processing of an image, such companies may require the help of computationally more powerful environments such as cloud servers. Companies again face with the dilemma mentioned in Sect. 5.1: outsourcing the images and the underlying image classification operations to an untrusted environment raise privacy issues since these images may contain sensitive information about individuals and more critically some malicious parties can gain access to various online systems using these individuals' images. Therefore, an extra layer of protection should be provided before outsourcing these images to the untrusted environment such as the use of advanced cryptographic techniques mentioned throughout the paper.

5.3 Performance Evaluation of Cryptographic Techniques on Arrhythmia Classification

5.3.1 The Neural Network Models

In order to evaluate the suitability and efficiency of the advanced cryptographic techniques mentioned in this paper, we propose a comparative study for neural network classification with the two previously described use cases, namely arrhythmia and image classification. To this aim, we build a small NN model for the arrhythmia classification and a deeper NN model for the image classification. These models are newly built in order to be compatible with the use of FHE and 2PC.

For the arrhythmia classification use case, the PhysioBank database[10] is employed for training and classification of the newly built NN model. The network consists of 2 fully-connected layers and 1 activation layer that uses x^2 with the input vector size 180, 40 hidden neurons and the output vector size 16 as defined in [42]. The model achieves 96.51% accuracy.

For the image classification use case, the MNIST database[11] that consists of handwritten digits is used to construct the NN model. The organization of the layers in the model are as follows: one convolution layer with 5 different 5×5 filters that have $(2, 2)$ strides, one x^2 activation layer, a flatten layer, a fully connected layer with 100 neurons, another x^2 activation layer and finally a fully connected layer with 10 neurons. This model achieves 97.39% accuracy.

5.3.2 Privacy Preserving Classifiers

Once these NN models are designed, the goal is to execute them over protected inputs. We have developed three different solutions that are based on the use of FHE, 2PC or Hybrid (see Sect. 4) for both NN models.

For the FHE-based solution, we use the CKKS scheme implemented in Microsoft SEAL 3.1. The CKKS scheme allows making computations on floating-point numbers, directly. We choose a precision of 15 bits after the point to ensure that the accuracy of the encrypted evaluation is the same as its evaluation in the non-encrypted version for both NN models. The network weights are not encrypted. Therefore, we mostly use operations between plaintexts and ciphertexts, instead of ciphertext and ciphertext, which improve performances the solution. We choose $m = 4096$ and $q = 2^{116}$ as parameters for the scheme in order to ensure 128-bit security.

In the 2PC-based solution, we propose to use the ABY framework [19] to realize the operations of the proposed NN models such as additions and multiplications. In particular, we propose to use arithmetic circuits in the ABY framework since the majority of the underlying operations are linear (matrix multiplications) and there are no comparisons. Moreover, we approximate all real numbers into integers by using a simple truncation method that consists of keeping only some digits of the fractional part (hence by multiplying them with 10^n). The resulting circuit for privacy preserving arrhythmia classifier has depth 5 and 127 arithmetic gates and for privacy preserving image classifier, the circuit has depth 7 and 37685 arithmetic gates. Both classifiers also allow for prediction in batches thanks to the use of the SIMD packing method.

For the Hybrid solution, we follow the approach in [36] and use the so-called Gazelle technique. In other words, we have implemented the linear operations such as vector/matrix multiplication using FHE and the non-linear ones such as operations in activation layer by using MPC. On the other hand, we use

[10] https://www.physionet.org/physiobank/database/mitdb.
[11] http://yann.lecun.com/exdb/mnist/.

HElib [1] as the homomorphic encryption library and BGV as the FHE tool. We also employ a truncation method to deal with the real numbers. This method is applied on the plaintext value before and after the classification such that floating point numbers are converted into integers before the classification and the result is converted into floating point number after the classification.

5.3.3 Experimental Results

In this section, we present the experimental results to compare the performance of the FHE-based, 2PC-based, and Hybrid solutions on the arrhythmia and image classifications. All the simulations were carried out using a computer which has six 4.0 GHz Intel Core i7-7800X processors, 128 GB RAM and 1 TB SSD disk. The experimental results are given in Table 1. We have performed two different tests for classifying of a single heartbeat/image and classifying heartbeats in batches of 2048 heartbeats/images.

Table 1. Performance Evaluation on ECG classification and image classification for FHE-based, 2PC-based and Hybrid solutions.

		1 Heartbeat	2048 Heartbeats	1 Image	2048 Images
FHE-based solution (with packing)	Comp Cost (ms.)	1253	1253	13570	13570
	Comm. Cost (MB)	0.0018	3.69	0.075	155
2PC-based solution (with packing)	Online Comp. Cost (ms.)	25.7	3792.7	205.4	369878
	Total Comp. Cost (ms.)	212.947	23092.4	1083.2	776778.3
	Comm. Cost (MB)	1.85	3801	3.5	82015
Hybrid solution (without packing)	Online Comp. Cost (ms.)	43	88064	1200	4505600
	Total Comp. Cost (ms.)	5638	11546624	6500	13312000
	Comm. Cost (MB)	15.5	31744	264	540672

For the arrhythmia classification, the 2PC-based solution seems to provide the lowest computational cost when compared with the other solutions for the classification of a single heartbeat. However, the FHE-based solution outperforms when classification is performed in batches. Additionally, the FHE-based solution has better advantage in terms of the communication cost since all the computations in this solution are realized at the server and there is no need for interaction except for the transfer of the input. Moreover, the NN model of arrhythmia classification, which only involves linear operations and one square operation, the FHE-based solution seems to be the most suitable one. Nevertheless, this may not be the case for deeper neural networks such as for the case of image classification. For the image classification, the 2PC-based solution and the hybrid solution outperforms the FHE based solution for the classification of a single image. On the other hand, the FHE-based solution again provides better results for the classification in batches. However, these two NN models are designed to be implemented in all proposed solutions and therefore they do not consist of any non-linear operations such as the max pooling layer and ReLU activation function which are not supported by FHE. The hybrid solution may be the most appropriate one in such a case as it combines the use of 2PC and FHE. Indeed, this particular solution is specifically designed to operate on large NN models and it can be seen that the increase rate on the computational cost is lower than in the case of other solutions. Furthermore, the current version of the hybrid solution does not support packing yet.

6 Summary

In this paper, we have presented the systematization of knowledge for privacy preserving Neural Network classifiers. We have first overviewed the NN operations and their approximations when needed for the cryptographic world. Later, we have introduced the definitions of advanced cryptographic techniques presented in this paper and with this aim, we have overviewed the existing solutions in the literature that utilize these techniques for NN classification. We further have developed particular NN models for the arrhythmia and image classification case studies, a small architecture for the first one and a deeper model for the second one and applied them on encrypted data using FHE-based, 2PC-based, and Hybrid solutions. We presented a performance evaluation to compare the FHE-based, 2PC-based, and Hybrid solutions. From the performance study, we can conclude that there is no single technique that outperforms. Moreover, in this study, the neural networks used are specifically selected by the compatibility with the FHE-based solution, which means these models do not consist of any non-linear operations. Therefore, for other usage scenarios, we may have to consider to use these operations, and this makes us deal with another dilemma between privacy, accuracy, and efficiency.

Acknowledgement. This work was partly supported by the PAPAYA project funded by the European Union's Horizon 2020 Research and Innovation Programme, under Grant Agreement no. 786767.

References

1. HElib: An Implementation of homomorphic encryption (2013). https://github.com/shaih/HElib
2. GDPR. Official Journal of the European Union (2016)
3. Ball, M., Carmer, B., Malkin, T., Rosulek, M., Schimanski, N.: Garbled neural networks are practical. Cryptology ePrint Archive, Report 2019/338 (2019)

4. Beaver, D.: Efficient multiparty protocols using circuit randomization. In: Feigenbaum, J. (ed.) CRYPTO 1991. LNCS, vol. 576, pp. 420–432. Springer, Heidelberg (1992). https://doi.org/10.1007/3-540-46766-1_34
5. Ben-David, A., Nisan, N., Pinkas, B.: FairplayMP: a system for secure multi-party computation. In: CCS (2008)
6. Boneh, D., Goh, E.-J., Nissim, K.: Evaluating 2-DNF formulas on ciphertexts. In: Kilian, J. (ed.) TCC 2005. LNCS, vol. 3378, pp. 325–341. Springer, Heidelberg (2005). https://doi.org/10.1007/978-3-540-30576-7_18
7. Bourse, F., Minelli, M., Minihold, M., Paillier, P.: Fast homomorphic evaluation of deep discretized neural networks. In: Shacham, H., Boldyreva, A. (eds.) CRYPTO 2018. LNCS, vol. 10993, pp. 483–512. Springer, Cham (2018). https://doi.org/10.1007/978-3-319-96878-0_17
8. Brakerski, Z., Gentry, C., Vaikuntanathan, V.: Fully homomorphic encryption without bootstrapping. Cryptology ePrint Archive, Report 2011/277 (2011)
9. Brakerski, Z., Gentry, C., Vaikuntanathan, V.: (Leveled) fully homomorphic encryption without bootstrapping. In: ITCS (2012)
10. Camgöz, N.C., Kındıroğlu, A.A., Akarun, L.: Sign language recognition for assisting the deaf in hospitals. In: Chetouani, M., Cohn, J., Salah, A.A. (eds.) HBU 2016. LNCS, vol. 9997, pp. 89–101. Springer, Cham (2016). https://doi.org/10.1007/978-3-319-46843-3_6
11. Canard, S., Carpov, S., Nokam, D., Sirdey, R.: Running compression algorithms in the encrypted domain: a case-study on the homomorphic execution of RLE (2017)
12. Chabanne, H., de Wargny, A., Milgram, J., Morel, C., Prouff, E.: Privacy-preserving classification on deep neural network (2017)
13. Chandran, N., Gupta, D., Rastogi, A., Sharma, R., Tripathi, S.: EzPC: programmable, efficient, and scalable secure two-party computation for machine learning. Euro S&P (2019)
14. Chen, L.F., Liao, H.Y.M., Ko, M.T., Lin, J.C., Yu, G.J.: A new LDA-based face recognition system which can solve the small sample size problem. Pattern Recogn. **33**, 1713–1726 (2000)
15. Cheon, J.H., Kim, A., Kim, M., Song, Y.: Homomorphic encryption for arithmetic of approximate numbers. In: Takagi, T., Peyrin, T. (eds.) ASIACRYPT 2017. LNCS, vol. 10624, pp. 409–437. Springer, Cham (2017). https://doi.org/10.1007/978-3-319-70694-8_15
16. Cheon, J.H., Stehlé, D.: Fully homomophic encryption over the integers revisited. In: Oswald, E., Fischlin, M. (eds.) EUROCRYPT 2015. LNCS, vol. 9056, pp. 513–536. Springer, Heidelberg (2015). https://doi.org/10.1007/978-3-662-46800-5_20
17. Chillotti, I., Gama, N., Georgieva, M., Izabachène, M.: Faster fully homomorphic encryption: bootstrapping in less than 0.1 seconds. In: Cheon, J.H., Takagi, T. (eds.) ASIACRYPT 2016. LNCS, vol. 10031, pp. 3–33. Springer, Heidelberg (2016). https://doi.org/10.1007/978-3-662-53887-6_1
18. Chillotti, I., Gama, N., Georgieva, M., Izabachène, M.: Faster packed homomorphic operations and efficient circuit bootstrapping for TFHE. In: Takagi, T., Peyrin, T. (eds.) ASIACRYPT 2017. LNCS, vol. 10624, pp. 377–408. Springer, Cham (2017). https://doi.org/10.1007/978-3-319-70694-8_14
19. Demmler, D., Schneider, T., Zohner, M.: ABY - a framework for efficient mixed-protocol secure two-party computation. In: NDSS (2015)
20. Ejgenberg, Y., Farbstein, M., Levy, M., Lindell, Y.: SCAPI: the secure computation application programming interface. Cryptology ePrint Archive, Report 2012/629 (2012)

21. ElGamal, T.: A public key cryptosystem and a signature scheme based on discrete logarithms. In: Blakley, G.R., Chaum, D. (eds.) CRYPTO 1984. LNCS, vol. 196, pp. 10–18. Springer, Heidelberg (1985). https://doi.org/10.1007/3-540-39568-7_2
22. Fan, J., Vercauteren, F.: Somewhat practical fully homomorphic encryption. Cryptology ePrint Archive, Report 2012/144 (2012)
23. Fan, J., Vercauteren, F.: Somewhat practical fully homomorphic encryption. IACR Cryptology ePrint Archive (2012)
24. Gentry, C.: Fully homomorphic encryption using ideal lattices. In: Mitzenmacher, M. (ed.) STOC (2009)
25. Gentry, C., Sahai, A., Waters, B.: Homomorphic encryption from learning with errors: conceptually-simpler, asymptotically-faster, attribute-based. In: Canetti, R., Garay, J.A. (eds.) CRYPTO 2013. LNCS, vol. 8042, pp. 75–92. Springer, Heidelberg (2013). https://doi.org/10.1007/978-3-642-40041-4_5
26. Gilad-Bachrach, R., Dowlin, N., Laine, K., Lauter, K.E., Naehrig, M., Wernsing, J.: CryptoNets: applying neural networks to encrypted data with high throughput and accuracy. In: ICML (2016)
27. Goldreich, O., Micali, S., Wigderson, A.: How to play any mental game or A completeness theorem for protocols with honest majority. In: STOC (1987)
28. Goldreich, O., Micali, S., Wigderson, A.: How to play any mental game or A completeness theorem for protocols with honest majority. In: ACM Symposium on Theory of Computing (1987)
29. Halevi, S., Shoup, V.: Algorithms in HElib. In: Garay, J.A., Gennaro, R. (eds.) CRYPTO 2014. LNCS, vol. 8616, pp. 554–571. Springer, Heidelberg (2014). https://doi.org/10.1007/978-3-662-44371-2_31
30. Halevi, S., Shoup, V.: Bootstrapping for `HElib`. In: Oswald, E., Fischlin, M. (eds.) EUROCRYPT 2015. LNCS, vol. 9056, pp. 641–670. Springer, Heidelberg (2015). https://doi.org/10.1007/978-3-662-46800-5_25
31. Hannun, A.Y., et al.: Cardiologist-level arrhythmia detection and classification in ambulatory electrocardiograms using a deep neural network. Nat. Med. **25**(1), 65 (2019)
32. Haralick, R.M., Shanmugam, K., Dinstein, I.: Textural features for image classification. IEEE Trans. Syst. Man Cybern. **6**, 610–621 (1973)
33. Henecka, W., Kögl, S., Sadeghi, A., Schneider, T., Wehrenberg, I.: TASTY: tool for automating secure two-party computations. In: ACM CCS (2010)
34. Hesamifard, E., Takabi, H., Ghasemi, M., Wright, R.N.: Privacy-preserving Machine Learning as a Service. PETS **2018**, 123–142 (2018)
35. Ibarrondo, A., Önen, M.: FHE-compatible batch normalization for privacy preserving deep learning. In: Garcia-Alfaro, J., Herrera-Joancomartí, J., Livraga, G., Rios, R. (eds.) DPM/CBT 2018. LNCS, vol. 11025, pp. 389–404. Springer, Cham (2018). https://doi.org/10.1007/978-3-030-00305-0_27
36. Juvekar, C., Vaikuntanathan, V., Chandrakasan, A.: Gazelle: a low latency framework for secure neural network inference. arXiv preprint (2018)
37. Lindell, Y., Pinkas, B.: Privacy preserving data mining. In: Bellare, M. (ed.) CRYPTO 2000. LNCS, vol. 1880, pp. 36–54. Springer, Heidelberg (2000). https://doi.org/10.1007/3-540-44598-6_3
38. Liu, J., Juuti, M., Lu, Y., Asokan, N.: Oblivious neural network predictions via MiniONN transformations. In: ACM CCS (2017)
39. Liu, J., Juuti, M., Lu, Y., Asokan, N.: Oblivious neural network predictions via MiniONN transformations. Cryptology ePrint Archive, Report 2017/452 (2017)
40. López-Alt, A., Tromer, E., Vaikuntanathan, V.: On-the-fly multiparty computation on the cloud via multikey fully homomorphic encryption. In: STOC (2012)

41. Malkhi, D., Nisan, N., Pinkas, B., Sella, Y.: Fairplay—a secure two-party computation system. In: USENIX (2004)
42. Mansouri, M., Bozdemir, B., Önen, M., Ermis, O.: PAC: privacy-preserving arrhythmia classification with neural networks. In: FPS (2019)
43. Aguilar Melchor, C., Kilijian, M.-O., Lefebvre, C., Ricosset, T.: A comparison of the homomorphic encryption libraries HElib, SEAL and FV-NFLlib. In: Lanet, J.-L., Toma, C. (eds.) SECITC 2018. LNCS, vol. 11359, pp. 425–442. Springer, Cham (2019). https://doi.org/10.1007/978-3-030-12942-2_32
44. Mohassel, P., Rindal, P.: ABY^3: a mixed protocol framework for machine learning. In: ACM CCS (2018)
45. Mohassel, P., Zhang, Y.: SecureML: a system for scalable privacy-preserving machine learning. In: S&P (2017)
46. Ohrimenko, O., et al.: Oblivious multi-party machine learning on trusted processors. In: USENIX (2016)
47. Orlandi, C., Piva, A., Barni, M.: Oblivious neural network computing via homomorphic encryption. EURASIP (2007)
48. Paillier, P.: Public-key cryptosystems based on composite degree residuosity classes. In: Stern, J. (ed.) EUROCRYPT 1999. LNCS, vol. 1592, pp. 223–238. Springer, Heidelberg (1999). https://doi.org/10.1007/3-540-48910-X_16
49. Rabin, M.O.: How to exchange secrets with oblivious transfer. Cryptology ePrint Archive, Report 2005/187 (2005)
50. Rivest, R.L., Shamir, A., Adleman, L.M.: A method for obtaining digital signatures and public-key cryptosystems. Commun. ACM **21**, 120–126 (1978)
51. Rouhani, B.D., Riazi, M.S., Koushanfar, F.: DeepSecure: scalable provably-secure deep learning. In: DAC (2018)
52. Sadegh Riazi, M., Weinert, C., Tkachenko, O., Songhori, E.M., Schneider, T., Koushanfar, F.: Chameleon: a hybrid secure computation framework for machine learning applications. arXiv e-prints (2018)
53. Singh, K., Sirdey, R., Artiguenave, F., Cohen, D., Carpov, S.: Towards confidentiality-strengthened personalized genomic medicine embedding homomorphic cryptography. In: ICISSP (2017)
54. Srinivasan, S., Latchman, H., Shea, J., Wong, T., McNair, J.: Airborne traffic surveillance systems: video surveillance of highway traffic. In: International Workshop on Video Surveillance & Sensor Networks (2004)
55. Wagh, S., Gupta, D., Chandran, N.: SecureNN: efficient and private neural network training. In: PETS (2019)
56. Wahab, A., Chin, S., Tan, E.: Novel approach to automated fingerprint recognition. IEE Proceedings - Vision, Image and Signal Processing (1998)
57. Wang, X., Malozemoff, A.J., Katz, J.: Faster secure two-party computation in the single-execution setting. Cryptology ePrint Archive, Report 2016/762 (2016)
58. Yao, A.C.C.: Protocols for secure computations (extended abstract). In: FOCS (1982)
59. Yao, A.C.C.: How to generate and exchange secrets (extended abstract). In: FOCS (1986)

Workshop on Privacy Challenges in Public and Private Organizations

Alessandra Bagnato[1(✉)], Paulo Silva[2], Ala Sarah Alaqra[3], and Orhan Ermis[4]

[1] SOFTEAM R&D Department, Paris, France
alessandra.bagnato@softeam.fr
[2] CISUC, Department of Informatics Engineering, University of Coimbra, Coimbra, Portugal
pmgsilva@dei.uc.pt
[3] Karlstad University, Karlstad, Sweden
alaa.alaqra@kau.se
[4] EURECOM, Sophia Antipolis, France
orhan.ermis@eurecom.fr

Abstract. Recent developments in information technology such as the Internet of Things and the cloud computing paradigm enable public and private organisations to collect large amounts of data to employ various data analytic techniques for extracting important information that helps improve their businesses. Unfortunately, these benefits come with a high cost in terms of privacy exposures given the high sensitivity of the data that are usually processed at powerful third-party servers. Given the ever-increasing of data breaches, the serious damage they cause, and the need for compliance to the European General Data Protection Regulation (GDPR), these organisations look for secure and privacy-preserving data handling practices. During the workshop, we aimed at presenting an approach to the problem of user data protection and control, currently being developed in the scope of the PoSeID-on and PAPAYA H2020 European projects.

Keywords: Privacy-enhancing dashboard · Privacy-preserving data analytics · GDPR

1 Introduction

Several European projects address the topical area of privacy, data protection, and digital identities. PAPAYA [1], and PoSeID-on [2], described next, are exploring ways of cooperation to enhance privacy assurances to end-users.

The goal of the PAPAYA project is to devise and develop a platform of privacy-preserving modules that protects the privacy of users on an end-to-end basis without sacrificing data analytics functionalities. The PAPAYA platform will integrate several privacy-preserving data analytics modules each of them dedicated to specific analytics operations and specific settings (e.g., single data

Published by Springer Nature Switzerland AG 2020
M. Friedewald et al. (Eds.): Privacy and Identity 2019, IFIP AICT 576, pp. 82–89, 2020.
https://doi.org/10.1007/978-3-030-42504-3_6

owner, multiple data owners). The platform aims to be usable in the sense that it also includes proper transparency and control mechanisms through a dashboard.

The PoSeID-on solution is based on innovative technologies such as blockchain, cloud and smart contracts. These technologies provide targeted benefits for end-users, enabling them to manage personal data and data access authorizations easily, securely and independently. This helps both public and private entities identify new business opportunities, be compliant with the GDPR while processing personal data, as well as undergo a substantial ICT-driven transformation, which will ensure higher security of end-users' data. PoSeID-on also impacts society as a whole, as it leads to increased trust in the digital single market, in addition to supporting fundamental rights in the digital society. The project will be evaluated through four different pilot studies (in Italy, France, Spain and Malta) that will test its functionalities in public, private and mixed contexts. Initially, pilots will involve a basic set of users to be enlarged during the evaluation months. The pilots (described in Deliverable 2.1[1]) will run in a controlled environment to simulate real-life services and conditions.

2 Motivation and Objectives

Recent advances in information technology such as the Internet of Things and/or the cloud computing paradigm enable public and private organisations to collect large amounts of data and use advanced techniques to infer valuable insights and improve their businesses. Unfortunately, these benefits come with a high cost in terms of privacy exposures given the high sensitivity of the data that are usually analysed/processed at powerful third-party servers. Given the ever-increasing of data breaches, the serious damage they cause, and the need for compliance to the European General Data Protection Regulation (GDPR), these organisations look for secure and privacy-preserving data handling practices.

Both projects decided it was worth to get feedback on the work done so far. Not only bilaterally, but also from all the workshop attendees. During the workshop, we presented our approach to the problem of user data protection and control that is currently being developed in the scope of the PoSeID-on H2020 European project. The presented solution complies with EU's General Data Protection Regulation and explores the use of Blockchain technology to provide data transactions protection and accountability, as well as full control of users over their data.

The goal of this workshop was to identify and present the privacy challenges related to data analysis by public and private organisations and to encompass research advances in the privacy-enhancing technologies that will enable privacy-preserving data management and GDPR compliance. Moreover, the workshop served as a discussion environment for those familiar with cryptographic tools, and discuss possible concerns and risks when it comes to applying such tools in different areas when data is critical and sensitive. We intended to understand

[1] https://www.poseidon-h2020.eu/documents/d2-1-use-case-analysis-and-user-scenarios/.

and shed light on the mental models, trust factors, and the possible risks and concerns when it comes to data analysis on the protected data and how these discussions might be used to foster collaboration among potential privacy-enhancing technology outputs of PoSeID-on and PAPAYA projects.

3 Workshop Format

The allocated time for the workshop was two hours. The objective was to first introduce both projects and then focus on the specific objectives and challenges. It was possible to present and discuss the previously agreed topics. The workshop was organized as follows:

- **Opening (5 min):**
  The workshop program presentation (Orhan Ermis).
- **PoSeID-on Project Presentation (20 min):**
  The presentation introduced the PoSeID-on H2020 EU project Project[2] (Alessandra Bagnato).
- **PoSeID-on Project Dashboard Demo (10 min):**
  Offered an insight about the dashboard and its functionalities (Paulo Silva).
- **Papaya Presentation (25 min):**
  The presentation introduced the Papaya Project[3] H2020 EU project (Orhan Ermis).
- **The End User requirements in PAPAYA Presentation (10 min):**
  Highlighted human aspects and results from end-user studies (Ala Sarah Alaqra).
- **Questionnaire Session on Human Factors (10 min):**
  Participants have provided feedback on PAPAYA e-Health use cases (Ala Sarah Alaqra).
- **Discussion Session (25 min):**
  Questions and debate regarding the use cases and respective approaches. Each speaker had a set of comments/questions to spark the discussion. Nevertheless, the discussion was naturally flowing in interventions from the audience.
- **PAPAYA-PoSeID-on collaboration (15 min):**
  A discussion with the participants from both projects to describe the needs of the dashboard and the needs of the analysis. More specifically, answering the questions "What are PAPAYA's needs with respect to the PoSeID-on's Dashboard?" and "What are PoSeID-on's needs with respect to Papaya's analytics?" was the objective of the discussion.

4 Lessons Learned from the PoSeID-on Project Perspective

In this session of the workshop, the PoSeID-on general objectives, the PoSeID-on architecture and the privacy challenges in one PoSeID-on Public and one PoSeID-on Private Organization were described.

[2] https://www.poseidon-h2020.eu/.
[3] https://www.papaya-project.eu.

4.1 Use Cases

The first presented PoSeID-on public organization use case and the respective challenge was related to the General Administration, Personnel and Services Department (DAG) of the Italian Ministry of Economy and Finance (MEF). The MEF is in charge of the management of payroll functions for approximately 2.1 million Italian public sector employees. Such service is provided through a unique payroll function, NoiPA, which annually manages more than €51B in payments. NoiPA is a portal created to manage administrative and economic data of central and peripheral Public Administration employees. Therefore, NoiPA has a big experience in personal data management and it could be very close to PoSeID-on project because this platform aims to collect the users' given authorizations (for sharing personal information) that are stored in the platform itself.

The second presented PoSeID-on Private sector organization use case was related to Softeam, a private French software vendor with about 1000 employees. Softeam develops a software called e-Citiz – a platform for Business Process Management for both e-government and companies, and which has been on the market since 2004. Softeam has a big experience in personal data management due to several business projects and some research projects. With the e-Citiz platform, Softeam proposes the SVE ("Saisine par Voie Electronique" which means Seizure by Electronic Way), an eService product allowing users to apply for a claim or any sort of demand to the company. The SVE pilot privacy challenge will imply the customization of the SVE product to integrate PoSeID-on solution to provide the users of the current SVE services with a single platform for personal data management, as well as to support SVE to be compliant with the GDPR.

The public and private challenges were discussed as well as the solutions proposed by the Papaya and PoSeID-on projects. The participants were particularly interested in the public aspects and challenges and potential help that the project results could give to European citizens. There were also questions regarding the trust issues that the PoSeID-on platform should provide for its users (for example the fact that it could be guaranteed by governmental organizations).

4.2 Dashboard and Data Analysis Modules

The team then presented the PoSeID-on Web Privacy Enhanced Dashboard as a web application giving Data Subjects access to the PoSeID-on functionalities. Access to the Web Dashboard is managed by national systems compliant with eIDAS (e.g., SPID, @firma, FRANCEconnect). Such systems guarantee users secured access to the digital services of Public Administrations. These "electronic Identities" are released by Identity Providers, accredited bodies that release the credentials (User ID and Password), after verifying the user's identity. The session was very positive and interactive. The dashboard mock-ups were well accepted and had positive feedback. The participants wondered about the dashboard compliance with W3C Standards, which is one of the aspects that PoSeID-on project takes into consideration.

The architecture of PoSeID-on was also presented and the audience was particularly interested in the PoSeID-on analytics capabilities, in particular, the Risk Management Module (RMM) and Personal Data Analyser (PDA). These components will be used to evaluate and manage a risk score as well as to monitor all personal data flow and usage in addition to related warnings generated, to detect and prevent anomalies and misbehaved transactions (data flow and usage).

The RMM leverages Apache Spark Streaming and associated machine learning library MLlib to build a data analysis pipeline in order to perform anomaly detection on incoming logs from system components. All data is stored in CassandraDB to guarantee throughput of writes and the partition of distributed data according to the deployment each RMM instance and Cassandra node.

The PDA explores the potentialities of Natural Language Processing (NLP), more precisely Named Entity Recognition (NER) to analyse personal information. It comprises an ensemble learning mechanism with the best-performing machine learning algorithms (e.g., Random Forest, Conditional Random Fields, Convolutional Neural Networks), NLP tools (e.g., Stanford CoreNLP or SpaCy) and regular expressions to provide an accurate analysis of data. Moreover, since it does not store any data and is open source, it is trustworthy and transparent.

With the Blockchain, Dashboard, RMM and PDA, the user could be aware of data privacy exposure and have control over his/her data. The audience particularly appreciated that PoSeID-on Users can be advised on which service they could eventually disable in case of anomalies or high exposure of their data to privacy risks. Moreover, the participants were also very interested in the way PoSeID-on Dashboard handles such amount of information.

One aspect that came out from the workshop discussion was that how much it is necessary to write a statement which clearly details and clarifies the provided transparency and add it to PoSeID-on Dashboard web site. It was also noticeable the interest in better understanding how we handle the data flowing through PoSeID-on. More specifically, over how the data is managed, kept, analysed by our modules. Therefore, this valuable input is going to be considered during the next stages of development as well as dissemination and presentation of the project.

5 Lessons Learned from the PAPAYA Project Perspective

In this session of the workshop, general objectives, use cases and underlying privacy-enhancing technologies were briefly explained. Moreover, the architecture and the planned dashboards for the players of PAPAYA were also explained. Another important benefit for the PAPAYA project is that we have the chance to discuss human aspects, particularly end-users, with privacy experts.

Workshop Participants on Human Aspects. The workshop also included a discussion regarding potential human aspects and privacy concerns for the privacy-enhancing technologies presented in the workshop. It was followed up by a small questionnaire, where we inquired about participants' perspectives of data privacy, key challenges and concerns, and factors to be considered in the use-case scenario presented. We chose to collect their opinions in a questionnaire format to allow some freedom of expressing their sharing concerns and opinions without revealing such information to fellow participants, especially in a privacy-aware group of experts, i.e., participants of the summer school.

Although we presented both of the eHealth use-cases of PAPAYA project [3] by **MediaClinics Italia** (project partner in the PAPAYA project), namely *UC1: Privacy preserving Arrhythmia Detection* and *UC2: Privacy preserving Stress Management* [7], we focused on UC1 during the presentation. The three different human actors in the use case scenario are the patient, pharmacist, and cardiologist. The eHealth use-case involves electrocardiogram (ECG) medical data that is to be collected from the patient via a wearable device and uploaded to the medical healthcare platform by the pharmacist. Data is then encrypted and send to the PAPAYA platform, where data is analyzed and sent back to the platform. Finally, the cardiologist receives the results of the analysis alongside raw data and other medical records to perform the diagnosis and send a report back to the patient.

In total, there were 11 participants in the workshop. We had optional demographic questions included in the questionnaire and the following are the responses for work/field, ages and genders for those who chose to answer those fields. There were only 5 who stated their field of work to be in research, academia, information technology, and computer science. There were 9 responses for age, of which 2 were in the age range of 21–30, 5 in the age range of 31–40, 1 in the age range of 41–50, and 1 preferred not to answer. There were 8 responses for genders, 2 were females, 4 males, and 2 preferred not to answer.

Workshop Feedback on Human Aspects. In human centered design approaches, end-users are taken into account early and throughout the development process of technologies. Following such approaches yield quality results that target human aspects throughout the development process by highlighting concerns and requirements which suit end-user's mental models [4,5]. In this workshop, we had the opportunity to engage privacy experts in our projects discussions. Below is a summary of the human factors part discussion of the PAPAYA.

Overview. Despite the limited number of participants, we had gathered varied inputs from participants, some included some UC2 points in their feedback sheets and were left-out from the summary. Main points raised by participants are summarized into their perspective/mental models on privacy, key challenges and concerns, communication requirements found in the following paragraphs.

Perspectives/Mental Models. In order to understand the mental model of participants, we inquired about their sharing behaviours and their opinions about privacy. We asked participants whether they are active on social media, personal (e.g., Facebook) and professional (e.g., Researchgate), so that we get an indication about their data sharing behaviour. Similarly in previous work, where it was shown that different stakeholders have different values for privacy that affect their privacy sharing behaviours [6]. 10 out of 11 participants indicated their responses, 6 indicated yes/occasionally on personal social media where 4 indicated no, whereas 7 indicated yes/occasionally on professional social media and 3 no. When asked whether privacy is not always the most important incentive on 5-point Likert scale from strongly disagree to strongly agree. 3 participants disagreed, 2 were neutral, 5 agreed, and 1 strongly agreed. A comment related to that question highlighted that privacy is the most important incentive just in research environment.

Key Challenges. Several challenges and concerns were highlighted by participants, they specified the need to address challenges with technologies, actors involved and trust factors for both use-case scenarios. It was indicated that technical challenges to PAPAYA for implementing and guaranteeing data accuracy should be addressed, as well as confidentiality, data security and privacy, and scope of data use and limitations. Additionally, considerations for ethical, legal and social factors pose trust challenges, such as the safety and the need for guarantees for patient's safety in the first use case scenario.

Communication Requirements. Communication with different actors and users was shown to be of significance by some participants; explanations of why and how data is being used, processed, stored, and who is involved. Also communicating procedure for safety, privacy and security, as in the case of data privacy breaches and how it is handled.

6 Conclusion

During the workshop, within the scope of PAPAYA and PoSeID-on project, we presented potential solutions to overcome privacy challenges related to data management and processing for public and private organizations. Additionally, we had the chance to discuss human aspects, particularly end-users, with privacy experts. Inputs from these experts was a helpful feedback for developing the privacy enhancing technologies of the use-cases particularly for the next stages of the project. In terms of PoSeID-on project, the most important inputs raised during discussions were "how the platform should provide trust to its users" and "how PoSeID-on project handles data flow, management and processing to achieve transparency". That is for sure, by considering these inputs, the consortium will have the chance to easily achieve the dissemination goals of the project. In terms of PAPAYA project, the most significant learned lessons was that we had the opportunity to get inputs on our questionnaire from privacy experts. These inputs are not only important for the dissemination activities of PAPAYA project but also plays an important role in the Data Subject Toolbox, which is mainly designed and will be used at the end of the project to explain privacy preserving data analytics techniques to data subjects.

Acknowledgement. This work was partly supported by the PAPAYA project and PoSeID-on project funded by the European Union's Horizon 2020 Research and Innovation Programme, under Grant Agreement no. 786767 and no. 786713, respectively.

References

1. Papaya Project - Platform for privacy preserving data analytics. www.papaya-project.eu. Accessed 17 Oct 2019
2. PoSeID-on Project - Protection and control of secured information by means of a privacy enhanced dashboard. www.poseidon-h2020.eu. Accessed 17 Oct 2019
3. Ciceri, E., Mosconi, M., Önen, M., Ermis, O.: PAPAYA: A Platform for Privacy Preserving Data Analytics (2019). https://ercim-news.ercim.eu/en118/special/papaya-a-platform-for-privacy-preserving-data-analytics. Accessed 22 Oct 2019
4. Abras, C., Maloney-Krichmar, D., Preece, J.: User-centered design. Bainbridge W. Encycl. Hum.-Comput. Interact. **37**(4), 445–56 (2004)
5. Anderson, N.S., Norman, D.A., Draper, S.W.: User centered system design: new perspectives on human-computer interaction. Am. J. Psychol. **101**(1), 148 (1988)
6. Alaqra, A.S., Wästlund, E.: Reciprocities or incentives? Understanding privacy intrusion perspectives and sharing behaviors. In: Moallem, A. (ed.) HCII 2019. LNCS, vol. 11594, pp. 355–370. Springer, Cham (2019). https://doi.org/10.1007/978-3-030-22351-9_24
7. Ciceri, E., Galliani, S., Mosconi, M., Azraoui, M., Canard, S.: D2.1: Use Cases and Requirements, PAPAYA Deliverable D2.1 (2019)

News Diversity and Recommendation Systems: Setting the Interdisciplinary Scene

Glen Joris[1,2](✉), Camiel Colruyt[3], Judith Vermeulen[4](✉), Stefaan Vercoutere[5], Frederik De Grove[1,2], Kristin Van Damme[1,2], Orphée De Clercq[3], Cynthia Van Hee[3], Lieven De Marez[1], Veronique Hoste[3], Eva Lievens[4], Toon De Pessemier[5], and Luc Martens[5]

[1] imec-mict-UGent, Ghent University, Ghent, Belgium
glen.joris@ugent.be
[2] Center for Journalism Studies, Department of Communication Sciences, Ghent University, Ghent, Belgium
[3] Language and Translation Technology Team, Department of Translation, Interpreting and Communication, Ghent University, Ghent, Belgium
[4] Law and Technology, Department of Interdisciplinary Law, Private Law and Business Law, Ghent University, Ghent, Belgium
judith.vermeulen@ugent.be
[5] imec-WAVES-UGent, Department of Information Technology, Ghent University, Ghent, Belgium

Abstract. Concerns about selective exposure and filter bubbles in the digital news environment trigger questions regarding how news recommender systems can become more citizen-oriented and facilitate – rather than limit – normative aims of journalism. Accordingly, this chapter presents building blocks for the construction of such a news algorithm as they are being developed by the Ghent University interdisciplinary research project #NewsDNA, of which the primary aim is to actually build, evaluate and test a diversity-enhancing news recommender. As such, the deployment of artificial intelligence could support the media in providing people with information and stimulating public debate, rather than undermine their role in that respect. To do so, it combines insights from computer sciences (news recommender systems), law (right to receive information), communication sciences (conceptualisations of news diversity), and computational linguistics (automated content extraction from text). To gather feedback from scholars of different backgrounds, this research has been presented and discussed during the 2019 IFIP summer school workshop on 'co-designing a personalised news diversity algorithmic model based on news consumers' agency and fine-grained content modelling'. This contribution also reflects the results of that dialogue.

Keywords: News personalisation · Algorithms · News recommender systems · Right to receive diverse information · News diversity · News content extraction · #NewsDNA

M. Friedewald et al. (Eds.): Privacy and Identity 2019, IFIP AICT 576, pp. 90–105, 2020.
https://doi.org/10.1007/978-3-030-42504-3_7

1 Introduction

In recent years, online news organisations – both web-editions of traditional news outlets and digital-only news sites – increasingly explore how recommender systems can be used to provide consumers with a tailor-made news offer. The New York Times, for example, uses a mix of editorial curation and algorithms to compose a newsletter tailored to each recipient [1]. In Belgium too, De Standaard and Het Nieuwsblad recently introduced a personal page that collects articles based on the reader's selected topics [2]. In The Netherlands, online newspapers nu.nl and Algemeen Dagblad both invested heavily in personalised news notifications [3]. Hence, news organisations are increasingly exploring implicit and explicit algorithmic news personalisation [4, 5], similarly to how companies such as Netflix and Amazon individualise content.

Personalisation typically relies on automated decision-making and recommender systems. However, in contrast to what people might think [6], these systems are not neutral as they primarily apply a commercial logic. More specifically, they produce recommendations based on the calculated relevance of news items vis-à-vis individual news consumers (for example taking into account selected fields of interests or past consumption patterns). Such practices contrast with the role of the media as 'a marketplace of ideas' in which citizens are confronted with a diverse array of ideas [7]. Although empirical research currently supports a more nuanced view [8], news recommender systems are argued to be a potential threat to an informed citizenry and the democratic processes between media, politics and audiences [9]. With these concerns in mind, several scholars have raised questions regarding how news recommender systems can be built in a more citizen-oriented way by maintaining the normative aims of journalism [10].

The current chapter presents research conducted by the interdisciplinary research project #NewsDNA at Ghent University (Belgium), which seeks to provide a possible answer in that regard. More specifically, it outlines building blocks for the construction of a recommender system that uses news diversity as a key driver for personalised news recommendations. As such, the deployment of artificial intelligence could support the media in providing people with information and stimulating public debate, rather than undermine their role in that respect. To gather feedback concerning this framework from scholars of different backgrounds, this research has been presented during the 2019 IFIP summer school workshop on 'co-designing a personalised news diversity algorithmic model based on news consumers' agency and fine-grained content modelling'. The results of this exercise are also included in this contribution.

As the current research builds on insights from multiple disciplines, the remainder of this paper is – as was the workshop – organised per discipline. First, we present a state-of-the-art overview of the most commonly used methods to design news recommender systems within computer sciences. Second, we explore the existence of a legal ground for receiving diverse news. Third, we expound on the conceptual meaning of news diversity by building on literature in communication sciences. Finally, we discuss the computational feasibility of news content extraction, provided by computational linguistics, to provide data to the aspired diversity-promoting news recommender system. Each part is followed by related questions we presented the workshop participants with as well as their answers in that regard.

2 News Recommendations Systems Today

In this section, we provide an overview of current news recommendation systems, from a computer science perspective. We present the two dominant approaches, being collaborative and content-based filtering, and outline the obstacles related to the development of a diversity-enhancing news recommender.

One of the most commonly used methods in the field of recommendation systems is *collaborative filtering*. Collaborative filtering assumes that people who had similar interests in the past are likely to have similar interests in the future [11]. As such, relevant news articles are predicted based on news articles read by so-called 'neighbours', other users who have historically had similar taste in news [12]. In essence, it is very similar to the concept of 'word of mouth': we often consult with our peers when gathering opinions about certain activities or decisions (e.g. interesting movies, tasty drinks). Especially for news, peer recommendations are still perceived as valuable [13].

Collaborative filtering methods have two major drawbacks when recommending news stories. First, news is quick and volatile, which exacerbates the first-rater problem [14, 15]: a new story cannot be recommended to users unless other users have read it before. This becomes problematic when trying to present the latest information in a timely manner, as it is not uncommon for collaborative filtering based methods to take several hours before sufficient clicks have been collected and a new item can be recommended. Generally, as an item gains more clicks, the system becomes more confident in its ability to recommend it. Hence, older and popular items dominate the recommendation process, which is not desirable for news recommendations. Second, there is the sparsity problem [16], which occurs when there is insufficient overlap between the consumption patterns of users. As the relevance of news stories sharply decreases over time, it is not unreasonable to assume little overlap between new and old users.

A second approach is *content-based filtering*, which does not have these shortcomings, and consequently is often used for news recommendations [15, 17]. Content-based systems use the news articles themselves to recommend similar news, both the content and its metadata. For example, the system looks at the topic of the news, the keywords or the broader classification (e.g. sports or domestic affairs), the author, word count, etc. This means that in contrast to collaborative filters, content-based systems treat recommendation as a single user classification problem.

However, this method also suffers from certain drawbacks. First, over-specialisation: content-based systems cannot provide recommendations outside the scope of what the user has already shown interest in. Within journalism, this is the trigger for concerns of filter bubbles and news personalisation [9]. Second, the performance of the system heavily depends on the quality of the content descriptions. In domains where the items consist of music or video, the extraction of a useful representation of the content can be very challenging. In journalism as well, news articles often do not have sufficient metadata, nor are metadata compatible across different news companies. Section 5 illustrates a few content dimensions that can be used in recommendation.

What both these techniques have in common is that they are based on similarity, either between users or items. The risk of such a recommendation strategy is that users are more likely to be exposed to a narrowing segment of popular items, as the focus lies on maximising the overlap between users' behaviour. As such, recommender systems today strive for news personalisation, which in fact contrasts with the aspired goal of a citizen-aware recommender system. This risk is compounded by the focus on metrics such as accuracy. Often, the performance of a recommender is solely measured in terms of its reconstructional capabilities (i.e. how precise the system is in predicting already consumed articles). All differences between the original and predicted user history are seen as losses in performance. When the lack of diversity is addressed, it is typically done as an adjunct to the standard procedures and through rudimentary means [18, 19]. Increasing diversity and novelty is only considered if it can be done without significantly compromising query similarity, and this application remains limited to aimlessly broadening of the coverage.

Having set the scene, the participants were asked whether they had ever encountered news personalisation in their daily lives and to share their perception of such practices in general. As regards the first question, one attendee indicated to have experienced personalisation at a news site whereby items presumed relevant to a particular consumer were highlighted. However, all news available remained accessible for all users in the same order. Another noted that as social media feeds are being personalised, the news you view on these platforms is so too. In response to the second question, concerns were raised as regards a presumed lack of awareness and transparency in relation to the existence of algorithmic selection processes as well as concerning the logic behind them. It was added that news consumers should be properly informed. One participant furthermore indicated to "dislike the feeling of being steered".

3 A Fundamental Right to Diverse Information

The importance of an easily available diverse news offer has been recognised in several recent policy documents. More specifically, it was argued that it has "the potential to make democratic processes more participatory and inclusive" and to foster public debate – which may ultimately secure democracy –, and could even "uncover, counterbalance, and dilute disinformation" [20, 21]. At EU level, therefore, "empower[ing] users with tools enabling a customised and interactive online experience so as to facilitate content discovery and access to different news sources representing alternative viewpoints" was set as a goal [21–23]. Interestingly, the EU Commission's independent High Level Expert Group on Fake news and Online Disinformation, when defining '[a]ctions in support of press freedom and pluralism' in the final report concerning their approach on disinformation, stated, amongst others, that public authorities must ensure the "protection of [a] basic right[…] to […] *diverse* information" (emphasis added) [22]. In that context, the question arises whether, and to what extent, the right to *freedom of expression and information*, laid down in both Article 10 of the European Convention on Human Rights ('ECHR') [24] – under which positive obligations[1] may arise [26] – and Article 11 of the Charter of Fundamental Rights of the European Union ('Charter') [27], indeed includes *a right to receive diverse information.* Its existence, including a corresponding responsibility for authorities to take affirmative action to ensure its effective exercise

[1] Negative obligations require States not to interfere in the exercise of rights, while positive obligations entail a duty to take the necessary measures to safeguard a right, or, more specifically, to adopt reasonable and suitable measures to protect the rights of individuals in [25].

[28], would enable citizens to *force* policymakers to adopt measures guaranteeing them access, potentially offline as well as online, to a diversity of information.

In its first paragraph, Article 10 ECHR puts forward that:

> "*Everyone* has the right to freedom of expression. This right shall include *freedom* to hold opinions and *to receive* and impart *information and ideas* without interference by public authority and regardless of frontiers [...]." (emphasis added)

On numerous occasions, the European Court of Human Rights ('ECtHR') has interpreted the 'freedom to receive information and ideas'. Indeed, already in the 1979 case of Sunday Times v. the UK, the Court stated that "[n]ot only do the [mass] media have the task of imparting [...] information and ideas [concerning matters of public interest]; *the public also has a right to receive them*" (emphasis added) [29]. In the Informationsverein Lentia and Others v. Austria judgment from 1993, it added that "[s]uch an undertaking cannot be successfully accomplished unless it is grounded in *the principle of pluralism, of which the State is the ultimate guarantor*" (emphasis added) and that "[t]his observation is especially valid in relation to audio-visual media, whose programmes are often broadcast very widely" [30]. On 8 July 1999, in the context of its decision in a number of cases against Turkey, all concerning the criminal convictions of the applicants in view of their involvement in the spread of separatist or pro-Kurdish propaganda [31], the ECtHR explicitly referred to "the public's right to be informed of a different perspective" and considered that the domestic authorities failed to sufficiently respect their negative obligation in that regard [32]. It furthermore concretised its by then settled Sunday Times case-law referred to above, by finding that "[i]t is [...] incumbent on the press to impart information and ideas on political issues, *including divisive ones*" (emphasis added), whilst the public is entitled to receive them [33]. In the Khurshid Mustafa and Tarzibachi v. Sweden case of 2008, which concerned (a prohibition of) the reception of information by means of a satellite dish, it was held in very clear terms that, "[i]n addition to the primarily negative undertaking of a State to abstain from interferences in Convention guarantees", "the genuine and effective exercise of freedom of expression under *Article 10 may require positive measures of protection, even in the sphere of relations between individuals*" (emphasis added) [34]. In the 2009 Times Newspapers LTD (Nos. 1 and 2) v. the United Kingdom case, it was considered that "the Internet plays an important role in enhancing the public's access to news and facilitating the dissemination of information in general" [35]. Late 2009, in the Manole and Others v. Moldova judgment, the Strasbourg Court ruled that "the State [must] ensure [...] that the public has access through television and radio to impartial and accurate information and *a range of opinion and comment, reflecting inter alia the diversity of political outlook within the country*" (emphasis added) [36]. Finally, in 2012, in Centro Europa 7 S.R.L. and Di Stefano v. Italy, it was clarified that, considering the sensitive nature of the audio-visual media sector, member States have a positive obligation to "put in place *an appropriate legislative and administrative framework to guarantee effective pluralism*" (emphasis added) [37].

Article 11 of the Charter provides that:

1. "*Everyone has the right to* freedom of expression. This right shall include *freedom* to hold opinions and *to receive* and impart *information and ideas* without interference by public authority and regardless of frontiers.
2. The freedom and *pluralism of the media shall be respected*." (emphasis added)

Article 52(3) of the Charter stipulates that "[i]n so far [the] Charter contains rights which correspond to rights guaranteed by the Convention for the Protection of Human Rights and Fundamental Freedoms, the meaning and scope of those rights shall be the same as those laid down in the European Convention on Human Rights". Clearly, such is the case for Article 11(1) [38, 39]. Moreover, Article 51(1) EU Charter puts forward that the provisions of the Charter are addressed to the institutions, bodies, offices and agencies of the Union and to its member states only when they are implementing Union Law. They should therefore "respect the rights, observe the principles and promote the application thereof in accordance with their respective powers and respecting the limits of the powers of the Union as conferred on it in the Treaties" [27 art 51(1)]. Leaving aside the discussions concerning the exact scope of application of EU fundamental rights in respect of actions of member states on the basis of Article 51(1) of the Charter [40], several scholars have argued that this provision confers on the Union an indirect power to adopt rules or measures protecting fundamental rights in the course of exercising its specific competences under the Treaties [40]. However, such a power would not allow the Union to take action if the protection of fundamental rights were to be the only or primary aim thereof [40]. It appears, for example, that by means of Article 15 as well as recitals 48 and 55 of the Audiovisual Media Directive [41], the EU legislator has sought to safeguard the right to receive information and to promote pluralism of the media by ensuring diversity in the production and programming of news in the EU, and therefore to respect the principles recognised by both paragraphs of Article 11 of the Charter [40]. In its 2013 Opinion in Sky Österreich, the Court of Justice of the European Union ('CJEU'), confirmed that the EU legislature was indeed 'entitled' – though not, on the basis of positive obligations, 'required' – to do so and to take measures to ensure public access to a diversity of information [42].

In conclusion, the ECtHR has clearly recognised a right of the public to be informed about different viewpoints concerning matters of general importance. The State is ultimately responsible for the effective exercise thereof, and this indisputably within the context of the audio-visual media sector. While the ECtHR so far has not recognised such a duty vis-à-vis States in relation to the public's right to receive information and ideas in the online environment, it indeed very well could, given its acknowledgment of the importance of the Internet in enhancing access to news and facilitating the dissemination of information [43]. The CJEU has also stressed the importance of media pluralism and diversity of information available to the public. Whereas the CJEU does not (yet) consider the Union – in view of its nature and competences [40] – to be directly responsible for taking positive action to that end, it found that the latter certainly *may* do so when exercising its attributed competences.

Considering the potential impact a right to diverse information could have in our contemporary society, we presented the participants with the following questions: one, do we need such a right, and two, should the government play a role in ensuring that citizens can access diverse information? The first question was collectively answered in the affirmative. One attendee in particular argued that one should have access to diverse information as it is enables him or her to make informed decisions. Thoughts in relation to the second question were, on the other hand, more varied. One group of participants noted that in countries where confidence in publicly-funded news is high, the government could also be trusted to guarantee diverse exposure to information. Others, however, stated that such involvement could very easily go wrong. They considered, more specifically, that it may lead to a situation in which people would only be shown content the authorities want them to see.

4 Unravelling News Diversity

As argued above, it becomes clear that today's news recommender systems do not take into account news diversity, even though the idea of receiving diverse information is an important prerequisite for maintaining a democratic society. This neglect also contrasts with the academic field of communication sciences in which news diversity has a long tradition in helping to understand and evaluate the role of news media in the public sphere [44, 45]. In fact, most research on news diversity date back to the arrival of audio-visual and digital media such as television and the web (i.e., 1995–2005, see e.g. [46–48]). Despite the existence of a significant body of literature around the concept of news diversity, however, communication scholars are still struggling with the question of what it means, and how it should be measured [48, 49]. Consequently, a wide range of diversity dimensions, assessments and assumptions are currently used to study news diversity [50].

The broad and ambiguous use of the concept is argued to have several academic and political implications. First, it endangers the broader validity and reliability of existing and future research, which is, in turn, essential for the organisation and application of scientific findings related to news diversity [51]. Second, and linked with the previous, there is a risk of formulating inadequate policy recommendations. For instance, with regard to the discussions on the existence of selective exposure or filter bubbles in the digital environment, the current literature is not able to present a clear overview on the state and outcome of diversity research in the digital environment. As a consequence, policy recommendations are rather limited to 'more research should be done' or 'insight into filter bubbles are indispensable' [9].

We argue that a clear description of what news diversity constitutes may be a first stepping stone to solve the above-mentioned issues. First, it may help scholars to map the current field and identify areas of ambiguity or neglect. Second, it enables news diversity scholars to make informed decisions when studying news diversity. This might be of particular importance for future diversity research, but also for the development of news recommenders.

In this section, we forward an approach to unravel the normative and conceptual assumptions underlying this concept ([52] for an extensive overview). These assumptions range from explicitly formulating the normative position to deciding on what kind of dimensions to measure (see Table 1). We will further elaborate on these assumptions by presenting three leading questions that enable the discussion on the meaning of news diversity.

Table 1. Distinguishing normative and conceptual assumptions of news diversity

Assumption	Leading question
Normative assumption: normative stance	Should news media reflect the diversity in society or should it treat all categories under study equally?
Conceptual assumption: sample selection	What or whom is studied: production, consumption or distribution?
Conceptual assumption: diversity dimensions	Which dimensions in news media content (e.g., gender, sentiment) or structure (e.g., ownership) are studied?

4.1 Normative Assumptions

The first leading question is concerned with the idea of *open and reflective diversity* [53, 54]. The former evaluates diversity as an equal media representation of all categories. The latter argues that media should reflect the diversity in society [55]. Take, for instance, research on the diversity of political opinions in the news. From an open point of view, diversity would be evaluated as an equal representation of all voices in the political spectrum. From a reflective viewpoint, evaluation of diversity would be based on the question to what extent these voices coincide with the current distribution of political opinions in society.

4.2 Conceptual Assumptions

A second question is related to *what or whom is studied*. Traditionally, this means a choice between the production side, in which news is made available, and the consumption side, in which people engage with news. However, in the current news environment, distribution actors such as search engines, recommendation systems, and aggregators could also be considered (e.g. [56]).

The third question deals with the most fundamental part of what constitutes news diversity: *the studied dimension(s) of diversity*. It concerns the focus of analysis, what researchers actually measure to make conclusions about news diversity. This might be centered on dimensions in the content or structure of news media. To name a few examples, we explain the content dimensions 'actor diversity' and 'party diversity'. The former refers to the affiliation or occupation of the actors who are quoted or paraphrased in the news [57]. The latter is concerned with the number of political parties across which a medium distributes its attention, either implicitly in terms of topics or explicitly in terms of party name [54].

To conclude, we want to emphasize that news diversity is a very broad concept, covering several aspects related to news, media and democracy. As such, news diversity remains an ambiguous concept when it is not accompanied by explications of the assumptions underlying this concept. Especially in the context of news recommendations systems, informed decisions on each of these assumptions as well as explicit statements should be made. Audiences, in the first place, but also other stakeholders such as policymakers should be aware of what kind of diversity is tweaked and which ideal is pursued.

During the workshop, the participants were asked to give their opinions in relation to the normative assumption related to news diversity. In particular, they were asked whether news media should reflect the diversity in society (i.e., reflective diversity) or should news media treat all categories under study equally (i.e., open diversity). Interestingly, a majority leaned towards 'open diversity'. Participants pointed out that while reflective diversity indeed mirrors society, it could, when taken too far, limit the forming of opinions. Instead, there could be a 'free market of diversity', in which an increase in reports on right-wing opinions would for example trigger a rise in the distribution of left-wing points of view (and vice versa). As such, a kind of equilibrium could be achieved. Other participants, moreover, pointed out that open diversity would allow niche opinions to grow and even become the majority. Therefore, it could also encourage change. For example, participants pointed to the idea that more coverage of female football, female scientists etc. could have a positive effect on the emancipation of women in society. Nonetheless, the participants discussed whether also the most extreme opinions should be allowed to circulate. Where one attendee stated to prefer to know of their existence, because this helps to assess one's own position on the spectrum, another considered that the right to access diverse information should be restricted in the same way as the freedom to express opinions. Accordingly, a diversity-enhancing news recommender should not promote content involving hate speech.

Concluding this discussion, we want to stress the importance of reflecting on the normative assumptions related to news diversity. As this workshop has shown, several (counter-)arguments may be used in favour or against open/reflective diversity. Future diversity research should focus on these (counter-)arguments to explore the consequences of each assumption.

5 Automated Extraction of Content Dimensions in Written News

This section zooms in on a number of diversity dimensions that can be detected in written news content. As indicated previously, several possibilities, such as actor diversity or the prominence of political parties, have been investigated in this field. These analyses often rely on manual coding of news items. Whereas manual analysis is powerful, it also practically restricts the number of media items that can be parsed, and easily leads to methodological differences between individual researchers. Automating the extraction of relevant dimensions, using techniques from the fields of computational linguistics and artificial intelligence, on the other hand, would allow for a standardised and fine-grained feeding of news algorithms on a large scale. The #NewsDNA research project focusses on the automated extraction of two possible content dimensions that can serve as building blocks for a diversity-driven news recommender: news topics and news events.

5.1 Topics

An intuitive analysis of the content of news articles is centred around news topics. In this context, topics are the general areas on which an article touches, such as politics, international news, or entertainment. News publishers use topics tags to organise their own news output. While some features of a topic taxonomy tend to recur, there is considerable variation between news outlets, making it hard to establish a mapping between them. Additionally, depending on the outlet, articles may belong exclusively to a single topic or to multiple topics. Variable tags based on current events, like *Brexit* or *immigration*, may be used alongside general tags. This lack of uniformity makes it generally impractical for the researcher to use outlet-provided topics tags for automatic analysis across publishers.

Some efforts exist to encourage consistent use of media topics in the news industry. An example of a topics framework promoted as a global standard is the IPTC Media Topics taxonomy [58]. It defines 17 top-level codes which hierarchically subdivide into subtopics up to five levels down. For example, the bottom-level code "housing and urban planning" can be traced back through "interior policy" and "government policy" to the top-level code "politics". The deeper into the tree, the more granular the topic definitions become.

5.2 Events

Topics provide a general idea of the content of an article by describing which aspects of society it touches on, but they do not say anything about its specific contents. More semantics-driven algorithms can shed light on the events described in the text. We briefly discuss one such technique applied in the #NewsDNA project and illustrate the research effort involved.

An attractive and little-explored dimension of analysis is that of news events, i.e. the real-world events which provide the material and context for news articles. For example, in a fictional example entitles "Russian spies arrested in England", the arrest of the Russian spies is the event that leads to the article being written. The goal of event extraction is to identify the real-world events referred to in news texts, as well as information on the actors, time, place, etc. involved in the event. In the example, the "Russian spies" are entities involved in the event and "England" is its stated location. Note that upstream technologies such as named entity extraction can play a role in discovering these participants [59].

An event extraction system, then, is an algorithm which takes as input a text and returns a number of event descriptions it has found in the text. Such a model is obtained through machine learning. First, a set of articles is prepared in which event descriptions have been manually annotated. Second, a machine-learning algorithm goes over this set and, through trial and error, learns to identify event descriptions matching the human-made gold standard. The system can then be run on previously unseen articles to produce new event descriptions.

Inevitably, to extract news events, we need to define what we consider to be a news event. Many different conceptions of "events" have been examined, some of which focus on the discovery of real events in text (see e.g. NewsReader [60], the ACE/ERE programs

[61, 62], RED [63]) and some of which focus on fine-grained text semantics (e.g. the FrameNet project [64]).

Typically, a taxonomy of event types is used, such that each event mention found in the text can be classified in a semantic category such as "Conflict-Attack" or "Transaction-TransferOwnership" (from ERE [62]). The advantage of a fixed taxonomy is that it naturally defines the scope of news events: events that cannot be classified are not recognised.

A sizeable body of work (around the previously cited research programs [61, 62, 65]) focuses on event extraction in a closed data context, where the corpus of articles is given and the event type taxonomy is fixed. This leads to systems that perform well at extracting those specific categories of events, but fail at handling unrestricted news text discussing a wider variety of events. In an open data context, an automatic system must capture all relevant events from in-coming news texts. Designing a taxonomy for this is a difficult balancing act: a small taxonomy will exclude many relevant events, while a taxonomy with many different types will suffer from data sparsity (i.e. some event types are so rare that they cannot be learned or extracted reliably). Additionally, a fixed taxonomy may not adapt well to a stream of news whose tone changes with time. This imposes a process of constant retraining of the algorithm, which is feasible provided that data are available. For instance, suppose that incoming news focuses on a certain terrorist attack one month, whereas the big story of the next month is centred around the question of immigration. A system trained on data from one period in time may be disadvantaged when dealing with news from another. A natural way to sidestep this limitation is to allow for events of type 'unknown' to be extracted, but even in that case care needs to be taken so that 'unknown' events remain a minority within the training data [66]. The prediction of events without type has not been fully explored, as the theoretical applications of this technology tend to presume event type prediction is a desirable feature, or, at least, useful for other downstream applications.

For the purposes of news recommendation, the appeal of event extraction lies in linking event descriptions across articles. Given two event descriptions, specialized systems can establish identity links between them; two mentions that refer to the same event are called co-referent. Co-reference links can be established within but also across articles. It has been thoroughly researched for nominal entities, but not for events, and even less across documents [67]. It allows us to link together articles based on a deep semantic interpretation. For instance, using a topic-based system, we are able to cluster articles based on tags such as politics or business, or if our system is capable of fine-grained topic analysis, more current tags such as Brexit or economic crisis. If we know the specific events that occur in the articles, and if we know how to establish co-reference links between events across articles, we can create clusters based on single events. For example, we could gather all articles discussing Theresa May's resignation in June, with far greater precision than using topic-based methods. In terms of addressing diversity, we could also use these clusters and links to broaden the scope of recommendations in a more organic way by, for example, recommending articles located at the edge of a cluster or from closely neighbouring ones.

Cross-document event recognition and co-reference is key to moving the state of the art in natural language understanding and personalised recommendation. While solutions based on dimensions such as topics and actors work well with recommender systems, we propose that a more granular semantic analysis based on events can further enhance the precision of news recommenders.

After having explained the difference between 'topics' and 'events, we asked our audience to think of other content dimensions which could be of relevance in the context of automated text analysis. A first participant considered it would be interesting to categorise new articles according to their 'level of argumentation'. Well-argued opinions could consequently be singled out and further discussed, which in turn may facilitate public debate. Another attendee suggested that content extraction techniques could be used to verify whether the title – rather than functioning as a 'clickbait' –matches the information contained in an article. The detection of 'viewpoints' also put forward as an option. In that regard, others put forward that structural elements, such as an author's affiliation or background, or his or her country of origin, could also serve as indicators of 'bias' in a news items.

6 Towards a Diversity-Promoting News Recommender

In this article, we addressed the conceptual development of a news diversity-enhancing recommendation system. To do so, we approached news recommendations from four different academic domains: computer sciences, law, communication sciences, and computational linguistics.

In the first section (i.e. computer sciences), we reviewed the state of the art of current news recommendation systems. In particular, we described two dominant methodologies – collaborative and content-based filtering – and unravelled their assumptions and drawbacks. We ended this section with a critique in that citizen – oriented concepts such as news diversity are currently underrepresented in these methodologies. Other concepts, such as accuracy or maximisation of the overlap between users' behavior, currently dominate the discourses in this field.

In the second discussion (i.e. law), we discussed the existence of a so-called 'right to diverse information'. By means of an analysis of relevant fundamental rights documents and case-law of the ECtHR and CJEU, we were able to support this statement. As a result, we concluded that governments carry the ultimate responsibility for the effective exercise of this right, in the past predominantly with respect to audio-visual media, but in the future potentially also with regard to the accessibility of online news.

In the third section (i.e. communication sciences), we explored the meaning of the mere notion of news diversity. As argued, diversity may function as an alternative, more citizen-oriented strategy to design news recommendation systems, yet the concept itself is characterized by ambiguity. As such, we started our discussion with the conceptual difficulties of this concept and their implications. Then, we presented an approach to unravel the normative and conceptual assumptions underlying this concept.

In the fourth discussion (i.e. computational linguistics), we explored how computational methods may enrich manual analysis in order to extract news content dimensions such as topics and events. We illustrated the usage of topic tags, and introduced automatic event extraction, citing applications, drawbacks and obstacles that emerge when these methods are set into practice.

Based on the feedback we received from the participants of the 2019 IFIP workshop on ‘co-designing a personalised news diversity algorithmic model based on news consumers’ agency and fine-grained content modelling’, we consider that ensuring and promoting diversity in information exposure is an important public policy goal. As future developments in the fields of recommender systems and automated content extraction may contribute to its achievement, further research into the conceptualization of a news-diversity enhancing algorithm will continue to be undertaken in the #NewsDNA project across the four disciplines. On the one hand, conceptual questions stemming from the fields of communication science and law will be considered on a fundamental level. This concerns questions such as ‘which dimensions should be selected to conceptualise news diversity?’ or ‘what is the optimal outcome of diversity to which audiences are steered?’ to which no unequivocal answers yet exist. On the other hand, the fields of computational linguistics and computer sciences, which enable such a recommender system, still carry operational questions and difficulties. Relevant content dimensions must be translated into content extraction algorithms, which is not a solved issue. The design of the recommendation algorithm must also be carefully considered, as the right balance has to be made between relevance and diversity.

References

1. All the news that’s fit for you: The New York Times. “Your Weekly Edition” is a brand-new newsletter personalized for each recipient. https://www.niemanlab.org/2018/06/all-the-news-thats-fit-for-you-the-new-york-times-your-weekly-edition-is-a-brand-new-newsletter-personalized-for-each-recipient/. Accessed 5 Dec 2019
2. Mis niets over uw favoriete thema’s via “Mijn dS”. https://www.standaard.be/cnt/dmf20190304_04229666. Accessed 9 Aug 2019
3. Gepersonaliseerd nieuws: matchmaker voor online media of journalistiek-ethisch mijnenveld? https://www.vn.nl/gepersonaliseerd-nieuws-matchmaker-of-mijnenveld/. Accessed 5 Dec 2019
4. Thurman, N., Moeller, J., Helberger, N., Trilling, D.: My friends, editors, algorithms, and I. Digital Journalism **7**, 447–469 (2019). https://doi.org/10.1080/21670811.2018.1493936
5. Thurman, N., Schifferes, S.: The future of personalization at news websites. Journal. Stud. **13**, 775–790 (2012). https://doi.org/10.1080/1461670X.2012.664341
6. Araujo, T.B., Helberger, N., Kruikemeier, S., de Vreese, C.H.: In AI we trust? Perceptions about automated decision-making by artificial intelligence. AI & Society (2020). https://doi.org/10.1007/s00146-019-00931-w
7. Mcquail, D.: McQuail’s Mass Communication Theory. SAGE Publications Ltd., London (2010)
8. Möller, J., Helberger, N., Makhortkh, M., van Dooremalen, S.: Filterbubbels in Nederland (2019). Commissariaat Voor de Media (2019)
9. Borgesius, F.J.Z., Trilling, D., Möller, J., Bodó, B., de Vreese, C.H., Helberger, N.: Should we worry about filter bubbles? Internet Policy Review **5**(1), 1–16 (2016)
10. Helberger, N.: On the democratic role of news recommenders. Digital Journalism **7**, 993–1012 (2019). https://doi.org/10.1080/21670811.2019.1623700
11. Recommendation Systems: General Collaborative Filtering Algorithm Ideas. http://www.cs.carleton.edu/cs_comps/0607/recommend/recommender/collaborativefiltering.html. Accessed 5 Dec 2019

12. Sarwar, B., Karypis, G., Konstan, J., Riedl, J.: Item-based collaborative filtering recommendation algorithms. In: Proceedings of the 10th International Conference on World Wide Web, pp. 285–295. ACM, New York (2001). https://doi.org/10.1145/371920.372071
13. Van Damme, K., Martens, M., Van Leuven, S., Vanden Abeele, M., De Marez, L.: Mapping the mobile DNA of news. Understanding incidental and serendipitous mobile news consumption. Digit. Journal., 1–20 (2019). https://doi.org/10.1080/21670811.2019.1655461
14. Good, N., Schafer, J.B., Konstan, J.A., Borchers, A., Sarwar, B., Herlocker, J., Riedl, J.: Combining collaborative filtering with personal agents for better recommendations. In: Proceedings of the Sixteenth National Conference on Artificial Intelligence and the Eleventh Innovative Applications of Artificial Intelligence Conference Innovative Applications of Artificial Intelligence, pp. 439–446. American Association for Artificial Intelligence, Menlo Park (1999)
15. Liu, J., Pedersen, E., Dolan, P.: Personalized news recommendation based on click behavior. In: 2010 International Conference on Intelligent User Interfaces (2010)
16. Chen, Y., Wu, C., Xie, M., Guo, X.: Solving the sparsity problem in recommender systems using association retrieval. JCP **6**, 1896–1902 (2011). https://doi.org/10.4304/jcp.6.9.1896-1902
17. Adnan, Md.N.M., Chowdury, M.R., Taz, I., Ahmed, T., Rahman, R.M.: Content based news recommendation system based on fuzzy logic. In: 2014 International Conference on Informatics, Electronics Vision (ICIEV), pp. 1–6 (2014). https://doi.org/10.1109/ICIEV.2014.6850800
18. Smyth, B., McClave, P.: Similarity vs. diversity. In: Aha, D.W., Watson, I. (eds.) ICCBR 2001. LNCS (LNAI), vol. 2080, pp. 347–361. Springer, Heidelberg (2001). https://doi.org/10.1007/3-540-44593-5_25
19. Lathia, N., Hailes, S., Capra, L., Amatriain, X.: Temporal diversity in recommender systems. In: Proceedings of the 33rd International ACM SIGIR Conference on Research and Development in Information Retrieval, pp. 210–217. ACM, New York (2010). https://doi.org/10.1145/1835449.1835486
20. United Nations (UN) Special Rapporteur on Freedom of Opinion and Expression, Organization for Security and Co-operation in Europe (OSCE) Representative on Freedom of the Media, Organization of American States (OAS) Special Rapporteur on Freedom of Expression, African Commission on Human and Peoples' Rights (ACHPR) Special Rapporteur on Freedom of Expression and Access to Information: Joint Declaration on Freedom of Expression and "Fake News", Disinformation and Propaganda (2017)
21. European Commission: Tackling online disinformation: a European Approach. Communication COM(2018) 236 final (2018)
22. High level Group on fake news and disinformation: A multi-dimensional approach to disinformation. Report of the independent High level Group on fake news and online disinformation. European Commission Directorate-General for Communication, Networks, Content and Technology (2018)
23. EU Code of Practice on Disinformation (2018)
24. European Convention for the Protection of Human Rights and Fundamental Freedoms. ETS No. 005 (1950)
25. Akandji-Kombe, J.-F.: Positive obligations under the European Convention on Human Rights. A guide to the implementation of the European Convention on Human Rights. Council of Europe, Strasbourg (2007)
26. Council of Europe/European Court of Human Rights: Positive obligations on member States under Article 10 to protect journalists and prevent impunity. European Court of Human Rights Case-law Research Reports (2011)
27. Charter of Fundamental Rights of the European Union. OJ C 326/391 (2000)

28. McGonagle, T.: Positive obligations concerning freedom of expression: mere potential or real power? In: Journalism at Risk: Threats, Challenges and Perspectives, pp. 9–35. Council of Europe, Strasbourg (2015)
29. The Sunday Times v. the United Kingdom (No. 1). ECtHR (1979)
30. Informationsverein Lentia and Others v. Austria. ECtHR (1993)
31. Voorhoof, D., van Loon, A., Vier, C.: IRIS Themes – Vol. III – Freedom of Expression, the Media and Journalists. Case-law of the European Court of Human Rights. European Audiovisual Observatory, Strasbourg (2017)
32. Erdoğdu and İnce v. Turkey. ECtHR (1999)
33. Sürek v. Turkey (No. 1). ECtHR (1999)
34. Khursid Mustafa and Tarzibachi v. Sweden. ECtHR (2008)
35. Times Newspapers LTD v. the United Kingdom (Nos. 1 and 2). ECtHR (2009)
36. Manole and Others v. Moldova. ECtHR (2009)
37. Centro Europa 7 S.R.L. and Di Stefano v. Italy. ECtHR (2012)
38. Explanations relating to the Charter of Fundamental Rights. OJ C 303/17 (2007)
39. Advocate General Jääskinen: Google Spain. CJEU (2013)
40. Beijer, M.: Limits of Fundamental Rights Protection by the EU: The Scope for the Development of Positive Obligations. Intersentia, Cambridge, Antwerp, Portland (2017)
41. Directive 2010/13/EU of the European Parliament and of the Council of 10 March 2010 on the coordination of certain provisions laid down by law, regulation or administrative action in Member States concerning the provision of audiovisual media services (Audiovisual Media Services Directive) (Text with EEA relevance) (2010)
42. Sky Österreich. CJEU (2013)
43. Animal Defenders International v. the United Kingdom. ECtHR (2013)
44. McQuail, D.: Media performance. In: The International Encyclopedia of Political Communication, pp. 1–9 (2015)
45. van der Wurff, R.: Do audiences receive diverse ideas from news media? Exposure to a variety of news media and personal characteristics as determinants of diversity as received. Eur. J. Commun. **26**, 328–342 (2011). https://doi.org/10.1177/0267323111423377
46. Day, A.G., Golan, G.: Source and content diversity in Op-Ed Pages: assessing editorial strategies in The New York Times and the Washington Post. Journalism Studies**6**, 61–71 (2005). https://doi.org/10.1080/1461670052000328212
47. Rodgers, R., Hallock, S., Gennaria, M., Wei, F.: Two papers in joint operating agreement publish meaningful editorial diversity. Newspaper Res. J. **25**, 104–109 (2004)
48. Voakes, P.S., Kapfer, J., Kurpius, D., Chern, D.S.-Y.: Diversity in the news: a conceptual and methodological framework. Journalism & Mass Communication Quarterly **73**, 582–593 (1996). https://doi.org/10.1177/107769909607300306
49. Raeijmaekers, D., Maeseele, P.: Media, pluralism and democracy: what's in a name? Media Cult. Soc. **37**, 1042–1059 (2015). https://doi.org/10.1177/0163443715591670
50. Napoli, P.M.: Deconstructing the diversity principle. Journal of Communication **49**, 7–34 (1999). https://doi.org/10.1111/j.1460-2466.1999.tb02815.x
51. Liu, P., Li, Z.: Task complexity: a review and conceptualization framework. Int. J. Ind. Ergon. **42**, 553–568 (2012). https://doi.org/10.1016/j.ergon.2012.09.001
52. Joris, G., De Grove, F., Van Damme, K., De Marez, L.: News diversity reconsidered: a systematic literature review unravelling the diversity in conceptualizations (submitted)
53. McQuail, D., Van Cuilenburg, J.J.: Diversity as a media policy goal: a strategy for evaluative research and a Netherlands case study. Gazette **31**, 145–162 (1983). https://doi.org/10.1177/001654928303100301
54. Takens, J., van Ruigrok, N., van Hoof, A., Scholten, O.: Old ties from a new(s) perspective: diversity in the Dutch press coverage of the 2006 general election campaign. Communications **35**, 417–438 (2010). https://doi.org/10.1515/comm.2010.022

55. McQuail, D.: Media Performance: Mass Communication and the Public Interest. Sage Publications, London (1992)
56. Möller, J., Trilling, D., Helberger, N., van Es, B.: Do not blame it on the algorithm: an empirical assessment of multiple recommender systems and their impact on content diversity. Information, Communication & Society **21**, 959–977 (2018). https://doi.org/10.1080/1369118X.2018.1444076
57. Masini, A., Van, A.P.: Actor diversity and viewpoint diversity: two of a kind? Communications **42**, 107–126 (2017). https://doi.org/10.1515/commun-2017-0017
58. Media Topics. https://iptc.org/standards/media-topics/
59. Colruyt, C., De Clercq, O., Hoste, V.: EventDNA: guidelines for entities and events in Dutch news texts (v1.0) (2019)
60. Vossen, P.: Newsreader Public Summary (2016)
61. Doddington, G., Mitchell, A., Przybocki, M., Ramshaw, L., Strassel, S., Weischedel, R.: The automatic content extraction (ACE) program – tasks, data, and evaluation. In: Proceedings of the Fourth International Conference on Language Resources and Evaluation (LREC 2004). European Language Resources Association (ELRA), Lisbon (2004)
62. Aguilar, J., Beller, C., McNamee, P., Van Durme, B., Strassel, S., Song, Z., Ellis, J.: A comparison of the events and relations across ACE, ERE, TAC-KBP, and FrameNet annotation standards. In: Proceedings of the Second Workshop on EVENTS: Definition, Detection, Coreference, and Representation, pp. 45–53. Association for Computational Linguistics, Baltimore (2014). https://doi.org/10.3115/v1/W14-2907
63. O'Gorman, T., Wright-Bettner, K., Palmer, M.: Richer event description: integrating event coreference with temporal, causal and bridging annotation. In: Proceedings of the 2nd Workshop on Computing News Storylines (CNS 2016), pp. 47–56. Association for Computational Linguistics, Austin (2016). https://doi.org/10.18653/v1/W16-5706
64. Ruppenhofer, J., Ellsworth, M., Schwarzer-Petruck, M., Johnson, C.R., Scheffczyk, J.: FrameNet II: Extended Theory and Practice. International Computer Science Institute (2006)
65. Vossen, P.: NewsReader at SemEval-2018 task 5: counting events by reasoning over event-centric-knowledge-graphs. In: Proceedings of the 12th International Workshop on Semantic Evaluation, pp. 660–666. Association for Computational Linguistics, New Orleans (2018). https://doi.org/10.18653/v1/S18-1108
66. Colruyt, C., De Clercq, O., Hoste, V.: Comparing event annotations: notes on the EventDNA corpus IAA study (2019)
67. Lu, J., Ng, V.: Event coreference resolution: a survey of two decades of research. In: Proceedings of the Twenty-Seventh International Joint Conference on Artificial Intelligence, pp. 5479–5486 (2018)

Language and Privacy

Ontology-Based Modeling of Privacy Vulnerabilities for Data Sharing

Jens Hjort Schwee(✉), Fisayo Caleb Sangogboye, Aslak Johansen, and Mikkel Baun Kjærgaard

University of Southern Denmark, Campusvej 55, 5230 Odense, Denmark
{jehs,fsan,asjo,mbkj}@mmmi.sdu.dk

Abstract. When several parties want to share sensor-based datasets it can be difficult to know exactly what kinds of information can be extracted from the shared data. This is because many types of sensor data can be used to estimate indirect information, e.g., in smart buildings a CO_2 stream can be used to estimate the presence and number of occupants in each room. If a data publisher does not consider these transformations of data their privacy protection of the data might be problematic. It currently requires a manual inspection by a knowledge expert of each dataset to identify possible privacy vulnerabilities for estimating indirect information. This manual process does not scale with the increasing availability of data due to the general lack of experts and the associated cost with their work. To improve this process, we propose a privacy vulnerability ontology that helps highlight the specific privacy challenges that can emerge when sharing a dataset. The ontology is intended to model data transformations, privacy attacks, and privacy risks regarding data streams. In the paper, we have used the ontology for modeling the findings of eight papers in the smart building domain. Furthermore, the ontology is applied to a case study scenario using a published dataset. The results show that the ontology can be used to highlight privacy risks in datasets.

Keywords: Open data · Data anonymization · Modeling methodologies · Data publishing · Data privacy · Privacy-Preserving Data Publishing

1 Introduction

Open data has a great potential for improving scientific practices in terms of transparency and efficiency [23]. Data is also the foundation for new intelligent data-driven software solutions in many application areas. The large sensor-networks installed in a modern smart building can monitor almost every aspect of the building and the occupants inside [15]. This includes data about how the building is being used by occupants and control systems. e.g., from occupancy counting sensors and the building's building management system (BMS).

Published by Springer Nature Switzerland AG 2020
M. Friedewald et al. (Eds.): Privacy and Identity 2019, IFIP AICT 576, pp. 109–125, 2020.
https://doi.org/10.1007/978-3-030-42504-3_8

Sharing such information with contractors who perform regular tasks – such as catering, cleaning, and facility management – would enable them to develop data-driven applications for these operations. Considering a scenario where a catering company knows how many people are actually on the premises, they would know exactly how many to cook for. If they also know who these people are, then they could have access to dietary needs and cook according to preferences.

The European Data Portal has made a guideline for how to create a strategy for sharing open data [8], which consists of:

1. ***Ambition.*** The strategy starts with setting the ambition for publishing data, including collecting a clear picture of the current data sharing situation in the organization. Furthermore, defining the intended situation.
2. ***Strategy.*** Create a strategy for an open data policy, which among others, includes identifying all data in the organization, aligning the legal aspects, and formalizing key performance indicators to measure the progress, both for sharing the data and the impact of sharing it.
3. ***Policy.*** Define the policy benefits for sharing the data. This includes defining, the scope, the goals, and the data types and quality of the data. Finally, the legal aspects, which include licensing, intellectual property, and privacy aspects.

Implementing such a strategy requires finding an appropriate data platform for the data, and making the data understandable for the recipient, as well as identifying the expected use from the open data community. Finally, a plan for how to maintain the data should be made, such that data release is up to date.

This process includes several stakeholders: The management of the company who develops the open data strategy; the data publisher, who is the one processing the data for sharing and who also is doing the privacy assessment, as well as maintaining the data; the data providers who either have actively participating in data collection (e.g., notes and sensor data from a smart watch) or passively being observed (e.g., data from a building-wide video analytics system); the data recipient, who are using the shared data.

When sharing data a data publisher needs to consider the privacy implications of sharing data. This is both to comply with privacy laws and regulations and to respect the interests of the data providers. One of the many challenges when sharing data is to identify the potential privacy implications of each part of the dataset, as well as combinations with available information through side channels. Combinations with other data can be the base of a privacy attack on the data which result in personal information about the data provider being revealed. Even more difficult is identifying privacy attacks, which can be achieved using artificial intelligence (AI) methods, as the AI area is a continuously moving target due to advancements in research. This makes it very difficult to keep an updated picture of the potential privacy risks. Here anonymizing the data and deleting any information related to the participants (e.g., name and height) is a method for privacy protection to limit such attacks on the data, but the current method also has their shortcomings. State-of-the-practice (SoP) methods for anonymization have, in recent research [26,28] been found insufficient to protect the released data. Indicating that the data publisher did not know the specific inference and data linkage possibilities, as well as the privacy risks related to the data release. Therefore, there is a need for solutions to address these problems.

Consider an example from the smart building domain, where a data publisher releases CO_2 data collected in a single room. The data publisher performs a privacy assessment for the CO_2 stream and identifies some privacy risks for the stream. Based on these, an anonymization strategy is selected for the data release. However, the data publisher needs also to be aware of the inference and data linkage possibilities using the stream, e.g., in the case of a CO_2 data stream an AI-based transformation model can estimate the number of occupants in the monitored area [4]. The combination of such inference possibilities and lack of privacy risks knowledge can lead to the data not being adequately protected for the data release. Furthermore, a data publisher must consider laws like the European General Data Protection Regulation (GDPR) [11], the EU's ePrivacy Directive [22], the California Consumer Privacy Act (CCPA) [7], and the Australian Privacy Principles (APPs) [20], which amongst others, defines the personally identifiable information that organizations are allowed to store and share, without reasonable cause or user consent. Many smart buildings are publicly accessible buildings. Therefore, an attacker can physically observe the building and combine ground truth with the published data to potentially infer complete knowledge about the building's use and conditions. This makes it challenging to select and create anonymization methods for smart building data.

In this paper, we address the very important privacy assessment step of the sharing process. The problem is that it can be very difficult for a data publisher to identify the potential privacy implications for a specific dataset. The data publisher might not have a full overview of the privacy risks related to each piece of data. This can lead to sharing of privacy problematic data which is a problem both for the individual data providers and for the publisher in terms of laws and regulations. Knowledge of the potential privacy risks is thus critical. Therefore, we present an ontology that for a given dataset can model privacy risks, privacy attacks, and possible data transformations. Previous work document a long range of possible transformations that can be applied to smart building datasets. However, it is impossible for data publishers to be experts and be up to date with the latest knowledge. In addition, the data publishers also need to consider that several transformations might be combined to reveal a certain piece of information from a dataset. Therefore, there is a need for a common format for modeling privacy risks. From a top-level view, the proposed ontology is illustrated in Fig. 1, including elements for data, transformations, privacy risks, and attacks, and the relationships among them. Using this ontology, the theoretical privacy risks associated with sharing some data can be modeled. The model can be used to consider privacy risks before sharing data. The paper will focus on the smart building domain, and will, therefore, propose a solution targeted this domain. However, the solution has a clear potential to be extended to other domains.

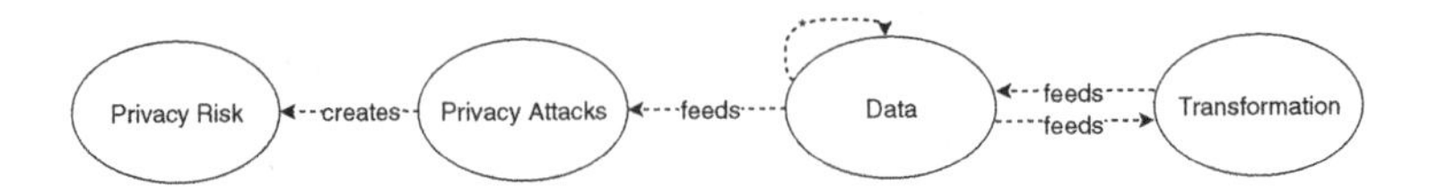

Fig. 1. The elements in the ontology. Includes classes and object relations.

The rest of the paper is structured as follows: Sect. 2 covers related work; Sect. 3 presents an analysis of a set of papers within the smart building domain to identify the concepts classes to be included in the ontology; Sect. 4 presents the design of the ontology; Sect. 5 presents an instance example of the ontology for each of the analyzed papers; Sect. 6 a case study combines all the individual models into a single instance of the ontology, which considers the privacy risks for a specific data release. Finally, Sects. 7 and 8 we will discuss and conclude on the paper.

2 Related Work

This section covers the related work including, including prominent privacy-preserving methods, e.g., methods for Privacy-Preserving Data Publishing (PPDP). The section also covers data transformations that can be performed on smart building data. Furthermore, the section also studies existing smart building ontologies, as well as the components necessary for creating instances of it and taxonomies for the sensitivity of attributes.

2.1 Protection Methods

A data publisher must consider if there is a need for protecting sensitive information in a dataset or the identity of monitored individuals using privacy-preserving data mining or PPDP methods, respectively [25]. The field of PPDP has developed several privacy models. Some of the more prominent ones include k-anonymity [30], l-diversity [18], and δ-presence [19] differential privacy [9]. These models protect against the privacy attacks of record-linkage, attribute-linkage, table-linkage, and probabilistic attacks, respectively.

More recently, Pappachan et al. [21], have developed a framework which can be used for configuring privacy settings of users and enforcing these when collecting and sharing user data in the smart building domain. The framework contains three elements: Internet of Things (IoT) resource registries, IoT assistants, and privacy-aware smart buildings. Resource registries are a collection of policies and sharing practices of IoT technologies. IoT assistants notify users about the policies of the IoT devices. Furthermore, the framework provides configuration for privacy settings, which enables the users to define which information they are willing to share with specific services. The operator of a smart building can also define which services are being used in the building, which can override the users' privacy settings. The privacy-aware smart building receives the privacy settings and uses them when collecting user data and sharing data.

Schwee et al. [28], have explored state-of-the-practice (SoP) PPDP methods on a published building dataset. They found that SoP methods like suppression were insufficient to protect the identity of rooms in the published dataset. As a result of this, identifying a potential privacy risk for such building data showing that occupancy data and room schedules can reveal the identity of a room.

2.2 Data Transformation

The recent advancements in AI have enabled data-driven creation of models that can by estimation transform a combination of input data to relevant information. A number of such transformation has been developed within the field of smart buildings: Arief-Ang et al. [4], have demonstrated that using nothing but CO_2 measurements, the amount of occupants in a room can be estimated. Furthermore, [3], presented how to generalize this method using domain-adaptation. Ardakanian et al. [1], have proposed an occupancy detection technique using fine-grained measurements from Variable Air Volume (VAV) systems. The results show that the estimated occupancy patterns can be used for per-zone schedules. Hence it can be used to estimate occupancy. In Sect. 3 more examples of such transformations are given when analyzing specific examples.

2.3 Ontologies and Taxonomies

Ontologies enable the structuring data from data collection and also to structure the description of the data collection. A number of ontologies have been proposed within the smart building field including Haystack, Industry Foundation Classes (IFC), and BRICK to describe the building and the installed data collection infrastructure. A prominent example is BRICK [5] which is an ontology to structure data about smart buildings. BRICK structures data based on the Resource Description Framework (RDF) [17] to define the relations between the elements (e.g. system or building elements) in the ontology as well as hierarchies of such elements. Having a BRICK model of a smart building enables the use of SPARQL [24] queries to discover control systems and the location of sensors. This enables applications to use the model to discover resources instead of hardcoding an application to the specific devices in a building. Furthermore, the model can be used to identify the available sensors at each location and explore how they are colocated.

Within the privacy field, a taxonomy for privacy sensitivity attributes can be found in [12]. It defines four types of attributes: Explicit identifiers, quasi-identifiers, sensitive attributes, and non-sensitive attributes. The explicit identifiers identify a record owner. The quasi-identifiers can, as a set, potentially identify a record owner. The sensitive attributes contain attributes which are person-specific information. The set of non-sensitive attributes is all attributes that do not fall into any of the others. The attributes in the Explicit identifier, quasi identifier, and sensitive attributes, need to be protected before they can be shared.

2.4 Summary

The mentioned privacy protection methods can be applied when it is known which potential privacy risks a dataset needs to be protected against. However, a data publisher has to identify the risks of the data, the inference opportunities, and the data linkage possibilities. The mentioned ontologies only model the physical relationships in the building, and thus leaves a gap when it comes to modeling data-sharing challenges. The privacy attribute taxonomy is in its current form difficult to apply on time-series data, which a lot of smart building data are. This is because the privacy risks for individuals often occur in a non-statically manner, e.g., outlines in the data streams. Therefore, to help data publishers there is a need for means for modeling privacy risks that incorporate knowledge created by the scientific community. Such models can, among others support data publishers in their privacy assessments for a specific dataset.

In this paper, we propose a method for modeling data, privacy risks, transformations, and privacy attacks and their relationships in a graph-based model using Resource Description Framework (RDF) triples. The graphs can be used for highlighting the potential privacy risk for a specific dataset before sharing the data. Thereby, giving the data publisher a mean to structure knowledge about the potential privacy risks and the ability to make better-informed decisions before sharing data.

3 Analysis of Domain Cases

In this section, we analyze existing work to identify the elements needed in an ontology for modeling privacy risks, privacy attacks, and possible data transformations. Six papers were selected to cover many types of sensor data, data transformations, and extracted information. Furthermore, all of them have been recently published in major conferences or journals in the field.

Electricity consumption measured by smart meters is a widely used sensor modality in buildings to capture electricity use and related occupant behavior. As an example, Sonta et al. [29], have proposed methods that for each desk as input use electricity consumption data for all available plugs. Their method can transform this data into desk-level activities of the occupant. In addition, the method can also produce a social network for the social interactions between the occupants in a whole office. Another example is Kleiminger et al. [16] that propose a method that from electricity consumption measured at a household granularity can estimate household occupancy. Furthermore, Beckel et al. [6] have created a model that from electricity consumption data at the household level can estimate several demographic parameters including the age of the residents, marital status, employment and the amount of bedrooms.

Door monitoring has been studied by many authors to capture room movement in buildings. As an example, Khalil et al. [14] propose a method that as input uses data from ultrasonic-based distance sensors in each door. Their method can from these measurements identify occupants passing through the door. Furthermore, the authors have proposed a method for detecting if the occupant passing the door is using a phone, holding a handbag, or a backpack.

Another relevant type of information is room occupation and number of occupants. As an example, Kjærgaard et al. [15] propose methods that from passive infrared movement or video-based sensors can estimate the presence and number of occupants. The methods can also via machine learning predict the future amount of occupants in an area. The spatial resolution of the predictions is similar to the monitored resolution, e.g., if monitoring private offices, then the prediction will be on the private office level. Likewise, Sangogboye et al. [27] have proposed a method that from passive infrared movement, temperature, and CO_2 sensor streams estimates the future amount of occupants in an area.

In a shopping context knowledge about the movement of shoppers and their intent is highly relevant to optimize shops. As an example, Kaur et al. [13] have proposed a method that from Wi-Fi logs estimates shoppers' physical movement. This physical information is then combined with shoppers' cyberbehvior based on weblogs to estimate the intent of the individual shoppers.

To study humans' health and wellbeing in indoor environments their activities and metabolic rate are too highly relevant types of information. As an example, Dziedzic et al. [10] have proposed an estimation model using the movement skeleton joints of the human body captured by a Microsoft Kinect installed in a household. The skeleton movement data is used to estimate an occupant's metabolic rate and activities. Furthermore, the paper highlights that using the data in conjunction with external data, one can estimate the weight of the occupant.

With the methods presented in the covered papers, one can extract several types of information from building-related data. An overview of what we have found in these analyses can be seen in Table 1. The table highlights which kind of information each of the methods in the papers needs as input to extract the specify type of information. Furthermore, the table highlights that the methods can extract many different types of information from building-related data. The papers use data in the form of time-series, graph, external, and metadata which an ontology must have classes to model.

4 Ontology-Based Modeling

In this section, we propose an ontology for modeling potential privacy attacks, privacy risks, and transformations. The ontology was developed based on our analyze results of existing work.

The analysis identified a number of different data types, namely: time-series, external, graph, and metadata. The analysis also highlighted a number of transformations that also need a concept in the ontology. Furthermore, the ontology has to model possible privacy risks and privacy attacks. All of the concepts are to be modeled using RDF. An overview of the ontology and the concepts can be found in Fig. 1. This gives the following concepts in the ontology:

Table 1. Overview over the papers analysed, and what information can be learned by the methods of each of the papers.

Paper	Sensor modality	Necessary resolution			Reveal information of type						
		Id	Spatial	Temporal	Relations	Behavior	Actions	Intent	Demographics	Health	Id
Sonta et al. [29]	Electricity consumption			X	X						
Khalil et al. [14]	Door distances		X	X		X					X
Kjærgaard et al. [15]	Door openings		X	X		X					
Kjærgaard et al. [15]	Occupant counts		X	X		X					X
Kjærgaard et al. [15]	PIR movement		X	X		X					X
Sangogboye et al. [27]	PIR movement & Temperature & CO_2		X	X		X					
Kaur et al. [13]	Web query Log & WiFi access-point association log & Shopping mall layout		X	X		X		X			
Kleiminger et al. [16]	Electricity consumption		X	X		X					
Beckel et al. [6]	Electricity consumption			X					X		
Dziedzic et al. [10]	Body skeleton joints		X	X			X	X		X	

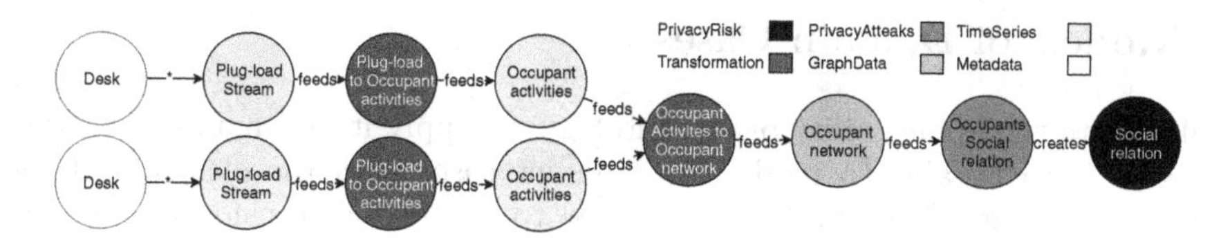

Fig. 2. Instance of the ontology, using the method presented by Sonta et al. [29].

- ***Data*** superclass, which is the superclass of all data types.
  - ***TimeSeries*** model any time-series data in relation to the data release.
  - ***External*** any external data which might be available in relation to the monitored area.
  - ***Metadata*** all relevant associated data that can be available about the data or the monitored context.
  - ***Graph*** is representing graphs data.
- ***Transformation*** is a concept class that models the transformation and inference possibilities, which can be performed on the data in the model. The class must have one or more feed relation from Data classes which represent the relations that are involved in providing input data to the transformation. Furthermore, the class has an outgoing feed relation to other Data classes to represent the output of transformations in terms of new types of data.
- ***PrivacyAttack*** is a class that models the potential attacks which can be performed on Data sources, which can be modeled using the feed relation. The attacks may only be executed if all of its inputs are available. The class can have relations to PrivacyRisk classes using a create relation.
- ***PrivacyRisk*** represents a potential privacy risk in relation to the data release. We use the term privacy risk to highlight that only if an attack is performed it results in an actual leak until then there is a risk that this might happen.

As an example, using the ontology, transformations, privacy attacks, and risk of the method by Sonta et al. [29] is shown in Fig. 2. We have modeled the context monitored as metadata and only two desks for clarity of the example. The streams of plug-load consumption measurements have been modeled as TimeSeries classes, these streams can be transformed into a time-series of occupant activities, using the proposed transformation. This is modeled as a transformation and produces a new TimeSeries. Based on these occupant activities, there is another transformation that can construct an occupant network, which can be represented by the Graph class. Finally, using the network, there can be performed PrivacyAttack estimating social relations among the occupants of the desks which is modeled as a PrivacyRisk. The instance of the ontology can be used to identify that if releasing plug-load consumption data, which is monitored on a desk level, this can potentially reveal social relations among the occupants.

5 Models of Domain Cases

To illustrate the applicability of the ontology we apply it in a number of cases. The goal is to be able to model the different cases with the elements of the proposed ontology. The ontology is applied to model the transformations and privacy risks identified in each of the case papers. In the models, we will only include elements that have been mentioned by the papers. Therefore, we do not include any potential other privacy risks or privacy attacks which could, in theory, be performed upon the data.

As the first case we consider the transformation methods described by Kjærgaard et al. [15]. We have modeled all of them in a single instance of the ontology, which can be found in Fig. 3. Using the model, it can be observed that the three time-series streams for CO_2, door opening measurements, and vision-based sensor count lines can be used to estimate the future presence and amount of occupants in the monitored zone.

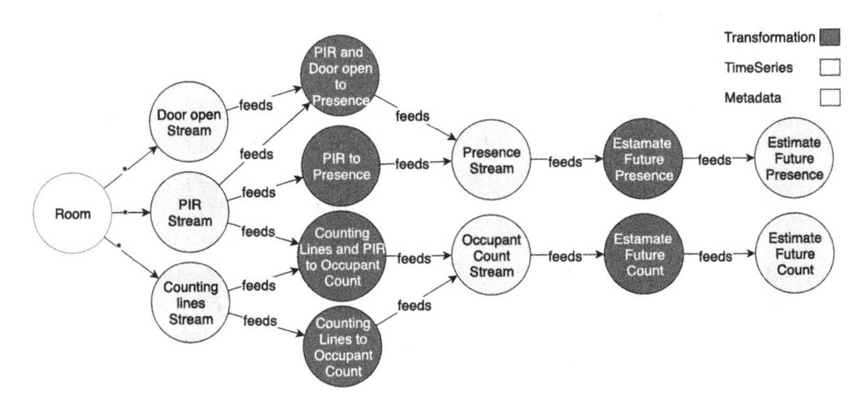

Fig. 3. Instance of the ontology, which models the findings of the paper Kjærgaard et al. [15]. We have chosen to model a single room as the monitored area.

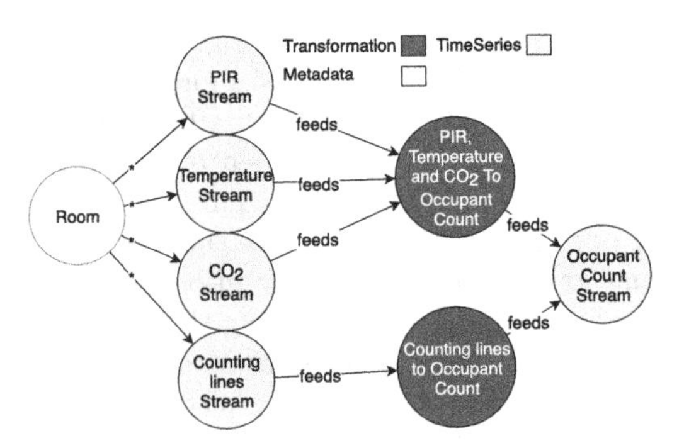

Fig. 4. Instance of the ontology, which models the findings of the paper Sangogboye et al. [27]. We have chosen to model a single room as the monitored area.

The instance for Sangogboye et al. [27], can be found in Fig. 4. The model highlights the possibility of estimating occupant counts using a combination of Passive infrared sensor (PIR) movement, CO_2, and temperature streams, in a monitored area.

The information found in Khalil et al. [14], has been modeled in the instance of the ontology found in Fig. 5. The model includes the privacy attack and risks on the user identity, which has been modeled as a potential privacy attack. The attack can be performed using the time-series streams about user behavior. Using the model it can be observed that based on the three distance sensors installed in a single door, it can be used to identify user behavior and in some cases identify the occupant passing through the door.

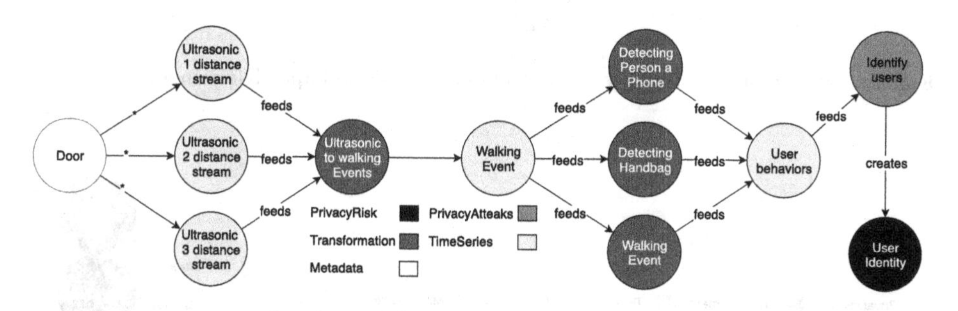

Fig. 5. Instance of the ontology, using the findings of the paper Khalil et al. [14].

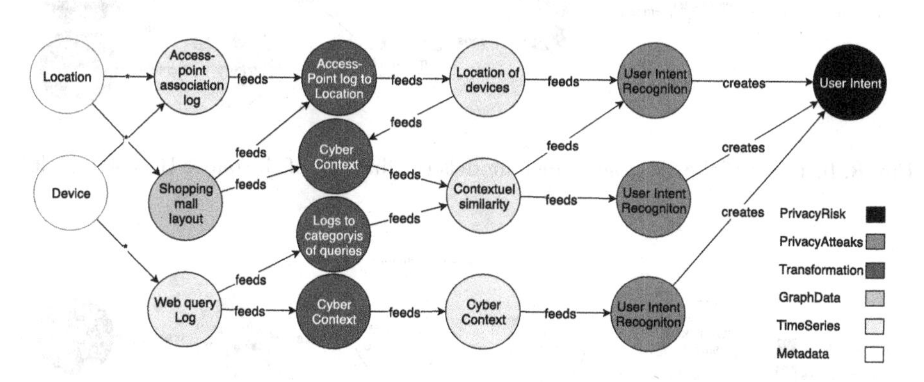

Fig. 6. Instance of the ontology, which models the findings of the paper Kaur et al. [13].

Figure 6 shows the instance of the ontology with the information found in [13]. The model includes the privacy attack and risks on the user intent, which was modeled as a potential privacy risk. The information can be exposed using three different privacy attacks, captured in the model.

The instance for the paper [16], can be seen in Fig. 7. This instance highlights that several methods can be used to detect occupancy presence using electricity consumption data from smart meters.

The instance for the paper [6], can be seen in Fig. 8. We have only modeled a part of the privacy risks for simplicity. In the paper, they have proposed two methods for the detection of household properties for the occupants living there, namely a regression model and a classification model. We have modeled each of them as a privacy attack with associated privacy risks.

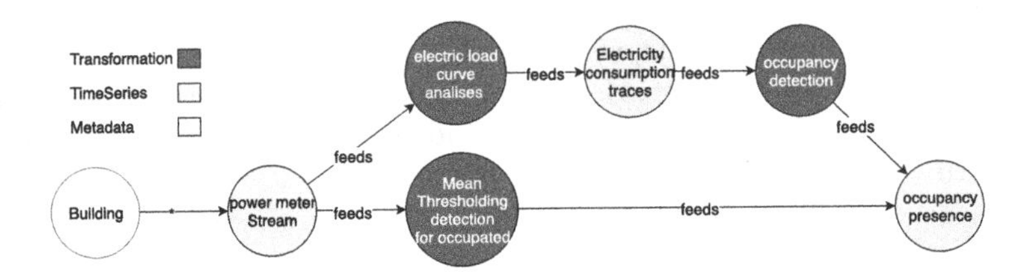

Fig. 7. Instance of the ontology, using the findings of the paper Kleiminger et al. [16].

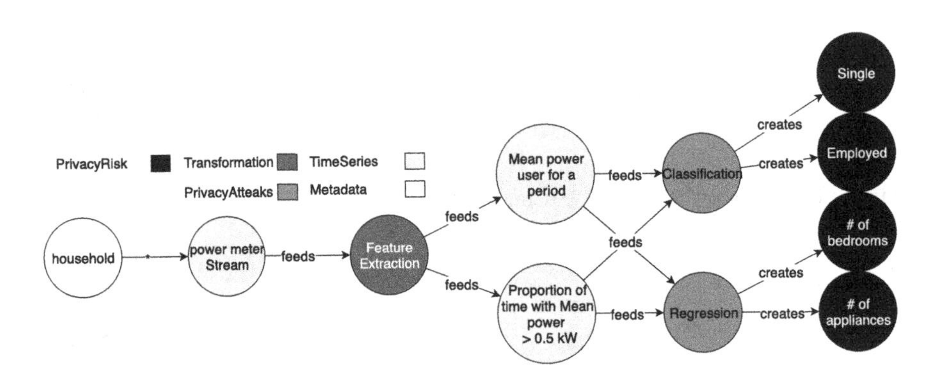

Fig. 8. Instance of the ontology, which models the findings of the paper Beckel et al. [6].

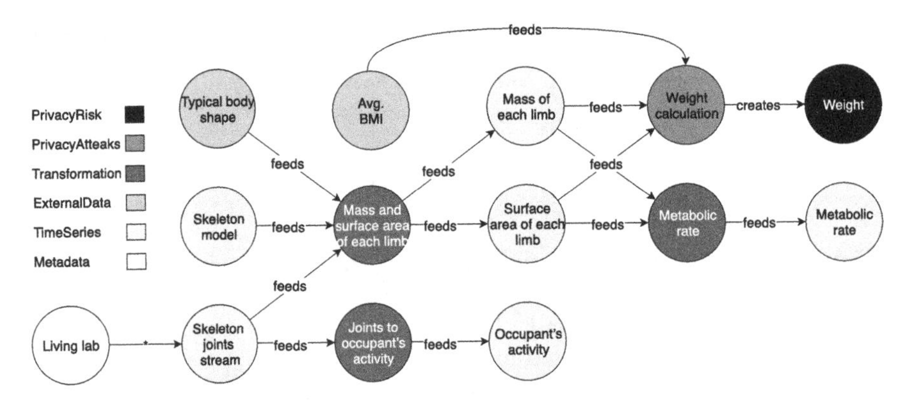

Fig. 9. Instance of the ontology, using the findings of the paper Dziedzic et al. [10].

The instance for the paper [10], can be seen in Fig. 9. In this instance, it can be observed that weight and metabolic rates for the monitored individual can be estimated by fusing body skeleton joint data with external data. Furthermore, the joint data can be used to determine the occupant's activities.

As shown the ontology was able to model the elements of the particular domain cases. This highlights the potential of the ontology. However, as discussed later the ontology has some limitations which should be explored in future work.

6 Modeling an Open Dataset

In this section, we present a model capturing expert knowledge from several papers [1,3,4] and apply it for privacy assessment for an open dataset. The model will combine the findings of the papers into a single instance of the ontology. We use the model to identify privacy risks for a real-world dataset already shared as open data [2]. The papers include the focus on data from ventilation systems in the form of VAV data which captures the inflow of air into a room and CO_2 which negatively correlate with oxygen content and thereby represent air quality. The papers propose methods to map these data sources to estimate occupant presence. This can, in the ontology, be directly modeled as a transformation from each of the data sources to occupancy presence.

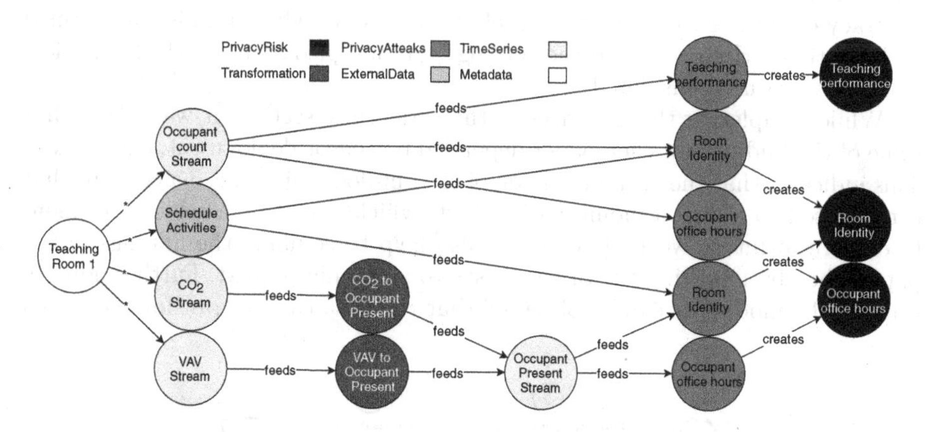

Fig. 10. Modeling case for an open dataset [2]. For simplicity the figure cover a single room from the released dataset with associated possible data transformations, privacy attacks, and privacy risks.

In terms of privacy attacks on occupant presence data, [28] highlighted several types of possible attacks. Firstly, they found that the room identity potentially can be identified using occupancy presence and counts combined with scheduled activities. Secondly, the occupancy presence and counts could also be used for identifying the working hours of an occupant working in the monitored area. Finally, it was found that the occupant counts for a teaching room, can be used to estimate teaching performance for each lecture in the facilities. These privacy attacks create three associated privacy risks. The model is shown in Fig. 10.

As a dataset for the case study, we have used [2]. We consider data from a single room with the following sensors: CO_2, VAV, and amount of occupants. These relations are modeled in Fig. 10. Here the metadata about room context and published data sources have been linked with the identified transformations, as well as the information about the room schedules that are publicly available [28].

With this model, a data publisher can for each type of data follow the graph and see what risks each data type result in. For instance, a data publisher can use the model to identify that occupant counts can potentially be used to estimate teaching performance and occupant office hours. The data publisher can also go the other way and from a privacy, risk backtrack what types of data result in this privacy risk. For instance, with the model one can observe that if the identity of a teaching room is to be hidden, there is a need for looking into how to anonymize the counts' data, as well as the presence data.

7 Discussion

The proposed ontology can model theoretical privacy risks in terms of how they relate to different data types. Therefore, it does not consider effects, such as the accuracy of data or the increasing level of uncertainty when data is transformed multiple times. However, since the ontology was designed to model the theoretical risks, this was not considered.

While completing the instances in the case study section, it was found that some of the findings for each paper depended on sensor deployment, e.g., in [15]. This indicates that the current version of the ontology might need a concept that can be used to model a monitored context, which would be a relevant element to explore in future work. This might also help to combine the findings in the papers, hereby modeling the privacy risks to a specific context. Furthermore, in some of the models, it can be observed that the data can be transformed into a

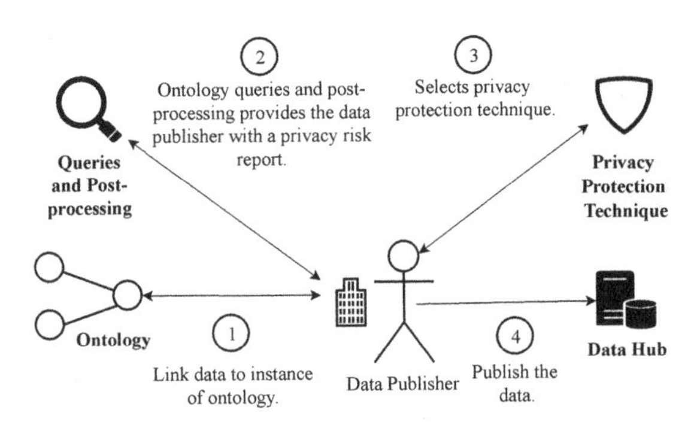

Fig. 11. The process for using the envisioned ontology for data publishing

number of time-series, which in some cases might pose a privacy risk, e.g., Fig. 9, where the data can be used to estimate occupant's activities or metabolic rates. However, this could be addressed by adding a privacy attack and privacy risk to each of them.

In this paper, we propose a privacy risk mapping ontology to improve the privacy assessment process for data publishing. However, we are aware that the ontology cannot stand on its own and would be hard for a non-domain expert to use. Therefore the ontology is only a piece of an envisioned tool-chain, which is sketched in Fig. 11. The process is as follows: (1) the data publisher maps the ontology instance, which has the privacy risks and attacks modeled, with the data considered for data publishing; (2) the ontology's queries and post-processing provide the data publisher with results of the potential inference, vulnerabilities, and attack vectors for each of the monitored contexts; (3) based on the output of the former step the data publisher selects a privacy protection technique; and (4) the final data is released. In this tool-chain, the ontology is to be used as a foundation and the queries and post-processing are to be reused between implementations for different datasets. The intended output of the tool-chain is privacy risks and possible methods for how to protect the dataset.

8 Conclusion

We set out to design an ontology enabling data publishers to make preemptive privacy assessments before sharing open datasets. To design the ontology we analyzed eight recent papers within the field of smart buildings to distill examples of data transformations and privacy risks. The papers cover many different forms of sensor modalities and how they can be transformed into information like occupant identity, actions, behaviors, and intents. The proposed ontology models data transformations, and associated privacy attacks and risks. We evaluated the ontology by constructing individual models for the methods of the eight papers. The result was that in all cases we could describe the datasets and methods of a paper by an instance of the ontology. Furthermore, we created a larger model based on related work which was applied to a particular dataset to be released. To support a privacy assessment the model could for this dataset highlight three privacy risks. In future work we plan that the ontology will be used as part of a larger tool-chain, helping data publishers perform privacy assessments for datasets before data sharing.

Acknowledgments. This study was supported by the HBODEx project (64018-0558). The authors are participating in IEA EBC Annex 79 and were support by EUDP (64018-0558).

References

1. Ardakanian, O., Bhattacharya, A., Culler, D.: Non-intrusive techniques for establishing occupancy related energy savings in commercial buildings. In: BuildSys 2016, pp. 21–30 (2016)

2. Arendt, K., et al.: Room-level occupant counts, airflow and CO_2 data from an office building. In: Proceedings of the First Workshop on Data Acquisition To Analysis, DATA 2018, pp. 13–14 (2018)
3. Arief-Ang, I.B., Hamilton, M., Salim, F.D.: A scalable room occupancy prediction with transferable time series decomposition of CO_2 sensor data. ACM Trans. Sens. Netw. **14**(3–4), 21:1–21:28 (2018)
4. Arief-Ang, I.B., Salim, F.D., Hamilton, M.: CD-HOC: indoor human occupancy counting using carbon dioxide sensor data. arXiv preprint arXiv:1706.05286 (2017)
5. Balaji, B., et al.: Brick: metadata schema for portable smart building applications. Appl. Energy **226**, 1273–1292 (2018)
6. Beckel, C., Sadamori, L., Staake, T., Santini, S.: Revealing household characteristics from smart meter data. Energy **78**, 397–410 (2014)
7. California State Legislature: California Consumer Privacy Act of 2018, June 2018. https://leginfo.legislature.ca.gov/faces/billTextClient.xhtml?bill_id=201720180AB375
8. Carrara, W., Oudkerk, F., Van Steenbergen, E., Tinholt, D.: Open data goldbook for data managers and data holders, February 2018
9. Dwork, C.: Differential privacy. In: Bugliesi, M., Preneel, B., Sassone, V., Wegener, I. (eds.) ICALP 2006, Part II. LNCS, vol. 4052, pp. 1–12. Springer, Heidelberg (2006). https://doi.org/10.1007/11787006_1
10. Dziedzic, J.W., Yan, D., Novakovic, V.: Real time measurement of dynamic metabolic factor (D-MET). In: Johansson, D., Bagge, H., Wahlström, Å. (eds.) CCC 2018. SPE, pp. 677–688. Springer, Cham (2019). https://doi.org/10.1007/978-3-030-00662-4_57
11. European Parliament and Council of the European Union: Regulations (EU) 2016/679 of the European Parliament and of the Council - general data protection regulation (GDPR). Official Journal of the European Union **L119**, 1–88 (2016). https://eur-lex.europa.eu/eli/reg/2016/679/oj
12. Fung, B.C.M., Wang, K., Chen, R., Yu, P.S.: Privacy-preserving data publishing: a survey of recent developments. ACM Comput. Surv. **42**(4), 14:1–14:53 (2010)
13. Kaur, M., Salim, F.D., Ren, Y., Chan, J., Tomko, M., Sanderson, M.: Shopping intent recognition and location prediction from cyber-physical activities via Wi-Fi logs. In: Proceedings of the 5th Conference on Systems for Built Environments, BuildSys 2018, pp. 130–139 (2018)
14. Khalil, N., Benhaddou, D., Gnawali, O., Subhlok, J.: Sonicdoor: scaling person identification with ultrasonic sensors by novel modeling of shape, behavior and walking patterns. In: BuildSys 2017, pp. 3:1–3:10 (2017)
15. Kjærgaard, M.B., Johansen, A., Sangogboye, F., Holmegaard, E.: Occure: an occupancy reasoning platform for occupancy-driven applications. In: CBSE 2016, pp. 39–48 (2016)
16. Kleiminger, W., Beckel, C., Santini, S.: Household occupancy monitoring using electricity meters. In: UbiComp 2015, pp. 975–986 (2015)
17. Klyne, G., Carroll, J., McBride, B.: RDF 1.1 concepts and abstract syntax, February 2014. https://www.w3.org/TR/rdf11-concepts/. Accessed 18 Oct 2019
18. Machanavajjhala, A., Gehrke, J., Kifer, D., Venkitasubramaniam, M.: L-diversity: privacy beyond k-anonymity. In: ICDE 2006, pp. 24–24, April 2006
19. Nergiz, M.E., Atzori, M., Clifton, C.: Hiding the presence of individuals from shared databases. In: SIGMOD 2007 (2007)
20. Office of the Australian Information Commissioner: Privacy Act 1988, Compilation No. 81, August 2019. https://www.legislation.gov.au/Details/C2019C00241

21. Pappachan, P., et al.: Towards privacy-aware smart buildings: capturing, communicating, and enforcing privacy policies and preferences. In: ICDCSW, pp. 193–198 (2017)
22. European Parliament, Council of the European Union: Directive 2002/58/EC of the European parliament and of the council of 12 July 2002 concerning the processing of personal data and the protection of privacy in the electronic communications sector (directive on privacy and electronic communications). Official Journal of the European Union **L201**, 37–47 (2002). https://eur-lex.europa.eu/legal-content/EN/TXT/?uri=OJ:L:2002:201:TOC
23. Pfenninger, S., DeCarolis, J., Hirth, L., Quoilin, S., Staffell, I.: The importance of open data and software: is energy research lagging behind? Energy Policy **101**, 211–215 (2017). https://doi.org/10.1016/j.enpol.2016.11.046
24. Prudhommeaux, E.: SPARQL query language for RDF (2008). http://www.w3.org/TR/rdf-sparql-query/
25. Rashid Asmaa, H., Mohd Yasin, N.: Privacy preserving data publishing: review. Int. J. Phys. Sci. **10**, 239–247 (2015)
26. Rocher, L., Hendrickx, J.M., de Montjoye, Y.A.: Estimating the success of re-identifications in incomplete datasets using generative models. Nat. Commun. **10**(1), 3069 (2019)
27. Sangogboye, F.C., Arendt, K., Singh, A., Veje, C.T., Kjærgaard, M.B., Jørgensen, B.N.: Performance comparison of occupancy count estimation and prediction with common versus dedicated sensors for building model predictive control. Build. Simul. **10**(6), 829–843 (2017)
28. Schwee, J., Sangogboye, F., Kjærgaard, M.: Evaluating practical privacy attacks for building data anonymized by standard methods. In: IoTSec 2019, April 2019
29. Sonta, A.J., Jain, R.K.: Inferring occupant ties: automated inference of occupant network structure in commercial buildings. In: BuildSys 2018, pp. 126–129 (2018)
30. Sweeney, L.: K-anonymity: a model for protecting privacy. Int. J. Uncertain. Fuzziness Knowl.-Based Syst. **10**(05), 557–570 (2002)

On the Design of a Privacy-Centered Data Lifecycle for Smart Living Spaces

Joseph Bugeja(✉) and Andreas Jacobsson(✉)

Internet of Things and People Research Center,
Department of Computer Science and Media Technology,
Malmö University, Malmö, Sweden
{joseph.bugeja,andreas.jacobsson}@mau.se

Abstract. Many living spaces, such as homes, are becoming smarter and connected by using Internet of Things (IoT) technologies. Such systems should ideally be privacy-centered by design given the sensitive and personal data they commonly deal with. Nonetheless, few systematic methodologies exist that deal with privacy threats affecting IoT-based systems. In this paper, we capture the generic function of an IoT system to model privacy so that threats affecting such contexts can be identified and categorized at system design stage. In effect, we integrate an extension to so called Data Flow Diagrams (DFD) in the model, which provides the means to handle the privacy-specific threats in IoT systems. To demonstrate the usefulness of the model, we apply it to the design of a realistic use-case involving Facebook Portal. We use that as a means to elicit the privacy threats and mitigations that can be adopted therein. Overall, we believe that the proposed extension and categorization of privacy threats provide a useful addition to IoT practitioners and researchers in support for the adoption of sound privacy-centered principles in the early stages of the smart living design process.

Keywords: IoT · Data lifecycle · Data Flow Diagrams · Data privacy · Privacy threats · Smart connected home · Smart living space · Facebook Portal

1 Introduction

IoT products are widely being deployed enabling the development of new applications. By combining the input of environmental and user activity information with control algorithms, activation mechanisms, and information feedback, traditional living spaces have been enhanced to smart living spaces. Smart living spaces involve different connected devices and services bringing up benefits such as improved convenience and experience for the users, optimized power efficiency, elderly telemonitoring, etc. [33]. The application of the IoT is broad, ranging from daily personal home applications to industrial automation applications or city transportation [13].

M. Friedewald et al. (Eds.): Privacy and Identity 2019, IFIP AICT 576, pp. 126–141, 2020.
https://doi.org/10.1007/978-3-030-42504-3_9

Despite their benefits, IoT technologies implemented in the home environment tend to generate and process a diverse amount of sensitive and personal data from users and making such data accessible to different entities almost instantaneously from anywhere there is an Internet connection. Data typically includes the geographical position of individuals, movement patterns, and sensed data that may reveal the physical conditions, behaviors, and activities of individuals. Improper usage of such data can lead to undesired privacy harms to an individual, group, and society. Some examples of consequences that can impair the users of IoT technologies include unsolicited advertisements, identity theft, discrimination, and more [5].

Our right to privacy as a concept is not a new idea, as early as 1890, Warren and Brandeis described privacy as the "the right to be let alone" [30]. They identified it as the right that enables individuals to have personal autonomy, freedom of association, moments of reserve, solitude, intimacy, and independence. Westin, as the computer era was emerging, described privacy as the "the claim of individuals, groups, or institutions to determine for themselves when, how, and to what extent information about them is communicated to others" [31]. While Westin's notion of privacy has been expanded upon (e.g., by Altman [2]) privacy remains a nebulous concept which is not amenable to precise definition [18]. This makes it especially challenging to enable a focused discussion about privacy in the IoT. In this paper, we focus on "information privacy" – a term that we adopt to signify the ability of users to exercise control over personal data about themselves while also minimize future privacy risks by protecting data after it is no longer under a user's direct control [28]. With the pervasiveness of IoT deployments it is essential to identify recent privacy threats and their related countermeasures.

Privacy concerns have motivated the development of legal regulation, such as the European General Data Protection Regulation (GDPR) [17], standards and frameworks for governing the processing of personal data, e.g., the privacy framework (ISO/IEC 29100:2011) [21], and as well engineering approaches such as "Privacy by Design (PbD)" [8] that aim to safeguard privacy since the beginning of the development process. Accordingly, software engineers are increasingly expected to give appropriate consideration to privacy and data protection issues throughout the development lifecycle [22]. However, the mentioned approaches alone are insufficient raising the need for a more proactive and integrative approach to safeguard the privacy of data subjects (i.e., the natural person to whom the personal data belongs to) [1]. In particular, tools are needed to better assist IoT practitioners to address the complexity and variability of privacy issues throughout the development process.

To this aim, in this paper we provide an understanding of: (i) what are the information privacy threats affecting a smart living space; (ii) which data lifecycle phase each privacy threats affects; and (iii) how can a privacy-preserving data lifecycle be modeled to reduce the exposure of end-users to such threats. Overall, the main contribution is an extension to a standard system modeling technique – Data Flow Diagrams (DFD) – with new processes and annotations

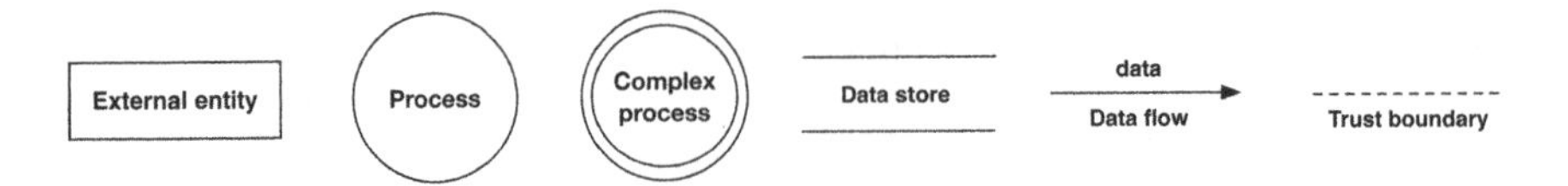

Fig. 1. DFD element types.

that can be used to incorporate privacy principles into the software design process. In summary, our contribution benefits IoT practitioners by helping bridge the gap between the technical design and concerns related to privacy compliance; and support the adoption of privacy principles in the early stages of the design process.

1.1 Data Flow Diagrams

DFDs are a visual notation that allows to model data flows in information systems in a structured way [29]. Graphically, the main DFD elements are represented in Fig. 1, and their basic elements are:

- *External entity*: Represents a person, organization, or services that are external to the system but interact with it.
- *Process*: Represents an activity or a function, e.g., updating user profile or consulting a cloud endpoint, that is performed for some specific business purpose.
- *Complex process*: These are a logical representation of a process, e.g., a mobile application or web application, that performs many distinct operations and thus can be represented or refined through separate DFDs.
- *Data store*: A collection of data that are at rest, e.g., databases, files, or cache information.
- *Data flow*: These are a single piece of data or a logical collection of several pieces of information, e.g., control, personal, or sensor data, that are being communicated.
- *Trust boundary*: Represents the border between untrustworthy and trustworthy elements or a delineation between data moving from low to high trust and vice versa.

DFDs are widely used during the system analysis phase to capture the requirements of a software system including the identification of security threats (e.g., in STRIDE) and privacy threats (e.g., in LINDDUN) [10].

1.2 IoT Entities, Processes, and Data Flows

An IoT system can be viewed as a dynamic and distributed networked system, capable of producing and consuming information [13]. At a high-level, it consists of connected devices (e.g., sensors), services (e.g., cloud-based analytics software), network infrastructure (e.g., protocols such as Bluetooth), and users (e.g., the smart living space inhabitants).

For an IoT system, the external entities can be generally grouped into data subject, data controller, and data user [15,24,32]. Data subject is the human entity that generates the original raw data and whose personal data are processed by the IoT system. Data controller (sometimes also referred to as "data holder," "data curator," or "data processor") is the person or organization that collects, stores, processes, and releases the data. Data users represent the entities that access the released data. Typically, data users are the data subjects however they can also be other devices or systems. At a general level, there are four main data phases in an IoT system [6,20,34]:

- *Data generation*: Represents the activity where the data subject interacts with the IoT system, directly or indirectly, to create personal data. Personal data are any information that is related to an identified or identifiable natural person. This interaction is typically, done through end-user devices such as smartphone applications with the help of services.
- *Data collection*: Collection represents the act of acquiring personal data from data subjects, including external sources. Typically, this is done through sensors embedded inside the connected device. Information at this stage may be stored in the IoT system. The longevity of this may range from temporary (transient storage), e.g., in memory buffers, to persistent (persistent storage), e.g., inside databases; and may occur automatically without involving directly the end-user.
- *Data processing*: The IoT analyses the data stored in the cloud data centers or inside the devices to provide the smart services. Due to shifting levels of autonomy, IoT devices are typically capable of making some decisions on their own, i.e., without human intervention. There are different processing models, including: cloud-centric, gateway-centric, and edge-centric data processing models.
- *Data disclosure*: Represents the act of disseminating, making available or transmitting personal data for external use by third-parties. Commonly, when the disclosure is done to the data user, this phase is typically referred to as data presentation. The output from this phase tends to range from notification to actuation.

This IoT data lifecycle, is also accompanied by a separate lifecycle, where the actual (physical) IoT device is initially deployed, then operated, and finally retired (decommissioned). However, for this study, we focus on the logical, i.e., the data lifecycle, as this corresponds to "information privacy."

1.3 Organization of the Paper

In Sect. 1, we provide an introduction of the research done, propose DFDs as a modeling tool, and describe the IoT data phases. Next, we provide a background overview and summarize the relevant related work. In Sect. 3, we identify the privacy protection goals, exemplify, and categorize the main IoT privacy threats directed to smart living spaces. Then, in Sect. 4, we provide a privacy-centered data lifecycle for smart living space applications using the proposed extension to DFDs. The proposed additions are applied in Sect. 5 to a smart connected home setup using Facebook Portal as its connected device. In Sect. 6, we discuss how the proposed extensions can help in addressing some of the privacy compliance requirements. Finally, in Sect. 7, we conclude this paper and identify some avenues for future work.

2 Background and Related Work

This work lies at the intersection of the research areas of privacy by design strategies, privacy threat modeling, and privacy threat analysis.

Privacy by Design Strategies. This is a design approach that aims to improve the overall privacy friendliness of IT systems [19]. PbD seeks to encourage the inclusion of privacy at the start rather than retrofitted into existing systems [8].

PbD was originally presented by Cavoukian and consists of seven foundational principles that divide privacy into high-level non-functional properties. Three principles covered in [9] are "privacy as the default," "end-to-end lifecycle protection," and "privacy embedded into design" [9]. These principles should be followed when developing privacy sensitive applications. Langheinrich [23] also proposed six principles for PbD. Three examples of protection principles mentioned in [23] are "choice & consent," "anonymity & pseudonymity," and "security". They were suggested for ubiquitous systems, but have found implementation in a range of fields. Recently, ENISA [11] summarized eight PbD strategies as derived by Hoepman [19] and applied them to the context of big data analytics. The strategies are: "minimize," "hide," "separate," "aggregate," "inform," "control," "enforce," and "demonstrate."

Privacy Threat Modeling. These approaches contribute to the realization of PbD strategies by providing a systematic, rigorous, and methodical approach towards privacy analysis.

LINDDUN [14] threat modelling framework uses a DFD to identify and model privacy threats. Privacy threats addressed by LINDDUN are: linkability, identifiability, non-repudiation, detectability, disclosure of information, unawareness, and non-compliance. Luna et al. [25] proposed QTMM (Quantitative Threat Modeling Methodology) to help draw objective conclusions about privacy-related attacks. It follows the same modeling steps as LINDDUN however it focuses on three privacy-specific threat categories: linkability, unawareness, and intervenability. Antignac et al. [3] proposed an extension to DFDs to support the inclusion of privacy concepts borrowed from the EU GDPR and ISO 29100 standard. This conceptual model is useful for making privacy concepts in the design model explicit.

Privacy Threat Analysis. Different scholars have proposed different taxonomies and catalogues to help identify and understand privacy threats.

The Antón-Earp taxonomy [4] identifies five privacy protection goals, e.g., "choice/consent," to safeguard the privacy of a customer's data, and seven corresponding categories of vulnerabilities, e.g., "information aggregation," that reflect a potential privacy violation. Solove's taxonomy [27] is a taxonomy of privacy harms that groups privacy threats, e.g., "surveillance," occurring at different system phases: information collection, information processing, information dissemination, and including invasion to the data subject. Ziegeldorf et al. [34] identified seven categories of privacy threats that affect an IoT-based system. Three examples of threats covered in [34] are identification, profiling, and linkage.

Main Observations. In reviewing the existing work, we observe that most of the mentioned approaches have not evolved to cater for IoT technologies. For instance, the Antón-Earp taxonomy [4] and Solove's taxonomy [27] while generic enough were created before the advancement of IoT technologies. This could mean that these taxonomies may be more applicable for studying web-based or traditional information systems but not necessarily IoT-based systems. Similarly, LINDDUN has not been validated on IoT systems [26]. Lastly, PbD while useful as an approach, there are still uncertainties about what it means in the context of IoT and especially how it can be implemented.

Our Contribution. Given that the focus of this paper is on IoT-based systems, we leverage the literature of Ziegeldorf et al. in [34] to identify privacy threats in smart living spaces. Different to that study, we organize the threats identified in that study according to the privacy protection goals being violated and identify corresponding data phases leading to those threats. This is similar to Solove's taxonomy [27] but with IoT specific phases. Furthermore, when it comes to modeling privacy we follow a similar conceptual modeling approach to Antignac et al. [3] but we focus on IoT-based systems and also propose different extensions and processes. These target both the data controller and data subject. Additionally, we suggest PbD strategies for mitigating the identified privacy threats at the different data lifecycle phases.

3 IoT Privacy Goals and Threats

In order to safeguard end-users right to privacy, three main privacy-specific protection goals have been proposed that articulate what is being protected and from who. These goals are: unlinkability, transparency, and intervenability [12,25,35].

Unlinkability is defined as the property that data processing is operated in such a way that the privacy-relevant data cannot be linked to any other set of privacy-relevant data outside of the domain. Transparency ensures that privacy-relevant data processing can be understood and reconstructed at any time. Intervenability ensures that the parties involved in any privacy-relevant data processing, including the data subject, have the possibility to intervene where necessary.

To the above, we also add the goals of confidentiality and detectability. Confidentiality represents the goal of preventing unauthorized access to information or systems. Detectability corresponds to the goal of preventing an attacker from sufficiently distinguishing if an item of interest exists or not. While these goals are somewhat related to the unlinkability goal, we added them separately since some IoT privacy threats are primarily affecting those.

Based on the work of Ziegeldorf et al. [34] and our observations, below we present a summary of IoT privacy threats applied to smart living systems and grouped according to the corresponding primary protection goal being violated by each. Table 1 outlines the different IoT privacy threats.

Table 1. Summary of the IoT privacy threats alongside the main privacy protection goal being violated by each and the corresponding data lifecycle phase during which each threat typically occurs. The symbol: ● indicates that the threat occurs often; ◐ indicates that the threat might occur; and ○ indicates that the threat rarely occurs.

Information privacy threats	Protection goals	Data generation	Data collection	Data processing	Data disclosure
Identification	Unlinkability	◐	◐	●	○
Localization and tracking	Unlinkability	◐	◐	●	○
Profiling	Unlinkability	○	◐	◐	●
Linkage	Unlinkability	○	◐	◐	●
Privacy-violating interaction and presentation	Confidentiality	●	○	○	●
Inventory attacks	Detectability	○	●	○	○
Lifecycle transitions	Transparency	○	●	○	○

Unlinkability Threats

- *Identification*: Identification characterizes the risk of associating a persistent identifier, e.g., name and address, with a data subject and thus revealing the identity of the individual [34]. Typically, all IoT control apps compel data subjects to identify themselves, e.g., by entering account details during system setup or dynamically through technologies such as facial recognition. This threat is dominant in the data processing phase but may also occur at the data generation and data collection phase [34].
- *Localization and tracking*: This threat allows for the recording of a person's location [34]. Different monitoring techniques are employed by IoT devices, e.g., built-in motion sensors, and as well reading the location directly from the smartphone. This threat is dominant in the data processing phase but may also occur at the data generation and data collection phase [34].
- *Profiling*: Represents the threat of collecting and correlating information about individual activities in order to generate new information from the original data [34]. For instance, a smart thermostat can collect temperature,

humidity, and ambient light data of the location where it is being used and then makes temperature adjustments for different situations accordingly. This may then be used to automatically infer that a person is home at a certain time and consequently offer targeted advertisements, e.g., food and drink adverts. Profiling threats mostly appear in the disclosure phase where information is forwarded to third-parties [34]. Nonetheless, these may also occur during data collection and processing.
- *Linkage*: This threat consists in linking different separated systems such that the combination of data sources reveals information that the subject did not disclose or intended to [34]. As an example, if an IoT device, such as a smart lock, is integrated with another device from a different manufacturer, e.g., a connected doorbell, through a cloud-based service such as IFTTT (if this then that), then the doorbell might acquire information, e.g., about successful/unsuccessful attempts, from the smart lock that it was not originally intended to process. This threat of linkage primarily appears in the data disclosure phase [34]. However, similar to profiling, it may also occur during data collection and processing.

Confidentiality Threats

- *Privacy-violating interaction and presentation*: Exposing personally identifiable information to individuals who are not supposed to have access to it [34]. As an example, information may be disclosed through a public medium, e.g., smart speaker system, to an unwanted audience, e.g., to temporary visitors in a smart home. This threat is dominant in the data generation and disclosure phase, in particular when information is presented to the users [34].

Detectability Threats

- *Inventory attacks*: These refer to the unauthorized collection of information about the existence and characteristics of personal things [34]. As an example, given the wireless nature of most of the smart devices, a malicious threat agent may deduce the presence of a certain medical device, e.g., insulin pump, and thereby inferring that a person is suffering from a certain medical condition, e.g., diabetes. This threat primarily occurs at the data collection phase [34].

Transparency Threats

- *Lifecycle transitions*: Occurs when users' private information collected during the IoT device's lifetime is disclosed during changes to the device's control spheres in their lifecycle [34]. As an example, sensors may collect private information about a user, then transmit it to the cloud for further analysis and returning the result. This transfer of data between different phases may disclose sensitive information about the user. This threat primarily occurs at the data collection phase [34].

4 Privacy-Centered Smart Living Data Lifecycle

In order to mitigate the identified threats, various mitigations have been proposed in scholarly literature, e.g., [19], and industry reports, e.g., [11].

For the unlinkability threats, we find in particular data-oriented strategies such as minimize and hide. Minimize, example by adopting a select-before-collect approach, reduces the amount of processed personal data to the minimal amount possible [11,19]. Hide, for example achieved through data encryption, anonymization, or attribute-based credentials, reduces personal data and their interrelations from plain view [11,19]. It is also useful as a strategy to reduce detectability threats. For the transparency threat, we refer to process-oriented strategies with protection strategies such as inform and control. As a strategy, the inform process notifies the data subjects whenever personal data are processed [11,19]. Acting as a counterpart to the inform strategy, is the control strategy that gives the data subjects agency over the processing of their personal data [11,19]. The inform strategy is also useful for mitigating the confidentiality threat. Additionally, the data-oriented strategy aggregate, is beneficial here. Aggregate, for example achieved through data provenance, can help restrict the amount of detail in the personal data that remains, e.g., before being sent or published as part of the data disclosure phase [11,19].

Data Subject Controls. Data subjects are the entities responsible for the data generation phase. For data subjects, new processes have to be added to offer them the capability of accessing, reviewing, and destroying stored personal data. Data access represents the act of specifying, retrieving, or consulting personal data values that are stored. Data review signifies the act of implementing the access right and rectifying personal data values by data subjects to ensure that their data is accurate, complete, and up-to-date. Data destroy characterizes the act of erasing, redacting, or disposing of personal data.

Furthermore, at the different lifecycle phases, it would be appropriate to have mechanisms in place that allow the data subjects the facility to be informed about data collection and use, and to have control over that. Overall, the successful implementation of these processes reduce the threat of identification, localization and tracking, and as well confidentiality threats.

Data Controller Controls. Data controllers are the entities responsible for the data collection, data processing, and data disclosure phase. For the data controllers, especially, at the data collection phase, the practice of data minimization is core to reduce personal data from being inappropriately disclosed to unauthorized entities. Strategies should be designed to minimize the amount of data types collected or requested by an IoT application, the volume and granularity of data stored by an IoT application, and raw data acquired by the system. The effective implementation of data minimization reduces the overall effect of the different IoT privacy threats.

Furthermore, when personal data need to be transmitted to the different external entities, in particular to data users as part of the data disclosure phase, these should be sent encrypted, ideally end-to-end starting from the data subject, in the network, and then off to the service provider backend infrastructure. Encrypted data communication reduces the potential privacy risks due to unauthorized access during data transfer between components. Encryption can also be applied when storing and processing data to reduce privacy violations due to malicious attacks and unauthorized access to personal data. Moreover, instances of personal data, especially if these are to be shared with another data controller, should ideally be transformed at source to anonymize the identity of the source or target device or user. Implemented properly these controls help reduce detectability and unlinkability threats.

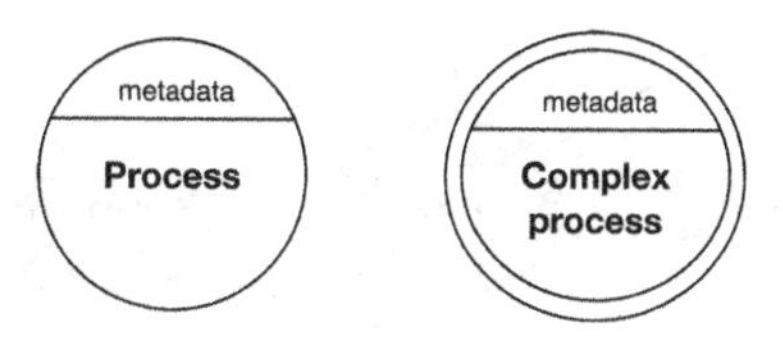

Fig. 2. A DFD extension to the process and complex process symbol. The metadata may represent a purpose statement indicating a reason for a process to collect and use personal data.

DFD Privacy-Centered Extensions. To capture the aforementioned tactics in the actual IoT system design, the identified data lifecycle phases have to be accordingly modified. With this, the IoT practitioner would be offered the possibility of introducing additional processes before and after each data lifecycle phase. This is to mitigate the different threats identified at each phase and to empower the data subject with control, choice, and flexibility over the processing of personal data.

We do so by extending the standard DFD notation with an annotation as shown in Fig. 2. The annotation allows for the direct specification of metadata. Through this, IoT practitioners can explicitly declare the purpose for collecting, processing, and disclosing the personal data of data subjects, and likewise to specify the duration for which personal data are stored in the smart living application. Overall, the different processes intended for the data subject and data controller are depicted in Fig. 3.

Collectively, when the proposed privacy protection measures are applied to the data collection, data processing, and data disclosure phases, we refer to these processes as: secure data collection, secure data processing, and secure data disclosure, respectively. These can be represented as complex processes; that are in the realm of the data controller; with privacy-preserving data being their output data flow. In terms of the data subject processes we aggregate these and represent them as a complex process – privacy manager. Essentially, this can be compared to an end-user privacy toolbox but it can also take the form of a separate physical device that is fully controllable by the data subject.

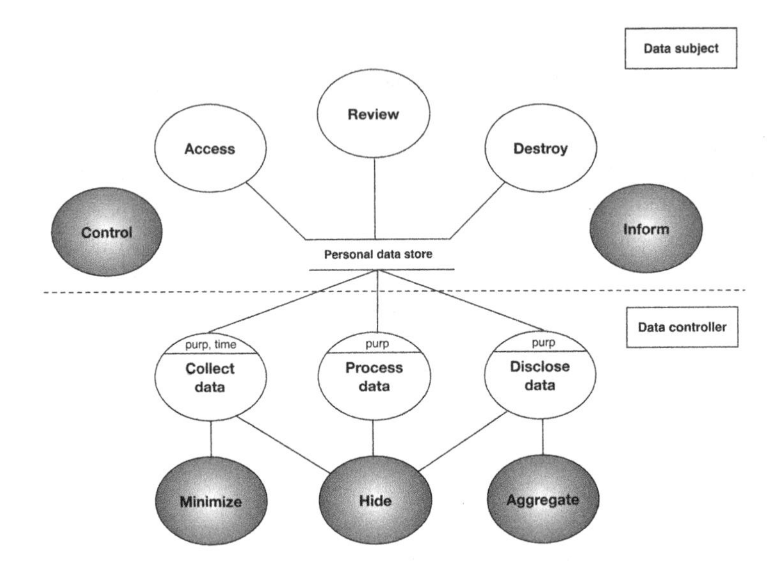

Fig. 3. Securing the IoT data lifecycle against IoT privacy threats. The data subject has the possibility to access, review, and remove any personal data retained by the system, and subsequently to get informed and have control over the data lifecycle processes. The data controller updates the IoT data lifecycle processes with functions to minimize, hide, and aggregate information, and to specify a purpose (*purp*) for each corresponding phase and duration (*time*) for data retention. Processes highlighted in gray colour represent PbD strategies.

5 Application of the Privacy-Centered Smart Living Data Lifecycle

Let us assume a smart connected home setup consisting of Facebook Portal (Portal)[1] as the main IoT device. Portal is primarily a smart screen and speaker device embedded with cameras, microphones, and Artificial Intelligence (AI), allowing for video-calling and advanced entertainment support.

Privacy Threat Identification. Using the previously established map of possible IoT privacy threats as detailed in Table 1 as a guide, together with Portal's privacy policy[2] and its feature list as supporting documentation, we identify a number of possible privacy threats in Portal.

Identification is a potential threat since an account is needed to be able to benefit from Portal. Localization and tracking is a possible threat as the data subject's location can be inferred from the smartphone device when enabled. Linkage is a potential threat especially since Amazon Alexa account can be connected to Portal for voice-based interactions. Privacy-violating interaction and presentation is a threat that is common to Portal and similar device types,

[1] https://portal.facebook.com [accessed December 13, 2019].
[2] https://portal.facebook.com/legal/data-policy [accessed December 13, 2019].

in particular as photos and videos may be pulled from Facebook and perhaps displayed unintentionally to guests or temporary visitors inside the smart living space. Inventory attacks may be a threat if an attacker is for instance able to detect Portal's presence, e.g., by passively observing the device's fingerprint or by inferring it from outgoing network flows. Profiling may be an eventual threat especially if the device is in future integrated with other connected systems such as WhatsApp and Instagram.

Privacy Threat Mitigation. In order to mitigate the identified privacy threats, we evolve the data lifecycle into a privacy-centered data lifecycle using the processes identified in Sect. 4. This version uses the secure processes: privacy manager, secure data collection, secure data processing, and secure data disclosure, which implemented properly reduce the discussed privacy threats.

More concretely, the privacy manager can be deployed to the data subject as a user-friendly application available over a smartphone. This application, might for instance give the data subject the access rights to view or delete their voice history, update applications permissions, and get notified about how Facebook (and Amazon for the voice-processing) use their data. Likewise, it could offer users ways to control, e.g., disable collection of personal data, when needed.

When it comes to the data controller, the secure data collection process might for instance refrain from asking data subjects certain particulars, for instance, about the geographical data where the device is installed in; secure data processing process might store data inside the actual device encrypted using a strong and approved cryptographic algorithm; and secure data disclosure process might obscure the real identity of the data subject before transmitting it, e.g., alongside the search criteria, to Amazon.

Graphically, the privacy-centered data lifecycle as applied to Portal is depicted in Fig. 4. Here, the proposed secure processes are indicated in the actual DFD, including being annotated with metadata indicating a specific purpose (e.g., collecting location data to only get calls when users are at home) and retention time (e.g., seven days) for processing activities.

6 Discussion

The successful implementation of the privacy-centered data lifecycle together with its secure processes helps data subjects attain different rights over their personal data. Based on the privacy principles and regulations of data processing in ISO/IEC 29100:2011 and GDPR, we outline the data subject's rights that the model is grounded upon:

- *Right to be forgotten:* The data subject shall have the right to obtain from the data controller the erasure of personal data concerning him or her without undue delay and the data controller shall have the obligation to delete personal data [GDPR Article 17, Clause 1; Individual participation and access (ISO/IEC 29100:2011)]. In our case, this right is implemented through the privacy manager, specifically through the data destroy process, and by having the retention period for data attributes specified explicitly in the secure data collection phase.

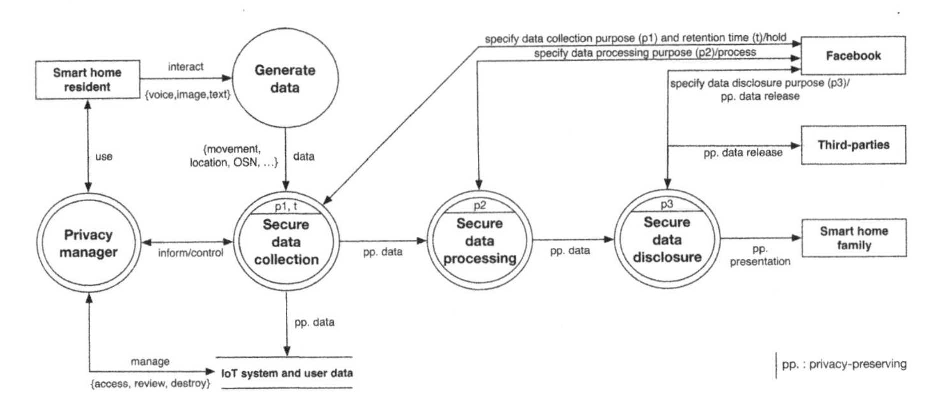

Fig. 4. Securing the IoT data lifecycle against IoT privacy threats using Facebook Portal as a use-case. As data is traveling across the different phases it is being outputted in a privacy-preserving (*pp*) manner. At the same time, the smart home residents can manage their collected personal data, and can get notified about it.

- *Right to restrict processing:* The data subject shall have the right to obtain from the data controller restriction of processing where one of the following applies: (a) the accuracy of the personal data; (b) the processing is unlawful; (c) the data controller no longer needs the personal data; (d) the data subject has objected to processing [GDPR Article 18, Clause 1; Use, retention and disclosure (ISO/IEC 29100:2011)]. In our case, this right is implemented through the privacy manager, through the data review process, and by having the data processing purpose specified explicitly as metadata in the secure data processing phase.
- *Right to data portability:* The data subject shall have the right to receive the personal data concerning him or her, that it provided to a data controller and has the right to transmit that data to another controller without hindrance [GDPR Article 20, Clause 1; Individual participation and access (ISO/IEC 29100:2011)]. In our case, this right is implemented through the privacy manager, through the data access process.
- *Right to object:* Consent should be given by a clear affirmative act establishing a freely given, specific, informed and unambiguous indication of the data subject's agreement to the processing of personal data relating to him or her [GDPR Recital 32; Consent and choice (ISO/IEC 29100:2011)]. In our case, this right is implemented through the privacy manager, through the inform and control strategy.

Through the use of the privacy-preserving processes and annotations, the choices of the IoT designer with regards to privacy are made more explicit in the design model. Nonetheless, in practice this needs to be accompanied by sound governance processes, standards, and practices that ensure that data controllers abide with privacy regulations, and thus safeguard individual's privacy rights accordingly. Here, we note that the GDPR, the forthcoming ePrivacy Regulation (ePR) [16], and the recent California Consumer Privacy Act (CCPA) [7], are key examples of good privacy regulations in this regard. Nevertheless, the fact that there are multiple regulations with overlapping sections raises the need for an international standard and consolidated guidelines for achieving data privacy.

When it comes to threat modeling DFDs provide a straightforward tool for doing so, but there are other threat modeling techniques that can be useful. For instance, Activity Diagrams (ADs) which are part of the Unified Modeling Language (UML) diagrams in UML 2.0. ADs are arguably more expressive than DFDs while retaining similar functionality towards threat modeling. Their expressiveness lies in the fact that they have a guard condition for the activity element, while DFD have no counterpart for this element. However, these extra elements while useful can make a diagram unnecessary complicated possibly resulting in a failure to identify threats in the system. Moreover, given that DFDs are focused on data, they provide a convenient choice for privacy analysis. Nonetheless, it would be useful to have a thorough evaluation of the proposed extensions. This could for instance be done by means of different empirical studies taking the perspective of requirements engineers and software architects.

Moving towards the system implementation phase, a data controller may leverage different technologies to realize the proposed mitigations. For instance, to address the threat of identification, localization and tracking, profiling, and linkage, data anonymization techniques or protocols, including data obfuscation, data encryption, data masking, and de-identification can be used. On the other hand, for reducing the threat of privacy-violating interaction and presentation, inventory attacks, and lifecycle transitions, access control with user-defined privacy policies or increasing user awareness are practical solutions. Additionally, there may be other PbD strategies that may be relevant depending on the particular context or use-case.

7 Conclusions and Future Work

IoT devices have brought added efficiencies and conveniences to smart living users. Nonetheless, connected devices bring unprecedented privacy threats to data subjects.

Recognizing this, we proposed a systematic data lifecycle approach that categorizes privacy threats affecting smart living spaces, mapping them to corresponding phases in the lifecycle. Moreover, we provided an extension to DFDs and a selection of PbD strategies to help mitigate the identified threats. Through the proposed privacy-centered lifecycle, a data controller can better plan in implementing privacy measures early-on in the software development process; and data subjects are empowered with improved control over their personal data.

For future work, it would be useful to develop a tool that provides IoT practitioners the facility to automatically elicit the privacy threats arising at each data lifecycle phase and how each can be mitigated through the use of the newly identified secure processes. This helps evaluate the presented DFD extensions but also from a privacy compliance perspective for instance serving as evidence that privacy measures have been thought through. Finally, it would be beneficial to develop an approach that can quantitatively measure the privacy risk exposure of an IoT-based system to different malicious threat agents. These can be represented as a new type of external entity, e.g., data privacy attacker, that aims to compromise the smart living privacy requirements. Possibly, after the system is modeled, the privacy risk exposure can be automatically calculated based on the agent's capabilities and the system's vulnerabilities.

Acknowledgments. This work has been carried out within the research profile "Internet of Things and People," funded by the Knowledge Foundation and Malmö University in collaboration with 10 industrial partners.

References

1. Alshammari, M., Simpson, A.: Privacy architectural strategies: an approach for achieving various levels of privacy protection. In: Proceedings of the 2018 Workshop on Privacy in the Electronic Society, pp. 143–154. ACM (2018)
2. Altman, I.: The environment and social behavior: privacy, personal space, territory, and crowding (1975)
3. Antignac, T., Scandariato, R., Schneider, G.: A privacy-aware conceptual model for handling personal data. In: Margaria, T., Steffen, B. (eds.) ISoLA 2016. LNCS, vol. 9952, pp. 942–957. Springer, Cham (2016). https://doi.org/10.1007/978-3-319-47166-2_65
4. Antón, A.I., Earp, J.B.: A requirements taxonomy for reducing web site privacy vulnerabilities. Requirements Eng. **9**(3), 169–185 (2004)
5. Bettini, C., Riboni, D.: Privacy protection in pervasive systems: state of the art and technical challenges. Pervasive Mob. Comput. **17**(PB), 159–174 (2015)
6. Bugeja, J., Jacobsson, A., Davidsson, P.: An empirical analysis of smart connected home data. In: Georgakopoulos, D., Zhang, L.-J. (eds.) ICIOT 2018. LNCS, vol. 10972, pp. 134–149. Springer, Cham (2018). https://doi.org/10.1007/978-3-319-94370-1_10
7. California Senate Judiciary Committee et al.: California consumer privacy act: Ab 375 legislative history (2018)
8. Cavoukian, A.: Privacy by design. Technical report (2009). http://www.ontla.on.ca/library/repository/mon/23002/289982.pdf
9. Cavoukian, A.: Privacy by design in law, policy and practice. A white paper for regulators, decision-makers and policy-makers (2011)
10. Chen, Y.T., Huang, C.C.: Determining information security threats for an iot-based energy internet by adopting software engineering and risk management approaches. Inventions **4**(3), 53 (2019)
11. D'Acquisto, G., Domingo-Ferrer, J., Kikiras, P., Torra, V., de Montjoye, Y.A., Bourka, A.: Privacy by design in big data: an overview of privacy enhancing technologies in the era of big data analytics. arXiv preprint arXiv:1512.06000 (2015)
12. Danezis, G., et al.: Privacy and data protection by design-from policy to engineering. arXiv preprint arXiv:1501.03726 (2015)
13. Miorandi, D., Sicari, S., De Pellegrini, F., Chlamtac, I.: Internet of things: vision, application areas and research challenges. Ad Hoc Netw. **10**, 1497–1516 (2012)
14. Deng, M., Wuyts, K., Scandariato, R., Preneel, B., Joosen, W.: A privacy threat analysis framework: supporting the elicitation and fulfillment of privacy requirements. Requirements Eng. **16**, 3–32 (2011)

15. Dwork, C., Roth, A.: The algorithmic foundations of differential privacy. Foundations Trends® Theor. Comput. Sci. **9**(3–4), 211–407 (2014)
16. European Commission: Proposal for a Regulation on Privacy and Electronic Communications (ePrivacy Regulation) (2017). https://eur-lex.europa.eu/legal-content/EN/TXT/?uri=CELEX:52017PC0010
17. European Union: Regulation (EU) 2016/679 of the European Parliament and of the Council of 27 April 2016. Technical report (2016). https://bit.ly/2Cxy5yP
18. Friedewald, M., Wright, D., Gutwirth, S., Mordini, E.: Privacy, data protection and emerging sciences and technologies: towards a common framework. Innov. Eur. J. Soc. Sci. Res. **23**(1), 61–67 (2010)
19. Hoepman, J.-H.: Privacy design strategies. In: Cuppens-Boulahia, N., Cuppens, F., Jajodia, S., Abou El Kalam, A., Sans, T. (eds.) SEC 2014. IAICT, vol. 428, pp. 446–459. Springer, Heidelberg (2014). https://doi.org/10.1007/978-3-642-55415-5_38
20. Hu, F., Jeyanthi, N.: Internet of Things (IoT) as Interconnection of Threats (IoT). In: Security and Privacy in Internet of Things (IoTs) (2016)
21. ISO: ISO 29100 Privacy Framework **2011**, 1–21 (2011)
22. Jacobsson, A., Boldt, M., Carlsson, B.: A risk analysis of a smart home automation system. Future Gener. Comput. Syst. **56**, 719–733 (2016)
23. Langheinrich, M.: Privacy by design—principles of privacy-aware ubiquitous systems. In: Abowd, G.D., Brumitt, B., Shafer, S. (eds.) UbiComp 2001. LNCS, vol. 2201, pp. 273–291. Springer, Heidelberg (2001). https://doi.org/10.1007/3-540-45427-6_23
24. Li, C., Palanisamy, B.: Privacy in internet of things: from principles to technologies. IEEE Internet of Things J. **6**, 1–18 (2018)
25. Luna, J., Suri, N., Krontiris, I.: Privacy-by-design based on quantitative threat modeling. In: 2012 7th International Conference on Risks and Security of Internet and Systems (CRiSIS), pp. 1–8. IEEE (2012)
26. Perera, C., Mccormick, C., Bandara, A.K., Price, B.A., Nuseibeh, B.: Privacy-by-design framework for assessing internet of things applications and platforms (2016)
27. Solove, D.J.: A taxonomy of privacy. U. Pa. L. Rev. **154**, 477 (2005)
28. Spiekermann, S., Cranor, L.: Privacy engineering. IEEE Trans. Softw. Eng. **35**(1), 67–82 (2009)
29. Tao, Y., Kung, C.: Formal definition and verification of data flow diagrams. J. Syst. Softw. **16**(1), 29–36 (1991)
30. Warren, S.D., Brandeis, L.D.: The Right to Privacy. Wadsworth Publishing Company, Belmont (1985)
31. Westin, A.F.: Privacy and freedom. Wash. Lee Law Rev. **25**(1), 166 (1968)
32. Yu, S.: Big privacy: challenges and opportunities of privacy study in the age of big data. IEEE Access **4**, 2751–2763 (2016)
33. Zhou, B., et al.: The carpet knows: identifying people in a smart environment from a single step. In: 2017 IEEE International Conference on Pervasive Computing and Communications Workshops (PerCom Workshops), pp. 527–532. IEEE (2017)
34. Ziegeldorf, J.H., Morchon, O.G., Wehrle, K.: Privacy in the internet of things: threats and challenges. Secur. Commun. Netw. **7**(12), 2728–2742 (2013)
35. Zwingelberg, H., Hansen, M.: Privacy protection goals and their implications for eID systems. In: Camenisch, J., Crispo, B., Fischer-Hübner, S., Leenes, R., Russello, G. (eds.) Privacy and Identity 2011. IAICT, vol. 375, pp. 245–260. Springer, Heidelberg (2012). https://doi.org/10.1007/978-3-642-31668-5_19

Language-Based Mechanisms for Privacy-by-Design

Shukun Tokas(✉), Olaf Owe(✉), and Toktam Ramezanifarkhani(✉)

Department of Informatics, University of Oslo, Oslo, Norway
{shukunt,olaf,toktamr}@ifi.uio.no

Abstract. The privacy by design principle has been applied in system engineering. In this paper, we follow this principle, by integrating necessary safeguards into the program system design. These safeguards are then used in the processing of personal information. In particular, we use a formal language-based approach with static analysis to enforce privacy requirements. To make a general solution, we consider a high-level modeling language for distributed service-oriented systems, building on the paradigm of active objects. The language is then extended to support specification of policies on program constructs and policy enforcement. For this we develop (i) language constructs to formally specify privacy restrictions, thereby obtaining a policy definition language, (ii) a formal notion of policy compliance, and (iii) a type and effect system for enforcing and analyzing a program's compliance with the stated polices.

Keywords: Privacy by design · Language-based privacy · Privacy compliance · Static analysis

1 Introduction

Advances in information technologies have often led to concerns about privacy. With the adoption of information and communication technology in our daily lives, the gathering and processing of personal information fundamentally increases the potential for privacy threats. In particular, privacy and data protection features are often ignored by conventional engineering approaches [1] or accommodated as an afterthought. Aligning the software ecosystem with the privacy-related requirements is an essential step towards better data protection. In order to endorse privacy as a first-class requirement and promote privacy compliance from the outset of product development, the *privacy by design* (PbD) requirement has been formally embedded in the GDPR regulations (Article 25 [2]). Article 25 [2] obliges the controllers to design and develop products with a built-in ability to demonstrate compliance towards the data protection obligations.

The main idea of privacy by design is to make privacy a key consideration in development of systems. Privacy by design is a framework consisting of seven

Published by Springer Nature Switzerland AG 2020
M. Friedewald et al. (Eds.): Privacy and Identity 2019, IFIP AICT 576, pp. 142–158, 2020.
https://doi.org/10.1007/978-3-030-42504-3_10

foundational principles: *(i)* proactive not reactive; preventive not remedial, *(ii)* privacy as default setting, *(iii)* privacy embedded into design, *(iv)* full functionality - positive-sum, not zero-sum, *(v)* end-to-end security - full lifecycle protection, *(vi)* visibility and transparency, and *(vii)* respect for user privacy - keep it user-centric. We focus on the *privacy embedded-into-design* principle, due to its potential connection with language mechanisms. We explore the idea of adding privacy requirements into programming/specification languages and use static analysis for enforcing such privacy requirements.

In this paper, we follow the privacy by design principle, by integrating necessary safeguards into the processing of personal information, using a language-based approach. In particular, we explore how to formalize fundamental privacy principles and to provide built-in abilities to fulfill data protection obligations. As a step towards this goal we develop a policy specification language that provide constructs for specifying privacy requirements on sensitive (personal) data. In particular, a policy is given by a set of triples that put restrictions on the *principals* that may access the information for certain *purposes* and the permitted *access rights*. Such policy statements are then linked with language constructs of a high-level modeling language oriented towards distributed and service-oriented systems. Policies are annotated with the *data types* and *methods*.

Certain aspects of privacy restrictions can be expressed by means of static concepts, while others can only be expressed at runtime, such as *data subject*, *consent*, and other user-defined changes. In this paper, we focus on statically declared policies and implicit consent (at compile time presence of policy implies consent). Changes in consent and policies are handled at runtime through predefined functionalities, which is beyond the scope of this article. In addition to *read* and *write* access, we consider *incremental* access (*incr*), allowing addition of sensitive information without read access and without modifying existing information. For instance, in a healthcare setting, a lab assistant may have incr access to treatment data, while a nurse may have both read and incremental access ($read \sqcup incr$), and a doctor may have full access ($read \sqcup write$). We formalize a notion of policy compliance, to develop a scheme of policy inheritance. Finally, to enforce policy compliance, we define a set of rules, i.e., the type and effect system that checks that the policies are respected when the sensitive information is accessed. The theory of the current work is presented in more details in [3].

In summary, the main idea is to provide language constructs that express privacy policy specifications capturing static aspects of privacy and use these to statically analyze a program's compliance with the policy specifications. We make the following contributions: (i) propose a policy language for specifying purpose, access and policy requirements (see Fig. 1), (ii) formalize a notion of policy compliance, (iii) show how the policy language can be used with an underlying object-oriented language, and (iv) develop a mechanic type and effect system for analyzing a program's compliance with the annotated privacy policies.

Paper Outline. The rest of the paper is structured as follows. Section 2 presents the formalization of privacy policies, including a policy definition language and a formalization of policy compliance. Section 3 introduces the core language, with support for the specification of privacy principles. Section 4 presents the type and effect system. Section 5 demonstrates the analysis on a small case study. Section 6 discusses related work, and Sect. 7 concludes the paper.

$$
\begin{array}{lll}
A & ::= \mathit{read} \mid \mathit{incr} \mid \mathit{write} \mid \mathit{self} & \text{basic access rights} \\
 & \quad \mid \mathit{no} \mid \mathit{full} \mid \mathit{rincr} \mid \mathit{wincr} & \text{abbreviated access rights} \\
 & \quad \mid A \sqcap A \mid A \sqcup A & \text{combined access rights} \\
\mathcal{P} & ::= (I, R, A) & \text{policy} \\
\mathcal{P}s & ::= \{\mathcal{P}^*\} \mid \mathcal{P}s \sqcap \mathcal{P}s \mid \mathcal{P}s \sqcup \mathcal{P}s & \text{policy set} \\
\mathcal{RD} & ::= \mathbf{purpose}\ R^+ & \\
 & \quad [\mathbf{where}\ \mathit{Rel}\ [\mathbf{and}\, \mathit{Rel}]^*] & \text{purpose declaration} \\
\mathit{Rel} & ::= R^+ < R^+ & \text{sub-purpose declaration}
\end{array}
$$

Fig. 1. BNF syntax definition of the policy language. I ranges over interface names and R over purpose names. The operators $\sqcup$ and $\sqcap$ denote join and meet, respectively.

2 Language Constructs for Policy Specification

Privacy policies are often described in natural language statements. To verify formally that the program satisfies the privacy specification, the desired notions of privacy need to be expressed explicitly. To formalize such policies, we define a policy specification language. Furthermore, to establish a link between policies and programming language constructs, we extend the syntax and semantics of a small core language (see Sect. 3). In our setting, a privacy policy is a statement that expresses permitted use of the sensitive information by the declared program entities. To support privacy-by-design, we define policies at the design level, and associate policies to data types and methods of interfaces and classes, such that the policies of a method in a class must comply with the corresponding policy in an interface of the class. In particular, a policy is given by a set of triples that put restrictions on: What *principals* may access the sensitive data, which *purposes* are allowed, and which *access-rights* are permitted. That being the case, a policy $\mathcal{P}$ is given by a triple (I, R, A), where *(i)* I ranges over interfaces, which are organized in an open-ended inheritance hierarchy, *(ii)* R ranges over purposes, which are organized in a hierarchy (reflecting specialization), and *(iii)* A ranges over access rights, which are organized in a lattice. Thus principals are expressed by the Interfaces, while new language constructs are added to represent purposes, access rights, and policies.

The language syntax for policies is summarized in Fig. 1, where [ ] is used as meta-parenthesis, and superscripts * and $^+$ denote general and non-empty repetition, respectively. Here we briefly discuss the specification constructs.

Principal describes the roles that can access sensitive information and is given by an interface. For instance for a call $x := o.m(\overline{e})$, where o is typed by an interface with policy (I, R, A), the caller object must support interface I. Interfaces are organized in an open-ended inheritance hierarchy, letting $I < J$ denote that I is a subinterface of J. For example,

$$Specialist < Doctor < HealthWorker$$

Any is predefined as the least specialized interface, i.e., the superinterface of all interfaces. We let $\leq$ denote the transitive and reflexive extension of $<$.

Purpose names are used to restrict usage of sensitive data to specific purposes. Such purpose names can be organized in a hierarchical structure, reflecting a *purpose hierarchy* [4]. We let purposes be organized in a directed acyclic graph reflecting specialization. Purpose names are defined by the keyword **purpose**. For instance, the declaration

$$\textbf{purpose}\ spl\_treatm,\ treatm\ \textbf{where}\ spl\_treatm\ <\ treatm$$

makes *spl_treatm* more specialized purpose than *treatm*. If data is collected for the purpose of *spl_treatm* then it cannot be used for *treatm*. However, if it is collected for the purpose of *treatm* then it can be used for *spl_treatm*. We let $\leq$ denote the transitive and reflexive extension of $<$.

Access-right describes permitted operations on sensitive data. Access rights are given by a lattice, with meet and join operations (see Fig. 2): *read* gives read access, *write* gives write access (without including read access), *incr* allows addition of new information but neither read nor write is included. The combination of *read* and *incr*, i.e., $read \sqcup incr$ is abbreviated *rincr* gives read and incremental access. Similarly, $write \sqcup incr$ is abbreviated as *wincr*, which gives write and incremental access. Full access is given by a combination of *read* and *write* (which includes incremental access), i.e., *full* is the same as $read \sqcup write$. These general access rights can be combined with access rights on self, i.e., access rights when the principal is the subject herself. (details are omitted). For instance, a nurse should be able to see treatment data of a patient and add new data, and needs *rincr* access, while a lab assistant may add lab data and needs only *incr* access. A patient should see data about herself, which requires $self \sqcap read$.

A single policy ($\mathcal{P}$) is given by a triple (I, R, A), and a policy set ($\mathcal{P}_s$) is given by a set of policy triples (with meet and join operation defined). For our purposes, we annotate methods with single policies while data types are annotated with policy sets reflecting the permitted usage by different principals.

Example. The example in Fig. 3 gives an illustration for declaring policies, and annotating methods and types with policies. The policy $(Doctor, treatm, rincr)$ restricts access to objects typed by the *Doctor* interface, for only *treatm* (treatment) purposes, and with *rincr* data access. This is checked by a type and effect system in Sect. 4. The policy set

$$\{(Doctor, treatm, full), (Doctor, treatm, rincr), (Nurse, treatm, read)\}$$

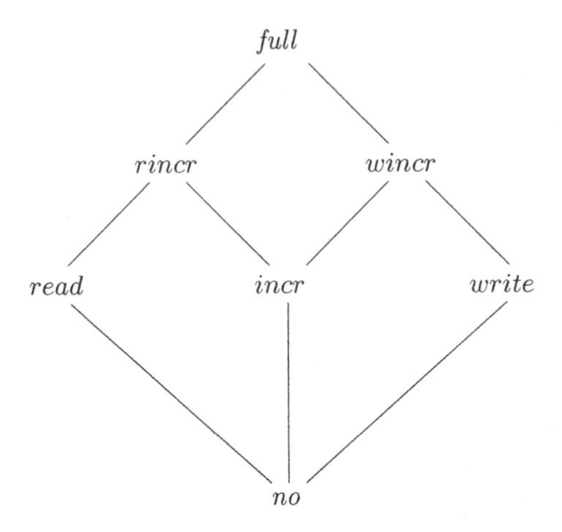

Fig. 2. The lattice for general access rights (without *self*). Note that *rincr* is the same as $read \sqcup incr$, *wincr* is the same as $write \sqcup incr$, and *full* is the same as $read \sqcup write$.

restricts access by these three policies. Here, the policy $(Doctor, treatm, rincr)$ is redundant since $(Doctor, treatm, rincr) \sqsubseteq (Doctor, treatm, full)$, and is colored grey to indicate that. Method *makePresc* has policy $(Doctor, treatm, rincr)$, meaning that this method must be called by a Doctor object (or a more specialized object), for purposes of treatment and with read and incremental access (but not write access). Thus a doctor can add new prescription, but not change or remove old ones. Method *getPresc* has policy $(Nurse, treatm, read)$, meaning that this method must be called by a Nurse object (or a more specialized object such as a Doctor object), for purposes of treatment, and with read-only access. These two methods, with associated policies, are inherited in interface *PatientData*.

2.1 Policy Compliance Definition

Here, we briefly present a few definitions needed to express policy compliance.

Definition 1 (Policy Compliance). *The sub-policy relation* $\sqsubseteq$, *expressing* policy compliance, *is defined by*

$$(I', R', A') \sqsubseteq (I, R, A) \triangleq I \leq I' \wedge R' \leq R \wedge A' \sqsubseteq A$$

(where the last $\sqsubseteq$ *operation is on access rights) with* $\bullet$ *as* bottom element, *representing non-sensitive information. It follows that* $\sqsubseteq$ *is a partial order.*

A policy $\mathcal{P}'$ complies with $\mathcal{P}$ if it has the same or larger interface, the same or more specialized purpose, and if the access rights of $\mathcal{P}'$ are the same or weaker than that of $\mathcal{P}$. In particular, the policy of the implementation of a method

```
purpose basic_treatm, treatm where basic_treatm < treatm

policy P_Doc = (Any, treatm, full)
policy P_AddPresc = (Doctor, treatm, rincr)
policy P_GetPresc = (Nurse, treatm, read)
policy P_Presc = {P_GetPresc, P_AddPresc, P_Doc}

type Presc == Patient * String :: P_Presc

interface Patient extends Subject {Void getSelfData() :: P_SelfPresc}
interface AddPresc {Void makePresc(Presc newp):: P_AddPresc}
interface GetPresc {Presc getPresc(Patient p) :: P_GetPresc}
interface PatientData extends AddPresc, GetPresc {}
interface Nurse extends Principal { Presc nurseTask() :: P_GetPresc}
interface Doctor extends Nurse{ Void doctorTask(Patient p) :: P_Doc}

class PATIENTDATA() implements PatientData {
  type PData = List[Presc] :: P_Presc
  PData pd = empty();
  Presc getPresc(Patient p){return last(pd/p)} :: P_GetPresc
  Void makePresc(Presc newp) {
     if newp ≠emptyString() then pd:+ newp fi } :: P_AddPresc }

class DOCTOR() extends NURSE implements Doctor{//inherits pd
  Void doctorTask(Patient p){
    Presc oldp = pdb.getPresc(p);
    String text = ...; //new presc using symptoms info and oldp
    Presc newp = (p, text); // here, new sensitive data is created!
    pdb!makePresc(newp)}:: P_Doc }
```

Fig. 3. Interface, class, type, and policy definitions for the Prescription Example. Grey policy specifications are implicit while underlined ones need to be explicitly stated. A class implementation of Nurse is omitted. The projection pd/p is the list of strings associated to patient p, and the function *last* gives the last element.

should comply with that of the interface. Note that $\bullet \sqsubseteq \mathcal{P}$ expresses that an implementation without access to sensitive information complies with any policy.

Moreover, the use of *self* in the access part allows us to distinguish between different kinds of self access for different purposes, such as $(Patient, all, read \sqcap self)$ and $(Patient, private\_settings, self)$. The latter gives full access to data about *self* for purposes of *private settings*, while the first gives read access to data about *self* for all purposes.

We define a lattice over sets of policies with meet and join operations, and generalize the definition of compliance to sets of policies:

Definition 2 (Compliance of Policy Sets).

$$\{\mathcal{P}'_i\} \sqsubseteq \{\mathcal{P}_j\} \triangleq \forall i \,.\, \exists j \,.\, \mathcal{P}'_i \sqsubseteq \mathcal{P}_j$$

This expresses that a policy set S' complies with a policy set S if each policy in S' complies with some policy in S. We define meet and join operations over policy sets by set union and a kind of intersection, respectively, adding implicitly derivable policies:

Definition 3 (Join and Meet over Policy Sets).

$$S \sqcup S' \triangleq closure(S \cup S')$$

$$S \sqcap S' \triangleq closure(\{P \mid P \sqsubseteq S \land P \sqsubseteq S'\})$$

where the closure operation *is defined by*

$$closure(S) \triangleq S \cup \{(I, R, A \sqcup A') \mid (I, R, A) \sqsubseteq S \land (I, R, A') \sqsubseteq S\}$$

We have a lattice with $\emptyset$ as the bottom element. The closure operation adds implicitly derivable policies, and ensures that $\{(I, R, A \sqcup A')\} \sqsubseteq \{(I, R, A)\} \sqcup \{(I, R, A')\}$. For instance, $\{(Doctor, treatm, read)\} \sqcup \{(Doctor, treatm, write)\}$ is the same as $\{(Doctor, treatm, full)\}$. These constructs are useful in specification of constraints and in capturing access to sensitive information with declared privacy policies. The meet operation typically reflects worst-case analysis.

Definition 4 (Implication on Policy Set). *We define the notation* $\mathcal{P}s' \Rightarrow \mathcal{P}s$ *(*$\mathcal{P}s'$ *implies* $\mathcal{P}s$*) by* $\{\bullet\} \Rightarrow \mathcal{P}s$ *and* $\mathcal{P}s \sqsubseteq \mathcal{P}s'$ *for* $\mathcal{P}s'$ *other than* $\{\bullet\}$.

Implication is used to check policy compliance of an actual parameter with respect to a formal parameter. If $\{\bullet\}$ is the policy on the actual parameter and $\mathcal{P}_{doc}$ the policy on the formal parameter, we will check $\{\bullet\} \Rightarrow \mathcal{P}_{doc}$.

Policies on Methods. Let $\mathcal{P}_{I,m}$ denote the policy of a method m given in an interface I, and $\mathcal{P}_{C,m}$ denote the policy of a method m given in a class C. We will require that the implementation of a method in a class (C) respects the policy stated in the interface (I), i.e., $\mathcal{P}_{C,m} \sqsubseteq \mathcal{P}_{I,m}$. And we also require that a method redefined in an interface (I) respects the policy of that method in a superinterface (J), i.e., $\mathcal{P}_{I,m} \sqsubseteq \mathcal{P}_{J,m}$. By transitivity of $\sqsubseteq$, a method implementation in a class that respects the policy given in an interface also respects the policy of the method given in a superinterface, i.e., $\mathcal{P}_{C,m} \sqsubseteq \mathcal{P}_{I,m}$ and $\mathcal{P}_{I,m} \sqsubseteq \mathcal{P}_{J,m}$ implies $\mathcal{P}_{C,m} \sqsubseteq \mathcal{P}_{J,m}$. For instance, consider an interface *GetPresc* with a method *getPresc*() with policy $(Nurse, treatm, read)$. An implementation of this method in a class must have a policy that complies with it, such as $(Any, treatm, read)$, $(Nurse, treatm, self \sqcap read)$, or $(Nurse, basic\_treatm, read)$. In contrast, the implementation cannot have policy $(Doctor, treatm, read)$, as this would not allow a $Nurse$ as the caller object, and also not $(Nurse, all, rincr)$, because this violates purpose and access restrictions.

Policies on Types. We let the policy of a type T, denoted $\mathcal{P}_T$, be a policy set. Let the policy set $\{(Doctor, treatm, rincr), (Nurse, treatm, read)\}$ be the policy set on type *Presc*. This allows the data of type *Presc* to be accessed based on these two policies, depending on the calling context. For instance, if the caller is a *Doctor* object and the purpose is *treatm* then *read* as well as *incr* access is allowed on data of type *Presc*. The policy set of an actual variable must imply the policy set of the type of the corresponding formal variable. Together, the policies on methods and types provide sufficient abstractions to control access to sensitive data.

In the next section we consider a high-level imperative language for service-oriented systems where policy specifications are integrated.

3 Embedding Policy with Program Constructs

We target object-oriented, distributed systems (OODS) and consider the active object programming paradigm [5], which is based on the actor model [6] and gives a high-level view of communication aspects in OODS. In the active object model, objects are autonomous and execute in parallel, communicating by so-called asynchronous method invocations. We assume interface abstraction, i.e., an object can only be accessed through an interface and remote field access is illegal. This allows us to focus on major challenges of modern architectures, without the complications of low-level language constructs related to the shared-variable concurrency model.

We propose a small core language, based on Creol [7], centered around a few basic statements. It has a compositional semantics which is beneficial to analysis [7,8]. The language is imperative and strongly typed, with data types for data structure locally inside a class. The data type sublanguage is side-effect-free. The motivation is that the language gives high-level descriptions of distributed systems and synchronous and asynchronous interaction based on methods, thereby avoiding shared variable access, and avoiding explicit signaling and notification. The BNF syntax of the language is summarized in Fig. 4. As before, optional parts are written in brackets (except for type parameters, as in $\mathsf{List}[T]$, where the brackets are ground symbols). Class parameters (z), method parameters (y) the implicit class parameter this and the implicit method parameter caller are read-only. A class may implement a number of interfaces, and for each method of an interface (of the class) it is required that the class defines the method such that policy of each method parameter and return value are respected. Additional methods may be defined in a class, but these may not be called from outside the class. The language supports single class inheritance and multiple interface inheritance (using the keyword **extends**). Below, we give BNF syntax for method and type declarations.

Definition 5 (Method Declaration Syntax).

$$T\ m([Y\ y]^*)\ [::\mathcal{P}]$$

where T is the result type and Y is the type of parameter y.

Pr	$::= [\mathcal{T} \mid \mathcal{RD} \mid In \mid Cl]^*$	program
$\mathcal{T}$	$::= \textbf{type}\ N\ [\overline{T}] =<type\_expression> [:: \mathcal{P}s]$	type definition
T	$::= I \mid \mathsf{Int} \mid \mathsf{Any} \mid \mathsf{Bool} \mid \mathsf{String} \mid \mathsf{Void} \mid \mathsf{List}[T] \mid N$	types
In	$::= \textbf{interface}\ I\ [\textbf{extends}\ I^+]\ \{D^*\}$	interface declaration
Cl	$::= \textbf{class}\ C\ ([T\ z]^*)$	class definition
	$[\textbf{implements}\ I^+]\ [\textbf{extends}\ C]$	support, inheritance
	$\{[T\ w\ [:= ini]]^*$	fields
	$[B\ [:: \mathcal{P}]]$	class constructor
	$[[\textbf{with}\ I]\ M]^*\}$	methods
D	$::= T\ m([T\ y]^*)\ [:: \mathcal{P}]$	method signature
M	$::= T\ m([T\ y]^*)\ [\{s\}]\ [:: \mathcal{P}]$	method definition
B	$::= \{[T\ x\ [:= rhs];]^*\ [s;]\ \textbf{return}\ rhs\}$	method blocks
v	$::= w \mid x$	assignable variable
e	$::= v \mid y \mid z \mid \mathsf{this} \mid \mathsf{caller} \mid \mathsf{void} \mid f(\overline{e})$	pure expressions
ini	$::= e \mid \textbf{new}\ C(\overline{e})$	initial value of field
rhs	$::= ini \mid e.m(\overline{e})$	right-hand sides
s	$::= \mathsf{skip} \mid s; s$	sequence
	$\mid v := rhs \mid v :+ rhs \mid e!m(\overline{e}) \mid I!m(\overline{e})$	assignment and call
	$\mid \textbf{if}\ e\ \textbf{then}\ s\ [\textbf{else}\ s]\ \textbf{fi}$	if statement
	$\mid \textbf{while}\ e\ \textbf{do}\ s\ \textbf{od}$	while statement

Fig. 4. BNF syntax of the core language. A field variable is denoted w, a local variable x, a method parameter y, a class parameter z, and list append is denoted $+$. The brackets in $[T]$ and $[\overline{T}]$ are ground symbols.

An inherited method m inherits the policy of m from the superinterface, unless the interface declares its own policy for m. However, the redefined policy of m (of interface I) cannot be more restrictive than that of the superinterface (J), i.e., $\mathcal{P}_{I,m} \sqsubseteq \mathcal{P}_{J,m}$, ensuring that a class implementation of m satisfying $\mathcal{P}_{I,m}$ also satisfies any declarations of m in a superinterface.

Definition 6 (Data Type Declaration Syntax and Sensitivity).

$$\textbf{type}\ N\ [TypeParameters] =< type\_definition > \ [:: \mathcal{P}s]$$

where the type parameters are optional. The predefined basic types (Nat, Int, String, Bool, Void) are non-sensitive. *A user-defined type is* sensitive *if a policy set is specified in the type definition.*

For example, a sensitive *String* type restricted by a policy $\mathcal{P}_s$ can be defined by

$$\textbf{type}\ Info\ = String :: \mathcal{P}_s$$

and encryption could go from Info to String, and decryption the other way.

We consider next *sensitive* functions, which create new sensitive data, for instance a product of individually non-sensitive data may be sensitive. Generator functions (here called constructors) are considered *sensitive* if they (i)

$$\text{(P-VAR)}\ \frac{read \sqsubseteq \Gamma[v] \sqcap (\mathcal{P}_{C,m}@(C,m))}{C,m \vdash [\Gamma]\ v ::\ \Gamma[v] \sqcap \Gamma[pc]}$$

$$\text{(P-FUNC)}\ \frac{\begin{array}{ll} C,m \vdash [\Gamma]\ e_i ::\ \mathcal{P} & \text{for each argument } e_i \text{ of a sensitive type} \\ write \sqsubseteq \mathcal{P}_T \sqcap (\mathcal{P}_{C,m}@(C,m)) & \text{if } f_T \text{ is a sensitive function} \end{array}}{C,m \vdash [\Gamma]\ f_T(\overline{e}) ::\ \mathcal{P}_T \sqcap \Gamma[pc]}$$

$$\text{(P-CALL)}\ \frac{\begin{array}{c} \mathcal{P}_{I,n} \sqsubseteq_{Co,R} \mathcal{P}_{C,m}@(C,m) \\ C,m \vdash [\Gamma]\ e :: \mathcal{P}' \\ C,m \vdash [\Gamma]\ e_i :: \mathcal{P}_i \quad \mathcal{P}_i \Rightarrow \mathcal{P}_{par(I,n)_i} \quad \text{for each i} \end{array}}{C,m \vdash [\Gamma]\quad e.n_I(\overline{e}) :: \mathcal{P}_{out(I,n)}}$$

Fig. 5. Policy rules for expressions and right-hand sides.

combine information about a subject with non-sensitive or sensitive information or (ii) use sensitive information. We assume that sensitive generators produce sensitive types (with some exceptions, such as constructors of encrypted data). Defined functions are *sensitive* if their type is sensitive and the definition directly or indirectly contains a sensitive application of a constructor. For instance we may (recursively) define a parameterized list type by $List[T] = empty()|append(List[T] * T)$ meaning that lists have the form $empty()$ or $append(l, x)$, where l is a list and x a value of type T. (We let the notation $l + x$ abbreviate $append(l, x)$.) The list is sensitive if T is sensitive, but the append constructor function is not sensitive. A pair product type can be defined by $PatientData = (Patient * String)$ where *Patient* is a interface representing a data *subject*. This type is sensitive (even though *String* is not), and the pair (*current_patient*, "*no health problems*") is a sensitive application of the product constructor. These examples suffice for our purposes here. It can be detected statically if a function is sensitive (further details are omitted). Applications of sensitive functions may create new sensitive data, something which require write access. This way the policy control is driven by the declared data types rather than variable declarations. Data types are reusable and therefore their policies are likely to more reliable and appropriate than one-time adhoc specification for program variables.

When the lawful basis of processing of personal information is performance of contract or other valid bases but not the consent, the policies must be formulated in a way that ensures that they are built into the system *by default*, i.e., no measures are required by the data subject in order to maintain his/her privacy. However, when consent is the basis of processing the data subjects, choices in privacy settings are captured at run time (as outlined in [9]).

We next show how to define static policy checking for our core language.

4 An Effect System for Privacy

We propose static policy checking defined by a set of syntax-directed rules, given as a type and effect system [10], but dealing with policies rather than types. We consider two kinds of judgments. For a statement s, the judgment

$$C, m \vdash [\Gamma]\ s\ [\Gamma']$$

expresses that inside a method body m and an enclosing class C, the statement(list) s when started in a state satisfying the environment Γ results in a state satisfying the environment Γ'. Here Γ is a mapping from program variable names to policy sets, such that the policy set of a variable in a given state gives an upper bound of the permitted operations. In order to deal with branches of if- and while-statements where the context policy is influenced by that of the if- and while-tests, Γ uses an additional variable pc (the program context) reflecting the current branching policy (as in [8]). Note that the rules are right-constructive in the sense that Γ' can be constructed from Γ and s.

For an expression or right-hand side e, the judgment

$$C, m \vdash [\Gamma]\ e :: \mathcal{P}s$$

expresses that the evaluation of e in a state satisfying Γ gives a value satisfying the policy set $\mathcal{P}s$, where m is the enclosing method and C the enclosing class.

Figure 5 defines the typing rules for expressions and right-hand sides, and Fig. 6 defines the typing rules for (selected) statements. We let $\mathcal{P}_{I,m}$ denote the policy of method m of interface I, $\mathcal{P}_{C,m}$ denote the policy of method m of class C, and $\mathcal{P}_T$ denote the policy associated with a type T. If no policy is specified for any declaration, we understand that there is no sensitive information, i.e., the policy is $\{\bullet\}$. Data types with sensitive constructors will be considered sensitive. A non-sensitive method would not be able to access or create sensitive data, and a non-sensitive type declaration would not allow assignment of sensitive information to variables of that type.

Rule P-VAR says that the policy of a variable v (a field, parameter, or local variable) is the one recorded in Γ for v, i.e., $\Gamma[v]$, combined with that of the program context pc. The premise states that there must be read access to v, both according to the policy set of the variable and according to the policy set of the enclosing method body. If the policy of the enclosing method m is (I, R, A), the *policy set of the method body* is defined by

$$(I, R, A)@(C, m) \triangleq (I, R, A) \cup (\cup_i \{(I_i, R, A)\})$$

where I_i ranges over all the interfaces of C that export m. Thus the policy set of the method body is that of the method and those where this object is the principle (as seen through one of the interfaces exporting m).

Rule P-FUNC says that the policy of a function application $f_T(\overline{e})$ is that of the resulting type T (detected by ordinary typing) combined with that of the program context pc. Sensitive arguments must be checked (which ensure read access to the variables occurring in these arguments), and in case f is a sensitive function application, there must be write access according to the policy of T and the policy of the method body. Constants (function without arguments), as well as object creation, have policy set $\{\bullet\}$.

Rule P-CALL says that the policy of a remote call $e.n_I(\overline{e})$ where I is the interface of the method (detected by ordinary typing), is the policy on the return type of the method (as given by the declaration of m in I). The first premise ensures that the policy of the called method complies with policy of the enclosing body. The second premise ensures that the callee expression has a valid policy, and the last premise ensures each actual parameter has a policy set that implies the policy set of the corresponding formal one.

The rule P-SKIP says that the environment is not changed. The rule for sequential composition says that the final environment of s_1 is used as the starting environment for the next statement s_2. The rules P-WRITE and P-LOCAL-WRITE say that the final environment is that of the right-hand side. Writing to a field requires write access, while writing to a local variable is always allowed. An incremental assignment $w : +e$ requires *incr* access, and the final environment is as for the assignment $w := w + e$. The premises for asynchronous call is as for P-CALL, and the resulting environment is unchanged (since no variable is changed).

$$\text{(P-SKIP)}\ \frac{}{C, m \vdash [\Gamma]\ skip\ [\Gamma]}$$

$$\text{(P-COMPOSITION)}\ \frac{C, m \vdash [\Gamma]\ s_1\ [\Gamma_1] \quad C, m \vdash [\Gamma_1]\ s_2\ [\Gamma_2]}{C, m \vdash [\Gamma]\ s_1; s_2\ [\Gamma_2]}$$

$$\text{(P-WRITE)}\ \frac{\begin{array}{c} C, m \vdash [\Gamma]\ rhs :: \mathcal{P} \\ write \sqsubseteq \Gamma_C[w] \sqcap (\mathcal{P}_{C,m}@(C, m)) \end{array}}{C, m \vdash [\Gamma]\ w := rhs\ [\Gamma[w \mapsto \mathcal{P}]]}$$

$$\text{(P-LOCAL-WRITE)}\ \frac{C, m \vdash [\Gamma]\ rhs :: \mathcal{P}}{C, m \vdash [\Gamma]\ x := rhs\ [\Gamma[x \mapsto \mathcal{P}]]}$$

$$\text{(P-INCR)}\ \frac{\begin{array}{c} C, m \vdash [\Gamma]\ rhs :: \mathcal{P} \\ incr \sqsubseteq \Gamma_C[w] \sqcap (\mathcal{P}_{C,m}@(C, m)) \end{array}}{C, m \vdash [\Gamma]\ w : +rhs\ [\Gamma[w \mapsto \Gamma[w] \sqcap \mathcal{P}]]}$$

$$\text{(P-ASYNCCALL)}\ \frac{C, m \vdash [\Gamma] \quad e.n_I(\bar{e}) :: \mathcal{P}_{out(I,n)}}{C, m \vdash [\Gamma] \quad e!n_I(\bar{e})\ [\Gamma]}$$

Fig. 6. Policy rules for statements.

Note that, if by mistake, no policy is specified due to forgetfulness, the static compliance checking would detect any use of sensitive information and the program would not pass the privacy checks. In particular, data types with constructors associating data to subjects will be considered sensitive. A non-sensitive method would not be able to access or create sensitive data, and a non-sensitive type declaration would not allow assignment of sensitive information to variables of that type.

We next show how to apply the static analysis on the Prescription case study.

5 Case Study

Consider the example from Fig. 3 where *Doctor*, *Nurse*, *Patient*, *PatientData*, *AddPresc*, *GetPresc* are interfaces. A *PatientData* object contains data for a number of patients, and can be accessed by doctors and nurses, based on different policies. Policies are declared by the keyword **policy**. Patient data *pd* of type *PData* (list of *Presc*) is labeled with polices: $\{\mathcal{P}_{GetPresc}, \mathcal{P}_{Doc}\}$, and an implicit policy $(Subject, all, self \sqcap read)$ is included in every policy set to allow read access when the principal is the data subject. This policy ($\mathcal{P}_{Presc}$) allows *(i)* a patient to access his/her own data, *(ii)* gives *full* (i.e., read, incr, write) access to the *Doctor* for *treatm* purposes, and *(iii)* gives *read-only* access to the *Nurse* for *treatm* purposes. The purpose *treatm* is declared by the keyword **purpose**. The policies need to be declared only once and then the effect system will keep track of the policies in a given program state. For example, the declaration of *makePresc*() includes the policy $\mathcal{P}_{AddPresc}$. Now we show an application of a few type rules, on the statements in the method $doctorTask()$ from Fig. 3.

1. $x := rhs$
   $String\ text = rhs$ //Apply P-LocalWrite
   The premise $rhs :: \bullet$ associates $\bullet$ with rhs, since it is a local variable and has no policy.
   $\Gamma[x \mapsto \mathcal{P}] \implies \Gamma[text \mapsto \bullet]$
   Gamma for *text* is updated with $\bullet$.
2. $Presc\ newp = (p, text)$; //Apply P-Func, P-LocalWrite
   (a) $read \sqsubseteq \Gamma[v] \sqcap (\mathcal{P}_{C,m}@(C,m))$
   $(\Gamma[p] \sqcap \Gamma[text]) \sqcap \mathcal{P}_{Presc}$
   $(\bullet \sqcap \bullet) \sqcap (\mathcal{P}_{C,m}@(C,m))$
   i.e., $(\bullet \sqcap \bullet) \sqcap \mathcal{P}_{Doc}$ since $(\mathcal{P}_{C,m}@(C,m)) = \mathcal{P}_{Doc}$
   which reduces to $\mathcal{P}_{Doc}$
   $read \sqsubseteq \mathcal{P}_{Doc}$ (i.e., $read \sqsubseteq \Gamma[v] \sqcap (\mathcal{P}_{C,m}@(C,m))$)
   which reduces to $read \sqsubseteq full$, using the notation $A \sqsubseteq (I, R, A')$ when $A \sqsubseteq A'$, and $A \sqsubseteq \{(I, R, A')_i\}$ when $A \sqsubseteq (I, R, A')_i$ for some i (i.e., $A \sqsubseteq A'_i$).
   (b) $write \sqsubseteq \mathcal{P}_T \sqcap (\mathcal{P}_{C,m}@(C,m))$, since the constructor $(\_, \_)$ is sensitive
   $write \sqsubseteq \mathcal{P}_{Doc} \sqcap \mathcal{P}_{Presc}$
   which reduces to $write \sqsubseteq full$, and the policy of $(p, text)$ is $\mathcal{P}_{Presc}$
   (c) $\Gamma[newp \mapsto \Gamma[(p, text)]]$
   $\Gamma[newp \mapsto \mathcal{P}_{Presc}]$ //since pc is empty here

   In the first statement, the policy set on *text* is $\{\bullet\}$ because it is not yet associated with a subject. But when non-sensitive *text* is combined with a subject identity, this is seen as construction of a sensitive data, and P-Func is used to ensure that the information can be read and constructed by the current context. The rest of the example can be checked in a similar way.

6 Related Work

Language-based mechanisms are techniques based on programming languages that are often used in developing secure applications. In particular, language-based security mechanisms are used in specification and enforcement of security policies. In recent years, various techniques (compilers, automated program analysis, type checking, program rewriting etc.) have been explored from the perspective of their applicability in enforcing security and privacy policies in programs. Privacy by Design (PbD) has been discussed and promoted from several viewpoints such as privacy engineering [1,11,12], privacy design patterns [13,14], and formal approaches [15–17]. Tschantz and Wing, in [17] and Métayer, in [15] discuss the significance of formal methods for foundational formalizations of privacy related aspects. In [16], Schneider discusses the main ideas of *Privacy by Design* and summarizes key challenges in achieving Privacy by Construction and probable means to handle these challenges. The paper calls for ways to ensure control of *purpose* integrated in programming languages. It is also indicated that in order to ensure that privacy-compliant code is sound and correct, formal methods would be helpful in proving soundness and completeness (with respect to a set of predefined privacy concepts). Privacy design strategies [13] focus on how to take privacy requirements into account from the beginning and make it a software quality attribute. The engineering aspects of privacy by design is addressed, but there is a lack on how to apply them in practice. In our work, we adhere to several privacy design strategies such as separating and hiding the data, and encapsulation in an object-oriented context.

Hayati and Abadi [4] describe a language-based approach based on information-flow control, to model and verify aspects of privacy policies in the Jif (Java Information Flow) programming language. In this approach data collected for a specific purpose is annotated with Jif principals and then the methods needed for a specific purpose are also annotated with Jif principals. Explicitly declaring purposes for data and methods ensures that the labeled data will be used only by the methods with connected purposes. Purposes are organized in a hierarchy, with sub-purposes. However, this representation of purpose is not sufficient to guarantee that principals will perform actions compliant with the declared purpose. But this can be checked statically in our approach, because the principal is restricted by a purpose-based access control.

Basin et al. [18] propose an approach that relates a purpose with a business process and use formal models of inter-process communication to demonstrate GDPR compliance. Process collection is modeled as data-flow graphs which depict the data collected and the data used by the processes. Then these processes are associated with a data purpose and are used to algorithmically (i) generate data purpose statements, (ii) detect violation of data minimization, and (iii) demonstrate compliance of some more aspects of GDPR. Since in GDPR, end-users should know the necessary purpose of data collection, some works such as [18] propose to audit logs and detect if a computer system supports a purpose. In a continuation of this work [19], Arfelt et al. show how such an audit can be automated by monitoring. Automatic audits and monitoring can be applied to a system like ours as a complementary step to verify how it complies with the GDPR. Besides, our work is more focussed on integrating such legal instruments during the design phase, using formal language semantics. In [20], Adams and Schupp consider black-box objects that communicate through messages. The approach is centered around algorithms that take as input an architecture and a set of privacy constraints, and output an extension of the original architecture that satisfies the privacy constraints. This work is complementary to ours in that it puts restrictions on the run-time message handling. In contrast to our work, the approach does not concern analysis of program code.

In [21], Ferrara and Spoto discuss the role of static analysis for GDPR compliance. The authors suggest combining taint analyses and backward slicing algorithms to generate reports relevant for the various actors (i.e., data protection officers, chief information officers, project managers, and developers) involved at various stages of GDPR compliance. In particular, taint analysis is performed on each program statement and then the data-flow of sensitive information is reconstructed using backward-slicing. These flows are then abstracted into the information needed by the compliance actors. However, they do not formalize nor check privacy policies (as we do).

In the sense of access control mechanisms such as RBAC that controls and restricts system access to authorized users, there are some common features. In addition to the hierarchies of roles and access rights supported by RBAC, our framework introduces hierarchies of purposes to control role access. However, our work uses static analysis while RBAC uses runtime analysis. Anthonysamy et al. [22] demonstrate a *semantic-mapping* approach to infer function specifications from semantics of natural language. This technique is useful in compliance verification as it aids in identification of program constructs that implements certain policies. The authors implement this technique in a tool, CASTOR, which takes policy statements (in natural language) and source code as input, and outputs a set of semantic mappings between policies and function specifications (function name, associated class, parameters etc.).

7 Conclusion

We have investigated challenges and opportunities in approaching privacy from the *by-design* perspective, i.e., embedding privacy design requirements into a language. We have considered a small core language supporting active objects, and extended it to integrate privacy policies. We chose three primary constituents of a privacy policy, i.e., *principal, purpose*, and *access right.* Policies are declared for methods and data types, and together restrict the usage of sensitive data.

We defined a language for formulating these policies, discussed static privacy polices, and formalized a concept of static privacy policies. We have formulated rules for policy compliance, given by an extended effect system. The problem of checking a program's compliance with privacy policies, reduces to efficient type-checking. The analysis is class-wise, which is a benefit in open object-oriented systems, and for scalability. Needless to mention that much work needs to be done, in terms of defining possibly new constructs and abstractions in order to formalize the essential data protection principles. In the future we would like to *(i)* extend the policy definition language, to express a wider range of privacy restrictions, *(ii)* work out a larger case study, and *(iii)* in particular focus on the dynamic policy and consent management.

Acknowledgments. The work is supported by the projects IoTSec no. 248113 and by SCOTT no. 283085. (its-wiki.no/wiki/IoTSec:Home and www.scott-project.eu).

References

1. Danezis, G., et al.: Privacy and data protection by design-from policy to engineering. arXiv preprint arXiv:1501.03726 (2015)
2. European Parliament and Council of the European Union: The General Data Protection Regulation (GDPR). https://eur-lex.europa.eu/eli/reg/2016/679/oj. Accessed 12 Dec 2019
3. Tokas, S., Owe, O., Ramezanifarkhani, T.: Static Checking of GDPR-Related Privacy Compliance for Object-Oriented Distributed Systems (2020, under review)
4. Hayati, K., Abadi, M.: Language-based enforcement of privacy policies. In: Martin, D., Serjantov, A. (eds.) PET 2004. LNCS, vol. 3424, pp. 302–313. Springer, Heidelberg (2005). https://doi.org/10.1007/11423409_19
5. Nierstrasz, O.: A tour of Hybrid - a language for programming with active objects. In: Advances in Object-Oriented Software Engineering, pp. 67–182. Prentice-Hall (1992)
6. Hewitt, C., Bishop, P., Steiger, R.: A universal modular ACTOR formalism for artificial intelligence. In: Proceedings of the Third International Joint Conference on Artificial Intelligence, IJCAI 1973, pp. 235–245. Morgan Kaufmann Publishers Inc. (1973)
7. Johnsen, E.B., Owe, O.: An asynchronous communication model for distributed concurrent objects. Softw. Syst. Model. **6**, 39–58 (2007)
8. Ramezanifarkhani, T., Owe, O., Tokas, S.: A secrecy-preserving language for distributed and object-oriented systems. J. Logic. Algebraic Methods Program. **99**, 1–25 (2018)
9. Tokas, S., Owe, O.: A formal framework for consent management. In: Proceedings of the 31st Nordic Workshop on Programming Theory, NWPT 2019, November 2019. https://doi.org/10.23658/taltech.nwpt/2019. ISBN 978-9949-83-520-1
10. Nielson, F., Nielson, H.R.: Type and effect systems. In: Olderog, E.-R., Steffen, B. (eds.) Correct System Design: Recent Insights and Advances. LNCS, vol. 1710, pp. 114–136. Springer, Heidelberg (1999). https://doi.org/10.1007/3-540-48092-7_6
11. Gürses, S., Troncoso, C., Diaz, C.: Engineering privacy by design reloaded. In: Amsterdam Privacy Conference, pp. 1–21 (2015)
12. Notario, N., et al.: PRIPARE: integrating privacy best practices into a privacy engineering methodology. In: 2015 IEEE Security and Privacy Workshops, pp. 151–158. IEEE (2015)
13. Hoepman, J.-H.: Privacy design strategies. In: Cuppens-Boulahia, N., Cuppens, F., Jajodia, S., Abou El Kalam, A., Sans, T. (eds.) SEC 2014. IAICT, vol. 428, pp. 446–459. Springer, Heidelberg (2014). https://doi.org/10.1007/978-3-642-55415-5_38

14. Colesky, M., Hoepman, J.-H., Hillen, C.: A critical analysis of privacy design strategies. In: 2016 IEEE Security and Privacy Workshops (SPW), pp. 33–40 (2016)
15. Le Métayer, D.: Formal methods as a link between software code and legal rules. In: Barthe, G., Pardo, A., Schneider, G. (eds.) SEFM 2011. LNCS, vol. 7041, pp. 3–18. Springer, Heidelberg (2011). https://doi.org/10.1007/978-3-642-24690-6_2
16. Schneider, G.: Is privacy by construction possible? In: Margaria, T., Steffen, B. (eds.) ISoLA 2018. LNCS, vol. 11244, pp. 471–485. Springer, Cham (2018). https://doi.org/10.1007/978-3-030-03418-4_28
17. Tschantz, M.C., Wing, J.M.: Formal methods for privacy. In: Cavalcanti, A., Dams, D.R. (eds.) FM 2009. LNCS, vol. 5850, pp. 1–15. Springer, Heidelberg (2009). https://doi.org/10.1007/978-3-642-05089-3_1
18. Basin, D., Debois, S., Hildebrandt, T.: On purpose and by necessity: compliance under the GDPR. Proc. Financ. Cryptogr. Data Secur. **18**, 20–37 (2018)
19. Arfelt, E., Basin, D., Debois, S.: Monitoring the GDPR. In: Sako, K., Schneider, S., Ryan, P.Y.A. (eds.) ESORICS 2019. LNCS, vol. 11735, pp. 681–699. Springer, Cham (2019). https://doi.org/10.1007/978-3-030-29959-0_33
20. Adams, R., Schupp, S.: Constructing independently verifiable privacy-compliant type systems for message passing between black-box components. In: Piskac, R., Rümmer, P. (eds.) VSTTE 2018. LNCS, vol. 11294, pp. 196–214. Springer, Cham (2018). https://doi.org/10.1007/978-3-030-03592-1_11
21. Ferrara, P., Spoto, F.: Static analysis for GDPR compliance. In: Proceedings of the Second Italian Conference on Cyber Security, Milan, No. 2058 in CEUR Workshop Proceedings (2018). http://ceur-ws.org/Vol-2058/paper-10.pdf
22. Anthonysamy, P., Edwards, M., Weichel, C., Rashid, A.: Inferring semantic mapping between policies and code: the clue is in the language. In: Caballero, J., Bodden, E., Athanasopoulos, E. (eds.) ESSoS 2016. LNCS, vol. 9639, pp. 233–250. Springer, Cham (2016). https://doi.org/10.1007/978-3-319-30806-7_15

Law, Ethics and AI

Aid and AI: The Challenge of Reconciling Humanitarian Principles and Data Protection

Júlia Zomignani Barboza(✉), Lina Jasmontaitė-Zaniewicz, and Laurence Diver

Vrije Universiteit Brussel, Brussels, Belgium
julia.zomignani.barboza@vub.be

Abstract. Artificial intelligence systems have become ubiquitous in everyday life, and their potential to improve efficiency in a broad range of activities that involve finding patterns or making predictions have made them an attractive technology for the humanitarian sector. However, concerns over their intrusion on the right to privacy and their possible incompatibility with data protection principles may pose a challenge to their deployment. Furthermore, in the humanitarian sector, compliance with data protection principles is not enough, because organisations providing humanitarian assistance also need to comply with humanitarian principles to ensure the provision of impartial and neutral aid that does not harm beneficiaries in any way. In view of this, the present contribution analyses a hypothetical facial recognition system based on artificial intelligence that could assist humanitarian organisations in their efforts to identify missing persons. Recognising that such a system could create risks by providing information on missing persons that could potentially be used by harmful actors to identify and target vulnerable groups, such a system ought only to be deployed after a holistic impact assessment has been made, to ensure its adherence to both data protection and humanitarian principles.

Keywords: Humanitarian action · Artificial intelligence · Facial recognition · Data protection · Humanitarian principles

1 Introduction

The use of artificial intelligence (hereafter AI) and biometrics systems is no longer the preserve of science fiction. On the contrary, a combination of advances in computing power and the vast amounts of data being generated by Internet-connected devices means that AI systems are now a truism in our everyday lives [1]. These systems are present, for example, in voice-activated digital assistants, biometric and facial recognition systems that unlock smartphones or allow access to buildings, traffic routing applications, purchase or viewing recommendations on online platforms, and many other features of smart devices. It is not surprising, therefore, to see that the humanitarian sector is also exploring how AI and biometrics tools can be applied to further the provision of humanitarian aid.

Published by Springer Nature Switzerland AG 2020
M. Friedewald et al. (Eds.): Privacy and Identity 2019, IFIP AICT 576, pp. 161–176, 2020.
https://doi.org/10.1007/978-3-030-42504-3_11

The use of AI and biometrics in the humanitarian sector, as in any other field, comes with many challenges. Some of these are specific to AI, while others are inextricably linked with the use of biometrics, particularly in relation to the protection of personal data. The most frequently cited challenge of AI systems relates to machine bias, exemplified by the controversial COMPAS algorithm used in the US to predict recidivism rates in criminal cases in order to assist judges in determining bail. The algorithm attracted criticism on the basis that it predicts black defendants as being almost twice as likely to reoffend as white defendants [2]. These predictions can be considered biased, depending on the technique for measuring fairness that is adopted [3]. Furthermore, the use of AI can have serious implications for individuals' rights to privacy and personal data protection, enshrined in international law,[1] especially considering the systems' increasing "capability of linking data or recognising patterns of data [that] may render non-personal data identifiable" [1; p. 11].

Apart from these general concerns about AI, its use in the humanitarian sector must also be reconciled with the humanitarian principles that govern how humanitarian organisations are supposed to act. In particular, organisations ought to comply with the principles of humanity and impartiality, of 'do no harm', of accountability, and the principled goals both of facilitating participation on the part of humanitarian beneficiaries and of building on local capacities [4].

Complying with such principles when deploying AI is particularly challenging given that humanitarian organisations usually lack the technical knowledge and resources to develop their own AI and biometric systems and must, therefore, rely on partnerships with private, for-profit technology companies. Such partners may have incentives which are incompatible with the humanitarian principles and which may in turn cause reputational damage to humanitarian organisations. As an illustrative example, the World Food Program (WFP) recently partnered with controversial data analytics firm Palantir to improve its food delivery and cash-based assistance programs. The WFP was heavily criticised as a result, leading it to issue a defensive public statement explaining how the partnership complied with the organisation's principles [5].

The use of AI systems in humanitarian aid raises numerous complementary issues from both the humanitarian and data protection perspectives, not least as to whether such use is compatible with either the principles pertaining to humanitarian assistance or the data protection principles implicated by the use of such technologies. Bearing these in mind, the challenge is how to deploy AI systems that comply with both data protection requirements and humanitarian principles. Taking into account the limited space available and the broadness of the field, we focus on one specific AI application, namely the use of facial recognition to identify missing persons.

The next section provides a brief overview of the evolution of humanitarian action, followed by Sect. 3 which analyses the use of AI and biometrics in the humanitarian sector. Sections 4 and 5 consider the challenges for privacy and data protection and the implementation of humanitarian principles, respectively.

[1] See Article 12 of the Universal Declaration of Human Rights, Article 17 of the International Covenant on Civil and Political Rights, and the Council of Europe Convention for the protection of individuals with regard to automatic processing of personal data, no. 108.

1.1 Terminology

For the purposes of this paper, we consider **artificial intelligence** to be "[a] set of sciences, theories and techniques whose purpose is to reproduce by a machine the cognitive abilities of a human being." [6] It includes the development of algorithms that improve their performance when completing a certain task with experience in the form of machine-readable data. The latter is usually referred to as **machine learning**, a subset of artificial intelligence more generally – this is the form of AI we focus on in this paper. We use the terms **humanitarian action**, **humanitarian assistance** and **humanitarian aid** interchangeably to refer to "any activity undertaken on an impartial basis to carry out assistance, relief and protection operations in response to a Humanitarian Emergency." [7; p. 8].

2 The Evolution of Humanitarian Action

Traditionally, humanitarian action was carried out by only a handful of dedicated organisations. In the present day, however, humanitarian action is provided by numerous actors, including states, international organisations, non-governmental organisations (NGOs), private companies, and private and technology-related philanthropic foundations (e.g. The Bill & Melinda Gates Foundation) [8; p. 7]. This shift toward the involvement of multiple parties has been partially driven by the increasing use of digital technologies to speed up the delivery of aid. The strive for (technology-based) efficiency has drawn in a range of new actors, e.g. mobile network operators and financial institutions involved in cash transfer programmes. This has increased the complexity of governing humanitarian action and has created more intricate data flows that include non-personal and personal data concerning both beneficiaries and humanitarian staff.

The presence of multiple actors in humanitarian action is not problematic per se, but the lack of a 'core set of shared values' can make such cooperation challenging [8], especially when the processing of beneficiaries' personal data is involved. To address these challenges, leading humanitarian organisations (e.g. the International Federation of Red Cross and Red Crescent Societies (IFRC), the International Committee of the Red Cross, and the United Nations High Commissioner for Refugees (UNHCR)) as well as smaller actors in the field (e.g. Terre des Hommes) have developed internal and external privacy policies. Many of these policies and their more recent revisions are intended to align such instruments with principles stemming from the EU's General Data Protection Regulation 2016/679 (GDPR). For example, the revised IFRC privacy policy allows data subjects to object to data processing, and provides for the reasonable expectation of notification should their personal data be disclosed to an unauthorised third party. It also foresees that in situations where "processing operations appear likely to result in a high risk to the rights or freedoms of a data subject", a data protection impact assessment should be carried out.

Taking into account the plurality of actors involved in humanitarian action and the fact that not all of them stand on the same footing, some organisations, such as the Harvard Humanitarian Initiative (HHI), have put forward guidelines and codes of ethics for those providing humanitarian assistance that are based largely on the principles of the

EU data protection framework.[2] Building on these observations it might be said that the GDPR represents the 'gold standard' for data protection, not least in the humanitarian sector. This is in spite of the fact that many humanitarian organisations are, due to their status in international law, not bound by the GDPR or any other national or regional data protection legislation. It should be noted that many of these organisations are based in Western countries that fall within the EU's sphere of influence and, therefore, the choice to follow EU standards may not be entirely neutral (this phenomenon is sometimes referred to as the 'Brussels effect' [9]). Regardless of the exact reasons behind the extraterritorial influence of the GDPR, it has in practice become the standard for data protection and, consequently, we rely on the principles and definitions that it sets out.

3 AI and Biometrics in Humanitarian Action

The use of AI and biometrics is not new to humanitarian organisations. For example, AI tools are being used to identify patterns and make predictions based on the analysis of social media platforms to detect disasters and identify the needs of affected populations [10]. Systems like the Artificial Intelligence for Disaster Response (AIDR) platform use AI "to automatically identify informative content on Twitter during disasters" [11].[3] In emergencies, AI also enables the automated mapping of disaster areas using satellite and aerial images [12]. AI can also assist in diagnosing disease, as well as in the early detection of pathogens, helping to avoid outbreaks and pandemics that may occur in the context of humanitarian crises (e.g. Microsoft's Project Premonition [13]).

Biometric systems, particularly those based on fingerprints and iris scans, have been used by humanitarian organisations "as part of their identification systems because of the benefits it can bring in efficiently identifying individuals and preventing fraud and/or misuse of humanitarian aid" [7; p. 98]. Such systems are often chosen by humanitarian organisations in the belief that they are typically "more difficult to counterfeit and, being digitally produced and stored, facilitate the efficient management of humanitarian aid in the field" [7; p. 99]. Some even argue that the use of biometric systems in humanitarian action "reveals a determined humanitarian focus on the individual" [14].

Despite the perceived advantages of using AI and biometrics in humanitarian action, the combination of the two technologies to facilitate the next generation of AI-based biometric systems (including facial recognition) raises important legal and ethical concerns. When combined, the two technologies could be (re)used for purposes and in contexts that humanitarian organisations may not fully anticipate or be able to control.

In this regard, the next subsection reflects on general trends emerging in the area of AI and facial recognition, with the aim of unpacking some of the issues that these pose in the broader context.

[2] For example, the HHI Signal Code includes the Right to Privacy and Security, according to which data of affected individuals should be (i) processed fairly and lawfully, and not further processed in a way incompatible with that purpose, (ii) adequate, relevant, and not excessive in relation to that purpose, (iii) accurate and, where necessary, kept up-to-date, and (iv) not kept longer than necessary to achieve the stated purpose under which informed consent and/or participation was obtained.

[3] This platform goes beyond simple keyword search, which its makers say can fail to identify over 50% of the textual content posted on Twitter that is relevant to disaster response.

3.1 General Trends in AI-Based Facial Recognition

Facial recognition from photographs is widely used by governments and private companies, and can be performed to a high degree of accuracy.[4] Such systems allow full identification of individuals and, with the rapid pace at which the technology is developing, facial recognition is possible even from live video (such examples are seen in China) in addition to static images.

A critical literature on facial recognition is beginning to emerge in tandem with these technological advancements, with a number of arguments being developed both in favour and against the technology. Some argue that more balanced and representative training datasets can facilitate accurate recognition across racial and gender boundaries, avoiding discrimination [16], while others warn that facial recognition is inherently dangerous and ought to be banned outright for any and all purposes (see [17, 18]).

There are multiple examples of facial recognition systems being used in practice and for various purposes. Traffic Jam [19], for example, uses facial recognition to assist law enforcement to locate victims of human trafficking and was estimated to have identified 3,000 victims of sex trafficking in 2018. Samsung provides facial recognition as a mechanism both for users to unlock their smartphones [20] and, in tandem with Diebold Nixdorf, to authenticate identity at ATMs [21].

Other uses of the technology currently being explored include identifying criminals, tracking school attendance, and seamless border crossing, as well as the identification of missing persons, which is further explained in the next section.

Considering the multiple and sometimes life-saving potential uses of the technology, it is to be expected that various public and private actors would seek to invest in the development of facial recognition systems. Recent revelations show, however, that some are willing to employ questionable methods in the development of facial recognition. The Ever smartphone application for example, which offered users free storage for photos, used the uploaded images to train and improve the company's facial recognition system. The only notification users received of these activities was a short and vague statement in the company's lengthy privacy policy [22]. Similar concerns have been expressed about other photo applications such as FaceApp (see [23–25]).

Besides concerns over consent and the longstanding debate over facial recognition systems' interference with privacy (see [26–30]), the potential of the technology to profile specific ethnic groups [31] and even purportedly to identify homosexuals [32] are worrisome, particularly if such systems are relied upon despite their inaccuracies. The harms that could arise from the use of facial recognition systems impact humanitarian organisations' compliance with humanitarian principles, as explained below in Sect. 5.

[4] Taigman et al. claim, for example, that Facebook facial recognition using the Deep Face method "reaches an accuracy of 97.35% on the Labeled Faces in the Wild (LFW) dataset, reducing the error of the current state of the art by more than 27%, closely approaching human-level performance". See [15].

3.2 The Prospective Use of Facial Recognition to Identify Missing Persons in the Humanitarian Sector

As mentioned in the introduction, taking into account the limits of this paper and the broadness of the field, we focus on the use of a hypothetical AI-based facial recognition system to identify missing persons. Various public actors have already deployed such systems: examples such as India's National Tracking System for Missing & Vulnerable Children are positively regarded; the system identified nearly 3,000 missing children within four days of launching a trial that matched photos of missing children with the faces of children throughout New Delhi [33]. To achieve that result, however, the technology processed the data of 45,000 children, raising concerns about interference with their privacy. Similarly, the British police is also preparing to use facial recognition systems to identify missing people in the UK by scanning CCTV footage [34].

In the humanitarian sector, the use of AI-based facial recognition will take a different shape. Humanitarian actors such as the International Committee of the Red Cross (ICRC) have a long history of identifying the fate of missing persons, restoring contact between family members, and facilitating family reunification [35]. In view of the significant inward migration to Europe in recent years, the ICRC together with National Red Cross Societies in Europe have developed the Trace the Face program [36], where those looking for a family member manually search an online database containing photos posted by individuals who are looking to be found.

For the purposes of this paper, we predict that in the near future AI-based facial recognition systems will be used to automate this search. In practice, this would mean that instead of manually going through the database, someone looking for a family member would upload a photo of their relative into the system to try and locate their family member in the database. This would be done upon entering into a contract with a humanitarian organisation providing access to such software. The system would then map the facial features in the photo uploaded, such as the distance between the eyes and the distance from forehead to chin, thus creating a facial 'signature'. This would then be compared to a database of known faces to look for a match (i.e. a photograph that contains the same facial features) [37].

In the posited case, then, the photo of the relative uploaded by the person seeking their family member would not be published publicly, but would instead pass through the facial recognition system in order to be compared with the public photos on the database. In the next sections, we analyse the potential challenges such use of facial recognition would pose to humanitarian organisations.

4 Privacy and Data Protection Concerns

As pointed out earlier, the deployment of AI-based facial recognition systems for identifying missing persons raises privacy and data protection concerns. As is the case with any technology with disruptive potential, especially when they are intended to process large amounts of personal data, the deployment of such systems should be preceded by a data protection impact assessment (DPIA), which is used to identify, evaluate, and address the risks to individuals and their personal data arising from a specific system.

Such assessments are typically done on the basis of principles stemming from the data protection framework, and include questions such as:

- Is the processing fair and lawful?
- Are the data subjects able fully to exercise their rights? If not, are the restrictions lawful and proportionate?
- How does the system operate?
- How is responsibility attributed among the involved parties?
- What are the risks posed by the chosen system to data subjects? How will each risk be treated?

Although some steps and characteristics are common to all impact assessments, these need to be tailored to the specificity and needs of a given project and its context. Humanitarian organisations, therefore, should develop DPIA guidance (if necessary with the assistance of external advisors and support of national data protection authorities) that takes the specific characteristics of humanitarian action into account. One might ultimately envision a standard emerging from the DPIA guidance documents of various humanitarian organisations.

In this regard, it is important to note that in the humanitarian sector compliance with data protection principles alone is not sufficient because of certain contradictions embedded in them. Take, for example, the principle of data minimisation. It can be said that the principle of data minimisation opposes the principle of accuracy, requiring data to be updated [38]. By minimising the amount of data processed, the controller can end up with applications that provide discriminatory results. Similarly, it is suggested that "data minimisation is not always be able to exclude privacy violating or discriminatory results given the redlining effect." In this regard, "[d]ata minimisation not only offers no adequate solution in this respect, it might also make it difficult to assess whether a rule is indirectly discriminatory or privacy violating" [38; p. 162].

Considering this background, we suggest that such assessments should be enriched by including humanitarian principles, which we consider further below in Sect. 5. For now, we maintain our focus on privacy and data protection. More specifically, the next sections will address the principles of purpose limitation, data security and fairness, as these principles are some of the most challenging ones to implement in the application foreseen by this paper.

4.1 Purpose Limitation

According to the purpose limitation principle, data must be collected for a clear and specific purpose and cannot be further processed for any purpose unrelated to the original. When it comes to AI, it is important to note that these systems 'learn' from the data that passes through them to improve their outcomes [39]. In the case of facial recognition, an AI system will learn from every photo that passes through it to better identify facial features in future photos. Thus it can be argued that despite the original purpose being the identification of missing persons, the photos will potentially be further processed to improve the system itself.

At this point, it is important to note that AI systems can be static or dynamic.[5] The former process (personal) data only to perform the task which was assigned to them (in this case, matching photos), while the latter process data both to perform the task (i.e. matching photos) and to refine their internal model to improve accuracy.

Consequently, even if humanitarian organisations deploy off-the-shelf AI systems developed by technology companies instead of developing their own, they may still have to deal with the fact that the data passing through the system (i.e. the photos of missing persons) is used to improve it. It is, therefore, essential that humanitarian organisations are aware of which type of system they are using and what the data protection implications are.

In this regard, it is worth noting that while it could be also argued that further processing the data to improve the system for humanitarian use is compatible with the initial purpose, since a better trained system has a better chance of identifying missing persons, humanitarian organisations should also be aware of the possibility of their partners using these data for commercial purposes. For example, when purchasing an off-the-shelf system, the technology company behind the AI may request that the data collected by the humanitarian organisation be used to improve the company's non-humanitarian systems that will later be commercialised by private or governmental entities.

4.2 Data Security

As the facial recognition system we envision processes biometric data – the features of an individual's face – to identify missing persons, such data must be subjected to a high level of security. This is because biometric data that is used to identify someone is considered to be a special category of personal data deserving of enhanced protection in a variety of legal regimes. According to the GDPR, for example, biometric data, which is defined as personal data relating to the physical, physiological or behavioural characteristics of a natural person which allow or confirm the unique identification of that natural person [41], will be considered a special category of data when collected "for the purpose of uniquely identifying a natural person" [42]. Uniquely identifying someone, especially someone belonging to a vulnerable group (which is often likely to be the case for those who have gone missing during emergency situations) can lead to stigmatisation and discrimination.

In China, for example, the government is said to be using facial recognition systems to track down and target Uighur Muslims [31]. While the system foreseen by this paper would not aim at categorising members of specific minorities, attacks against such systems (mentioned below) could alter them or retrieve the biometric data that passes through them, which could later be used to identify members of certain groups. Furthermore, if technology companies manage to use data gathered by humanitarian organisations to improve their commercial systems (as foreseen above), such systems could be trained to accurately identify the traits of certain ethnicities and later be sold to

[5] Static models will not change over time and will always apply the model developed with the training data. This allows the programmer to maintain full control of the model but stops the system from refining itself over time. Dynamic models, on the other hand, avail themselves of input data to adjust to changes and refine their outputs, for more see [40; p. 10].

harmful actors that wish to target them. In this regard, it is important to note that when using biometrics,

> "[g]iven the potentially harmful consequences for the persons concerned, more stringent requirements will have to be met in the impact assessment process of any measure interfering with an individual's dignity in terms of questioning the necessity and proportionality as well as the possibilities of the individual to exercise his right to data protection in order for that measure to be deemed admissible." [43; p. 15]

When determining which security measures should be put in place, humanitarian organisations should take into consideration that harmful actors may try to conduct deliberate attacks that aim at (i) revealing information about the data that passed through the system (model inversion), (ii) undermining the utility of the system by adding noise to the input data or inserting bad data that will induce the system to misread the information or emphasise the wrong features (poisoning attack) or (iii) gaining unauthorised access to the system (backdoor attack) and modifying it after it has been trained. Such deliberate attacks can further decrease the quality of outcomes, sometimes leading to false positives and negatives. This is particularly relevant because even data that would not be publicly available (such as the photos of missing relatives uploaded to the system, as in the example above) might potentially be revealed through these types of attacks. The need for a high level of security is thus evident.

4.3 Fairness and Bias

The principle of fairness requires that all processing activities respect data subjects' interests and that data controllers take action to prevent arbitrary discrimination against individuals [40]. The risk of discriminatory bias in AI systems is widely known and may be a result of, for example, using biased datasets to train the system, systemic biases in society that are reflected in the data, or even choices of the programmer when deciding which features to assign more value to in each dataset. Without delving into the intricacies of information theory and machine learning research design, it can at a minimum be said that AI developers should assess the quality, nature and origin of the personal data used to develop the system as well as considering the potential risks to individuals and groups of using de-contextualised data, which can create de-contextualised results [44].

In the case at hand, the matching of the photo of a relative to that of a missing person in a database should not determine the fate of an individual in the same way the recidivism algorithm mentioned in the introduction did; however, when one considers that most facial recognition systems perform better with male faces than female faces and on lighter skinned faces than on darker skinned faces [16], there is a high probability that the system will not perform as well on certain minority groups, inevitably leading to mismatches. Such results are likely to exacerbate an already stressful situation for vulnerable beneficiaries who are relying on such systems to find their missing relatives. Conversely, systems that are very efficient in identifying those belonging to a certain minority can be used by harmful actors to target them, as is the concern with the Uighur in China.

It is essential therefore that those deploying such systems (considered to be data controllers) carry out frequent assessments on both the system itself and the data used to develop and improve it, in order to address any possible form of bias or discrimination [45]. The consequences of not taking such measures are not only legal, but also have an impact on humanitarian principles, as will be explained below.

5 Application of Humanitarian Principles

As mentioned above, data protection impact assessments can be used to identify risks associated with the processing of individuals' personal data, including the potential negative effects of a specific technology on a variety of fundamental rights as well as the ethical and social consequences of the data processing [46]. Considering humanitarian organisations' mandate of providing humanitarian assistance, it is essential that humanitarian principles are also considered and, therefore, included in such assessments. Furthermore, it should be noted that principles setting requirements for personal data protection and humanitarian action are complementary and reinforce each other (see [47]). Indeed, the two frameworks are built with individuals' dignity and empowerment at their core. In the following sub-sections we elaborate on some the most topical humanitarian principles. In this regard, humanity and impartiality are the core two principles of humanitarian action. They embody the idea that humanitarian action should aim to prevent and alleviate human suffering wherever it may be found, and that it should be provided to anyone in need, regardless of nationality, race, religious beliefs, class, or political opinion [4].

In this regard, the concept of discriminatory bias in AI is once again relevant. Studies have shown that facial recognition systems perform better in the population of the region where they were developed [16]. Considering that many of these systems are developed in Western countries, this is one of the explanations of why they tend to perform better on light skinned males (see above in Sect. 4.2). Databases such as the ICRC's Trace the Face are based in Europe and, most likely, will make use of Western-developed systems, even though the database itself contains photos from different regions and ethnic groups. It would be necessary therefore to scrutinise whether a facial recognition system applied in such a case is able correctly to identify those coming from other regions or belonging to certain minority groups that are underrepresented in the data.

Considering that the impartiality principle requires that humanitarian action benefits everyone in need in a non-discriminatory manner, deploying a system that is likely to benefit only those belonging to a majority Western group may not be appropriate. Furthermore, because such systems can be less accurate within minority groups, using them to identify these groups might lead to a high number of false positives or false negatives, magnifying the detrimental emotional effects on an already-vulnerable class of user.

It is also important to consider that deploying ineffective systems may negatively affect an organisation's reputation and local acceptability, hindering their access to local population and, potentially, their ability to operate in a certain area, thus obstructing the humanitarian mission.

5.1 'Do No Harm'

The 'do no harm' principle is part of humanitarian organisations' commitment to ensure "[c]ommunities and people affected by crisis are not negatively affected and are more prepared, resilient and less at-risk as a result of humanitarian action" [48; p. 59]. In other words, humanitarian organisations should not leave a negative footprint in the contexts where they act.

Organisations may face various dilemmas in which the consequences of their actions might be unclear. For example, whether they would have to share data they gather to identify missing persons with law enforcement authorities and other public or state actors. This is particularly relevant considering that if harmful actors were somehow to acquire such data it may allow them to identify vulnerable groups with the intention of harming them. The ICRC highlighted this potential issue in its Restoring Family Links (RFL) strategy for 2020–2025:

> "The sharing of potentially sensitive information on affected people with other entities, such as States, that wish to use such data for non-humanitarian purposes might expose individuals to new risks, such as profiling, discrimination, arrest or even exclusion from humanitarian assistance. This would negatively impact the safety of the very people humanitarian action is trying to help, contradict the "do no harm" principle and be incompatible with the Fundamental Principles, especially neutrality, impartiality and independence" [49; p. 5]

When making decisions around such dilemmas, it is recommended that humanitarian organisations make "sufficient enquiries about the ethics, interests, risks and professional reliability of individuals and organizations in your agency's political, commercial and humanitarian supply chain and delivery network; and acting upon information received" [4; p. 109].

5.2 Participation by Beneficiaries and Building on Local Capacity

In essence, the principle of building on local capacity implies that humanitarian organisations should act in a way that brings beneficiaries into the process as far as possible, such that disaster responses are not simply 'imposed' upon them [4; p. 82]. This principle, however, extends beyond participation. As rightly noted by former ICRC President Jakob Kellenberger, humanitarians' work should allow the re-establishment of physical and mental resilience in affected populations. In particular, it should aim to assist those affected to regain their autonomy, in order to better "cope with the shock and trauma caused" [50; p. 987]. This notion of autonomy is also present in the data protection framework, linking closely to the notion of informational self-determination [51; pp. 8–9]. We recognise that the two notions are being used in different contexts, but ultimately they each serve to respect and empower the individual. At the same time, Western individualistic notions of autonomy and rational decision-making may not be appropriate across the spectrum of geographical and cultural contexts in which humanitarian assistance is provided. At any rate, if the goal of humanitarian action is to save and protect life in order to facilitate its subsequent flourishing, it is important that this goal is not undermined by data processing that denies the individual similar opportunities. Squaring that circle may prove to be extremely challenging in practice, where AI systems that rely on personal data are used in contexts where beneficiaries have little power to contest the processing, and perhaps not even the correct frame of reference to conceptualise it.[6]

In the hypothetical system introduced above, following the rationale of the building on local capacity principle, several questions arise. One might ask to what extent the "technical wisdom that local knowledge often knows best" could be applied within the scope of such a project. Furthermore, it might be queried whether the skills and understanding of the local population are such that the application is in practice useful to them [4; p. 80]. Finally, it must be considered whether or not the affected beneficiaries can provide meaningful consent to the processing of personal data and, perhaps more importantly, whether they comprehend the risks that the processing may constitute in the case of an AI-facilitated biometric system. In the end, it may well be that respecting the principles of beneficiary participation and local capacities entails the conclusion that AI ought not to be used.

5.3 Accountability: Value for Money or Humanitarian Effectiveness

The Code of Conduct for the International Red Cross and Red Crescent Movement and NGOs in Disaster Relief [52] states that "[w]e [those who adhere to it] hold ourselves accountable to both those we seek to assist and those from whom we accept resources." This commitment can prove challenging, however, when resources come from national or regional authorities which, as mentioned above, may pressure humanitarian organisations or NGOs into sharing the biometric data sets they have collected (in this case, the photos of missing persons and facial patterns found by the system), "with the risk of the data being used for purposes other than strictly humanitarian purposes (e.g. law enforcement, security, border control or monitoring migration flows)" [7; p. 99]. In this case, being accountable to donors can compromise accountability to beneficiaries, since acquiescing to donors' requests to share data might endanger beneficiaries (e.g. in case such data is used to target minorities such as the Uighur, mentioned above).

Furthermore, beyond addressing the risks of their own activities (including the sharing of beneficiaries' data), humanitarian organisations also need to consider the actions of their technology partners, who might have divergent interests and whose commercial imperative might be at odds with the humanitarian programs they engage with, for example the improvement of their technology for later re-use in non-humanitarian contexts (recall the example of Ever above). Vulnerable groups and individuals whose data are processed for the purposes of providing aid might find themselves inducted into surveillance-capitalist paradigm that is not of their choosing, particularly where the social and legal norms of (Western) data protection law do not easily translate across cultural and developmental boundaries. These concerns are often raised when humanitarian organisations partner with the private sector, as was the case in WFP's partnership with Palantir, mentioned above. When partnering with the private sector, therefore, it is

[6] But compare [51], detailing how the WFP halted aid where local beneficiaries refused to permit their biometric data to be harvested, reportedly for reasons of sovereignty (as opposed to data protection).

essential that humanitarian organisations can clearly present to partners what is needed and what is expected in the project, especially given that technology companies might have different ways of working in terms of budget, timelines, reporting etc. Moreover, humanitarian organisations need to assess the risks of working with external partners, possibly through an impact assessment, to ensure that affected populations and their data are not endangered by the partnership.

Such risks reinforce the utmost importance of humanitarian organisations taking a precautionary approach. They should conduct thorough impact assessments before opting to use facial-recognition techniques in their programs and should use such technologies only after taking all feasible precautions to protect those they seek to help, in the humanitarian spirit that is their very *raison d' être*. In this regard, and similar to the accountability principle in data protection, accountability requires evidence. More specifically, "[a]gencies need to be able to 'know and show' what they have intended, decided, done, and the results that have flowed from their actions." [4; p. 99].

6 Outlook for Future Engagement

Compliance with data protection principles by themselves does not suffice for the legitimate deployment of new technologies by humanitarian organisations. Due to the humanitarian sector's mission to carry out assistance, relief and protection operations in an impartial manner that does not cause harm to beneficiaries, it is essential that the humanitarian principles are also complied with. Before deciding on the use of a certain technology, therefore, humanitarian organisations should conduct a holistic assessment of compliance of a specific system with both legal rules and humanitarian principles. In the case of using AI-based facial recognition to identify missing persons, the principles identified above are particularly relevant and unless all of them can be complied with, such a system should not be implemented.

Further research and analysis will be necessary to fully tease out the interplay between the humanitarian and data protection principles. However, by drawing attention to the complementariness of the two regimes, humanitarian organisations can be sensitised to the multidimensional concerns that AI-based facial recognition poses in the context of providing aid. The laudable aim of saving and protecting vulnerable human lives must not be inadvertently undermined by the use of AI systems that infringe fundamental rights and/or disrespect local culture and custom.

References

1. Centre for Information Policy Leadership: First Report: Artificial Intelligence and Data Protection in Tension (2018)
2. Angwin, J., Larson, J., Mattu, S., Kirchner, L.: Machine Bias. ProPublica (2016). https://www.propublica.org/article/machine-bias-risk-assessments-in-criminal-sentencing. Accessed 2 Aug 2019
3. Corbett-Davies, S., Pierson, E., Feller, A., Goel, S.: A computer program used for bail and sentencing decisions was labeled biased against blacks. It's actually not that clear. Washington Post (2016). https://www.washingtonpost.com/news/monkey-cage/wp/2016/10/17/can-an-algorithm-be-racist-our-analysis-is-more-cautious-than-propublicas/. Accessed 23 Oct 2019

4. Slim, H.: Humanitarian Ethics - A Guide to the Morality of Aid in War and Disaster, 1st edn. Oxford University Press, New York (2015)
5. World Food Program: A statement on the WFP-Palantir partnership (2019). https://insight.wfp.org/a-statement-on-the-wfp-palantir-partnership-2bfab806340c. Accessed 2 Aug 2019
6. Council of Europe, Glossary on Artificial Intelligence: https://www.coe.int/en/web/artificial-intelligence/glossary. Accessed 5 Aug 2019
7. Kuner, C., Marelli, M.: Handbook on Data Protection in Humanitarian Action, 1st edn. International Committee of the Red Cross, Geneva (2017)
8. Scott, R.: Imagining more effective humanitarian aid a donor perspective. 18 OECD Development Co-operation Directorate 34 (2014)
9. Bradford, A.: The Brussels Effect. Northwest. Univ. Law Rev. **107**(1), 1–68 (2012)
10. Hashtag Standards for Emergencies (2014). https://www.unocha.org/sites/unocha/files/Hashtag%20Standards%20for%20Emergencies.pdf. Accessed 2 Aug 2019
11. iRevolutions: AIDR: Artificial Intelligence for Disaster Response (2013). https://irevolutions.org/2013/10/01/aidr-artificial-intelligence-for-disaster-response/. Accessed 2 Aug 2019
12. Meier, P.: Digital Humanitarians: How Big Data Is Changing the Face of Humanitarian Response, 1st edn. CRC Press, Boca Raton (2015)
13. Project Premonition website. https://www.microsoft.com/en-us/research/project/project-premonition/. Accessed 2 Aug 2019
14. Slim, H.: Eye Scan Therefore I am: The Individualization of Humanitarian Aid. European University Institute (2015). https://iow.eui.eu/2015/03/15/eye-scan-therefore-i-am-the-individualization-of-humanitarian-aid/. Accessed 2 Aug 2019
15. Taigman, Y., Yang, M., Ranzato, M.A., Wold, F..: DeepFace: closing the gap to human-level performance in face verification. In: 2014 IEEE Conference on Computer Vision and Pattern Recognition, pp. 1701–1708. IEEE, Columbus (2014). https://research.fb.com/wp-content/uploads/2016/11/deepface-closing-the-gap-to-human-level-performance-in-face-verification.pdf. Accessed 23 Oct 2019
16. Buolamwini, J., Gebru, T.: Gender shades: intersectional accuracy disparities in commercial gender classification. Proc. Mach. Learn. Res. **81**, 1–15 (2018)
17. Hartzog, W.: Facial Recognition Is the Perfect Tool for Oppression. Medium (2018). https://medium.com/s/story/facial-recognition-is-the-perfect-tool-for-oppression-bc2a08f0fe66. Accessed 27 May 2019
18. Golumbia, D.: Do You Oppose Bad Technology, or Democracy? Medium (2019). https://medium.com/@davidgolumbia/do-you-oppose-bad-technology-or-democracy-c8bab5e53b32. Accessed 23 Oct 2019
19. Traffic Jam website. https://www.marinusanalytics.com/traffic-jam. Accessed 2 Aug 2019
20. Samsung: Use Facial recognition security on your Galaxy phone. https://www.samsung.com/us/support/answer/ANS00062630/. Accessed 2 Aug 2019
21. Samsung: Diebold ATM Samsung SDS Nexsign – Digital Banking (2017). https://www.samsungsds.com/global/en/about/news/1196788_1373.html. Accessed 2 Aug 2019
22. Vincent, J.: A photo storage app used customers' private snaps to train facial recognition AI – Photo app Ever pivoted its business without informing users. The Verge (2019). https://www.theverge.com/2019/5/10/18564043/photo-storage-app-ever-facial-recognition-secretly-trained-ai. Accessed 2 Aug 2019
23. Collie, M.: 'Just walk away from it': The scary things companies like FaceApp can do with your data. GlobalNews (2019). https://globalnews.ca/news/5653531/faceapp-data-mining/. Accessed 5 Aug 2019
24. Carman, A.: FaceApp is back and so are privacy concerns. The Verge (2019). https://www.theverge.com/2019/7/17/20697771/faceapp-privacy-concerns-ios-android-old-age-filter-russia. Accessed 5 Aug 2019

25. Olavario, D.: FaceApp: are security concerns around viral app founded? Euronews (2019). https://www.euronews.com/2019/07/17/faceapp-are-security-concerns-around-viral-app-founded-thecube. Accessed 5 Aug 2019
26. Milligan, C.S.: Facial recognition technology, video surveillance, and privacy. South. Calif. Interdiscip. Law J. **9**, 295–334 (1999)
27. Bowyer, K.W.: Face recognition technology: security versus privacy. IEEE Technol. Soc. Mag. **23**(1), 9–19 (2004)
28. Privacy International: The police are increasingly using facial recognition cameras in public to spy on us (2019). https://privacyinternational.org/feature/2726/police-are-increasingly-using-facial-recognition-cameras-public-spy-us. Accessed 1 Aug 2019
29. Frew, J.: How Facial Recognition Search Is Destroying Your Privacy. Make Use Of (2019). https://www.makeuseof.com/tag/facial-recognition-invading-privacy/. Accessed 1 Aug 2019
30. Curran, D.: Facial recognition will soon be everywhere. Are we prepared? The Guardian (2019). https://www.theguardian.com/commentisfree/2019/may/21/facial-recognition-privacy-prepared-regulation. Accessed 1 Aug 2019
31. Mozur, P.: One Month, 500,000 Face Scans: How China Is Using A.I. to Profile a Minority. The New York Times (2019). https://www.nytimes.com/2019/04/14/technology/china-surveillance-artificial-intelligence-racial-profiling.html. Accessed 2 Aug 2019
32. Lewis, P.: 'I was shocked it was so easy': meet the professor who says facial recognition can tell if you're gay. The Guardian (2018). https://www.theguardian.com/technology/2018/jul/07/artificial-intelligence-can-tell-your-sexuality-politics-surveillance-paul-lewis. Accessed 2 Aug 2019
33. Cuthbertson, A.: Indian Police Trace 3,000 Missing Children in Just Four Days Using Facial Recognition Technology. Independent (2018). https://www.independent.co.uk/life-style/gadgets-and-tech/news/india-police-missing-children-facial-recognition-tech-trace-find-reunite-a8320406.html. Accessed 2 Aug 2019
34. Bernal, N.: Facial recognition to be used by UK police to find missing people. The Telegraph (2019). https://www.telegraph.co.uk/technology/2019/07/16/facial-recognition-technology-used-uk-police-find-missing-people/. Accessed 2 Aug 2019
35. Restoring Family Links website. https://familylinks.icrc.org/en/Pages/home.aspx. Accessed 2 Aug 2019
36. Trace the Face – Migrants in Europe website. https://familylinks.icrc.org/europe/en/Pages/Home.aspx. Accessed 2 Aug 2019
37. Norton: How does facial recognition work? https://us.norton.com/internetsecurity-iot-how-facial-recognition-software-works.html. Accessed 5 Aug 2019
38. Gellert, R.: Understanding the risk based approach to data protection: an analysis of the links between law, regulation, and risk. Ph.D. thesis, Vrije Universiteit Brussel (2017)
39. Burrell, J.: How the machine "thinks": understanding opacity in machine learning algorithms. Big Data Soc. 1–12 (2016)
40. The Norwegian Data Protection Authority: Artificial intelligence and privacy (2018). https://www.datatilsynet.no/globalassets/global/english/ai-and-privacy.pdf. Accessed 2 Aug 2019
41. GDPR: Art. 4(14)
42. GDPR: Art. 9
43. Article 29 Data Protection Working Party: Opinion 3/2012 on developments in biometric technologies (2012). https://www.pdpjournals.com/docs/87998.pdf. Accessed 2 Aug 2019
44. Council of Europe: Guidelines on artificial intelligence and data protection (2019). https://rm.coe.int/guidelines-on-artificial-intelligence-and-data-protection/168091f9d8. Accessed 2 Aug 2019
45. Article 29 Data Protection Working Party: Guidelines on Automated individual decision-making and Profiling for the purposes of Regulation 2016/679 (2018). https://ec.europa.eu/newsroom/article29/item-detail.cfm?item_id=612053. Accessed 2 Aug 2019

46. Mantelero, A.: Artificial intelligence and big data: a blueprint for a human rights, social and ethical impact assessment. Comput. Law Secur. Rev. **34**(4), 754–772 (2018)
47. Kuner, C., Svantesson, D.J.B., Cate, F.H:, Lynskey, O., Millard, C.: Data protection and humanitarian emergencies. Int. Data Priv. Law **7**(3), 147–148 (2017)
48. Sphere: The Sphere Handbook – Humanitarian Charter and Minimum Standardsin Humanitarian Response (2018). https://spherestandards.org/handbook/editions/. Accessed 2 Aug 2019
49. ICRC: Restoring Family Links and Data Protection – background document (2019). https://rcrcconference.org/app/uploads/2019/06/33IC-RFL-background-document_en.pdf. Accessed 2 Aug 2019
50. McGoldrick, C.: The future of humanitarian action: an ICRC perspective. Int. Rev. Red Cross **93**(884), 965–991 (2011)
51. Martin, A., Taylor, L.: Biometric Ultimata — what the Yemen conflict can tell us about the politics of digital ID systems. Global Data Justice (2019). https://globaldatajustice.org/2019-06-21-biometrics-WFP/
52. Code of Conduct. https://media.ifrc.org/ifrc/who-we-are/the-movement/code-of-conduct/. Accessed 5 Aug 2019

Get to Know Your Geek: Towards a Sociological Understanding of Incentives Developing Privacy-Friendly Free and Open Source Software

Oğuz Özgür Karadeniz[1] and Stefan Schiffner[2(✉)]

[1] KU Leuven, Leuven, Belgium
oguzozgur.karadeniz@kuleuven.be
[2] Université du Luxembourg, Esch sur Alzette, Luxembourg
Stefan.Schiffner@uni.lu

Abstract. In this paper we sketch a road map towards a sociological understanding of software developers' motivations to implement privacy features. Although there are a number of studies concerning incentives for developers, these accounts make little contribution to a comprehensive sociological understanding, as they are either based on a simplistic view of FOSS development in terms of altruism vs. utilitarianism, or are focused on individual psychological factors, leaving room for research that takes into account the complex social context of FOSS development. To address this gap, we propose a mixed methods approach, incorporating the strengths of qualitative and quantitative techniques for a comprehensive understanding of FOSS development as a social field. We then sketch how we envision developing a game theoretic approach based on the gathered data to analyze the situation in the field with respect to privacy features and propose relevant changes in policy and best practices.

1 Introduction

Motivation. The GDPR [1], and the European Commission's Ethics Guidelines for Trustworthy AI [3] indicate the importance of data privacy in contemporary society. With GDPR, developers are now legally obliged to implement Privacy by Design (PbD), or in the lingua of the legal text, data protection by design, in their development process. However, privacy breaches continue to occur, despite policy initiatives and guidelines that legally incentivize or nudge developers towards privacy-friendly features. In fact, as pointed out by ENISA's threat landscape [17], threats continue to rise as the landscape becomes more and more complex.

Moreover, as pointed in the Ethics Guidelines for Trustworthy AI [3], legal compliance, e.g. with the GDPR, does not guarantee that a system meets any ethical standards. In order to meet privacy standards, software developers need

Published by Springer Nature Switzerland AG 2020
M. Friedewald et al. (Eds.): Privacy and Identity 2019, IFIP AICT 576, pp. 177–189, 2020.
https://doi.org/10.1007/978-3-030-42504-3_12

to move beyond checklist compliance towards a continuous critical approach concerning the product/service and its social impact. Hence software developers need to continually prioritize and reassess the implementation of secure features, rather than a superficial compliance to law.

In this paper, we seek to pave the way for a better sociological understanding of software developers' motivations when taking implementation decisions for privacy-friendly and secure software. In order to reduce direct monetary and other external motives, we focus on free and open source software development (FOSS). In the first part of the paper, we present a literature review indicating the lack and importance of a sociological understanding of this issue. Following this, we offer a methodological approach based on mixed methods, which is not only appropriate for gaining a better understanding of FOSS developers' motives, but also for bringing about positive social transformation.

Background. FOSS development represents an immense contribution to technology and, through this, to society. Open source software has often contributed to more secure and privacy-friendly software either as its main goal, such as the development of PGP[1] for email encryption and Tor[2] for anonymity protection, or by setting high standards, such as the Linux kernel[3], and the Firefox web browser[4]. Some open source software has been developed and maintained for decades, and has thus become a practical alternative for commercial products. At other times the contribution has been more indirect, as in cases where the success of certain superior aspects of open source products led to improvements in the same direction in closed source competitors, such as end-to-end encryption for messenger apps.

Many open source licensing models allow commercial applications to make use of them. This is arguably a double-edged sword, since it adds a potential layer of vulnerability to every service that uses them. As code is open to the public in FOSS projects, it is easier to find vulnerabilities, but due to their dependence on voluntary contributions, maintenance is often more difficult, potentially exposing applications and services that use them. On the other hand, FOSS developers also have the potential to facilitate the development of privacy-friendly and secure software, such as programs to identify vulnerabilities and assess risks [20], or by setting high standards as mentioned above. Finally, open source software usually comes with 'no warranty' despite popular belief in their superiority in this respect, which can be erroneous at times [8,15]. This puts FOSS in a very critical role in terms of privacy and security, since, although it represents a very important contribution and potential improvement, its perceived security can be misleading given the reluctance among developers [5,32] to implement PbD.

Our Contribution. In this paper, we argue that in order to bring about social transformation with regards to the production of privacy-friendly and secure

[1] https://www.openpgp.org.
[2] https://www.torproject.org.
[3] https://www.kernel.org.
[4] https://www.mozilla.org/.

software and services, a more formal model of the motivations of FOSS developers is necessary. We detail the steps towards a model that allows for a better sociological understanding of the issue, which can be used to render advice for policy changes. Firstly, we point to the areas where legal incentives fell short of preventing breaches or the development of needlessly privacy invasive services. Secondly, we present the literature on developer motivations to highlight the need for a sociological understanding of the motivations of FOSS developers. We will then offer our own methodological approach for a sociological inquiry into this problem based on mixed methods, combining the strengths of different quantitative and qualitative techniques for a more comprehensive view of the cultural context being investigated [16]. Following this, we describe how game theory can be used to construct a model that draws from the findings of a sociological research into this field. Finally, we offer a thought experiment that illustrates our approach.

2 Related Work

FOSS Development and Developer Motivation. In his seminal work "The Cathedral and the Bazaar" [28], Raymond compares the FOSS development scene to a crowded bazaar where individuals with multiple underlying agendas interact. According to him, the gift culture that characterizes the motivations of FOSS developers is ultimately self-interested rather than altruistic. Despite competition and the self-interest of the participants, FOSS development culture results in a harmonious system and better and more secure software than commercial development [23,28]. Raymond's work can be considered as a step towards an understanding of FOSS as a social field with its own relatively autonomous dynamics, as it identifies motivations and goals in FOSS development other than altruism, such as recognition and gaining rank within the community. However, as Lin argues, Raymond's account is a long way from capturing the diversity of the field that it acknowledges or addressing how individual efforts can work harmoniously to develop sophisticated software within this complexity [23].

Raymond's work had a tremendous influence on academic and popular discourses on FOSS. The work led to the popular belief that open source is in many ways superior to commercial, epitomized by the phrase "Given enough eyeballs, all bugs are shallow", commonly attributed to Linus Torvalds [8]. Coining the term "vulgar Raymondism", Bzrukov severely criticizes this idealization for ignoring many problematic aspects of FOSS development ranging from developer burnout, hyper-inflated egos among developers and disagreements within the community [7,8].

With the influence of Raymond's work, many researchers in the field tried to situate motivations and incentives in the field in terms of self-interest and altruism. For example, Lerner and Tirole [21,22] formulate motivations underlying contribution to open source in terms of career advancement and reputation. According to them, contribution to open source is connected with an expectation of delayed return in career and economic terms due to gain in reputation and an associated ego boost. Bonaccorsi and Cristina also mention a combination of altruistic or self-interested motives, but they also find that these motivations differ between firms and individuals. According to their survey, individual programmers have a tendency to have social and altruistic motives, such as contribution to the movement and belief in non-proprietary code [9]. Others such as Hars and Ou, and Roberts et al. differentiate between intrinsic and extrinsic motivations, as they associate intrinsic motivations with societal contribution and extrinsic motivations with economic and reputational gains [4,6,29].

While the research Raymond's work inspired sought to shed more light on motivations to participate in open source, research focusing on motivations in implementations of PbD features is more sparse. More recent work by Spiekermann [32] and Bednar et al. [5] further indicate the need for research and intervention in the field by documenting the reluctance to implement PbD. Using both qualitative and quantitative data, their findings indicate organizational and individual components to the low motivation for ethical system development and implementing privacy features [5,32].

FOSS and Non-monetary Incentives. Lin points out that the reductionist instrumentalist approach to developer motivations is inadequate, since FOSS development is not characterized by a universal activity: the actors in FOSS development assume different identities (e.g. leader, follower, developer, user, hacker) and competitive and cooperative strategies that change between different social circles [23]. Moreover, as Gabriella Coleman notes, individual motivations and ethical commitments do not remain constant, but change over the course of projects, adding another level of complexity to the mosaic of ethical and motivational parameters [15]. In addition, there are other actants in the field such as corporations, NGOs, legal administrative bodies and platforms such as GitHub[5] and SourceForge[6] which further complicate matters.

Following from these criticisms, it can be argued that an account of developer motivations needs to move beyond the assumption that monetary and legal incentives will work universally as sources of motivation, and take into account the diverse social contexts of FOSS developers and sources of motivation specific to them. In this regard, following Bourdieu [14], recognizing the non-monetary forms of capital in the field of FOSS development can be valuable in understanding motivations. For example, recognition and reputation, which are forms of symbolic capital, have been referred as an alternative to monetary capital by Roberts and other researchers following him [4,6,29]. This type of symbolic capital is not necessarily antagonistic to commercial interests in FOSS development[7], since the reputation and experience gained in open source projects are reasonably aligned with developers' career progression goals. However, this brings the pitfall of adopting an overly reductionist approach which assumes that every decision in the field can be explained in terms of career goals and

[5] https://github.com.
[6] https://sourceforge.net.
[7] Unlike, for example, *avant-garde* art. See Bourdieu [13].

ignoring the rest of the cultural aspects of FOSS. As, e.g., Gabriella Coleman argues in Coding Freedom [15], participation in FOSS development and hacking has important ethical and political aspects. Similarly, Linus Torvalds's autobiography 'Just for Fun' [33] can be seen as a narrative on how 'fun' can be considered a source of motivation in FOSS. Bednar et al. [5] find that developers describe implementation of privacy features using terms such as 'inconvenient', 'not pleasing', 'challenging', 'nightmare', or 'interesting', 'exciting' and 'satisfying' respectively. As these findings show, the forms of capital and sources of motivation in software development involve diverse elements, including symbolic capital (reputation, recognition), politics, ethics, excitement, convenience and fun.

3 Research Agenda

As stated in the previous sections, the field of FOSS development has many types of non-monetary rewards, incentives, values and goals related to developers' motivation. Due to this diversity and dependence on social context, an inquiry into motivations in this field needs to address these diverse and field-specific parameters. The research we propose hopes to bring about a social change towards the adoption of PbD principles in open source developer communities, based on a theoretical framework combining game theory and field theory. Our work examines FOSS development as a relatively autonomous field, with its specific rules, goals, values, and forms of capital, in order to understand the motivations of FOSS developers in implementing software features, and to effectively model objective functions based on them to bring about change towards a more favorable equilibrium.

3.1 Model Assumptions

Game theory assumes an individual that develops rational strategies consisting of actions oriented towards a goal [30]. In sociology, the notion of rational individual is used not as an accurate and realistic representation, but as a simplification that allows for accounting for complex mechanisms [2]. Despite this acknowledgment, this assumption is criticized for being too narrow, and rarely coinciding with social reality, ultimately defeating its purpose [2].

Goldthorpe [19] (as cited in Glaesser [18]), who favors rational action theory over the concept of habitus for large-scale data, acknowledges that actors do not act purely on rational terms in individual cases. He puts forth a notion of subjective rationality based on the idea that

> [...] actors may hold beliefs, and in turn pursue courses of action, for which they have 'good reasons' in the circumstances in which they find themselves, even though they may fall short of the standard of rationality that utility theory would presuppose. [18]

This view contextualizes actors' courses of action in terms of their social circumstances, which also includes the information available to them. As actors do not necessarily have the resources or disposition to access the 'best information imaginable', they usually act on the available information [18]. They may be aware or unaware of their lack of information or the costs of accessing additional information and thus act on incomplete information [18].

The assumption of rational choice is similarly criticized within game theory since rational action is sometimes dependent on extra-game circumstances including 'history of past interactions, existing customs and practices, or contingently salient features of the particular instantiation of the game' [30]. This echoes the criticisms from sociology, as it implies that a simplified and universalized model of an individual acting optimally on the best information ignores the diversity of individual and social contexts.

The complex social circumstances of FOSS development culture warrants taking these views into account. On one hand, the field of FOSS development is diverse, with various developer agents and other actants having a large spectrum of backgrounds and motivations. Secondly, as previous research on the issue attests, goals, norms and practices in the field are varied and contextual [23]. To factor this complexity, brought about by the diversity of individual cognitive structures and social contexts in FOSS development, we make an assumption based on Pierre Bourdieu's concepts of field and habitus [10,11]. Our assumption is that actors in the field are limited in their capacity of making rational and optimal choices by their *habitus* and *specific rules and goals of the subfield.*

Habitus. Bourdieu defines habitus as "systems of durable, transposable dispositions, [...] principles which generate and organize practices and representations" [11]. Habitus is a system of cognitive and motivating structures associated with past individual and class experience. Although Habitus can result in mastery in actions towards goals without conscious motivation, it also includes strategic thinking involving "an estimation of chances presupposing transformation of the past effect into an expected objective" in the light of present "objective potentialities" [11].

As Bourdieu's concept of habitus does not exclude rational choice but contextualizes it within past experience and present conditions, it has the potential to make up for the shortcomings of a purely rational model of individual in studying goal-oriented choices and motivation. One example is a study on education choices in Germany and England by Glasser and Cooper, who find that young individuals make their educational choices rationally and oriented towards goals, yet these choices are also informed by their class and the culture of their educational institute, confirming habitus as a factor that affects the decision space and actions available to individuals [18].

To summarize, despite the tendency to represent habitus and rational choice as polar opposites or mutually exclusive concepts, habitus is inclusive of rational choice, although based on limited information or class/individual experience. Consequently, we take both habitus and rational choice as important components of the cultures that we chose to study, informing and conditioning each other.

Specific Rules and Goals of the Subfield. According to Bourdieu, the social world is composed of semi-autonomous systems called fields, which are not completely independent but rather have their own laws and forms of capital [10]. It follows that a field, such as literature, can have its own, non-monetary forms of capital such as recognition, prestige and autonomy [13]. Fields have their own specific logic by which individuals can take positions within them or forms of capital such as monetary capital that can be converted into other forms and vice-versa [10,13].

The goals, forms of capital, the logic of their accumulation and conversion, and the rules of a field do not necessarily coincide with other fields and society as a whole. This difference affects the capacity of the agents to take rational actions in it, as well as complicating our means of interpreting these actions in purely rational psychological terms. Therefore, our second assumption is that the specific conditions, rules and goals of the field leading to courses of action do not necessarily coincide with rational choice.

Thus, our ability to analyze developer motivations in FOSS development as actions oriented towards goals, and formulate them as objective functions require taking into account the specific cognitive structures of these agents and the rules specific to the field. This context-specific information cannot be derived from macro-sociological accounts or overly-general assumptions about individual cognition. Our research addresses this issue by integrating a qualitative field study aimed towards understanding the habituses of the developers in this field, and the rules specific to the field. This insight into the contextualized motivations enable us to use field-specific variables in our objective functions, rather than generalized notions of utilitarian or altruistic choices.

3.2 Data Collection

The research we propose uses mixed methods, i.e., combining qualitative and quantitative techniques. Due to their suitability for research projects that aspire to practical application and social transformation, mixed methods have been associated in social sciences with pragmatism [16,27,31]. The qualitative phase of the research is an exploration of FOSS development communities as a social field. A field consists of a structured set of positions; these positions differ from spatial positions in that they include alignments based on interpersonal relations and common goals [12,24]. This first phase is intended to gather insight into the values, goals and forms of capital specific to the field of FOSS development, as well as the agents that take positions within it. This will be achieved through a combination of in-depth and focus-group interviews and participant observation in developer meet-up events such as conferences. While interviews will be useful in understanding individual and group motivations and shared goals, observations in developer conferences will supplement this knowledge to shed more light no interactions. The results of the first phase will facilitate selection of relevant parameters to be measured and thus will shape the quantitative inquiries that follow.

The second part of the research consists of developing questionnaires guided by feature vectors built according to the findings of the previous stage. At this stage, the individual data collected from developers will be used to examine the field using local network regression [25,26], and to visualize it with dimensionality reduction techniques. After mapping the field as clusters based on individual features, we will be able to determine different subfields with their respective motivating goals, capital forms and habituses. This would enable us to distill objective functions to model a two-player game (developers and users) in order to change the equilibria of the subfields to a state favorable to PbD. Our aim at this stage is to be able to address the diversity of the subfields, avoiding the pitfalls of previous research and legal incentives, while keeping the model as simple and elegant as possible. After the dissemination of tools based on our game model, the final quantitative stage of the research is an assessment of the impact of our intervention, and proposal of strategies that encourage PbD based on our findings.

Sample. Our proposed study focuses on European FOSS mobile game developers. Mobile software is significant for the issue of privacy and security as it is a platform that is very susceptible to exploitation. As mobile games are easily installed, and are used almost ubiquitously by people of all ages and backgrounds, they represent an important risk with respect to privacy and security. For our field study, we strive for maximum variation in our sample of FOSS developers in terms of age, career stage, educational background and project size. This variation will facilitate exploration of a wide variety of subfields that include developers from diverse social backgrounds.

3.3 Engineer the Game: Mechanism Design

The modeling of the FOSS field is intended to describe and understand the current situation. A mere understanding, as detailed as it might be, does not bring about change. However, the aim of our research is to produce actionable advice for policy makers. To do this, we use an approach based on mechanism design. Here we will introduce the basic notions of mechanism design and present a thought experiment how to analyse the resulting games.

Mechanism design originated in economic theory or, more precisely, in game theory. It takes an engineering approach to design mechanisms, i.e. incentives, toward desired objectives. By informing this approach with the qualitative and quantitative results of our field research, as sketched in Fig. 1, we acknowledge the general critique of game theory, i.e. unrealistic assumptions about the rationality of agents and over-simplification [2].

Games: Objective Functions and Equilibria – Two Examples. For our thought experiment, let's assume that we find in the analyses of our initial interviews and observations that our participants have different occupational statuses and organizational roles (e.g. junior, senior, student), and have different roles in the decision-making processes of their FOSS projects. They also have various

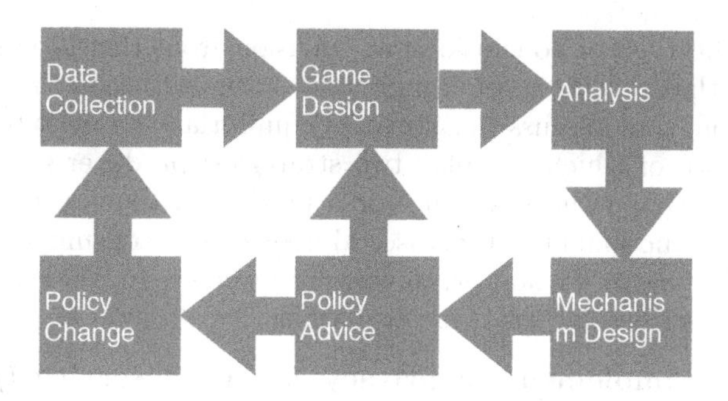

Fig. 1. Our experimental setup

motives for participating in FOSS (e.g. to build reputation for themselves or their organization, to have a project to show in job applications, to gain skills and knowledge, to contribute to society). They formulate the impact of their project in terms of their individual careers, contribution to the FOSS world, and society in general. Finally, let us say that we find that all of our participants think PbD features are important, but all are reluctant to implement them for differing reasons: Some state that PbD features take too much time and hinder the project, some think their skills are not sufficient for their implementation and are concerned that it will harm their reputation if they fail, and some do not want the ethical responsibility of implementing them correctly in case they fail.

Looking at these preliminary results, we develop our questionnaires. Our data can be roughly organized under habitus and forms of capital (education and skills, career, role in project or organization, reputation, money), values and goals (e.g. reputation, skills, contribution), strategies and positions (e.g. organizational and project hierarchy, career plans), and motivational factors (e.g. importance placed on PbD, time and resources or lack thereof, fear of losing reputation, wishing to contribute to society vs. fear of doing damage by developing insecure software). With the collected data and using local network regression [26] we will be able to cluster individuals and observe subfields formed on the basis of these features. This will help us develop a game model addressing the diversity of these clusters while remaining simple and elegant. Based on this model, we can develop tools to shift the equilibrium towards a more sensible state, followed by our assessment of impact.

Table 1. Payoff matrix of our example game: the privacy battle.

	$c = pet$	$c = \overline{pet}$
$p = pet$	2,1	0
$p = \overline{pet}$	0	1,2

To illustrate the type of conclusions we envision from this process, we briefly sketch a game that reflects a very rough view on the situation that we set out in our introduction, then discuss the expected equilibria. The most basic game has two players, both of which can play two strategies: producer p, which can opt to provide a privacy-friendly product or a product without privacy protection but with equal functionality otherwise. Moreover, the consumer c might chose to buy a privacy-aware or non-privacy-aware product. In Table 1, we illustrate this game. It describes the following situations:

The producer implements a privacy-enhanced service ($p = pet$). In doing so, the service becomes more expensive, which might have a negative impact on the profit margin.

If the consumer requires privacy ($c = pet$), the producer will reach some satisfaction, but at higher implementation costs, while the consumer will be fully satisfied. Hence the consumer payoffs are relatively higher.

If the consumer is willing to give up privacy ($c = \overline{pet}$) for a lower priced product, both consumer and producer might be unsatisfied, because the consumer might not be willing to pay the higher price of the privacy-enhanced product.

The producer implements a service without privacy ($p = \overline{pet}$). By doing this, the service becomes less expensive, which might have a positive impact on the profit margin.

If the consumer requires privacy ($c = pet$), both consumer and producer might be unsatisfied, because the consumer might not be willing to give up privacy.

If the consumer is willing to give up privacy ($c = \overline{pet}$) for a lower-priced product, the producer will be fully satisfied but the consumer will need to surrender some of their revenue in the form of privacy.

From today's market observation, we expect a payoff matrix similar to that described above. That is, if both players opt for the more privacy-aware option c, will get their highest payoff and p some payoff lower than that. If, on the other hand, both opt for a less privacy-friendly option, p will get their highest payoff and c some lower payoff. In the remaining cases, no player will get any payoff since either the consumer is not willing to pay the higher price or is not willing to give up their privacy. Further research is needed to determine more precise figures. Note in the fundamental game theory literature, games of this form are often introduced as "battle of the sexes games".

From everyday observation, which indicates a lack of privacy-friendly services, we venture to construct the following hypothetical analysis of the game: we know that "battle of the sexes" type of games have three Nash equilibria namely two pure, where both players opt for the same privacy preference, and one mixed. While the pure strategies are utility optimal, i.e. both ensure that the sum of all revenues is maximized, both are unfair: one party will always receive the higher revenue. On the other hand, the mixed strategy is highly inefficient. From the above, we conclude that consumers settle for a lower overall revenue due to the lack of better options.

While this simple game may broadly approximate the current situation, we do not assume that such a simple game will be sufficient to describe the current motives of FOSS developers. Hence we need to collect data to come up with a somewhat more complex game.

4 Conclusion

In this paper, we have attempted to provide a road map for research on FOSS developer motivations, with the aim of bringing about social change towards privacy-friendly features, while taking into account and learning from the shortcomings of previous research. We argued that the problems in previous research on this topic stem from the fact that it is mostly based on assumptions about developers that do not reflect the specific conditions of the FOSS development scene. More specifically, the research we see in the literature typically assumes that the developers are hedonistic rational individuals that make self-interested choices, or assumes that they take part in FOSS development due to altruistic motives. Neither of these models successfully explains the reluctance to implement privacy features or the failure of legal incentives obliging the implementation of PbD. More recent research, such as Spiekermann [32] and Bednar et al. [5], go beyond such limited explanations and point to organizational and individual reasons underlying this reluctance, indicating the necessity for research that takes into account the specific individual and social contexts surrounding developer motivations.

To address this, we have proposed an approach combining field theory and game theory into a mixed methods research endeavour with the aim of gaining deeper insight into FOSS communities and of suggesting relevant policy changes. From field theory, we take the concept of field and habitus in order to replace the general/universal models of individual with a historically and socially contextualized understanding of FOSS developers. This approach enables us to explore the field of FOSS development and the developer motivations with respect to field specific rules, goals, forms of capital and the collective past experience of the individuals therein. The findings will be used to develop games based on social context rather than dehistoricized assumptions about developers. The insights from the field study and game model will then be used to develop guidelines, tools and policy suggestions aimed at social change towards privacy-friendly software.

References

1. Regulation (EU) 2016/679 of the European Parliament and of the Council of 27 April 2016 on the protection of natural persons with regard to the processing of personal data and on the free movement of such data, and repealing directive 95/46/EC (General Data Protection Regulation). Official J. **L 119/1**. http://eur-lex.europa.eu/legal-content/EN/TXT/PDF/?uri=CELEX:32016R0679&from=DE

2. Adams, J.: The unknown James Coleman. Contemp. Sociol.: A J. Rev. **39**(3), 253–258 (2010). https://doi.org/10.1177/0094306110367907
3. High-Level Expert Group on Artificial Intelligence: Ethics Guidelines for Trustworthy AI (2019). https://ec.europa.eu/digital-single-market/en/news/ethics-guidelines-trustworthy-ai
4. Hars, A., Ou, S.: Working for free? Motivations for participating in open-source projects. Int. J. Electron. Commerce **6**(3), 25–39 (2002). https://doi.org/10.1080/10864415.2002.11044241
5. Bednar, K., Spiekermann, S., Langheinrich, M.: Engineering privacy by design: are engineers ready to live up to the challenge? Inf. Soc. **35**(3), 122–142 (2019). https://doi.org/10.1080/01972243.2019.1583296
6. Benbya, H., Belbaly, N.: Understanding developers' motives in open source projects: a multi-theoretical framework. Commun. Assoc. Inf. Syst. **27** (2010). https://doi.org/10.17705/1CAIS.02730
7. Bezroukov, N.: Open source software development as a special type of academic research: critique of vulgar Raymondism. First Monday **4**(10) (1999). https://doi.org/10.5210/fm.v4i10.696
8. Bezroukov, N.: A second look at the cathedral and the bazaar. First Monday **4**(12) (1999). https://doi.org/10.5210/fm.v4i12.708
9. Bonaccorsi, A., Rossi, C.: Altruistic individuals, selfish firms? The structure of motivation in open source software. First Monday **9**(1) (2004). https://doi.org/10.5210/fm.v9i1.1113
10. Bourdieu, P.: Distinction: A Social Critique of the Judgement of Taste. Harvard University Press, Cambridge (1984)
11. Bourdieu, P.: The Logic of Practice. Stanford University Press, Stanford (1990)
12. Bourdieu, P.: Sociology in Question. Sage Publications, Thousand Oaks (1993)
13. Bourdieu, P.: The Rules of Art: Genesis and Structure of the Literary Field. Stanford University Press, Stanford (1996)
14. Bourdieu, P.: The forms of capital. In: Readings in Economic Sociology, pp. 280–291. Blackwell (2002). https://doi.org/10.1002/9780470755679.ch15
15. Coleman, E.G.: Coding Freedom: The Ethics and Aesthetics of Hacking. Princeton University Press, Princeton (2012)
16. Creswell, J.W., Creswell, J.D.: Research Design: Qualitative, Quantitative, and Mixed Methods Approaches. Sage Publications, Thousand Oaks (2017)
17. ENISA: ENISA threat landscape report 2018—15 top cyberthreats and trends (2018)
18. Glaesser, J., Cooper, B.: Using rational action theory and Bourdieu's habitus theory together to account for educational decision-making in England and Germany. Sociology **48**(3), 463–481 (2013). https://doi.org/10.1177/0038038513490352
19. Goldthorpe, J.H.: Rational action theory for sociology. Br. J. Sociol. **49**(2), 167 (1998). https://doi.org/10.2307/591308
20. Kilani, J.: Will the GDPR affect open source projects? (2020). https://www.termsfeed.com/blog/gdpr-open-source/
21. Lerner, J., Tirole, J.: The open source movement: key research questions. Eur. Econ. Rev. **45**(4–6), 819–826 (2001). https://doi.org/10.1016/S0014-2921(01)00124-6
22. Lerner, J., Tirole, J.: Some simple economics of open source. J. Ind. Econ. **50**(2), 197–234 (2002). https://doi.org/10.1111/1467-6451.00174
23. Lin, Y.: The future of sociology of FLOSS. First Monday (Special Issue 2) (2005). https://doi.org/10.5210/fm.v0i0.1467

24. Martin, J.L.: The Explanation of Social Action. Oxford University Press (2011). https://doi.org/10.1093/acprof:oso/9780199773312.001.0001
25. Martin, J.L., Slez, A., Borkenhagen, C.: Some provisional techniques for quantifying the degree of field effect in social data. Socius: Sociol. Res. Dyn. World **2**, 1–18 (2016). https://doi.org/10.1177/2378023116635653. 237802311663565
26. McMahan, P., Slez, A., Martin, J.L.: Local network regressions. Socius: Sociol. Res. Dyn. World **5**, 1–7 (2019). https://doi.org/10.1177/2378023119845758. 237802311984575
27. Morgan, D.L.: Paradigms lost and pragmatism regained: methodological implications of combining qualitative and quantitative methods. J. Mixed Methods Res. **1**(1), 48–76 (2007). https://doi.org/10.1177/2345678906292462
28. Raymond, E.: The cathedral and the bazaar. Knowl. Technol. Policy **12**(3), 23–49 (1999). https://doi.org/10.1007/s12130-999-1026-0
29. Roberts, J.A., Hann, I.H., Slaughter, S.A.: Understanding the motivations, participation, and performance of open source software developers: a longitudinal study of the Apache projects. Manag. Sci. **52**(7), 984–999 (2006). https://doi.org/10.1287/mnsc.1060.0554
30. Sensat, J.: Game theory and rational decision. Erkenntnis (1975-) **47**(3), 379–410 (1997). http://www.jstor.org/stable/20012813
31. Shannon-Baker, P.: Making paradigms meaningful in mixed methods research. J. Mixed Methods Res. **10**(4), 319–334 (2015). https://doi.org/10.1177/1558689815575861
32. Spiekermann, S., Korunovska, J., Langheinrich, M.: Inside the organization: why privacy and security engineering is a challenge for engineers. Proc. IEEE **107**(3), 600–615 (2019). https://doi.org/10.1109/jproc.2018.2866769
33. Torvalds, L., Diamond, D.: Just for Fun: The Story of an Accidental Revolutionary. HarperCollins, New York (2001)

Recommended for You: "You Don't Need No Thought Control". An Analysis of News Personalisation in Light of Article 22 GDPR

Judith Vermeulen(✉)

Ghent University, Ghent, Belgium
judith.vermeulen@ugent.be

Abstract. More and more often personalisation at news websites is being introduced. In order to be capable of providing suggestions, recommender systems create and maintain a user profile per individual consumer that captures the latter's preferences over time. This requires the processing of personal data, and more specifically its collection and use for both profiling purposes and the eventual delivery of recommendations. In view of the fact that providing users with individualised news recommendations is usually realised by solely automated means and may in some cases violate people's absolute and non-derogable rights to freedom of thought and freedom of opinion, it could be prohibited on the basis of Article 22(1) of the EU General Data Protection Regulation. Online newspapers ideally abstain from becoming personalisation-only forums and give users some form of control as regards the determination of their reading preferences and interests.

Keywords: News personalisation · News websites · Automated individual decision-making · Article 22 GDPR · Fundamental rights · Freedom of thought · Freedom of opinion

1 Introduction

A banner titled "your privacy" pops up when visiting the website of The Guardian: "We use cookies to improve your experience. To find out more, read our privacy and cookie policy" [1]. Such 'improvement', so a (rare) consultation of the latter reveals, includes amongst others "showing readers journalism that is relevant to them" [2]: a bespoke offer based on what you like. Which proportions said customisation takes on, how such impacts user interfaces, and to what extent items are still distributed to everyone in the same way, is not clarified.

The reinvention by this newspaper of its service of providing people with news as well as its way of announcing it appears to be no stand-alone example, on the contrary. Thus seemingly unnoticed – hardly anyone actually runs through those policies before pressing the prominent button "I'm OK with that" –, news *personalisation* by online news providers, both web-editions of traditional news outlets and digital-only news

Published by Springer Nature Switzerland AG 2020
M. Friedewald et al. (Eds.): Privacy and Identity 2019, IFIP AICT 576, pp. 190–205, 2020.
https://doi.org/10.1007/978-3-030-42504-3_13

sites, appears to become commonplace. Simultaneously, the era of *mass* distribution of information and opinions by the press to all citizens alike may be coming to an end.

In what follows, first, the phenomenon of news personalisation at news websites will be described; second, it will be analysed whether this practice falls within the scope of Article 22 GDPR, which contains a general prohibition on automated individual decision-making having legal or similarly significant effects, and if so, what would be the consequences thereof; finally, there will be a conclusion.

2 News Personalisation at News Websites

Neil Thurman, Professor of Computational Journalism, has been studying the phenomenon of news 'personalisation', 'customisation' or 'individualisation' already since 2011 [3, 4]. He, together with Steve Stifferes, Professor of Financial Journalism, defines it as "a form of user-to-system interactivity that uses a set of technological features to adapt the content, delivery, and arrangement of a communication to individual users' explicitly registered and/or implicitly determined preferences" [5, 6]. They consider it to be the result of a set of "complex interactions that exist between computer algorithms and those behind their logic, data about individual and aggregated user behaviour, decisions on classification and indexing, explicit user choices, editorial and journalistic decisions, and user profiles, demographics and location" [6]. From their definition, it is moreover clear that these authors make a distinction between two different ways to determine user preferences: explicit and implicit personalisation [6]. In the former case, users actively register their interests [3, 6]. They could, for example, be allowed to select their favourite topics or columns and be enabled to always easily access them via a "My Page" or "My News Feed"-function [7]. The latter technique on the other hand covers situations in which user preferences are inferred "from data collected, for example, via a registration process or via the use of software that monitors user activity" [3, 6]. There, the goal could be to personalise what stories people see on the homepage [8].

Personalisation at news websites could also take the form of recommendations as to what to read next in a sidebar or below articles, a customised page besides a homepage, mobile editions or apps offering customisation options etc. [9, 10].

From a computer science perspective, research regarding news recommender systems has for the most part concerned either content-based filtering, collaborative filtering or hybrid approaches [11, 12]. The first-mentioned method consists in the analysis of a set of items by means of feature extraction techniques in order to represent their content in a form suitable for the next processing steps; the subsequent building of a model comprising a structured representation of user interests based on the features of the items 'rated' (see *infra*) by that user; and, finally, the recommendation of new interesting items using these profiles [13]. Collaborative filtering, on the other hand, relies on user-item responses or 'ratings' (see *infra*) of an individual user u as well as those of other users in the system [14]. Recommendations are made by finding correlations among users of a recommendation system [15]: "[t]he key idea is that the rating of u for a new item i is likely to be similar to that of another user v if u and v have rated other items in a similar way. Likewise, u is likely to rate two items i and j in a similar fashion, if other users have given similar ratings to these two items" [14]. User-item ratings can, moreover, either be used directly to predict ratings for new items, or employed to learn a predictive model [14]. Hybrid systems combine several filtering techniques, such as the aforementioned ones, in order to further improve their results [16, 17] (Fig. 1).

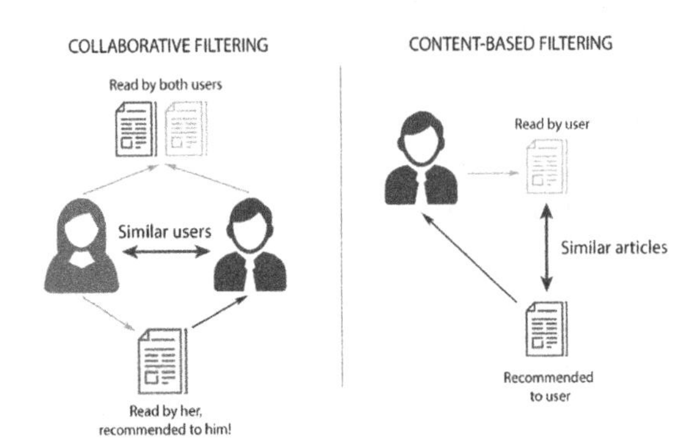

Fig. 1. Collaborative filtering and content-based filtering [18].

In order to be capable of providing suggestions, recommender systems, in any event, need to create and maintain a user profile per individual consumer that captures the latter's preferences over time [11]. Each type of recommender system however has its own approach to learning user profiles as they use different kinds of data and represent information in a different manner [15]. Often, the content of read news items is used to deduct consumer interest (content-based filtering) and/or user features are being considered in comparison to those of others (collaborative filtering) [11]. Some approaches, in addition, consider people's demographic information, for example, to overcome the so-called cold-start problem, which refers to situations where either little is known about a user's preferences or an item has not sufficiently been viewed or rated [11, 19].

In as far as the provision of customised reading suggestions to consumers is based on 'information relating to an identified or identifiable individual' – thus leaving aside item-only data used in content-based filtering –, the data used qualifies as 'personal data' within the meaning of Article 4(1) of the EU's General Data Protection Regulation ('GDPR') [20]. In terms of personal data 'processing', which the GDPR defines as "any operation or set of operations [that] is performed on personal data or on sets of personal data" [20, art 4(2)], news personalisation essentially evolves in three stages: collection, profiling and recommending [21].

First of all, news providers, gather several data points per reader. Hence, they keep record of information provided by users when creating an account (e.g.: gender, age, location) or for receiving newsletters (e.g.: email address), as well as of so-called user 'ratings' [22]. The latter take a variety of forms as they can be either scalar/numerical (e.g.: 1–5 stars), ordinal (e.g.: agree, neutral, disagree), binary (e.g.: like/dislike), or unary in case the interaction of a user with an item is implicitly captured (e.g.: time spent browsing it). The amassing of the latter can take place through the use of cookies, which can be described as small pieces of data a site asks your browser to store on your computer, tablet or mobile phone [23]. They allow the recognition of your device upon return and remember, depending on their specific function, certain of your actions and preferences over time [23]. In a second step, the collected material is evaluated – where appropriate in combination with data relating to other users – using statistics and artificial intelligence, to predict a news consumer's preferences or interests, and consequently, what news items are most suitable to recommend to him or her [16, 24]. This process coincides with the GDPR-definition of 'profiling' [20, art 4(4)]. Finally, users are offered recommendations – which can be reflected in user interfaces in several manners (see *supra*) – in relation to specific items when (re-)visiting news sites.

3 A Prohibition Based on Article 22 GDPR and Freedom to Form Thoughts and Opinions Concerning Matters of General Interest

In view of the foregoing, the question arises whether Article 22 GDPR on 'automated decision-making, including profiling' applies to news personalisation as described and if so, what would be the implications thereof.

The first paragraph of said provision provides that:

> *"The data subject shall have the right not to be subject to a decision based solely on automated processing, including profiling, which produces legal effects concerning him or her or similarly significantly affects him or her".*

According to the Article 29 Data Protection Working Party, the previous independent European advisory body on data protection and privacy and predecessor of the European Data Protection Board, Article 22(1) contains a general prohibition and is accordingly relevant regardless of whether a potentially affected individual takes any action [25]. Instead, its applicability depends on the fulfilment of two requirements: a decision involving a natural person must be made 'solely' as a result of automated processing, including profiling, *and* affect a data subject's legal rights or produce an effect concerning him or her that is equivalent or similarly significant in its impact [25].

3.1 Generating News Recommendations by Solely Automated Means

As regards the first criterion, the Article 29 Group explains that "[a]n automated process produces what is in effect a recommendation concerning a data subject" [25]. As becomes clear from the descriptions above, in case of news personalisation, a set of operations is performed on personal (user), and in case of content-based filtering also non-personal (item), data which results in (a list of) recommended content [16]. For the purposes of Article 22, each individual suggestion can be considered a 'decision' based on the 'processing' of personal data, and more specifically, on its collection for learning user profiles and predicting individuals' (lack of) interest in certain items, and use to eventually offer news consumers individualised suggestions [10]. Recommendations are moreover generated by 'solely automated means', which entails that an algorithm decides which articles will be proposed and that their delivery to the reader happens automatically, without any prior meaningful assessment by a human [25]. Recommender systems, indeed, "provide *automated* and personalised suggestions [...] to consumers" (emphasis added) [14]. They "have evolved to fulfil the natural dual need of buyers and sellers by *automating* the generation of recommendations based on data analysis" (emphasis added) [26 at "Recommender Systems"].

3.2 Limitations on the Basis of the Freedom to Form Thoughts and Opinions Concerning Matters of General Interest

To fall within the scope of Article 22 GDPR, decision-making furthermore needs to produce either 'legal' or 'similarly significant' effects concerning the data subject [25]. We argue that news personalisation may potentially interfere with – and thus violate (see *infra*) – a person's fundamental rights to freedom of thought and freedom of opinion and could therefore indeed affect his or her legal rights in breach of the prohibition laid down by said provision [25].

Freedom of thought and freedom of opinion are respectively contained in two consequent provisions of the European Convention on Human Rights ('ECHR') [27]. More specifically, Article 9(1) ECHR provides that "everyone has the right to freedom of thought" – as well as of "conscience and religion" –, while Article 10(1) ECHR, on freedom of expression, grants individuals "freedom to hold opinions". According to the majority opinion in legal doctrine, these two freedoms essentially overlap in scope [28, 29], which is why we consider them as one and the same for the purposes of the analysis in this contribution. Statements concerning either one of them will therefore also be considered true in relation to the other. These freedoms involve the area often referred to as the *forum internum* [30]. It is crucial to note that they are absolute and unconditional and can therefore not be restricted or interfered with without being violated [28, 31, 32]. Indeed, Article 9(2) ECHR only allows limitations in relation to the freedom to *manifest* one's religion or beliefs. Whereas Article 10(2) ECHR, on the other hand, in general terms provides that the exercise of the freedoms listed in its first paragraph (see *infra*) may be subject to restrictions, the Council of Europe's Committee of Ministers has stated that allowing any in relation to the freedom to hold opinions "will be inconsistent with the nature of a democratic society" [33].

The right to freedom of thought and opinion, apart from freedom to *have* and *hold* thoughts and opinions (first dimension), also guarantees freedom to *form* and *develop* them (second dimension) [31, 34]. This latter liberty in particular constitutes the subject of further analysis in the context of this contribution. In the past, States have not found it difficult to allow people freedom to think [35]. Some even considered it impossible to restrict this freedom as it protects the realm of the mind (*forum internum*) [29]. This may explain the lack of case-law specifically dealing with the freedoms of thought and opinion up to now. In view of the emergence of new technologies such as those described above, their value may nonetheless become more tangible [36–39]. Importantly, the guarantee of the right to freely form and develop thoughts and opinions depends on the right to receive information and ideas. Indeed, the free formation of thoughts and opinions is only possible when one is properly informed [29]. People are *influenced* by the information they are presented with [40]. When such influencing reaches an unacceptable level, the right to freedom of thought and opinion will however be considered *interfered with* [28, 41–43]. The freedom of opinion formation, in turn, constitutes a prerequisite for the realisation of a person's right to freely *express* his or her thoughts and opinions, for these have to be conceived before the can be voiced [31]. This latter right as well as the freedoms to receive and impart information and ideas are also encompassed in Article 10(1) ECHR. It may be pointed out that these, unlike the freedom of thought and opinion, *can* justifiably be restricted, provided that the conditions thereto laid down by Article 10(2) ECHR in that regard are fulfilled [29].

At this point, it is important to note that the freedom of expression rights serve a dual purpose. More specifically, they are directed at both the *individual* and the *citizen* [29]. In the first place, freedom of expression, as other liberal rights, enables *individuals* to ward off interferences by the State [28]. As such it delineates a sphere of *freedom* or autonomy for their personal development [29]. People are *free to form* inner – purely private – *thoughts and opinions* as well as *to express them* publicly in the "market place of ideas" without fear of repression [28]. While the latter takes place in public, it is nonetheless considered an essential component of the individual's privacy being the externalization of one's personality [28, 29]. Next to 'liberté-autonomie', the freedom of expression is also conceived as 'liberté-participation' [28]. As such, it aims at the social integration of a person in society rather than at warding off interferences [28]. In accordance with settled case-law of the European Court of Human Rights ('ECtHR'), *citizens* have, more specifically, the right to *express their opinions* on political matters as well as regarding issues of public interest more generally and, accordingly, to take part in public debate [44, 45]. The Court considers, more generally, that "freedom of political debate" – to be understood as including debate on all matters of public interest [46] – "is at the very core of the concept of a democratic society which prevails throughout the Convention" [47]. Accordingly, it ruled on several occasions that it requires "very strong reasons for justifying restrictions on such debate" [45, 48]. The twofold function of Article 10 of the European Convention was famously underscored by the ECtHR for the first time in its *Handyside v. the UK* judgment where it stated that "[f]reedom of expression constitutes one of the essential foundations of [a democratic] society, one of the basic conditions for its progress and for the development of every man" [49]. Both purposes are moreover complementary. Whereas 'uninfluenced' – in as far as that is possible – opinions of *individuals* form the basis of democratic debate, taking part therein as a *citizen* constitutes the best method to safeguard the autonomous sphere [29].

Importantly, the Court in Strasbourg views the press as the most important enabler for the realisation of the public debate freedom [50]. This explains both why this actor is attributed special protection (which will however not be the subject of further discussion here) and assigned certain responsibilities under Article 10 of the Convention [50]. In *Leander v. Sweden*, it was decided that a person, in general, has a right to receive information and ideas that others wish or may be willing to impart to him or her [51]. However, in cases such as *Lingens v. Austria* and *Şener v. Turkey*, the Court held that the press has the *task* of imparting information and ideas on political issues just as on those in other areas of general interest and that the public has a right to receive them [47, 52]. This was explained by the fact that freedom of the press affords the *public* one of the best means of discovering and forming an opinion [47, 52]. In the case of *Khursid Mustafa v. Sweden*, which concerned a conflict between two private parties, the Court confirmed that also *individual applicants* are free to receive "political and social news" or "reports on events of public concern" as they constitute "the most important information protected by Article 10" [53]. The freedom to receive information and ideas can thus be considered dependent on the freedom to impart information and ideas [33].

Freedom to receive information and ideas moreover entails that people are free to consume available content and undertake the necessary steps to inform themselves [29]. They may gather information through all possible lawful sources [33]. As such, everyone has the right to buy books, magazines or newspapers [29]. Whereas *publishing* can be considered to take place in the public sphere, *reading* on the other hand is in essence a private activity [50]. In that regard, it can be noted that the EU High-Level Expert Group on Artificial Intelligence identified 'respect for human autonomy' as the first of four ethical principles which should be respected in order to ensure that AI systems are developed, deployed and used in a trustworthy way [54]. Rooted in human rights theory, this norm puts forward that "humans interacting with AI systems must be able to keep full and effective self-determination over themselves, and be able to partake in democratic process" [54]. As explained above, the participation of citizens in society indeed necessitates individual autonomy (and vice versa) [29] (Fig. 2).

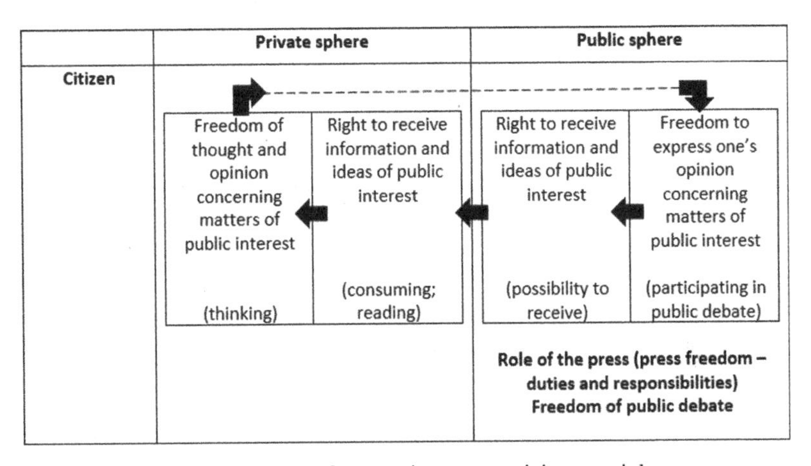

Fig. 2. Freedom of expression as a participatory right.

The right to freedom of thought and opinion – in its second dimension – is however only *interfered with* when one's thoughts and opinions are being influenced to an extent their free formation could be corrupted and such thus exceeds a certain threshold [28, 41–43]. The question is, therefore, whether the (lack of) information one receives as a result of the customisation of his or her online news offer at a news website can reasonably be considered to have a 'sufficiently' significant effect on the formation or development of one's inner ideas concerning matters of general interest. Rather than evaluating the (in)existence or dangers of filter bubbles [55, 56], the aim is thus to assess whether news personalisation could impede citizens from being adequately informed and, as a consequence thereof, from forming their opinions. In view of the fact that establishing the actual impact of phenomena such as news personalisation – in all its forms – requires extensive and continuously updated empirical findings [57], this contribution presents a theoretical framework for analysis and is based on legal research.

It can be noted that the media, irrespective of their intentions with respect to personalisation, presume an important gatekeeper role [50]. They function as intermediaries between speakers and sources of information on the one hand, and the public on the other [50]. However, as explained, the ECtHR expects the press to at least inform people *on events of interest* [28, 47, 52]. In *Lingens v. Austria*, the Court added that it could not accept the opinion of the Austrian domestic authorities "to the effect that the task of the press was to impart information, the interpretation of which had to be left primarily to the reader" [47]. The Court, in the case of *Stoll v. Switzerland*, in a similar vein, pointed to the *influence* wielded by the media in contemporary society: "not only do they inform, they can also suggest by the way in which they present the information how it is to be assessed" [58]. Provided that journalists "are acting in good faith and on an accurate factual basis and provide "reliable and precise" information in accordance with the ethics of journalism" [58], it is thus accepted for media outlets to be influencing their audience by means of *how* they report on issues of public concern. Manfred Nowak, Professor of International Human Rights [59], considered, however, that "[i]t is not easy to delineate between the (impermissible) interference with freedom of opinion and (permissible) daily influencing of the formation of our opinions [...] by all the news, commentary and information disseminated by the mass media in our communications and information society" [28]. This seems to suggest that the latter, distributing news items to every consumer alike, are nonetheless at the limits of what can be considered *acceptable influence*. Therefore, and having regard to the freedom to consume published information – particularly relating to matters of public interest –, the individualisation of people's feeds might go too far. In that regard, it may be interesting to note that according to the New York Times even its most active users only read a handful of the 200 news stories it puts out each day [8].

Moreover, in line with what David Kaye, the current Special Rapporteur on the promotion and protection of the right to freedom of opinion and expression, stated in his 2018 Annual Report, online content curation by means of technology may negatively impact an individual's capacity to form opinions in view of the fact that algorithmic ranking may induce users to believe that items recommended to them are particularly worth reading at a given moment in time [43, point 24–25]. He noted moreover that "media outlets elevate particular stories to the front page with the intention of shaping and influencing individual knowledge about significant news of the day" [60, point 24]. Having regard to the case-law of the ECtHR referred to above, this seems indeed what the structuring or placement of articles by news media, both print and electronic, first and foremost should lead to.

Therefore, we argue that news personalisation at news websites, to the extent it would impede people from having the possibility of being informed (1a) and/or informing themselves about matters of public concern (1b), and/or impair their intellectual capacity to assess the general importance of the news items they are being recommended (2), surpasses the limits of influencing acceptable in light of the freedom of thought and opinion.

When online newspapers offer recommendations only *in addition* and next to the main news items of the day, provided that the latter can be distinguished from the former, are bundled together and displayed in order of importance, neither of these issues (1a, 1b and 2) seem to be present. In such circumstances, readers wishing to be informed concerning matters of interest to accordingly form their opinions are not hindered in that respect (1) and are enabled to assess the respective public interest value of the items (2). Personalisation would as such not negatively affect public debate. News providers could for example maintain a general page and as a supplementary service offer their audience personalised reading suggestions – by means of sidebars, below articles, via an additional personal page etc. Considering personalisation in such circumstances only concerns articles lacking public interest value, the fact whether or not a user has an explicit say regarding the determination of his or her preferences (see *supra*) is of secondary importance in the context of this analysis. The same cannot be said in case news personalisation at news websites takes the form of a *personalised page* without there being available a designated section regarding important items. Instead, such articles would be shown more or less prominently depending on the individual profile of the reader concerned. If generated recommendations would moreover be based on one's *implicitly inferred interests*, a reader might not only miss out on articles concerning issues of general interest (1a), but would at the same time be deprived of the freedom to consume the news items amongst those of importance which they themselves consider the most relevant (1b). Considering people, in case of implicit personalisation, would be offered a list of recommendations in accordance with their presumed preferences, it can moreover be assumed that their ability to assess their respective societal weightiness would be affected as well (2). Explicit personalisation on the contrary empowers (1b) users to a certain extent as it allows them to ban certain types of content from their feed while giving them the opportunity to 'adjust' their preferences and perhaps even 'select all' (1a) available predetermined topics, events, etc. In the latter case they should however again somehow be enabled to distinguish between information of general interest and other news (2) (Fig. 3).

Personalisation	Additional	-only
Explicit	1a, 1b, 2	1a(?), 1b, 2(?)
Implicit	1a, 1b, 2	×

Fig. 3. Permissibility of personalisation at news websites.

In order to put this analysis into perspective, the potential effects of news personalisation can be compared to those of (political) micro-targeting practices, which are also considered to violate people's freedom of opinion in some case [36, 38]. As explained, the former could influence the formation of one's thoughts on issues of general interest in view of the fact that it may impede a person from receiving information about such matters and from assessing the respective importance of the articles he or she is recommended. The latter, on the other hand, is problematic when it amounts to manipulation and affects our cognitive autonomy [36]. A manipulator seeks to insinuate himself in a target's decision-making process and thus deprives the individual of the authorship over his or her ideas [57]. Whereas micro-targeting could hinder the free *development* of thoughts (second dimension), it might also affect the freedom to *have* a personal opinion as its purpose is to have people *adopt* another point of view (first dimension).

3.3 Explicit Consent as an Exception

Considering the fact that Article 22(1) GDPR sets out a general prohibition, online newspapers should in principle not engage in a type of news individualisation that could violate Articles 9 and 10 ECHR [25]. However, Article 22(2) contains three exceptions to this rule. Considering its problematic nature, the following analysis will in that respect focus on implicit personalisation-only news websites. First of all, point (a) of Article 22(2) sets out that paragraph 1 shall not apply if the decision "is necessary for entering into, or performance of, a contract between the data subject and a data controller". The EDPB recently, in version 2.0 of its guidelines on the processing of personal data under Article 6(1)(b) GDPR in the context of the provision of online services to data subjects, acknowledged that "personalisation of content may (but does not always) constitute an intrinsic and expected element of certain online services, and therefore may be regarded as necessary for the performance of the contract with the service user in some cases" [61]. In the draft text, it stated that online news aggregator websites, whose services consist of "providing users with tailored content from multiple online sources through a single interface", may rely on contract as a legal basis for processing "because the function of the service directly relates to personalised content" [62]. In view of the fact that the press itself – meaning (online) news providers (see *supra*) – carries a great responsibility in informing the public so as to facilitate public debate, it cannot be accepted that their primary purpose could also be to customise their readers' news offer in accordance with the latter's preferences. The example concerning news aggregators has moreover been deleted in the second version of the EDPB guidelines.

Therefore, we consider that online newspapers may not use Article 22(2) (a) GDPR to deploy a personalisation-only news service that relies on implicit profiling techniques. Secondly, point (b) provides that automated individual decision-making would be permissible in case its use "is authorised by Union or Member State law". As such is not the case for news personalisation, this exception too is not relevant here.

Of particular importance in this context however is third one – point (c) –, which stipulates that the prohibition does not apply if the decision-making "is based on the data subject's explicit consent". To be validly expressed, consent must (amongst others) be 'informed' [20, art 4(11)]. According to recital 42 GDPR, this entails that the data subject should be aware of the purposes of the processing for which the personal data are intended. Article 29 Working Party has explained in that regard that "where the processing involves profiling-based decision making" it must be made clear to the person concerned "that the processing is for the purposes of both (a) profiling and (b) making a decision based on the profile generated" [25]. Moreover, data subjects must "understand exactly what they are consenting to" and "have enough relevant information about the envisaged use and consequences of the processing to ensure that any consent they provide represents an informed choice" [25]. In view of the analysis above, this means that online news providers engaging in implicit personalisation-only type of services should explain to potential consumers that the articles in their future news feed will be ranked in accordance with their own preferences rather than taking into account the societal interest of the items. The GDPR also requires a data subject's consent regarding the processing of his or her personal data to be 'freely given' [20, art 4(11)]. As stated in recital 42 of the GDPR, such is not the case "if the data subject has no genuine or free choice or is unable to refuse or withdraw consent without detriment". It has been argued that this requirement inherently conflicts with so-called "take it or leave it"-practices [63] which lead to the *de facto* exclusion of data subjects from being provided with the service concerned in case they do not consent. Indeed, forcing users to accept the placement of so-called tracking cookies, which are used to collect user behaviour information (see *supra*), due to the fact that a refusal entails that a website simply cannot be accessed, is therefore often considered unlawful [64–66]. This would imply that news personalisation at a specific news website should in fact never be mandatory, though, in contrast, always ought to be optional. In this context consent must moreover be 'explicit' and therefore manifested by means of an *express* statement [67]. When processing entails significant risks, a high level of control over personal data is thus deemed appropriate [25]. In the digital or online context, the Article 29 Working Party explains that "a data subject may be able to issue the required statement by filling in an electronic form, by sending an email, by uploading a scanned document carrying the signature of the data subject, or by using an electronic signature" [67]. A two stage verification of consent, whereby a person is requested to reply to an email in which is asked for consent by "I agree" and subsequently receives a verification link or code that must be clicked to confirm agreement, can also suffice [67]. In such circumstances, consenting through cookie banners (see *supra*) cannot be allowed.

In the context of their transparency obligations, Article 13(2)(f) GDPR moreover requires controllers [20, art 4(7)], at the time when personal data are obtained from the data subject – as we assume in this paper they would be –, to provide them with specific information in case they engage in processing activities as referred to in Article 22 GDPR. They must, first of all, inform readers about the fact that they do so, which in case the processing is based on consent they already should have (see *supra*). Secondly, and "at least in those cases, meaningful information about the logic involved" must be provided. According to the Article 29 Working Party this requires finding "simple ways to tell the data subject about the rationale behind, or the criteria relied on in reaching the decision" [25]. Finally, controllers should explain "the significance and the envisaged consequences of [the] processing". This requirement comes down to clarifying "how the automated decision-making might affect the data subject" [25]. As explained above, also this information should normally be obtained by the latter also earlier on and namely prior to agreeing to the processing.

Article 22(3) GDPR, in addition, states that "in the case [...] referred to in point [...] (c) of paragraph 2, the data controller shall implement suitable measures to safeguard the data subject's rights and freedoms", *in casu* being his or her right to freedom of thought and opinion. At the least, a person should be granted the right to obtain human intervention on the part of the controller, to express his or her point of view and to contest the decision [20, art 22(3)]. While in case of news personalisation the implications of these assurances are, admittedly, not straightforward, some ideas may nonetheless be posited in that respect. Considering that the Working Party put forward that any review must be carried out by someone who has the capability to *change* the decision, data subjects might be able to request the adjustment of their profile – and with that the recommendations based thereon – in case they feel impeded from receiving information and ideas of general importance. However, bearing in mind that user preferences are determined by algorithmic means (see *supra*) and therefore not readily adaptable by human beings, the only practical solution might be to completely reset a user profile – something which is said to be also possible on the basis of the right to erasure ('right to be forgotten') laid down in Article 17 of the GDPR [10]. The relevance of being able to express one's opinion and to contest the decision might be similar.

Lastly, it can be noted that Article 22(4) GDPR in principle prohibits decisions referred to in paragraph 2 to be based on special categories of personal data unless "the data subject has given explicit consent to the processing of those personal data for one or more specified purposes" or "processing is necessary for reasons of substantial public interest, on the basis of Union or Member State law" [20, art 22(4), 9(1), 9(2)(a), (g)]. Article 9(1) GDPR specifies that such categories (amongst others) include data revealing political opinions and philosophical beliefs. Therefore, if a news consumer would, as a result of implicit personalisation, be recommended more articles concerning either left or right-wing political parties or touching upon topics associated with progressive rather than conservative convictions or vice versa, this could be problematic. In the context of this analysis, we consider moreover that only the first-mentioned exception can be applied.

4 Conclusion

More and more often, personalisation at news websites is being introduced. It can be done on the basis of explicitly registered or implicitly inferred user preferences, take several forms and rely on different types of (implicit) filtering techniques. Customising consumers' news offer, however, requires the processing of personal data, and more specifically its collection as well as use for profiling and the eventual delivery of recommendations. In view of the fact that providing users with individualised news recommendations is usually realised by solely automated means and has the potential to violate people's absolute and non-derogable rights to freedom of thought and freedom of opinion, it could be prohibited on the basis of Article 22(1) of the EU General Data Protection Regulation. The adaptation of news feeds according to implicitly inferred user profiles without there being available a designated section regarding important news does, in any event, not appear acceptable. However, news consumers could, as foreseen by Article 22(2) GDPR, nonetheless explicitly agree to such employment of their personal information, though only provided that they would be offered an informed and genuine choice in that regard. In such circumstances, online newspapers must also provide their readers with simple and meaningful information concerning the rationale behind the algorithm used and the recommendations it generates. The latter can moreover only be based on a person's political opinions or philosophical beliefs in case he or she gave its explicitly consented thereto.

Ideally, online newspapers abstain from becoming personalisation-only forums and give users some form of control as regards the determination of their reading preferences and interests.

References

1. News, sport and opinion from the Guardian's global edition. https://www.theguardian.com/international. Accessed 06 May 2019
2. Cookies on the Guardian website. https://www.theguardian.com/info/cookies. Accessed 06 May 2019
3. Thurman, N.: Making 'The Daily Me': technology, economics and habit in the mainstream assimilation of personalized news. Journalism **12**, 395–415 (2011). https://doi.org/10.1177/1464884910388228
4. Website of Dr Neil Thurman, Professor of Communication. LMU Munich. https://neilthurman.com/. Accessed 31 July 2019
5. Professor Steve Schifferes. https://www.city.ac.uk/people/academics/steve-schifferes. Accessed 01 Aug 2019
6. Thurman, N., Schifferes, S.: The paradox of personalization: the social and reflexive turn of adaptive news. In: The Handbook of Global Online Journalism, pp. 373–391. Wiley (2012). https://doi.org/10.1002/9781118313978.ch20
7. Mis niets over uw favoriete thema's via "Mijn dS". https://www.standaard.be/cnt/dmf20190304_04229666. Accessed 09 Aug 2019
8. Slefo, G.P.: "New York Times" plans to invest heavily in AI (2018). https://www.crainsnewyork.com/technology/new-york-times-plans-invest-heavily-ai
9. Kunert, J., Thurman, N.: The form of content personalisation at mainstream, transatlantic news outlets: 2010–2016. Journal. Pract. **13**, 759–780 (2019). https://doi.org/10.1080/17512786.2019.1567271
10. Eskens, S.: A right to reset your user profile and more: GDPR-rights for personalized news consumers. Int. Data Priv. Law. https://doi.org/10.1093/idpl/ipz007
11. Karimi, M., Jannach, D., Jugovac, M.: News recommender systems – survey and roads ahead. Inf. Process. Manage. **54**, 1203–1227 (2018). https://doi.org/10.1016/j.ipm.2018.04.008
12. de Souza Moreira, G., Ferreira, F., da Cunha, A.M.: News session-based recommendations using deep neural networks. In: Proceedings of the 3rd Workshop on Deep Learning for Recommender Systems - DLRS 2018, pp. 15–23 (2018). https://doi.org/10.1145/3270323.3270328
13. Lops, P., de Gemmis, M., Semeraro, G.: Content-based recommender systems: state of the art and trends. In: Ricci, F., Rokach, L., Shapira, B., Kantor, Paul B. (eds.) Recommender Systems Handbook, pp. 73–105. Springer, Boston (2011). https://doi.org/10.1007/978-0-387-85820-3_3
14. Desrosiers, C., Karypis, G.: A comprehensive survey of neighborhood-based recommendation methods. In: Ricci, F., Rokach, L., Shapira, B., Kantor, Paul B. (eds.) Recommender Systems Handbook, pp. 107–144. Springer, Boston (2011). https://doi.org/10.1007/978-0-387-85820-3_4
15. Pazzani, M.J.: A framework for collaborative, content-based and demographic filtering. Artif. Intell. Rev. **13**, 393–408 (1999)
16. Ricci, F., Rokach, L., Shapira, B., Kantor, P.B. (eds.): Recommender Systems Handbook. Springer, Boston (2011). https://doi.org/10.1007/978-0-387-85820-3
17. Vozalis, M., Margaritis, K.G.: Collaborative filtering enhanced by demographic correlation. In: Proceedings of the AIAI Symposium on Professional Practice in AI, Part of the 18th World Computer Congress, pp. 293–402 (2004)
18. Mohamed, M.H., Khafagy, M.H., Ibrahim, M.H.: Recommender systems challenges and solutions survey. In: 2019 International Conference on Innovative Trends in Computer Engineering (ITCE), pp. 149–155 (2019). https://doi.org/10.1109/ITCE.2019.8646645

19. Lika, B., Kolomvatsos, K., Hadjiefthymiades, S.: Facing the cold start problem in recommender systems. Expert Syst. Appl. **41**, 2065–2073 (2014). https://doi.org/10.1016/j.eswa.2013.09.005
20. Regulation (EU) 2016/679 of the European Parliament and of the Council of 27 April 2016 on the protection of natural persons with regard to the processing of personal data and on the free movement of such data, and repealing Directive 95/46/EC (General Data Protection Regulation) (Text with EEA relevance) (2016)
21. Verdoodt, V., Lievens, E.: Targeting children with personalised advertising: how to reconcile the (best) interests of children and advertisers. In: Data Protection and Privacy Under Pressure: Transatlantic Tensions, EU Surveillance, and Big Data, pp. 313–341. Maklu (2017)
22. Overeenkomst voor gegevensgebruik ('agreement for data use') Mediahuis. https://www.mediahuis.be/overeenkomst-voor-gegevensgebruik/. Accessed 24 May 2019
23. European Commission: Cookies. http://ec.europa.eu/ipg/basics/legal/cookies/index_en.htm#section_2. Accessed 09 May 2019
24. European Data Protection Supervisor: Towards a new digital ethics. Data, dignity and technology. Opinion 4/2015 (2015)
25. Article 29 Data Protection Working Party: Guidelines on Automated individual decision-making and Profiling for the purposes of Regulation 2016/679. WP251 (2017)
26. Sammut, C., Webb, G.I. (eds.): Encyclopedia of Machine Learning. Springer, Boston (2011). https://doi.org/10.1007/978-0-387-30164-8
27. European Convention for the Protection of Human Rights and Fundamental Freedoms. ETS No. 005 (1950)
28. Nowak, M.: U.N. Covenant on Civil and Political Rights: CCPR Commentary. N.P. Engel, Kehl (2005)
29. Velaers, J.: De beperkingen van de vrijheid van meningsuiting. Maklu Uitgevers, Antwerpen (1991)
30. Murdoch, J.: Freedom of thought, conscience and religion. A guide to the implementation of Article 9 of the European Convention on Human Rights. Council of Europe, Belgium (2007)
31. Schabas, W.A.: The Universal Declaration of Human Rights: The Travaux Préparatoires. Cambridge University Press, Cambridge (2013)
32. UN Human Rights Committee (HRC): General comment no. 34, Article 19, Freedoms of opinion and expression (2011)
33. Macovei, M.: Freedom of expression. A guide to the implementation of Article 10 of the European Convention on Human Rights. Council of Europe (2004)
34. Murdoch, J.: Protecting the right to freedom of thought, conscience and religion under the European Convention on Human Rights. Council of Europe, Strasbourg (2012)
35. Scheinin, M.: Article 18. In: Eide, A., Alfredsson, G., Melander, G., Rehof, L.A., Rosas, A., with the collaboration of Theresa Swinehart (eds.) The Universal Declaration of Human Rights: A commentary. Scandinavian University Press; Distributed world-wide excluding Scandinavia by Oxford University Press, Oslo, Oxford (1992)
36. Committee of Ministers: Declaration on the manipulative capabilities of algorithmic processes. Decl(13/02/2019)1 (2019)
37. European Data Protection Supervisor: Online manipulation and personal data. Opinion 3/2018 (2018)
38. European Data Protection Board: The use of personal data in the course of political campaigns. Statement 2/2019 (2019)
39. Amnesty International: Surveillance giants: how the business model of google and facebook threatens human rights (2019)
40. Bowman v. the United Kingdom. ECtHR (1998)

41. Bossuyt, M.J.: Guide to the "Travaux Préparatoires" of the International Covenant on Civil and Political Rights. Martinus Nijhoff Publishers, Dordrecht (1987)
42. Vermeulen, J.: Permissibility of news personalisation in view of the freedoms of thought and opinion. In: Presented at the BILETA Belfast: Back to the Futures? (2019)
43. Special Rapporteur on the promotion and protection of the right to freedom of opinion and expression: Report on the nature and scope of the right to freedom of opinion and expression, and restrictions and limitations to the right to freedom of expression (1994)
44. Feldek v. Slovakia. ECtHR (2001)
45. Sürek v. Turkey (No. 1). ECtHR (1999)
46. Mendel, T.: Freedom of Expression: A Guide to the Interpretation and Meaning of Article 10 of the European Convention on Human Rights. Centre for Law and Democracy. https://rm.coe.int/16806f5bb3
47. Lingens v. Austria. ECtHR (1986)
48. Savva Terentyev v. Russia. ECtHR (2018)
49. Handyside v. the United Kingdom. ECtHR (1976)
50. van Hoboken, J.V.J.: Search engine freedom: on the implications of the right to freedom of expression for the legal governance of Web search engines (2012). https://dare.uva.nl/search?metis.record.id=392066
51. Leander v. Sweden. ECtHR (1987)
52. Şener v. Turkey. ECtHR (2000)
53. Khursid Mustafa and Tarzibachi v. Sweden. ECtHR (2008)
54. High-Level Expert Group on Artificial Intelligence: Ethics Guidelines for Trustworthy AI. European Commission (2019)
55. Zuiderveen Borgesius, F.J., Trilling, D., Möller, J., Bodó, B., de Vreese, C.H., Helberger, N.: Should we worry about filter bubbles? Internet Policy Rev. **5**, 16 p. (2016). https://doi.org/10.14763/2016.1.401
56. Möller, J., Helberger, N., Makhortkh, M., van Dooremalen, S.: Filterbubbels in Nederland. Commissariaat voor de media (2019)
57. Susser, D., Roessler, B., Nissenbaum, H.: Online manipulation: hidden influences in a digital world. Geo. L. Tech. Rev. **4**(1), 45 p. (2019). https://doi.org/10.2139/ssrn.3306006
58. Stoll v. Switzerland. ECtHR (2007)
59. Professor Manfred Nowak. https://www.chr.up.ac.za/world-moot-currently-confirmed-judges-in-the-final-round/95-moot-courts/world-moot-court/judges/1648-professor-manfred-nowak. Accessed 08 Aug 2019
60. Special Rapporteur on the promotion and protection of the right to freedom of opinion and expression: Report on Artificial Intelligence technologies and implications for the information environment. E/CN.4/1995/32 (1994)
61. European Data Protection Board (EDPB): Guidelines on the processing of personal data under Article 6(1)(b) GDPR in the context of the provision of online services to data subjects (Version 2.0). Guidelines 2/2019 (2019)
62. European Data Protection Board (EDPB): Guidelines on the processing of personal data under Article 6(1)(b) GDPR in the context of the provision of online services to data subjects (Version for public consultation). Guidelines 2/2019 (2019)
63. GDPR: noyb.eu filed four complaints over "forced consent" against Google, Instagram, WhatsApp and Facebook. https://noyb.eu/4complaints/?lang=nl, https://noyb.eu/4complaints/. Accessed 25 Mar 2019
64. Commission nationale de l'informatique et des libertés (CNIL): The CNIL's restricted committee imposes a financial penalty of 50 Million euros against GOOGLE LLC. https://www.cnil.fr/en/cnils-restricted-committee-imposes-financial-penalty-50-million-euros-against-google-llc. Accessed 09 Aug 2019

65. Autoriteit Persoonsgegevens: Websites moeten toegankelijk blijven bij weigeren tracking cookies. https://www.autoriteitpersoonsgegevens.nl/nl/nieuws/websites-moeten-toegankelijk-blijven-bij-weigeren-tracking-cookies. Accessed 09 Aug 2019
66. Autoriteit Persoonsgegevens: Normuitleg AP over cookiewalls (2019)
67. Article 29 Data Protection Working Party: Guidelines on consent under Regulation 2016/679. WP259 rev.01 (2018)

Biometrics and Privacy

Data Privatizer for Biometric Applications and Online Identity Management

Giuseppe Garofalo(✉), Davy Preuveneers, and Wouter Joosen

imec - DistriNet, KU Leuven, Celestijnenlaan 200A, 3001 Heverlee, Belgium
{giuseppe.garofalo,davy.preuveneers,wouter.joosen}@cs.kuleuven.be

Abstract. Biometric data embeds information about the user which enables transparent and frictionless authentication. Despite being a more reliable alternative to traditional knowledge-based mechanisms, sharing the biometric template with third-parties raises privacy concerns for the user. Recent research has shown how biometric traces can be used to infer sensitive attributes like medical conditions or soft biometrics, e.g. age and gender. In this work, we investigate a novel methodology for private feature extraction in online biometric authentication. We aim to suppress soft biometrics, i.e. age and gender, while boosting the identification potential of the input trace. To this extent, we devise a min-max loss function which combines a siamese network for authentication and a predictor for private attribute inference. The multi-objective loss function harnesses the output of the predictor through adversarial optimization and gradient flipping to maximize the final gain. We empirically evaluate our model on gait data extracted from accelerometer and gyroscope sensors: our experiments show a drop from 73% to 52% accuracy for gender classification while loosing around 6% in the identity verification task. Our work demonstrates that a better trade-off between privacy and utility in biometric authentication is not only desirable but feasible.

1 Introduction

Biometrics have become a prevalent form of authentication. A broad spectrum of services, with their own unique security requirements, uses some form of biometric authentication, e.g. messaging applications or banking services. Biometrics are preferred over traditional knowledge-based systems, such as PINs and passwords, due to their ease of use, robustness and uniqueness. Moreover, the wide availability of mobile sensors allows for the deployment of near frictionless multi-modal systems.

Sensor based gait recognition is regarded as a promising approach towards unobtrusive user authentication [9,17,29,30]. Despite being less robust than well-established biometrics, motion data takes advantage of body worn sensors that are widely implemented in modern devices and require little to no effort by the

M. Friedewald et al. (Eds.): Privacy and Identity 2019, IFIP AICT 576, pp. 209–225, 2020.
https://doi.org/10.1007/978-3-030-42504-3_14

user. By enabling continuous user authentication, gait authentication is a natural candidate for multi-modal settings, e.g. by combining face recognition to walking data [17]. In this way, we not only improve accuracy, but also strengthen our system against forging and spoofing attacks [9]. Moreover, gender recognition is a relevant topic to be addressed in gait recognition for future applications in healthcare [17,30]. The ever-improving resilience of continuous authentication systems based on accelerometer and gyroscope measurements clashes with the lack of a comprehensive assessment in terms of sensitive data leakage, demanding for techniques to protect a user's privacy against sensitive inferences. In this work, we explore gait authentication and soft biometric recognition, i.e. gender and age, as a use case for our adversarial framework for privacy.

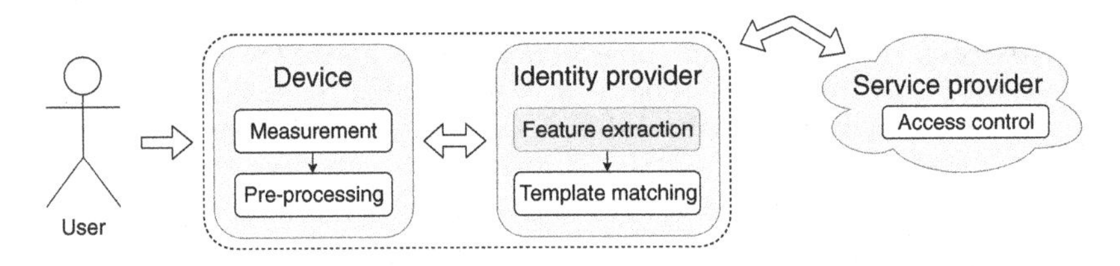

Fig. 1. Biometric authentication pipeline for online identity management. The user interacts with the identity provider and the service provider through his/her device.

As shown in Fig. 1, biometric authentication systems typically involve three entities [26]: a device equipped with sensors, a service provider that authorizes access, and an identity provider that verifies the identity. The authentication pipeline is composed of three steps. During step one, the user device collects the biometric signal. The latter is then cleaned and prepared for feature extraction, which is the second step in the pipeline. Consequently, the pre-processed signal is transformed into a set of relevant features that can be matched with a stored user template. For example, a face image may be turned into a vector of numeric features, while a gait trace could become a 2D image. Herein, the feature extraction and matching scheme are implemented by the identity provider. The final step consists of sending the output of template matching to the service provider, which grants access to its services based on proper access control policies. These three blocks can be incorporated as parts of the user's device or exist in isolation. In alternative, hybrid implementation are possible, e.g feature are extracted in the user's device while the templates are matched remotely. The latter scenario, i.e. online authentication, requires the user to send sensitive data over an unreliable network, exposing him/her to potential privacy leaks. Handling biometric data, including storing and processing templates, calls for additional security and privacy guarantees.

Misuse of biometric templates leads to severe privacy leakages for the end-user. Recent work has shown the presence of sensitive data in biometric traces, including medical conditions and soft biometrics [2,18,22]. If the user consented for his/her biometric template to be stored on a third-party server, he/she has to be aware of the potential disclosure of such sensitive data. For example, a *curious* service provider might want to learn more about its customers to advertise them with tailored products and increase its sales. Even in the unrealistic hypothesis that the user can blindly trust his/her recipient, adversaries might steal the user's template by impersonating the trusted party or attacking remote servers [32]. However, in many cases the third-party will need access to relevant information to keep its services alive. This calls for techniques which enable the sharing of the least amount of sensitive identifying information, e.g. *private biometric features.*

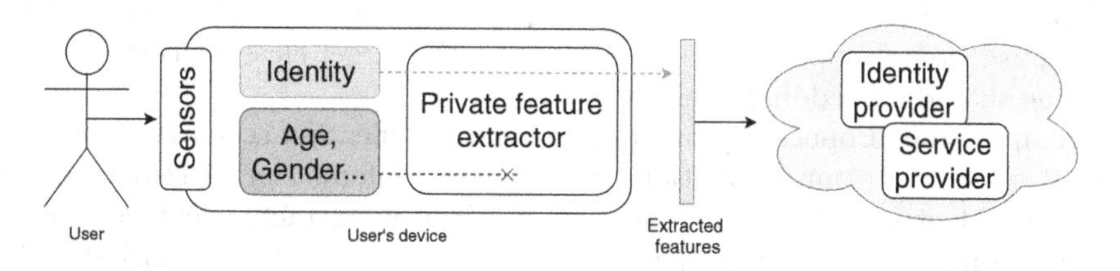

Fig. 2. Pipeline for private feature extraction in a biometric sharing scenario. The central element is the extractor, which is shared with the user.

Recent history testifies that algorithms are prone to discriminate based on racial or sex attributes [4,5], leading to discussions on how to mitigate bias in AI [13]. Typical sources of bias are the *training dataset*, directly reflecting unfair patterns in the external world, and a *flawed machine learning (ML) process*, not accounting for discrimination or, even worse, deliberately fueling inequalities. By suppressing highly-sensitive information like gender or age, we propose a novel representation of the biometric trace which discourages differences based on group membership.

Our work is also motivated by the General Data Protection Regulation (GDPR) [8], which tackles private data collection and processing problems. Article 25 puts the emphasis on the scope limitation by defining the *data minimization principle* which explicitly requires third-parties to limit their data collection to what is useful for their purposes, upon consent. However, storing data such as biometrics clashes with this principle because of its intrinsic repurposable nature, which has been theorized and validated thoroughly in recent years. Hence, there is a need to design data reduction processes for sharing sensitive information in order to protect the users against unprompted attribute inferences. By sharing only what is needed for a predefined task, we can also address the discrimination problem, bringing fairness and transparency in the ML pipeline.

Figure 2 shows our pipeline for private biometric sharing. The *user's device* is in charge of collecting motion data and extracting features to be shared with an external authentication provider. However, the user typically lacks the necessary resources, both hardware and data, to train a feature extractor him/her-self. Thus, we embrace a data-driven approach to derive a private feature extractor on the server side. This approach exploits a publicly available dataset to train the model, allowing to derive only one feature extractor for all the users willing to authenticate. In particular, our adversarial framework is composed of three entities closely working together: a (private) feature extractor, an (adversarial) siamese identity verifier and a predictor for the private task. During training, the feature extractor will iteratively adapt to changes in the classifier, which in turn will challenge the extractor. This procedure models the mutual information between the identity and the private attributes, guaranteeing protection against sensitive inferences. Eventually, the feature extractor is published by the identity/service provider and becomes available for local usage on the user's device, as depicted in Fig. 2. By harnessing our feature extractor, neither third parties nor channel eavesdroppers can accurately infer the target classification attribute from the shared embedding.

We apply our proposed framework in a gait verification scenario using fixed inertial sensors. In our evaluation, we compare different privatized traces to assess the identifiability of the users while the new extracted features cannot be used to infer the user's gender or age, which are our private attributes. We emulate our adversary's ability to infer the sensitive attributes by means of transfer learning, i.e. training an unseen classifier for the private task on the private features. Our main contributions include:

1. Devising an adversarial framework exploiting a novel loss function to train a private feature extractor starting from variable-length gait traces.
2. Evaluating the privacy-utility trade-off w.r.t soft biometrics privacy in the gait authentication domain.
3. Using the biggest known inertial sensors dataset, which includes almost 500 users and 5 different activities.

The rest of the paper is organized as follows. We identify the gap and differences with related work in Sect. 2. We present our framework for privacy-preserving feature extraction in Sect. 3. The experimental protocol and results are presented in Sect. 4. Section 5 concludes our work.

2 Related Work

Mordini and Ashton [22] have performed an extensive study of medical pattern retainment in biometric templates: psychiatric conditions can be inferred from gait traces, chromosomal diseases can be accurately guessed from face images or fingerprints, while neurological pathologies have been associated to a broad range of behavioural biometrics. The same leakage potential holds true for electrocardiogram (ECG) signals [18], iris recognition [2] and other bio or behavio-metrics [6]. Similarly, soft biometrics like age, gender or race are linked to physiological or behavioural traits of the user. In a recent work, we proved the feasibility of age and gender estimation from gait traces in the frame of the *OU-ISIR Wearable Sensor-based Gait Challenge: Age and Gender* (GAG 2019) competition at the 12th IAPR International Conference on Biometrics[1] [28]. The goal of this

[1] http://www.am.sanken.osaka-u.ac.jp/GAG2019/.

competition was to improve the state-of-the-art in soft biometric prediction from accelerometer and gyroscope traces. Even without crucial information on sensors position, we were able to achieve ~76% accuracy for gender classification and a mean absolute error of ~6 years for age estimation, eventually obtaining the best result among all contestants. Our model is inspired by [28] as follows: we harness temporal convolutional networks (TCNs) for feature extraction and few dense layers for soft biometric prediction. On top of the extracted features, we have built a siamese network for user verification and we have plugged a gradient reversal layer for attribute privatization.

Several works tackled the problem of discrimination in the ML pipeline [1,7]. Typically, they focus on the output of the decision function and how to make it independent from a particular group membership. In contrast to previous work, we address this problem indirectly, aiming to achieve **soft biometric privacy**. By suppressing information deemed to be private, we discourage discriminative attributes to influence the learning process, thereby representing a source of bias. For example, by minimizing the information about the gender in motion data, we encourage the building of a gender-agnostic gait verification system. In the soft biometric privacy landscape, our work is the first one focusing on time sequences and, specifically, gait authentication.

The approaches to protect user's privacy divide into context-free and context-aware techniques. Context-free techniques, like differential privacy (DP), model worst-case adversaries regardless of his/her real capabilities and discarding relevant contextual information, i.e. about the problem to be solved. DP provides strong privacy guarantees, delivering a shrinking in data usefulness. Context-aware strategies, on the other hand, incorporate the retainment of task-specific utility by selectively adding noise where it matters. This advantage comes at the expenses of a formal characterization of the relationship between public variables, i.e. what we aim to share, and private variables, i.e. what we aim to protect, which is rarely available in practice.

Data-driven optimization has been recently proposed as a mean to achieve context-aware privacy. By exploiting recent advances in adversarial optimization, it is possible to model the joint distribution between shared and private variables. Generative adversarial networks (GANs) have been recently proposed as an effective tool to achieve this goal [12]. They model a min-max game between a generator and a discriminator, where the former tries to fool the latter in an iterative learning process. This concept has been first adapted to the privacy domain by Huang et al. who define the generative adversarial privacy (GAP) framework [14]. Inspired by their work, we harness adversarial training to obtain a private feature extractor, representing our feature generator. This generator is used to obtain a compressed template representation for a specific, measurable and limited purpose, while also minimizing sensitive disclosure.

Morales et al. [21] recently proposed a method to reduce gender and race information in latent representations of face images. Their method is based on a modification of the triplet loss function, which is commonly employed in face verification scenarios [27]. Our work differs from theirs for two reasons: first, we exploit adversarial optimization to maximize the privacy-utility trade-off; second, our feature extractor is designed to model temporal dependencies in the input data, making it more suitable for gait samples than face images. Similarly, Mirjalili et al. proposed a framework to impart gender privacy to face images [19,20]. Drawing from their work, we empirically evaluate the privacy and generalizability of our approach by training several models, which simulate the ability of a malicious entity. As before, they focus on face recognition systems rather than temporal data.

Malekzadeh et al. [16] have considered motion data and gait authentication in a different min-max optimization scenario: perturbing identity while preserving task-specific utility. Their classification task is activity recognition, which has been extensively studied in the gait literature in addition to being arguably a private variable. Moreover, we shift the focus towards building a private extractor to be used by end-users instead of generating a privatized trace in the input domain. By compacting the trace in a latent space representation, we reduce the interpretability of the shared sample while minimizing the risks of sensitive inference. Similarly, Ossia et al. [24] investigates the use of siamese networks for privatizing the user's identity while preserving gender classification accuracy. Besides the different learning goal, they focus on fine-tuning existing, pre-trained networks. In our framework, the minimization of sensitive attribute is embedded in the learning process itself. By simply applying fine-tuning to the last layers of the feature extractor, we would discourage the achievement of a better sub-optimal solution for the min-max optimization problem.

3 Private Feature Extraction Framework

We propose a novel framework for protecting sensitive variables when sharing biometric data in an online authentication scenario. In this section, we tackle 4 key aspects which characterize our framework: (i) the *main steps*, stakeholders, and threat model, (ii) the nominal *privacy-preserving loss function*, (iii) the designed *architecture*, i.e. the neural networks to approximate the nominal loss, (iv) and the architecture *min-max optimization strategy*.

3.1 High-Level Framework and Threat Model

Our framework faces the problem of sharing biometric data without exposing user's private information. This process requires the interaction between two entities: the user and the service provider. The user is willing to share what is needed to accomplish the main task but he is worried that certain information might leak along the way. Let us assume user A wants to be authenticated towards service B, then sharing the raw data will reveal attributes which A might consider private, e.g. A's gender. As discussed in Sect. 2, traditional techniques can be employed to solve this problem, however they come with several limitations such as expensive computations at the edge or having to trust external entities. Instead, we propose the use of a contained set of *private features* extracted from biometric data on the user side. These features are carefully optimized during the training phase, with a twofold purpose: (1) to preserve the information which identifies the user and (2) to suppress a specific private variable, e.g. the gender of the user.

Our proposed framework requires three actions by the involved actors: (i) the identity/service provider trains a private feature extractor for authentication purposes; (ii) the feature extractor is published, which allows the user to extract authentication features and assess its privacy guarantees; (iii) by following the authentication protocol, the authentication features are shared with the service provider which grants access to the system based on given access control policies. This three-step procedure protects the user against unprompted sensitive inferences: the service provider will not be able to improve its knowledge about the user w.r.t. to the selected private variables, thus we protect against *function creep* [31]. This is inherent to the local feature extraction step (ii), which suppresses private variables within the user's device. Moreover, sharing the private extractor enables any external entities, like the user, to assess and analyze the privacy of the model, which is only trained on *public data*. It is worth noticing that the authentication-party might want to train a ML model to authenticate the user based on the received features, i.e. step (iii), and this part of the model has to be kept private in order to guarantee users' privacy. In addition, the feature extractor could be trained by different parties than the third-party, but we rely on the realistic hypothesis that the features are especially crafted for the main task. Therefore, the service provider is better suited to design the feature extraction step. If we assume that the features are intended for different uses, then we can assign step (i) to another external, mediating entity without affecting the presented framework.

While our solution overcomes traditional techniques limitations, several challenges regarding privacy estimation arise. By delegating the training phase to a cloud-based service, we cut down the computational power which is requested to the user. Thus, only feature extraction of a pre-trained model is performed on the user side. In addition to energy consumption, by extracting the biometric representation locally, we avoid the need for an external mediator. Therefore, we free the user from trusting an external entity. However, unlike traditional techniques like DP, we can only provide empirical privacy guarantees of the shared representation. This is due to a different sharing scenario, which involves single temporal traces as opposed to large databases of many users.

We evaluate the privacy of the extracted representation by looking at the performance of a newly trained ML classifier. Having fixed the discernment of a discrete sensitive variable as the learning goal, and provided the extracted features as the model input, we derive an empirical definition of privacy: the better the classifier performs, the higher the sensitive leakage. In practice, the classifier mimics the capabilities of a *curious service provider* willing to obtain valuable information about its users. The provider has access to the public dataset used to train the feature extractor (step (i)), and it also knows the training details as well as the trained model weights. Thus, the third-party is able to use public data

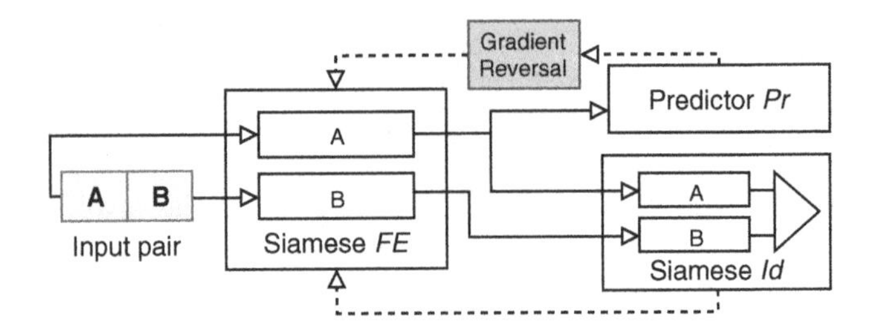

Fig. 3. Our framework with its three challenging entities: the siamese feature extractor (FE), the soft biometric predictor, and the siamese identity verifier (Id).

to train a classifier discerning the private variable from the extracted features. We test the classifier against a group of test users, which simulates the sharing of features during regular usage. The service provider acts as the most powerful adversary for knowledge and resources, so we indirectly test the privacy of the extractor against any, less or equally powerful, external adversaries trying to infer sensitive information from the shared representation. In conclusion, if the service provider is not able to train a classifier which discerns the private variable from the shared representation, we consider the features to be *safe* against inferences. However, it has to be stressed that a more powerful estimator which is able to extract private information from the shared template might exist, which is a realistic assumption drawn from recent work [25].

3.2 Privacy-Preserving Training Objective

We present here our nominal loss function for soft biometric privacy, i.e. the ideal training objective to be approximated via neural networks optimization. Given an input X, we search for the optimal feature extractor $FE(\cdot)$, which outputs the private embedding $Z = FE(X)$. We define $D(\cdot, \cdot)$ as a measure of the dependency between two variables, such that $D(Z, T)$ measures the dependency between the private embedding and the classification task T we aim to suppress, while $D(Z, I)$ describes the usefulness of the latent representation for authentication purposes, being I the identity of the user. The nominal loss to be minimized becomes

$$NL = \alpha * D(Z, I) - \beta * D(Z, T) \tag{1}$$

where α and β regulate the importance of each term, which purpose is to fine-tune the privacy-utility trade-off. As mentioned before, we make use of deep neural networks to approximate $D(\cdot, \cdot)$. Therefore, an estimation of this measure is embedded in the weights of the models after training. We estimate it by extracting features and analyzing what newly created ML models are capable of learning from these features.

3.3 Neural Networks Architecture

Our framework is composed of three competing blocks: a private extractor (Fig. 4), an identity verifier (Fig. 5b) and a task-dependent predictor (Fig. 5a). As mentioned earlier, the classification pipeline is inspired by [28] with the addition of a siamese block for gait verification. An overview of the interacting components is presented in Fig. 3.

The input of the model is 3-dimensional raw accelerometer and 3-dimensional raw gyroscope measurements recorded from inertial sensors. In our implementation, we account for variable-length input by simply stacking sensors measurements (6-dimensional measurements) without pruning the obtained sequences.

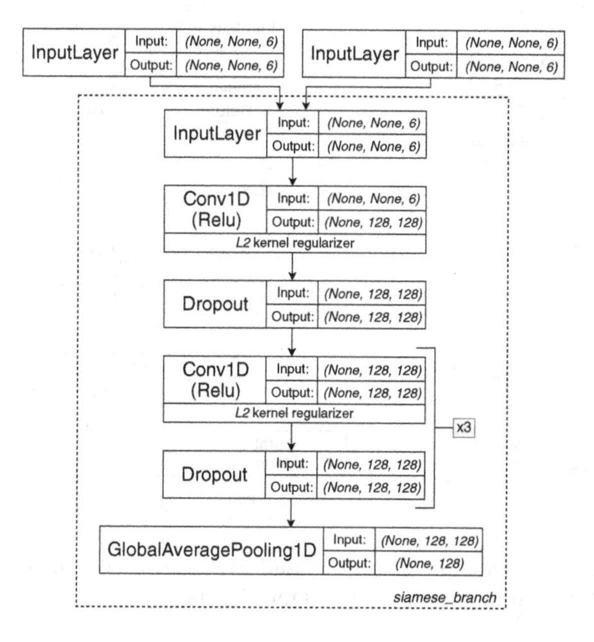

Fig. 4. Implementation of *Siamese FE*. The sub-model *siamese_branch* is shared among the two input layers and contains a quintessential element, namely *Conv1D*.

Input data is fed to a *siamese feature extractor FE*, which details are presented in Fig. 4. Every layer is presented along with its input and output shapes, where *None* represents either a variable length trace or a variable number of samples, i.e. a batch. As first proposed by Bai et al. [3], we perform dynamic feature extraction by harnessing temporal convolutional networks (TCNs). This family of networks have been demonstrated to achieve state-of-the-art performance when dealing with temporal data, behaving equal to or better than recurrent neural networks. TCNs potential is mainly due to *dilated* convolutions, which address both complexity of the network and low-level spatial accuracy. Convolutional layers are intertwined with *dropout* layers to prevent over-fitting the training set, thus acting as regularizers by zeroing-out random filters which are re-activated when testing the model. Following best practices, $L2$ regularization is also introduced to penalize complex models through the loss function. Finally, a *global average layer* flattens the output of the extractor to obtain 128 features, which are the objective of the privatization of our optimization framework. They are then conveyed into two branches: a *soft biometric predictor* and a *siamese identity verifier.*

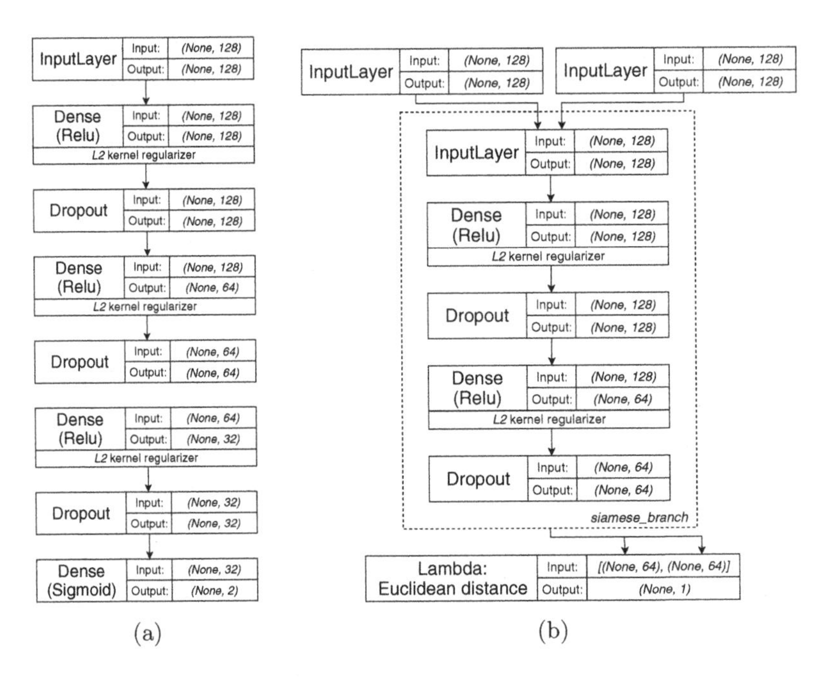

Fig. 5. Implementations of *Predictor* (a) and *Siamese Id* (b). In (b), the sub-model *siamese_branch* is shared among the two input layers.

The first branch is the soft biometric predictor *Pr*, which details are depicted in Fig. 5a. This model is composed of several fully connected layers with *ReLU* activations. As before, dropout and regularization techniques are employed to improve generalizability. The model is designed once but tuned and optimized separately for gender and age. During optimization, we minimize *binary cross-entropy* to maximize our sensitive variable prediction accuracy.

The second branch is a siamese neural network for identity verification *Id*, which details are shown in Fig. 5b. This model is composed of two stacked fully connected layers which are duplicated into two parallel branches sharing their weights (in Fig. 5b, only one branch is shown). These branches converge into a distance-based function, the *contrastive loss*, which is typically used in a verification scenario to increase the similarity, i.e. decrease the distance in the latent space, of samples belonging to the same class while driving away dissimilar pairs.

In order to obtain two feature vectors, the feature extractor is turned into a siamese model as follows: we duplicate the FE, obtaining two branches which share their weights and allowing to feed a pair of gait traces into the input layer.

By combining our three blocks, we end up with the following multi-objective loss function

$$G(\theta, \phi) = min_{\theta,\phi}(\alpha * Id(\phi) + \beta * Pr(\theta)) \tag{2}$$

where θ and ϕ are the weights of *PR* and *Id*, respectively, and α and β weight each term. By minimizing G, however, the predictor accuracy is maximized, thus we need to reverse this trend to protect the latent representation against inferences of the sensitive variable. First introduced in domain adaptation by Ganin et al. [10], we plug a *gradient reversal layer* between *FE* and *Pr* which minimizes the *Pr* accuracy when training *FE*. This leads to a min-max loss function which is the approximated realization of Eq. 1.

3.4 Optimization of the Networks

As common in adversarial training, the three blocks are optimized separately, via *strict alternation*. Figure 3 shows the interaction among the elements, which single iteration works as follows. *Id* and *Pr* are fed with features extracted from *FE* and trained for one epoch to minimize the *contrastive loss* and the *cross-entropy*, respectively. *FE* is later trained for a single epoch to optimize Eq. 2 by freezing the weights of the other networks. By reversing the gradient of *Pr*, *FE* will adapt to its changes becoming more and more resilient to sensitive inferences. This challenge approximates the dependency measure D (cfr. Eq. 1) which is key to achieve a satisfactory sub-optimal trade-off between privacy and utility.

By alternating the gradient updates among the three networks, we enforce dynamic adaptation to future updates. Suppose *Pr* is trained w.r.t. the true labels until a acceptable sub-optimal solution is found. *FE* can be trained to beat *Pr*, hiding the sensitive variable from the learnt representation. However, this holds true only for one sub-optimal solution of *Pr* and does not prevent inferences from future re-training. Instead, by letting *Pr* adapt to changes in *FE*, and vice-versa, both models converge to a more satisfactory solution in terms of privacy vs. utility. Hyper-parameters play an important role in network convergence, impeding *Pr* to win over *FE*; yet, we can only empirically estimate the best hyper-parameters for our task.

Our feature extraction strategy differs from traditional dimensionality reduction techniques because we actively trigger the reaction of a discriminator, exploiting its input to refine our final output. However, approaches like Principal Component Analysis (PCA) or noise addition at the bottleneck of the network could still be applied as a complementary, simple aid to achieve a better sub-optimal solution.

4 Evaluation

In this section, we evaluate the ability of our framework to hinder classification of gender from motion features while preserving the identity of the user. We present the experimental setup, followed by the evidence we found.

4.1 Experimental Setup

We define our networks using Keras 2.2.4 with a Tensorflow backend, which runs on a machine with a 3.4 GHz i5-7500 CPU, 16 GB RAM, and an NVIDIA Titan V GPU. As in [28], we choose the OU-ISIR labelled gait action dataset [23] to train our model. Angular velocity and acceleration in the 3 dimensions are collected from sensors fixed on a belt, at a sampling rate of 100 Hz. Every trace is associated to age, gender and current activity of the user.

In order to generate training pairs, we undertake several steps. We first divide users into training and test sets following a 80%–20% proportion, after which we use the same ratio to split the training set into training and validation sets. Then we generate fixed length sequences through windowing with overlapping: a user is selected and his/her trace is divided into several traces of length 2.56 s with 20% overlap, i.e. 0.5 s at the end and 0.5 s at the beginning of the trace. Finally, we create pairs to feed our model with through an iterative procedure in two steps: (1) for each user, the current trace is coupled with the subsequent one and a *similar pair label* is assigned to the pair; (2) the second trace of the previous pair is coupled with a random trace from a different user and a *dissimilar pair label* is assigned to this couple. Since the first branch of the siamese *FE* (i.e., branch A in Fig. 3) is responsible for feeding *Pr*, a label with the gender of the first user is associated to the pair.

We empirically select the best hyper-parameter configuration for our networks. We select the Nadam optimizer to train the models, following the procedure explained in Sect. 3.4. Every model is trained on mini-batches within the set [5,25,50], while the number of epochs varies between 15 and 50. We fix α and β to 1. Intuitively, one can expect α and β to have a predictable impact on the privacy-utility trade-off when training the features extractor. Due to the adversarial nature of the training procedure, however, tuning these variables proved to be highly sensitive w.r.t. the given setting and selected hyper-parameters. We argue that this effect can be associated to the training of the predictors, which are independent from α and β in our implementation. Nevertheless, as underlined before, trivial noise addition and dimensionality reduction could be employed as a better, more stable alternative for privacy-utility trade-off tuning.

After training, the sensitive variable predictor and the identity verifier are re-trained on the privatized features. As suggested by previous work [24], the privacy of a sensitive variable can be evaluated via transfer learning, i.e. freezing our pre-trained feature extractor and training a soft biometric predictor from scratch. In order for a thorough evaluation of the generalizability of our approach, unseen classifiers have to be taken into consideration. Hence, we define two models: (1)

Table 1. Scores for baseline siamese verifier trained without predictor feedback.

Epochs	Verification accuracy	f1-score (SVM)
200 (early stopping)	90.93% ±0.15	72.58%

Table 2. Re-train f1-score for the predictors and accuracy for the siamese identity verifier after adversarial privatization.

Epochs	Batch size	Verification accuracy	f1-score	f1-score (SVM)
15	25	82.14% ± 0.89%	51.15% ± 0.70%	60.97%
15	50	87.15% ± 0.38%	50.68% ± 0.57%	65.26%
25	25	84.47% ± 0.53%	50.20% ± 0.15%	63.28%
25	50	85.28% ± 0.48%	50.10% ± 0.00%	**52.99%**

a DNN-based predictor resembling the one we used in our adversarial framework, which ensures protection against our target model; (2) a Support Vector Machine (SVM) which is typically used downstream of DNN for feature extraction. The SVM is optimized by applying grid search, which exhaustively searches for the hyper-parameter combination with the best score. $L2$-normalization is also performed to maximize our accuracy metric, i.e. *f1-score*.

We repeat each experiment 10 times, reporting the f1-score for the gender and the average verification accuracy for the verifier. All the results, besides the f1-score which results from a deterministic search, are presented with their standard deviation.

4.2 Experimental Results

Table 1 shows the results for our baseline model: a siamese feature extractor for identity verification. After training, the extracted features are used to infer the gender of the user, resulting in a f1-score of 72.58% for the SVM in the best configuration. This underlines the retainment of soft biometric information in the authentication features, especially if we compare this figure to the state-of-the-art gender prediction accuracy presented in [28], i.e. 75.77%.

We compare our baseline with our proposed approach for feature privatization, which is summarized in Table 2. A high variability in the results can be observed, which is mainly due to the instability of the adversarial learning procedure. The *f1-score* shows how our privatization mechanism protects the features against possible re-training of our target predictor in each setting. However, we have to take into account generalization and we must be able to protect against unseen classifiers. SVM *f1-score* proves that we are able to achieve a nearly optimal result (50%) by carefully tuning our hyper-parameters, i.e. number of epochs and batch size. This comes at the expense of a nearly 6% loss in verification accuracy. For a batch size of 50, and increase in the number of epochs corresponds with a slight decline in verification accuracy. This drop indicates how letting the network train for a larger number of epochs improves the empirical privacy (see smaller f1-scores in Table 2) while decreasing the utility, even by just a tiny fraction.

We identify two main limitations for this work which are linked to our privacy evaluation and chosen dataset. First, we empirically evaluate the privacy of our framework by re-training different models on the extracted features. Future work could derive a formal evaluation of the privacy guarantees from an information-theory point of view. Second, the OU-ISIR dataset provides us with sufficient data for our scopes but its data is collected in a constrained environment by sensors fixed on a belt. In a real world scenario, we deal with different orientations of mobile devices, and its sensors, carried by the users. A more realistic dataset is needed to properly evaluate and compare solutions in the gait domain.

As a future direction, we aim to tackle the linkability of templates across services while hiding different private variables for one user. Since we are not delivering a full-fledged biometric template protection (BTP) scheme, we do not directly address linkability of traces, assessing instead the retainment of private information for a specific use-case, i.e. gender classification. Hence, our framework alone does not fulfill the two requirements of the standard on biometric data protection ISO/IEC 24745 (2011) [15], i.e. *irreversibility* and *unlinkability*. However, by tackling the data minimization problem we aim to address problems which are complementary to BTP schemes: (1) we exclude unnecessary data from transmission and processing, possibly improving privacy and performance of crypto schemes, and (2) we help preventing or fighting back algorithmic bias by feeding algorithms with more neutral and task-specific data. We envision a hybrid system where BTP schemes and adversarial training maximize the utility for a specific task without compromising users privacy. Future directions include exploring age or race prediction to evaluate cross-task linkability, and analyzing the advantage of applying a biometric crypto scheme on top of our minimization framework for compliance with existing requirements for private and secure biometric data processing and management. To this extent, Barrero et al. [11] have proposed a metric to evaluate the local and global linkability of biometric templates.

5 Conclusion

In this work, we demonstrated the effectiveness of an adversarial learning technique towards privatization of biometric features from sequential data. We evaluated our approach on the gender estimation use-case, inspired by a recent work. Our evaluation supported our approach, showing a dip in the f1-score from 73% to 52.99% in the best case, which is very close to random guess (50%).

Further evaluation is needed to assess the effectiveness of our approach against different use-cases, but our results show that a solution to the long standing problem of data-minimization for biometrics is possible. Data-driven techniques have the potential to achieve the optimal trade-off between privacy and utility, something traditional techniques usually struggle with. We advocate for new tools for the user to manage his own identity and the amount of sensitive information which is shared with third-parties.

Acknowledgement. This research is partially funded by the Research Fund KU Leuven. Work for this paper was supported by the European Commission through the H2020 project CyberSec4Europe (https://www.cybersec4europe.eu/) under grant No. 830929. We gratefully acknowledge the support of NVIDIA Corporation with the donation of the Titan V GPU used for this research.

References

1. Alvi, M., Zisserman, A., Nellåker, C.: Turning a blind eye: explicit removal of biases and variation from deep neural network embeddings. In: Leal-Taixé, L., Roth, S. (eds.) ECCV 2018. LNCS, vol. 11129, pp. 556–572. Springer, Cham (2019). https://doi.org/10.1007/978-3-030-11009-3_34
2. American Academy of Ophthalmology: Evidence mounts that an eye scan may detect early Alzheimer's disease (2018). https://www.aao.org/newsroom/news-releases/detail/evidence-eye-scan-may-detect-early-alzheimers. Accessed 14 May 2019
3. Bai, S., Kolter, J.Z., Koltun, V.: An empirical evaluation of generic convolutional and recurrent networks for sequence modeling. arXiv:1803.01271 (2018)
4. Buolamwini, J., Gebru, T.: Gender shades: intersectional accuracy disparities in commercial gender classification. In: Proceedings of the 1st Conference on Fairness, Accountability and Transparency. Proceedings of Machine Learning Research, vol. 81, pp. 77–91. PMLR, New York, 23–24 February 2018
5. Cohn, J.: Google's algorithms discriminate against women and people of colour (2019). http://theconversation.com/googles-algorithms-discriminate-against-women-and-people-of-colour-112516. Accessed 14 May 2019
6. Dantcheva, A., Elia, P., Ross, A.: What else does your biometric data reveal? A survey on soft biometrics. IEEE Trans. Inf. Forensics Secur. **11**(3), 441–467 (2016)
7. Das, A., Dantcheva, A., Bremond, F.: Mitigating bias in gender, age and ethnicity classification: a multi-task convolution neural network approach. In: Leal-Taixé, L., Roth, S. (eds.) ECCV 2018. LNCS, vol. 11129, pp. 573–585. Springer, Cham (2019). https://doi.org/10.1007/978-3-030-11009-3_35
8. European Parliament: Regulation (EU) 2016 of the European Parliament and of the Council, on the protection of natural persons with regard to the processing of personal data and on the free movement of such data, and repealing Directive 95/46/EC (General Data Protection Regulation) (2016)
9. Gafurov, D.: A survey of biometric gait recognition: approaches, security and challenges. In: NIK Conference (2007)
10. Ganin, Y., et al.: Domain-adversarial training of neural networks. J. Mach. Learn. Res. **17**(59), 1–35 (2016)
11. Gomez-Barrero, M., Galbally, J., Rathgeb, C., Busch, C.: General framework to evaluate unlinkability in biometric template protection systems. IEEE Trans. Inf. Forensics Secur. **13**(6), 1406–1420 (2018)
12. Goodfellow, I., et al.: Generative adversarial nets. In: Advances in Neural Information Processing Systems 27, pp. 2672–2680. Curran Associates, Inc. (2014)
13. Hao, K.: This is how AI bias really happens—and why it's so hard to fix (2019). https://www.technologyreview.com/s/612876/this-is-how-ai-bias-really-happensand-why-its-so-hard-to-fix/. Accessed 14 May 2019

14. Huang, C., Kairouz, P., Chen, X., Sankar, L., Rajagopal, R.: Context-aware generative adversarial privacy. Entropy **19**(12), 656 (2017)
15. Information technology - Security techniques - Biometric information protection. Standard, International Organization for Standardization (2011)
16. Malekzadeh, M., Clegg, R.G., Cavallaro, A., Haddadi, H.: Mobile sensor data anonymization. In: Proceedings of the International Conference on Internet of Things Design and Implementation, IoTDI 2019, pp. 49–58. ACM (2019)
17. Marsico, M.D., Mecca, A.: A survey on gait recognition via wearable sensors. ACM Comput. Surv. **52**(4), 86:1–86:39 (2019)
18. Matovu, R., Serwadda, A.: Your substance abuse disorder is an open secret! Gleaning sensitive personal information from templates in an EEG-based authentication system. In: 2016 IEEE 8th International Conference on Biometrics Theory, Applications and Systems (BTAS), pp. 1–7, September 2016
19. Mirjalili, V., Raschka, S., Ross, A.: Gender privacy: an ensemble of semi adversarial networks for confounding arbitrary gender classifiers. In: 2018 IEEE 9th International Conference on Biometrics Theory, Applications and Systems (BTAS), pp. 1–10, October 2018
20. Mirjalili, V., Raschka, S., Ross, A.: FlowSAN: privacy-enhancing semi-adversarial networks to confound arbitrary face-based gender classifiers. IEEE Access **7**, 99735–99745 (2019)
21. Morales, A., Fiérrez, J., Vera-Rodríguez, R.: SensitiveNets: learning agnostic representations with application to face recognition. CoRR abs/1902.00334 (2019)
22. Mordini, E., Ashton, H.: The transparent body: medical information, physical privacy and respect for body integrity. In: Mordini, E., Tzovaras, D. (eds.) Second Generation Biometrics: The Ethical, Legal and Social Context. The International Library of Ethics, Law and Technology, vol. 11, pp. 257–283. Springer, Dordrecht (2012). https://doi.org/10.1007/978-94-007-3892-8_12
23. Ngo, T.T., Makihara, Y., Nagahara, H., Mukaigawa, Y., Yagi, Y.: Similar gait action recognition using an inertial sensor. Pattern Recogn. **48**(4), 1289–1301 (2015)
24. Ossia, S.A., Shamsabadi, A.S., Taheri, A., Rabiee, H.R., Lane, N.D., Haddadi, H.: A hybrid deep learning architecture for privacy-preserving mobile analytics. CoRR abs/1703.02952 (2017)
25. Pittaluga, F., Koppal, S., Chakrabarti, A.: Learning privacy preserving encodings through adversarial training. In: 2019 IEEE Winter Conference on Applications of Computer Vision (WACV). IEEE (2019)
26. Rui, Z., Yan, Z.: A survey on biometric authentication: toward secure and privacy-preserving identification. IEEE Access **7**, 5994–6009 (2019)
27. Schroff, F., Kalenichenko, D., Philbin, J.: FaceNet: a unified embedding for face recognition and clustering. In: 2015 IEEE Conference on Computer Vision and Pattern Recognition (CVPR), pp. 815–823, June 2015
28. Van hamme, T., Garofalo, G., Argones Rúa, E., Preuveneers, D., Joosen, W.: A systematic comparison of age and gender prediction on IMU sensor-based gait traces. Sensors **19**(13), 2945 (2019)
29. Van hamme, T., Preuveneers, D., Joosen, W.: Improving resilience of behaviometric based continuous authentication with multiple accelerometers. In: Livraga, G., Zhu, S. (eds.) DBSec 2017. LNCS, vol. 10359, pp. 473–485. Springer, Cham (2017). https://doi.org/10.1007/978-3-319-61176-1_26
30. Wan, C., Wang, L., Phoha, V.V.: A survey on gait recognition. ACM Comput. Surv. **51**(5), 89:1–89:35 (2018)

31. Winner, L.: Autonomous Technology: Technics-Out-of-Control as a Theme in Political Thought. MIT Press, Cambridge (1977)
32. Zeitz, C., et al.: Security issues of internet-based biometric authentication systems: risks of man-in-the-middle and BioPhishing on the example of BioWebAuth. In: Security, Forensics, Steganography, and Watermarking of Multimedia Contents, p. 68190R (2008)

What Does Your Gaze Reveal About You? On the Privacy Implications of Eye Tracking

Jacob Leon Kröger[1,3](✉), Otto Hans-Martin Lutz[1,2,3], and Florian Müller[1,3]

[1] Technische Universität Berlin, Straße Des 17. Juni 135, 10623 Berlin, Germany
kroeger@tu-berlin.de
[2] Fraunhofer Institute for Open Communication Systems, Berlin, Germany
[3] Weizenbaum Institute for the Networked Society, Berlin, Germany

Abstract. Technologies to measure gaze direction and pupil reactivity have become efficient, cheap, and compact and are finding increasing use in many fields, including gaming, marketing, driver safety, military, and healthcare. Besides offering numerous useful applications, the rapidly expanding technology raises serious privacy concerns. Through the lens of advanced data analytics, gaze patterns can reveal much more information than a user wishes and expects to give away. Drawing from a broad range of scientific disciplines, this paper provides a structured overview of personal data that can be inferred from recorded eye activities. Our analysis of the literature shows that eye tracking data may implicitly contain information about a user's biometric identity, gender, age, ethnicity, body weight, personality traits, drug consumption habits, emotional state, skills and abilities, fears, interests, and sexual preferences. Certain eye tracking measures may even reveal specific cognitive processes and can be used to diagnose various physical and mental health conditions. By portraying the richness and sensitivity of gaze data, this paper provides an important basis for consumer education, privacy impact assessments, and further research into the societal implications of eye tracking.

Keywords: Eye tracking · Gaze · Pupil · Iris · Vision · Privacy · Data mining · Inference

1 Introduction

Being an important part of visual perception and human behavior, eye movements have long been a subject of research interest. The first approaches to measure a person's gaze direction date back to the early 1900s [74]. Until recently, these technologies were severely limited by the cost of the equipment required, a lack of precision, and poor usability and were only used in very specific niches of research. Over the last few years, however, with rapid advances in sensor technology and data processing software, eye tracking solutions have become easy to use, lightweight, efficient, and affordable and found increasing adoption in many fields, including gaming, marketing, automotive technology, military, and healthcare [26].

M. Friedewald et al. (Eds.): Privacy and Identity 2019, IFIP AICT 576, pp. 226–241, 2020.
https://doi.org/10.1007/978-3-030-42504-3_15

While alternatives[1] exist, the most popular method today is video-based eye tracking, where mathematical models are used to calculate a person's gaze direction from video recordings, for example based on the shape and position of pupil and iris, or based on light reflection patterns in the eyes [2]. This method can not only be used in head-mounted devices, such as smart glasses and virtual reality headsets, but also through built-in cameras in laptops, tablets, and smartphones without requiring any additional hardware [45, 56]. With further improvements in cost and performance, eye tracking may soon be included as a standard feature in various consumer electronics, moving us towards a "pervasive eye tracking world" [58].

The many beneficial uses and enormous potentials of the rising technology have to be acknowledged and should be embraced. However, a more ubiquitous use of eye tracking will also raise serious privacy concerns – not only because gaze data may be collected and shared in non-transparent ways, but also because such data can unexpectedly contain a wealth of sensitive information about a user.

Drawing from a broad range of scientific disciplines, including neuroscience, human-computer interaction, medical informatics, affective computing, experimental economics, psychology, and cognitive science, this paper provides a structured overview and classification of sensitive pieces of information that can be disclosed by analyzing a person's eye activities. According to the reviewed literature, eye tracking data may reveal information about a user's biometric identity (Sect. 2.1), mental activities (Sect. 2.2), personality traits (Sect. 2.3), ethnic background (Sect. 2.4), skills and abilities (Sect. 2.5), age and gender (Sect. 2.6), personal preferences (Sect. 2.7), emotional state (Sect. 2.8), degree of sleepiness and intoxication (Sect. 2.8), and physical and mental health condition (Sect. 2.9). In order to take rapidly evolving technology trends and newly emerging privacy threats into account, we will consider not only proven and established approaches but also inference methods that are subject to ongoing research. Limitations of the presented methods and their practical applicability will be reflected upon in Sect. 3, followed by a conclusion in Sect. 4.

2 Inference of Personal Information from Eye Tracking Data

With reference to published research, filed patents, and existing commercial products, this section presents and categorizes personal information that can be inferred from eye tracking data. As a basis for potential inferences, eye tracking devices can record a large variety of gaze parameters.

Some of the most commonly measured eye movements are fixations, saccades, and smooth pursuit eye movements [85]. During a fixation, the eyes are relatively stable and focused on a specific position, allowing for information to be acquired and processed. Saccades are rapid eye movements from one fixation point to another, lasting 30 to 80 ms [87]. Smooth pursuit movements are performed when eyes are closely following a moving visual target. In addition to the spatial dispersion, duration, amplitude, acceleration, velocity, and chronological sequence of such eye movements, many eye trackers capture various other eye activities, including eye opening and closure (e.g., average distance

[1] For an overview of existing types of eye tracking, refer to [2].

between the eyelids, blink duration, blink frequency), ocular microtremors, pupil size, and pupil reactivity [19, 58]. Furthermore, most eye trackers videotape parts of the user's face and may thereby capture additional information, such as the number and depth of wrinkles, and a user's eye shape and iris texture [40]. Therefore, these parameters were also considered in our investigation into the richness and sensitivity of eye tracking data. Fig. 1 provides an introductory overview of common eye tracking measures and the categories of inferences discussed in this paper.

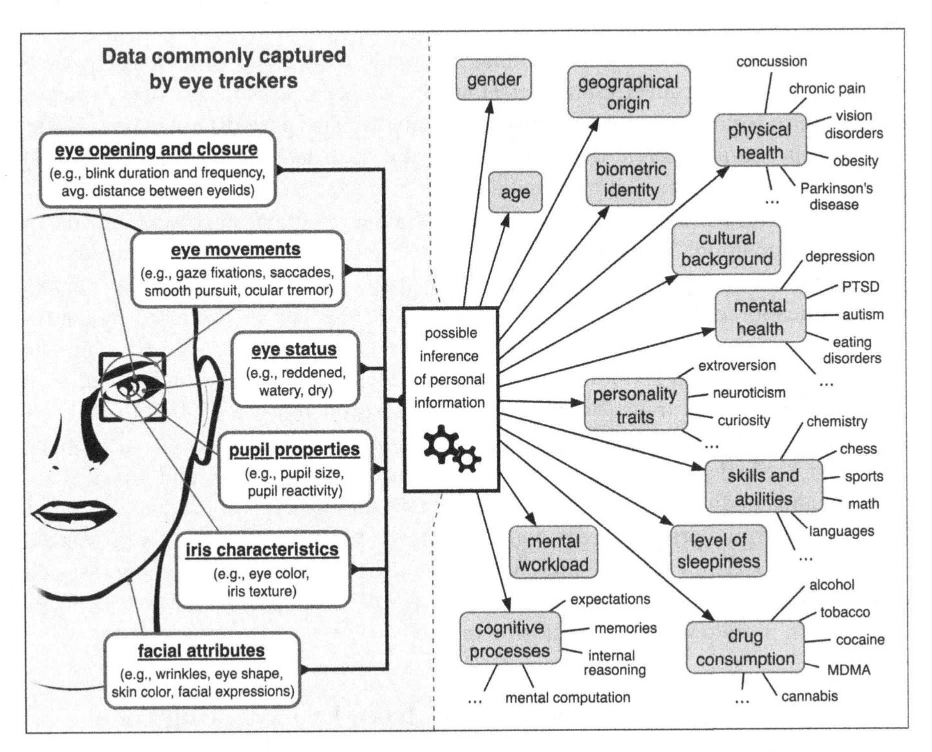

Fig. 1. Overview of sensitive inferences that can be drawn from eye tracking data.

2.1 Biometric Identification

Due to differences in physical oculomotor structure and brain functioning, certain gaze characteristics are unique for every individual, similar to fingerprints, and can thus be exploited for biometric identification [9, 74, 76]. Among other methods, people can be told apart based on distinct patterns of pupil reactivity and gaze velocity [9], or by comparing their eye movement trajectories when they focus on a moving target – even if the eye activity is only recorded through an ordinary smartphone camera [55].

Aside from such gaze-based measures, the complex textures and color patterns in a person's iris are also suitable for biometric identification. This approach, called *iris recognition*, is being used in a variety of real-world security and surveillance applications and has been recognized as "one of the most powerful techniques for biometric identification ever developed" [64]. Even though their iris scanning capability is usually not advertised, it should be understood that commodity eye trackers often record and process high-resolution images of the user's iris, which can not only be used to uniquely identify the user but also to deceive iris-based authentication mechanisms and thereby steal the user's identity [40].

In cases where a unique identification of an individual is not possible (e.g., because the person is not registered in the recognition system database), other attributes inferred from eye tracking data, such as age and gender (see Sect. 2.6), health condition (see Sect. 2.9), or ethnicity (see Sect. 2.4), can still help to classify the target person into a specific demographic group and thereby approximate the identity [74].

2.2 Monitoring of Mental Workload and Cognitive Processes

Certain patterns in eye movement, pupil dilation, and eye blinking have been recognized as reliable indicators of mental workload in people of any age [19, 63], sometimes offering higher accuracy than conventional methods like Electroencephalography [8]. Through eye tracking, it is also possible to distinguish a user's moments of awareness from moments of distraction and mind wandering [31, 84].

Apart from detecting a user's mental presence and measuring the mere intensity of cognitive processing, eye tracking can also provide insights into specific conscious and unconscious thought processes in a large variety of contexts. Among other mental tasks and activities, ocular measures have been used to study memory retrieval [19, 31], problem solving [31, 75], learning processes [44, 69], the formation of expectations [19, 27], internal reasoning [19], and mental computations [19, 31].

Eye tracking data can not only – to a certain extent – reveal what we remember, imagine, expect, and think about, but also our specific decision-making strategies [19, 28] and cognitive styles, i.e., individual differences in the way we acquire, process, and interpret information [72]. For example, people can be classified as field-dependent vs. field-independent (people of the latter type pay more attention to detail and exhibit a more analytical approach to processing visual information) [72], or as verbalizers vs. visualizers (people of the latter type can process visual information, such as images and diagrams, better than textual information) [44]. The gaze-based inference of such cognitive styles is feasible and can achieve high accuracies, as has been confirmed in a recent study by Raptis et al. [72].

Researchers from the field of cognitive science and experimental psychology have suggested that eye tracking data will not only be used for the real-time analysis but also for the prediction of human decisions and behavior [28].

2.3 Inference of Personality Traits

Experimental research has shown that it is possible to automatically infer personality traits from eye tracking data [34, 35, 42]. For example, gaze patterns captured during everyday tasks can be used to evaluate users along the so-called Big Five traits, namely openness to experience, conscientiousness, extroversion, agreeableness, and neuroticism

[34, 42]. The gaze-based assessment of personality traits is possible not only in binary form (high vs. low) but also in the form of ranges. In [35], for instance, eye movement analysis was used for the automatic recognition of different levels of curiosity.

Besides the Big Five traits and curiosity, gaze metrics were found to be associated with various other personality traits, including emotional intelligence [54], indecisiveness [36], the tendency to ruminate [21], trait anxiety [42], sexual compulsivity [87], boredom susceptibility [70], and general aggressiveness [6]. Eye tracking has even been used to investigate people's attachment styles in interpersonal relationships (e.g., secure, withdrawn, fearful, enmeshed) [81].

Based on data from 428 study participants, Larsson et al. [53] also suggest that some personality traits, including tendermindedness, warmth, trust, and impulsiveness, are genetically linked to certain iris characteristics, offering – besides gaze behavior – another potential ocular biomarker to analyze people's personalities.

2.4 Inference of Cultural Affiliation and Ethnicity

It is widely agreed that culture fundamentally shapes human cognitive processing and behavior [11]. Studies have shown that intercultural differences are reflected in certain gaze characteristics [12, 24, 41, 61]. For example, people of different cultural background were found to exhibit discriminative eye-movement patterns when seeking information on search engine results pages [61], when exploring complex visual scenes [12, 24], and when viewing videos of actors performing cultural activities [41]. Some cultural biases in visual processing are so pronounced that they can still be measured when external stimuli draw attention in an opposite manner to the respective bias [24].

Additionally, eye movements can reveal a person's knowledge of certain cultural practices. For instance, in an eye tracking study by Green et al. [27], Chinese infants exclusively predicted the goal of eating actions performed by an actor with chopsticks, whereas European infants only anticipated that food would be brought to the mouth when eating actions were performed with Western cutlery, as indicated by their predictive gaze shifts towards the actor's mouth.

Some studies have also investigated how people of different "race"[2] differ in their viewing behavior [25, 33, 88]. Apart from the fact that video-based eye trackers can directly record the eye color, eye shape, and skin color of a user, it has been observed in eye tracking studies that test subjects view "other-race faces" differently than faces of their "own race" in terms of the facial features scanned (e.g., initial focus and greater proportion of fixation time on the eyes vs. nose and mouth) [25, 88]. Furthermore, researchers have observed characteristic changes in pupil size, which are attributed to elevated cognitive effort during face recognition, when people look at "other-race faces" [88]. Such differences have been reported, for example, between "Black and White observers" [33] and between "Western Caucasian and East Asian observers" [25]

[2] The authors share the UNESCO's position [60] that the classification of human populations into "races" is inadequate and obsolete. Nevertheless, it is important to monitor the state of research in this field, especially because any information indicative of a person's ethnic background can serve as a basis for racist discrimination. All terms related to the concept of "race" in this paper are cited from external sources and do not reflect the authors' views.

and could potentially allow inferences about the genetic and ethnic background of eye tracking users.

Eye tracking data may also allow inferences about a user's native language. For instance, considerable differences in eye movement patterns during reading can be observed between native and non-native speakers of English [39]. Eye tracking can even be used to determine which specific words are difficult to understand for a person [51]. Among other things, such information could help in estimating a subject's nationality or geographical origin.

2.5 Skill Assessment

Eye tracking has been used extensively in the study of human expertise and to discriminate between performance levels in a variety of areas [30, 31, 69, 75]. For example, gaze behavior can be analyzed to assess reading and listening comprehension skills [10, 92]. During a corresponding task or scenario, eye tracking can also be used to distinguish between experts and novices in chess [75], several sports [46], chemistry [69], mathematics [31], school teaching [14], and various medical skills, including surgery, nursing, anesthesia, and radiology [30].

Among other gaze characteristics, expertise is often associated with systematic eye movement patterns reflecting a specific task strategy [31], with the targeted inspection of important regions and task-relevant information [30, 75], and with more consistent gaze patterns over consecutive trials of a task [46].

In some fields, eye tracking has not only been used as a tool to discriminate between people of different skill levels, but also to predict people's task performance and learning curves [52, 69] and to examine specific learning disabilities, such as mathematical difficulties and dyslexia [31, 85].

2.6 Age and Gender Recognition

Just like physical shape, skin texture, and cognitive abilities, human eyes and visual behavior are fundamentally affected by the aging process [20, 36]. For example, eye tracking studies found age-related differences in people's visual explorativeness, pupil reactions to certain visual stimuli, and error rates in eye movement tasks [36, 42].

Furthermore, detailed frontal face images, which are typically required for video-based eye tracking, have already been used for automated age estimation, for instance based on wrinkles in the eye area [15]. Dynamic facial expressions, such as smiles, may also be analyzed to infer the age of test subjects [17]. Other parameters utilized for computerized age-group recognition include iris size and iris texture [20].

As with age, a person's gender can be reflected in certain eye tracking measures. For instance, studies found systematic gender differences in people's fixation distribution while viewing natural images (e.g., stills from romance films or wildlife documentaries) [68], during online shopping [38], when playing video games [42], and when viewing sexual stimuli [87]. Researchers have already used such differences in visual behavior to automatically classify the sex of test subjects [68].

2.7 Inference of Preferences and Aversions

Eye tracking is widely employed to investigate people's interests, likes, and dislikes. Spontaneous attention to specific objects in a visual scene (e.g., in terms of frequency, duration, and sequence of gaze fixations) is regarded as a natural indicator of interest [19, 74, 87]. For data presentation and analysis, gaze fixations are commonly aggregated into heat maps to quickly identify potential regions and objects of interest [74].

Besides the focus of visual attention, other eye parameters, such as pupil dilation and blink properties, can also be used to analyze a person's degree of interest and to distinguish between positive, neutral, and negative responses to visual stimuli [55]. Emotion detection from gaze data, which can assist in analyzing a user's interests and preferences [55, 83], will be discussed in Sect. 2.8.

Among other things, eye tracking has been used to examine preferences for certain types of gambling [65], mobile apps [56], activities of daily living [86], types of food [32], colors, geometric shapes, and product designs [3], pieces of clothing, animals, video game characters, and items of furniture [83]. Beyond mere interest, existing research even suggests that people's patterns of visual attention reflect their consumption and purchasing behavior [91].

Eye tracking has also been used extensively in the study of love and sexual desire. For example, researchers have analyzed pupillary responses and the allocation of visual attention to measure levels of sexual arousal and to investigate mating preferences towards specific facial characteristics, age groups, body shapes, body parts, and signs of social dominance [3, 87].

Apart from positive interests, visual attentional biases captured by eye trackers can also reflect a person's phobias and aversions (e.g., fear of spiders) [3, 37]. Some interests and preferences can already be inferred from eye tracking data with high accuracy [56, 73, 87] and several patents have been filed in this field [3, 83].

2.8 Detection of Short- and Medium-Term User States

Moods and Emotions. Eye tracking is increasingly used in the interdisciplinary field of affective computing, where systems are developed to automatically recognize human emotions based on physiological signals and behavioral cues [73, 83]. It has been shown that various ocular measures, including pupil size, blink properties, saccadic eye movements, and specific biases of visual attention, can contain information about a person's emotional state [4, 23, 55, 59].

Gaze data can reflect emotional arousal and the valence of emotions (positive, negative, neutral) [19, 55] as well as more specific affective states, such as happiness and enthusiasm [83], acute stress and worry [59], humorous moods and disgust [73], curiosity [4], distress, nervousness, and hostility [23], fear, anger, sadness, and surprise [55].

Eye tracking can not only be used to detect emotions with high accuracy [73] but also to estimate the intensity of emotions [55, 83]. Based on gaze parameters, existing methods can even distinguish whether a user's emotional response to a given stimulus is rational or purely instinctive [55].

Fatigue and Sleepiness. For over two decades, there have been approaches to automatically derive a person's level of sleepiness from certain ocular measures, such as blink rate, blink duration, average distance between the eyelids, fixation durations, and velocity of eye movements [57]. Recent studies have confirmed the suitability of eye tracking measures as indicators for sleepiness and fatigue [63, 89]. Corresponding methods have already been patented and achieve high accuracies – not only while the user is working on specific cognitive tasks, but also during everyday natural-viewing situations [57, 89].

Intoxication. The consumption of alcohol and other recreational drugs can have measurable effects on various eye and gaze properties, such as decreased accuracy and speed of saccades, changes in pupil size and reactivity, and an impaired ability to fixate on moving objects [29, 67, 85].

Apart from alcohol, significant abnormalities in oculomotor functioning were found in people under the influence of nicotine, 3,4-methylenedioxymethamphetamine ("MDMA"), and tetrahydrocannabinol ("THC") [29, 70].

Researchers have demonstrated the ability to differentiate between drug-impaired and sober subjects with high accuracy based on eye tracking data [29]. The magnitude of some ocular effects is closely associated with the amount of drugs consumed [85] and certain effects can even be detected at non-intoxicating doses [77]. In addition to pupillary changes and eye movement impairments, an attentional bias towards drug-related visual stimuli has been observed among intoxicated test subjects [67].

Not only a state of intoxication, but also an acute state of drug deprivation and craving can have a distinct effect on certain eye tracking parameters [29, 70].

2.9 Health Assessment

Physical Health. Many diseases and medical conditions directly affect the eyes, or parts of the brain that are responsible for oculomotor function, and thereby cause gaze impairments [3, 19, 30]. Characteristic eye movement patterns were found, for example, in people suffering from concussion [43], fetal alcohol syndrome [3], irregular growth [3], chronic pain [22], neurocognitive impairment due to preterm birth [82], multiple sclerosis [3], Alzheimer's disease [30, 43], Tourette syndrome [19], Parkinson's disease [30], and various vision disorders (e.g., myopia, farsightedness, and blind spots) [3, 43].

As filed patents and published experimental studies show, eye movement analysis can be used to diagnose, monitor, prognose, and sometimes even predict various health disorders [30, 43] which can be subsumed under the umbrella term ETDCC ("Eye Tracking-Relevant Diseases, Conditions, and Characteristics") [3].

Research has further demonstrated that certain patterns in gaze orientation and pupil reactivity to food-related stimuli (e.g., high vs. low calorie food images) can be indicative of overweight and obesity [32].

Mental Health. Abnormal eye movements can be used as behavioral biomarkers for the diagnosis of various mental health problems [1, 5, 29]. Oculomotor dysfunctions and gaze peculiarities are found, for example, in sufferers of anxiety disorder [29], depression [1], bipolar disorder [30], borderline personality disorder [6], schizophrenia [5], obsessive–compulsive disorder [13], binge-eating disorder [79], ADHD [7], mild cognitive impairment [30], autism [43], and posttraumatic stress disorder [66].

Some common symptoms of mental disorders are irregularities in blink rate and blink duration [19], abnormal stability and dispersal of gaze fixations during free viewing [5], unusual biases of visual attention [66], impaired smooth pursuit eye-movement performance [85], eye contact avoidance, and abnormal distance between the eyelids [1].

Certain mental illnesses, including depression and schizophrenia, can already be detected automatically via eye tracking [1, 5, 30] and corresponding methods have been filed as patents [43]. Besides the possibility of binary classification (suffering vs. not suffering), some ocular measures are associated with the severity of mental disorders [19]. Not only acute disorders can be reflected in gaze data, but also past mental health issues and even the personal risk of future outbreaks [71, 78]. For example, researchers have observed characteristic gaze patterns in previously depressed individuals [78] and found biases in visual attention that were predictive of future depression scores at a delay of more than two years [71].

Substance Use Disorders. Apart from acute states of intoxication (which we have discussed in Sect. 2.8), eye tracking data may contain information about a user's longer-term drug consumption habits and addictions. Numerous eye tracking studies have reported a strong attentional bias towards drug-related visual cues in addicts of cocaine [16], alcohol [67], cannabis [90], and tobacco [18, 70].

Among other possible methods, such attentional biases can be detected by measuring how quickly, how often, and for how long a person's eyes fixate on corresponding stimuli in comparison to neutral stimuli, or by testing the person's ability to look away from drug-related stimuli on command [16, 18]. Significant biases have not only been observed in long-term addicts but also in habitual drug users without clinical symptoms of dependency [18, 67]. The strength of attentional biases towards drug-related visual cues was found to be correlated with scores on drug use scales, such as the Obsessive Compulsive Cocaine Scale [16] and with self-reported lifetime drug consumption [62]. Research has also shown that certain biases in visual attention can be predictive of craving and even relapse in drug addiction [16].

3 Discussion and Implications

As shown in the previous section, various kinds of sensitive inferences can be drawn from eye tracking data. Among other categories of personal data, recorded visual behavior can implicitly contain information about a person's biometric identity, personality traits, ethnic background, age, gender, emotions, fears, preferences, skills and abilities, drug habits, levels of sleepiness and intoxication, and physical and mental health condition. To some extent, even distinct stages of cognitive information processing are discernable from gaze data. Thus, devices with eye tracking capability have the potential to implicitly capture much more information than a user wishes and expects to reveal. Some of the categories of personal information listed above constitute *special category data*, for which particular protection is prescribed by the EU's General Data Protection Regulation (Art. 9 GDPR).

Of course, drawing reliable inferences from eye tracking data is not a trivial task. Many situational factors can influence eye properties and gaze behavior in complex ways, making it difficult to measure the effect of a particular action, internal process, or personal characteristic of the user in isolation [55]. Seemingly identical ocular reactions can result from completely different causes. For example, an intensive gaze fixation on another person's face may indicate liking, aversion, confusion, recognition, and much more. Similarly, a sudden change in pupil size can be indicative of many different feelings or internal states, including physical pain, sexual arousal, interest, happiness, anger, or simply be a reaction to ambient events and conditions, such as noise or varying lighting [19, 55].

In spite of existing challenges and limitations, the reviewed literature demonstrates that there is considerable potential for inferences in many areas and that numerous research projects, patented systems, and even commercial products have already taken advantage of the richness of eye tracking data to draw inferences about individuals with high accuracy.

It should be acknowledged that many of the cited inference methods were only tested under controlled laboratory conditions and lack evaluation in real-world scenarios [4, 18, 27, 52, 65, 67, 69, 86, 88]. On the other hand, it may reasonably be assumed that some of the companies with access to eye tracking data from consumer devices (e.g., device manufacturers, ecosystem providers) possess larger sets of training data, more technical expertise, and more financial resources than the researchers cited in this paper. Facebook, for example, a pioneer in virtual reality and eye tracking technology, is also one of the wealthiest and most profitable companies in the world with a multi-billion dollar budget for research and development and a user base of over 2.3 billion people [93]. It seems probable that the threat of unintended information disclosure from gaze data will continue to grow with further improvements of eye tracking technology in terms of cost, size, and accuracy, further advances in analytical approaches, and the increasing use of eye tracking in various aspects of daily life.

In assessing the privacy implications of eye tracking, it is important to understand that, while consciously directed eye movements are possible, many aspects of ocular behavior are not under volitional control – especially not at the micro level [19, 55]. For instance, stimulus-driven glances, pupil dilation, ocular tremor, and spontaneous blinks mostly occur without conscious effort, similar to digestion and breathing. And even for those eye activities where volitional control is possible, maintaining it can quickly become physically and cognitively tiring [58] – and may also produce certain visible patterns by which such efforts can be detected. Hence, it can be very difficult or even impossible for eye tracking users to consciously prevent the leakage of personal information.

Though this paper focuses on privacy risks, we do not dispute the wide-ranging benefits of eye tracking. Quite the opposite: we believe that it is precisely the richness of gaze data and the possibility to draw insightful inferences from it that make the rising technology so valuable and useful. But to exploit this potential in a sustainable and socially acceptable manner, adequate privacy protection measures are needed.

Technical safeguards have been proposed to prevent the unintended disclosure of personal information in data mining, including specialized solutions for eye tracking data [58, 80]. These comprise the fuzzing of gaze data (i.e., inserting random noise into the signal before passing it down the application chain) and the utilization of derived parameters (e.g., aggregated values instead of detailed eye fixation sequences) [58]. Experiments have already shown that approaches based on differential privacy can prevent certain inferences, such as user re-identification and gender recognition, while maintaining high performance in gaze-based applications [80]. In addition to approaches at the technical level, it should also be examined whether existing laws provide for sufficient transparency in the processing of gaze data and for proper protection against inference-based privacy breaches. The promises and limitations of existing technical and legal remedies are beyond the scope of this paper but deserve careful scrutiny and will be considered for future work.

Even though eye tracking is a demonstrative example, the threat of undesired inferences is of course much broader, encompassing countless other sensors and data sources in modern life [47]. In other recent work, we have examined sensitive inferences that can be drawn from voice recordings [49] and accelerometer data [48, 50], for instance. In our view, the vast possibilities of continuously advancing inference methods are clearly beyond the understanding of the ordinary consumer. Therefore, we consider it to be primarily the responsibility of technical experts, technology companies, and governmental agencies to inform consumers about potential consequences and protect them against such covert invasions of privacy. Also, since it is unlikely that companies will voluntarily refrain from using or selling personal information that can be extracted from already collected data, there should be strong regulatory incentives and controls.

4 Conclusion

While the widespread adoption of eye tracking holds the potential to improve our lives in many ways, the rising technology also poses a substantial threat to privacy. The overview provided in this paper illustrates that, through the lens of advanced data analytics, eye tracking data can contain a rich array of sensitive information, including cues to a user's biometric identity, gender, age, ethnicity, personality traits, drug consumption habits, moods and emotions, skills, preferences, cognitive processes, and physical and mental health condition. Since inference methods are often based on hidden patterns and correlations that are incomprehensible to ordinary consumers, it can be impossible for them to understand and control what information is revealed.

Although there is extensive literature on the analysis of eye tracking data, we believe that many possible inferences have not yet been investigated. Keeping track of the evolving possibilities of data mining methods in this field is therefore an important avenue for future research. This paper represents a crucial first step towards understanding the sensitivity of eye tracking data from a holistic perspective. The findings compiled herein are significant enough to warrant a warning to users whose privacy could be affected, as well as a call for action to the public and private actors entrusted with protecting user privacy in consumer electronics. Considering the rapid proliferation of eye tracking technology, existing technical and legal safeguards urgently need to be assessed regarding their ability to avert undesired inferences from gaze data, or to at least prevent the misuse of sensitive inferred information.

References

1. Alghowinem, S., et al.: Eye movement analysis for depression detection. In: IEEE International Conference on Image Processing, pp. 4220–4224 (2013)
2. Al-Rahayfeh, A., Faezipour, M.: Eye tracking and head movement detection: a state-of-art survey. IEEE J. Transl. Eng. Health Med. **1**, 2100212 (2013)
3. Avital, O.: Method and System of Using Eye Tracking to Evaluate Subjects (Patent No.: US20150282705A1) (2015)
4. Baranes, A., et al.: Eye movements reveal epistemic curiosity in human observers. Vis. Res. **117**, 81–90 (2015). https://doi.org/10.1016/j.visres.2015.10.009
5. Benson, P.J., et al.: Simple viewing tests can detect eye movement abnormalities that distinguish schizophrenia cases from controls with exceptional accuracy. Biol. Psychiatry **72**(9), 716–724 (2012). https://doi.org/10.1016/j.biopsych.2012.04.019
6. Bertsch, K., et al.: Interpersonal threat sensitivity in borderline personality disorder: an eye-tracking study. J. Pers. Disord. **31**(5), 647–670 (2017)
7. Blazey, R.N., et al.: ADHD Detection by Eye Saccades (Patent No.: US6652458B2) (2003)
8. Borys, M., et al.: An analysis of eye-tracking and electroencephalography data for cognitive load measurement during arithmetic tasks. In: 10th International Symposium on Advanced Topics in Electrical Engineering (ATEE), pp. 287–292 (2017)
9. Cantoni, V., et al.: Gaze-based biometrics: an introduction to forensic applications. Pattern Recogn. Lett. **113**, 54–57 (2018). https://doi.org/10.1016/j.patrec.2016.12.006
10. Chita-Tegmark, M., et al.: Eye-tracking measurements of language processing: developmental differences in children at high risk for ASD. J. Autism Dev. Disord. **45**(10), 3327–3338 (2015). https://doi.org/10.1007/s10803-015-2495-5
11. Chizari, S.: Exploring the role of culture in online searching behavior from cultural cognitive perspective: case study of American, Chinese and Iranian Graduate Students. In: iConference Proceedings. iSchools, Philadelphia (2016)
12. Chua, H.F., et al.: Cultural variation in eye movements during scene perception. Proc. Natl. Acad. Sci. **102**(35), 12629–12633 (2005). https://doi.org/10.1073/pnas.0506162102
13. Cludius, B., et al.: Attentional biases of vigilance and maintenance in obsessive-compulsive disorder: an eye-tracking study. J. Obsessive Compuls. Relat. Disord. **20**, 30–38 (2019). https://doi.org/10.1016/j.jocrd.2017.12.007
14. Cortina, K.S., et al.: Where low and high inference data converge: validation of CLASS assessment of mathematics instruction using mobile eye tracking with expert and novice teachers. Int. J. Sci. Math. Educ. **13**(2), 389–403 (2015)
15. Dehshibi, M.M., Bastanfard, A.: A new algorithm for age recognition from facial images. Signal Process. **90**(8), 2431–2444 (2010)
16. Dias, N.R., et al.: Anti-saccade error rates as a measure of attentional bias in cocaine dependent subjects. Behav. Brain Res. **292**, 493–499 (2015)
17. Dibeklioğlu, H., et al.: A smile can reveal your age: enabling facial dynamics in age estimation. In: Proceedings of the 20th ACM International Conference on Multimedia, pp. 209–218. ACM Press, Nara (2012). https://doi.org/10.1145/2393347.2393382
18. DiGirolamo, G.J., et al.: Breakdowns of eye movement control toward smoking cues in young adult light smokers. Addict. Behav. **52**, 98–102 (2016)
19. Eckstein, M.K., et al.: Beyond eye gaze: what else can eyetracking reveal about cognition and cognitive development? Dev. Cogn. Neurosci. **25**, 69–91 (2017)
20. Erbilek, M., et al.: Age prediction from iris biometrics. In: 5th International Conference on Imaging for Crime Detection and Prevention (ICDP), pp. 1–5 (2013)
21. Fang, L., et al.: Attentional scope, rumination, and processing of emotional information: an eye-tracking study. Emotion **19**(7), 1259–1267 (2018)

22. Fashler, S.R., Katz, J.: Keeping an eye on pain: investigating visual attention biases in individuals with chronic pain using eye-tracking methodology. J. Pain Res. **9**, 551–561 (2016). https://doi.org/10.2147/JPR.S104268
23. Gere, A., et al.: Influence of mood on gazing behavior: preliminary evidences from an eye-tracking study. Food Qual. Prefer. **61**, 1–5 (2017)
24. Goh, J.O., et al.: Culture modulates eye-movements to visual novelty. PLoS ONE **4**(12), e8238 (2009). https://doi.org/10.1371/journal.pone.0008238
25. Goldinger, S.D., et al.: Deficits in cross-race face learning: insights from eye movements and pupillometry. J. Exp. Psychol. Learn. Mem. Cogn. **35**(5), 1105–1122 (2009)
26. Grand View Research: Global Eye Tracking Market Size By Type, Industry report. https://www.grandviewresearch.com/industry-analysis/eye-tracking-market. Accessed 25 Oct 2019
27. Green, D., et al.: Culture influences action understanding in infancy: prediction of actions performed with chopsticks and spoons in Chinese and Swedish infants. Child Dev. **87**(3), 736–746 (2016)
28. Guazzini, A., et al.: Cognitive dissonance and social influence effects on preference judgments: an eye tracking based system for their automatic assessment. Int. J. Hum Comput Stud. **73**, 12–18 (2015). https://doi.org/10.1016/j.ijhcs.2014.08.003
29. Hall, C.A., Chilcott, R.P.: Eyeing up the Future of the Pupillary Light Reflex in Neurodiagnostics. Diagnostics **8**(1), 1–20 (2018). https://doi.org/10.3390/diagnostics8010019
30. Harezlak, K., Kasprowski, P.: Application of eye tracking in medicine: a survey, research issues and challenges. Comput. Med. Imag. Graph. **65**, 176–190 (2018)
31. Hartmann, M., Fischer, M.H.: Exploring the numerical mind by eye-tracking: a special issue. Psychol. Res. **80**(3), 325–333 (2016). https://doi.org/10.1007/s00426-016-0759-0
32. Hendrikse, J.J., et al.: Attentional biases for food cues in overweight and individuals with obesity: a systematic review of the literature. Obes. Rev. **16**(5), 424–432 (2015)
33. Hills, P.J., Pake, J.M.: Eye-tracking the own-race bias in face recognition: revealing the perceptual and socio-cognitive mechanisms. Cognition **129**(3), 586–597 (2013)
34. Hoppe, S., et al.: Eye movements during everyday behavior predict personality traits. Front. Hum. Neurosci. **12**, 1–8 (2018). https://doi.org/10.3389/fnhum.2018.00105
35. Hoppe, S., et al.: Recognition of curiosity using eye movement analysis. In: International Conference on Pervasive and Ubiquitous Computing, pp. 185–188 (2015)
36. Horsley, M. (ed.): Current Trends in Eye Tracking Research. Springer, Cham (2013). https://doi.org/10.1007/978-3-319-02868-2
37. Huijding, J., et al.: To look or not to look: an eye movement study of hypervigilance during change detection in high and low spider fearful students. Emotion **11**(3), 666–674 (2011). https://doi.org/10.1037/a0022996
38. Hwang, Y.M., Lee, K.C.: Using an eye-tracking approach to explore gender differences in visual attention and shopping attitudes in an online shopping environment. Int. J. Hum. Comput. Interact. **34**(1), 15–24 (2018)
39. Ito, A., et al.: Investigating the time-course of phonological prediction in native and non-native speakers of English: a visual world eye-tracking study. J. Mem. Lang. **98**, 1–11 (2018). https://doi.org/10.1016/j.jml.2017.09.002
40. John, B., et al.: EyeVEIL: degrading iris authentication in eye tracking headsets. In: ACM Symposium on Eye Tracking Research & Applications (ETRA), pp. 1–5. ACM Press, Denver (2019). https://doi.org/10.1145/3314111.3319816
41. Kardan, O., et al.: Cultural and developmental influences on overt visual attention to videos. Sci. Rep. **7**(1), 11264 (2017). https://doi.org/10.1038/s41598-017-11570-w
42. Kaspar, K., König, P.: Emotions and personality traits as high-level factors in visual attention: a review. Front. Hum. Neurosci. **6**, 321 (2012)
43. Kempinski, Y.: System and Method of Diagnosis Using Gaze and Eye Tracking (Patent No.: US20160106315A1) (2016)

44. Koć-Januchta, M., et al.: Visualizers versus verbalizers: effects of cognitive style on learning with texts and pictures – an eye-tracking study. Comput. Hum. Behav. **68**, 170–179 (2017). https://doi.org/10.1016/j.chb.2016.11.028
45. Krafka, K., et al.: Eye tracking for everyone. In: 2016 IEEE Conference on Computer Vision and Pattern Recognition (CVPR), pp. 2176–2184. IEEE, Las Vegas (2016)
46. Kredel, R., et al.: Eye-tracking technology and the dynamics of natural gaze behavior in sports: a systematic review of 40 years of research. Front. Psychol. **8**, 1–15 (2017)
47. Kröger, J.: Unexpected inferences from sensor data: a hidden privacy threat in the Internet of Things. In: Strous, L., Cerf, V.G. (eds.) IFIPIoT 2018. IAICT, vol. 548, pp. 147–159. Springer, Cham (2019). https://doi.org/10.1007/978-3-030-15651-0_13
48. Kröger, J.L., et al.: Privacy implications of accelerometer data: a review of possible inferences. In: Proceedings of the 3rd International Conference on Cryptography, Security and Privacy (ICCSP). ACM, New York (2019). https://doi.org/10.1145/3309074.3309076
49. Kröger, J.L., et al.: Privacy implications of voice and speech analysis - information disclosure by inference. In: Fricker, S., et al. (eds.) Privacy and Identity 2019. IFIP AICT, vol. 576, pp. 242–258. Springer, Cham (2020). https://doi.org/10.1007/978-3-030-42504-3_16
50. Kröger, J.L., Raschke, P.: Is my phone listening in? On the feasibility and detectability of mobile eavesdropping. In: Foley, S.N. (ed.) DBSec 2019. LNCS, vol. 11559, pp. 102–120. Springer, Cham (2019). https://doi.org/10.1007/978-3-030-22479-0_6
51. Kunze, K., et al.: Towards inferring language expertise using eye tracking. In: CHI 2013 Extended Abstracts on Human Factors in Computing Systems, pp. 217–222. ACM Press, Paris (2013). https://doi.org/10.1145/2468356.2468396
52. Lallé, S., et al.: Prediction of users' learning curves for adaptation while using an information visualization. In: International Conference on Intelligent User Interfaces, pp. 357–368. ACM Press, Atlanta (2015)
53. Larsson, M., et al.: Associations between iris characteristics and personality in adulthood. Biol. Psychol. **75**(2), 165–175 (2007). https://doi.org/10.1016/j.biopsycho.2007.01.007
54. Lea, R.G., et al.: Trait emotional intelligence and attentional bias for positive emotion: an eye tracking study. Pers. Individ. Differ. **128**, 88–93 (2018)
55. Lemos, J.: System and Method for Determining Human Emotion by Analyzing Eye Properties (Patent No.: US20070066916A1) (2007)
56. Li, Y., et al.: Towards measuring and inferring user interest from gaze. In: International Conference on World Wide Web Companion, pp. 525–533. ACM Press, Perth (2017). https://doi.org/10.1145/3041021.3054182
57. Liang, C.-C., et al.: System for Monitoring Eyes for Detecting Sleep Behavior (Patent No.: US5570698A) (1996)
58. Liebling, D.J., Preibusch, S.: Privacy considerations for a pervasive eye tracking world. In: International Joint Conference on Pervasive and Ubiquitous Computing: Adjunct Publication, pp. 1169–1177 ACM Press, New York (2014)
59. Macatee, R.J., et al.: Attention bias towards negative emotional information and its relationship with daily worry in the context of acute stress: an eye-tracking study. Behav. Res. Ther. **90**, 96–110 (2017). https://doi.org/10.1016/j.brat.2016.12.013
60. Mader, G.: Declaration of Schlaining Against Racism, Violence and Discrimination. Austrian Commission for UNESCO, Vienna (1995)
61. Marcos, M.-C., et al.: Cultural differences on seeking information: an eye tracking study. In: CHI 2013: Workshop Many People, Many Eyes. ACM, Paris (2013)
62. Marks, K.R., et al.: Fixation time is a sensitive measure of cocaine cue attentional bias. Addict. Abingdon Engl. **109**(9), 1501–1508 (2014). https://doi.org/10.1111/add.12635
63. Martins, R., Carvalho, J.: Eye blinking as an indicator of fatigue and mental load—a systematic review. In: Arezes, P., et al. (eds.) Occupational Safety and Hygiene III, pp. 231–235. CRC Press (2015). https://doi.org/10.1201/b18042-48

64. Matey, J.R., et al.: Iris on the move: acquisition of images for iris recognition in less constrained environments. Proc. IEEE **94**(11), 1936–1947 (2006)
65. McGrath, D.S., et al.: The specificity of attentional biases by type of gambling: an eye-tracking study. PLoS ONE **13**(1), e0190614 (2018)
66. Milanak, M.E., et al.: PTSD symptoms and overt attention to contextualized emotional faces: evidence from eye tracking. Psychiatry Res. **269**, 408–413 (2018)
67. Miller, M.A., Fillmore, M.T.: Persistence of attentional bias toward alcohol-related stimuli in intoxicated social drinkers. Drug Alcohol Depend. **117**(2), 184–189 (2011)
68. Moss, F.J.M., et al.: Eye movements to natural images as a function of sex and personality. PLoS ONE **7**(11), e47870 (2012). https://doi.org/10.1371/journal.pone.0047870
69. Peterson, J., Pardos, Z., Rau, M., Swigart, A., Gerber, Colin, McKinsey, J.: Understanding student success in chemistry using gaze tracking and pupillometry. In: Conati, C., Heffernan, N., Mitrovic, A., Verdejo, M.F. (eds.) AIED 2015. LNCS (LNAI), vol. 9112, pp. 358–366. Springer, Cham (2015). https://doi.org/10.1007/978-3-319-19773-9_36
70. Pettiford, J., et al.: Increases in impulsivity following smoking abstinence are related to baseline nicotine intake and boredom susceptibility. Addict. Behav. **32**(10), 2351–2357 (2007). https://doi.org/10.1016/j.addbeh.2007.02.004
71. Price, R.B., et al.: From anxious youth to depressed adolescents: prospective prediction of 2-year depression symptoms via attentional bias measures. J. Abnorm. Psychol. **125**(2), 267–278 (2016). https://doi.org/10.1037/abn0000127
72. Raptis, G.E., et al.: Using eye gaze data and visual activities to infer human cognitive styles: method and feasibility studies. In: Conference on User Modeling, Adaptation and Personalization (UMAP), pp. 164–173. ACM Press, Bratislava (2017)
73. Raudonis, V., et al.: Evaluation of human emotion from eye motions. Int. J. Adv. Comput. Sci. Appl. **4**(8), 79–84 (2013). https://doi.org/10.14569/IJACSA.2013.040812
74. Ravi, B.: Privacy Issues in Virtual Reality: Eye Tracking Technology. Bloomberg Law, Arlington County (2017)
75. Reingold, E., Sheridan, H.: Eye movements and visual expertise in chess and medicine. In: Liversedge, S.P., Gilchrist, I.D., Everling, S. (eds.) The Oxford Handbook of Eye Movements, pp. 528–550. Oxford University, Oxford (2011)
76. Rigas, I., et al.: Biometric recognition via eye movements: saccadic vigor and acceleration cues. ACM Trans. Appl. Percept. **13**(2), 1–21 (2016)
77. Roche, D.J.O., King, A.C.: Alcohol impairment of saccadic and smooth pursuit eye movements: impact of risk factors for alcohol dependence. Psychopharmacology **212**(1), 33–44 (2010). https://doi.org/10.1007/s00213-010-1906-8
78. Sears, C.R., et al.: Attention to emotional images in previously depressed individuals: an eye-tracking study. Cogn. Ther. Res. **35**(6), 517–528 (2011)
79. Sperling, I., et al.: Cognitive food processing in binge-eating disorder: an eye-tracking study. Nutrients **9**(8), 903 (2017). https://doi.org/10.3390/nu9080903
80. Steil, J., et al.: Privacy-aware eye tracking using differential privacy. In: ACM Symposium on Eye Tracking Research & Applications, pp. 1–9 (2019). https://doi.org/10.1145/3314111.3319915
81. Szymanska, M., et al.: How do adolescents regulate distress according to attachment style? A combined eye-tracking and neurophysiological approach. Prog. Neuropsychopharmacol. Biol. Psychiatry **89**, 39–47 (2019). https://doi.org/10.1016/j.pnpbp.2018.08.019
82. Telford, E.J., et al.: Preterm birth is associated with atypical social orienting in infancy detected using eye tracking. J. Child Psychol. Psychiatry **57**(7), 861–868 (2016)
83. Thieberger, G., et al.: Utilizing Eye-tracking to Estimate Affective Response to a Token Instance of Interest (Patent No.: US9569734B2) (2017)
84. Tobii: Tobii Pro wearable eye tracking for driver safety. https://www.tobiipro.com/fields-of-use/psychology-and-neuroscience/customer-cases/audi-attitudes/. Accessed 13 Sept 2019

85. Vidal, M., et al.: Wearable eye tracking for mental health monitoring. Comput. Commun. **35**(11), 1306–1311 (2012). https://doi.org/10.1016/j.comcom.2011.11.002
86. Wang, C.-Y., et al.: Multimedia recipe reading: predicting learning outcomes and diagnosing cooking interest using eye-tracking measures. Comput. Hum. Behav. **62**, 9–18 (2016)
87. Wenzlaff, F., et al.: Video-based eye tracking in sex research: a systematic literature review. J. Sex Res. **53**(8), 1008–1019 (2016)
88. Wu, E.X.W., et al.: Through the eyes of the own-race bias: eye-tracking and pupillometry during face recognition. Soc. Neurosci. **7**(2), 202–216 (2012)
89. Yamada, Y., Kobayashi, M.: Fatigue detection model for older adults using eye-tracking data gathered while watching video: evaluation against diverse fatiguing tasks. In: 2017 IEEE International Conference on Healthcare Informatics (ICHI), pp. 275–284 (2017). https://doi.org/10.1109/ICHI.2017.74
90. Yoon, J.H., et al.: Assessing attentional bias and inhibitory control in cannabis use disorder using an eye-tracking paradigm with personalized stimuli. Exp. Clin. Psychopharmacol. (2019). https://doi.org/10.1037/pha0000274
91. Zamani, H., et al.: Eye tracking application on emotion analysis for marketing strategy. J. Telecommun. Electron. Comput. Eng. **8**(11), 87–91 (2016)
92. Zhan, Z., et al.: Online Learners' reading ability detection based on eye-tracking sensors. Sensors **16**(9), 1457 (2016). https://doi.org/10.3390/s16091457
93. Fourth Quarter and Full Year 2018 Results. Facebook, Inc., Menlo Park, USA (2019)

Privacy Implications of Voice and Speech Analysis – Information Disclosure by Inference

Jacob Leon Kröger[1,2(✉)], Otto Hans-Martin Lutz[1,2,3], and Philip Raschke[1]

[1] Technische Universität Berlin, Straße des 17. Juni 135, 10623 Berlin, Germany
{kroeger,philip.raschke}@tu-berlin.de
[2] Weizenbaum Institute for the Networked Society, Berlin, Germany
[3] Fraunhofer Institute for Open Communication Systems, Berlin, Germany

Abstract. Internet-connected devices, such as smartphones, smartwatches, and laptops, have become ubiquitous in modern life, reaching ever deeper into our private spheres. Among the sensors most commonly found in such devices are microphones. While various privacy concerns related to microphone-equipped devices have been raised and thoroughly discussed, the threat of unexpected inferences from audio data remains largely overlooked. Drawing from literature of diverse disciplines, this paper presents an overview of sensitive pieces of information that can, with the help of advanced data analysis methods, be derived from human speech and other acoustic elements in recorded audio. In addition to the linguistic content of speech, a speaker's voice characteristics and manner of expression may implicitly contain a rich array of personal information, including cues to a speaker's biometric identity, personality, physical traits, geographical origin, emotions, level of intoxication and sleepiness, age, gender, and health condition. Even a person's socioeconomic status can be reflected in certain speech patterns. The findings compiled in this paper demonstrate that recent advances in voice and speech processing induce a new generation of privacy threats.

Keywords: Audio · Voice · Speech · Microphone · Privacy · Inference · Side channel

1 Introduction

Since the invention of the phonograph in the late 19th century, it has been technically possible to record and reproduce sounds. For a long time, this technology was exclusively used to capture pieces of audio, such as songs, audio tracks for movies, or voice memos, and for the telecommunication between humans. With recent advances in automatic speech recognition, it has also become possible and increasingly popular to interact via voice with computer systems [96].

Microphones are ubiquitous in modern life. They are present in a variety of electronic devices, including not only phones, headsets, intercoms, tablet computers, dictation machines and baby monitors, but also toys, household appliances, laptops, cameras, smartwatches, cars, remote controls, and smart speakers.

M. Friedewald et al. (Eds.): Privacy and Identity 2019, IFIP AICT 576, pp. 242–258, 2020.
https://doi.org/10.1007/978-3-030-42504-3_16

There is no question that microphone-equipped devices are useful and important in many areas. It is hard to imagine a future, or even a present, without them. However, as a growing proportion of audio recordings is disseminated through insecure communication networks and processed on remote servers out of the user's control, the ubiquity of microphones may pose a serious threat to consumer privacy. Research and public debates have addressed this concern, with published reports looking into technical and legal aspects regarding data collection, processing, and storage, as well as access and deletion rights of the data subjects [18, 32, 96]. Yet, the recent privacy discourse has paid too little attention to the wealth of information that may unexpectedly be contained in audio recordings.

Certain characteristics of human speech can carry more information than the words themselves [94]. With the help of intelligent analysis methods, insights can not only be derived from a speaker's accent, dialect, sociolect, lexical diversity, patterns of word use, speaking rate and rhythms, but also from acoustic properties of speech, such as intonation, pitch, perturbation, loudness, and formant frequencies. A range of statistics can be applied to extract hundreds or even thousands of utilizable speech parameters from just a short sequence of recorded audio [19, 80].

Based on literature of diverse scientific disciplines, including signal processing, psychology, neuroscience, affective computing, computational paralinguistics, speech communication science, phonetics, and biomedical engineering, Sect. 2 of this paper presents an overview of sensitive inferences that can be drawn from linguistic and acoustic patterns in audio data. Specifically, we cover inferences about a user's biometric identity (Sect. 2.1), body measures (Sect. 2.2), moods and emotions (Sect. 2.3), age and gender (Sect. 2.4), personality traits (Sect. 2.5), intention to deceive (Sect. 2.6), sleepiness and intoxication (Sect. 2.7), native language (Sect. 2.8), physical health (Sect. 2.9), mental health (Sect. 2.10), impression made on other people (Sect. 2.11), and socioeconomic status (Sect. 2.12). Additionally, we examine information that can be extracted from the ambient noise and background sounds in a voice recording (Sect. 2.13). Section 3 provides a discussion of the presented findings with regard to their limitations and societal implications, followed by a conclusion in Sect. 4.

2 Inference of Personal Information from Voice Recordings

Based on experimental studies from the academic literature, this section presents existing approaches to infer information about recorded speakers and their context from speech, non-verbal human sounds, and environmental background sounds commonly found in audio recordings. Where available, published patents are also referenced to illustrate the current state of the art and point to potential real-world applications.

Figure 1 provides an introductory overview of the types of audio features and the categories of inferences discussed in this paper.

2.1 Speaker Recognition

Human voices are considered to be unique, like handwriting or fingerprints [100], allowing for the biometric identification of speakers from recorded speech [66]. This has been

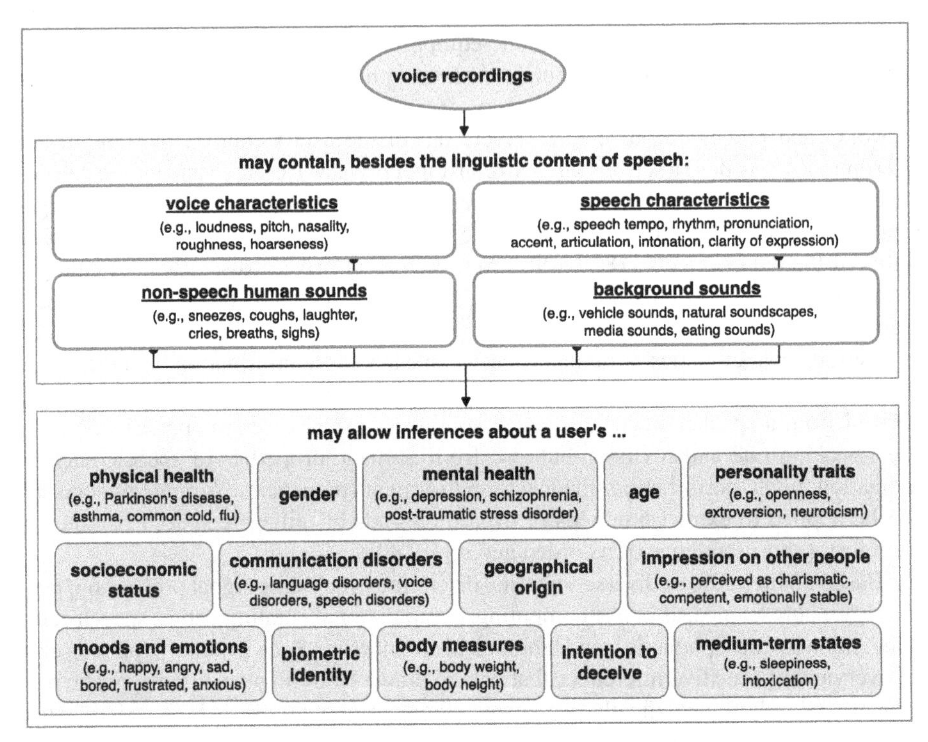

Fig. 1. Overview of some sensitive attributes discernable from speech data.

shown to be possible with speech recorded from a distance [71] and with multi-speaker recordings, even under adverse acoustic conditions (e.g., background noise, reverb) [66]. Voice recognition software has already been transferred into patents [50] and is being applied in practice, for example to verify the identity of telephone customers [40] or to recognize users of virtual assistants like Amazon Alexa [1].

Mirroring the privacy implications of facial recognition, voice fingerprinting could be used to automatically link the content and context of sound-containing media files to the identity of speakers for various tracking and profiling purposes.

2.2 Inference of Body Measures

Research has shown that human listeners can draw inferences about body characteristics of a speaker based solely on hearing the target's voice [42, 55, 69]. In [42], voice-based estimates of waist-to-hip ratio (WHR) of female speakers predicted the speaker's actual WHR, the estimated shoulder-to-hip ratio (SHR) of male speakers predicted the speaker's actual SHR measurements. In another study, human evaluators estimated the body height and weight of strangers from a voice recording almost as well as they did from a photograph [55].

Various attempts have been made to identify the acoustic voice features that enable such inferences [25, 29, 69]. In women, relationships were discovered between voice parameters, such as subharmonics and frequency pertubation, and body features, including weight, height, body mass index, and body surface area [29]. Among men, individuals with larger body shape, particularly upper body musculature, are more likely to have low-pitched voices, and the degree of formant dispersion in male voices was found to correlate with body size (height and weight) and body shape (e.g., waist, chest, neck, and shoulder circumference) [25].

Although research on the speech-based assessment of body configuration is not as advanced as other inference methods covered in this paper, corresponding algorithms have already been developed. For instance, researchers were able to automatically estimate the body height of speakers based on voice features with an accuracy of 5.3 cm, surpassing human performance at this task [69].

Many people feel uncomfortable sharing their body measurements with strangers [12]. The researchers who developed the aforementioned approach for speech-based body height estimation suggest that their algorithm could be used for "applications related to automatic surveillance and profiling" [69], thereby highlighting just some of the privacy threats that may arise from such inference possibilities.

2.3 Mood and Emotion Recognition

There has been extensive research on the automatic identification of emotions from speech signals [21, 23, 53, 95, 99]. Even slight changes in a speaker's mental state invoke physiological reactions, such as changes in the nervous system or changes in respiration and muscle tension, which in turn affect the voice production process [20]. Besides voice variations, it is possible to automatically detect non-speech sounds associated with certain emotional states, such as crying, laughing, and sighing [4, 23].

Some of the moods and emotions that can be recognized in voice recordings using computerized methods are happiness, anger, sadness, and neutrality [86], sincerity [37], stress [95], amusement, enthusiasm, friendliness, frustration, and impatience [35], compassion and sarcasm [53], boredom, anxiousness, serenity, and astonishment [99]. By analyzing recorded conversations, algorithms can also detect if there is an argument [23] or an awkward, assertive, friendly, or flirtatious mood [82] between speakers.

Automatic emotion recognition from speech can function under realistic noisy conditions [23, 95] as well as across different languages [21] and has long been delivering results that exceed human performance [53]. Audio-based affect sensing methods have already been patented [47, 77] and translated into commercial products, such as the voice analytics app *Moodies* [54].

Information about a person's emotional state can be valuable and highly sensitive. For instance, Facebook's ability to automatically track emotions was a necessary precondition for the company's 2014 scandalous experiment in which the company observed and systematically manipulated mental states of over 600,000 users for opaque purposes [14].

2.4 Inference of Age and Gender

Numerous attempts have been made to uncover links between speech parameters and speaker demographics [26, 34, 48, 92]. A person's gender, for instance, can be reflected in voice onset time, articulation, and duration of vowels, which is due to various reasons, including differences in vocal fold anatomy, vocal tract dimensions, hormone levels, and sociophonetic factors [92]. It has also been shown that male and female speakers differ measurably in word use [26]. Like humans, computer algorithms can identify the sex of a speaker from a voice sample with high accuracy [48]. Precise classification results are achieved even under adverse conditions, such as loud background noise or emotional and intoxicated speech [34].

Just as the gender of humans is reflected in their anatomy, changes in the speech apparatus also occur with the aging process. During puberty, vocal cords are thickened and elongated, the larynx descends, and the vocal tract is lengthened [15]. In adults, age-related physiological changes continue to systematically transform speech parameters, such as pitch, formant frequencies, speech rate, and sound pressure [28, 84].

Automated approaches have been proposed to predict a target's age range (e.g., child, adolescent, adult, senior) or actual year of birth based on such measures [28, 85]. In [85], researchers were able to estimate the age of male and female speakers with a mean absolute error of 4.7 years. Underlining the potential sensitivity of such inferred demographic information, unfair treatment based on age and sex are both among the most prevalent forms of discrimination [24].

2.5 Inference of Personality Traits

Abundant research has shown that it is possible to automatically assess a speaker's character traits from recorded speech [3, 79, 80, 88]. Some of the markers commonly applied for this purpose are prosodic features, such as speaking rate, pitch, energy, and formants [68] and characteristics of linguistic expression [88].

Existing approaches mostly aim to evaluate speakers along the so-called "Big Five" personality traits (also referred to as the "OCEAN model"), comprising openness, conscientiousness, extroversion, agreeableness, and neuroticism [88]. The speech-based recognition of personality traits is possible both in binary form (high vs. low) and in the form of numerical scores [79]. High estimation accuracies have been achieved for all OCEAN traits [3, 80, 88].

Besides the Big Five, voice and word use parameters have been correlated with various other personality traits, such as gestural expressiveness, interpersonal awkwardness, fearfulness, and emotionality [26]. Even culture-specific attributes, such as the extent to which a speaker accepts authority and unequal power distribution, can be inferred from speech data [101].

It is well known that personality traits represent valuable information for customer profiling in various industries, including targeted advertising, insurance, and credit risk assessment – with potentially harmful effects for the data subjects [17, 18]. Some data analytics firms also offer tools to automatically rate job applicants and predict their likely performance based on vocal characteristics [18].

2.6 Deception Detection

Research has shown that the veracity of verbal statements can be assessed automatically [60, 107]. Among other speech cues, acoustic-prosodic features (e.g., formant frequencies, speech intensity) and lexical features (e.g., verb tense, use of negative emotion words) were found to be predictive of deceptive utterances [67]. Increased changes in speech parameters were observed when speakers are highly motivated to deceive [98].

Speech-based lie detection methods have become effective, surpassing human performance [60] and almost reaching the accuracy of methods based on brain activity monitoring [107]. There is potential to further improve the classification performance by incorporating information on the speaker's personality [2], some of which can be inferred from voice recordings as well (as we have discussed in Sect. 2.5).

The growing possibilities of deception detection may threaten a recorded speaker's ability to use lies as a means of sharing information selectively, which is considered to be a core aspect of privacy [63].

2.7 Detection of Sleepiness and Intoxication

Medium-term states that affect cognitive and physical performance, such as fatigue and intoxication, can have a measurable effect on a speaker's voice. Approaches exist to automatically detect sleepiness from speech [19, 89]. There is even evidence that certain speech cues, such as speech onset time, speaking rate, and vocal tract coordination, can be used as biomarkers for the separate assessment of cognitive fatigue [93] and physical fatigue [19].

Similar to sleepiness and fatigue, intoxication can also have various physiological effects, such as dehydration, changes in the elasticity of muscles, and reduced control over the vocal apparatus, leading to changes in speech parameters like pitch, jitter, shimmer, speech rate, speech energy, nasality, and clarity of pronunciation [5, 13]. Slurred speech is regarded as a hallmark effect of excessive alcohol consumption [19].

Based on such symptoms, intoxicated speech can be automatically detected with high accuracy [89]. For several years now, systems have been achieving results that are on par with human performance [13]. Besides alcohol, the consumption of other drugs such as ±3,4-methylenedioxymethamphetamine ("MDMA") can also be detected based on speech cues [7].

2.8 Accent Recognition

During childhood and adolescence, humans develop a characteristic speaking style which encompasses articulation, phoneme production, tongue movement, and other vocal tract phenomena and is mostly determined by a person's regional and social background [64]. Numerous approaches exist to automatically detect the geographical origin or first language of speakers based on their manner of pronunciation ("accent") [9, 45, 64].

Research has been done for discriminating accents within one language, such as regional Indian accents in spoken Hindi (e.g., Kashmiri, Manipuri, Bengali, neutral Hindi) [64] or accents within the English language (e.g., American, British, Australian, Scottish, Irish) [45], as well as for the recognition of foreign accents, such as Albanian,

Kurdish, Turkish, Arabic and Russian accent in Finnish [9] or Hindi, Russian, Italian, Thai, and Vietnamese accent in English [9, 39].

By means of automated speech analysis, it is not only possible to identify a person's country of origin but also to estimate his or her "degree of nativeness" on a continuous scale [33]. Non-native speakers can even be detected when they are very fluent in the spoken language and have lived in the respective host country for several years [62]. Experimental results show that existing accent recognition systems are effective and have long reached accuracies comparable to human performance [9, 39, 45, 62].

Native language and geographical origin can be sensitive pieces of personal information, which could be misused for the detection and discrimination of minorities. Unfair treatment based on national origin is a widespread form of discrimination [24].

2.9 Speaker Pathology

Through indicative sounds like coughs or sneezes and certain speech parameters, such as loudness, roughness, hoarseness, and nasality, voice recordings may contain rich information about a speaker's state of health [19, 20, 47]. Voice analysis has been described as "one of the most important research topics in biomedical electronics" [104].

Rather obviously, recorded speech may allow inferences about communication disorders, which can be divided into language disorders (e.g., dysphasia, underdevelopment of vocabulary or grammar), voice disorders (e.g., vocal fold paralysis, laryngeal cancer, tracheoesophageal substitute voice) and speech disorders (e.g., stuttering, cluttering) [19, 88].

But also conditions beyond the speech production can be detected from voice samples, including Huntington's disease [76], Parkinson's disease [19], amyotrophic lateral sclerosis [74], asthma [104], Alzheimer's disease [27], and respiratory tract infections caused by the common cold and flu [20]. The sound of a person's voice may even serve as an indicator of overall fitness and long-term health [78, 103].

Further, voice cues may reveal a speaker's smoking habit: A linear relationship has been observed between the number of cigarettes smoked per day and certain voice features, allowing for speech-based smoker detection in a relatively early stage of the habit (<10 years) [30]. Recorded human sounds can also be used for the automatic recognition of physical pain levels [61] and the detection of sleep disorders like obstructive sleep apnea [19].

Computerized methods for speech-based health assessment reach near-human performance in a variety of recognition and analysis tasks and have already been translated into patents [19, 47]. For example, Amazon has patented a system to analyze voice commands recorded by a smart speaker to assess the user's health [47].

The EU's General Data Protection Regulation classifies health-related data as a *special category of personal data* for which particular protection is warranted (Art. 9 GDPR). Among other discriminatory applications, such data may be used by insurance companies to adjust premiums of policyholders according to their state of health [18].

2.10 Mental Health Assessment

Speech abnormalities are a defining characteristic of various mental illnesses. A voice with little pitch variation, for example, is a common symptom in people suffering from schizophrenia or severe depression [36]. Other parameters that may reveal mental health issues include verbal fluency, intonation, loudness, speech tempo, semantic coherence, and speech complexity [8, 31, 36].

Depressive speech can be detected automatically with high accuracy based on voice cues, even under adverse recording conditions, such as low microphone quality, short utterances, and background environmental noise [19, 41]. Not only the detection, but also a severity assessment of depression is possible using a speech sample: In men and women, certain voice features were found to be highly predictive of their HAMD (Hamilton Depression Rating Scale) score, which is the most widely used diagnostic tool to measure a patient's degree of depression and suicide risk [36]. Researchers have even shown that it is possible to predict a future depression based on speech parameters, up to two years before the speaker meets diagnostic criteria [75].

Other mental disorders, such as schizophrenia [31], autism spectrum conditions [19], and post-traumatic stress disorder [102], can also be detected through voice and speech analysis. In some experiments, such methods have already surpassed the classification accuracy of traditional clinical interviews [8].

In common with a person's age, gender, physical health, and national origin, information about mental health problems can be very sensitive, often serving as a basis for discrimination [83].

2.11 Prediction of Interpersonal Perception

A person's voice and manner of expression have a considerable influence on how he or she is perceived by other people [44, 51, 88, 90]. In fact, a single spoken word is enough to obtain personality ratings that are highly consistent across independent listeners [10]. Research has also shown that personality assessments based solely on speech correlate strongly with whole person judgements [88]. Conversely, recorded speech may reveal how a speaker tends to be perceived by other people.

Studies have shown, for example, that fast talkers are perceived as more extroverted, dynamic, and competent [80], that individuals with higher-pitched voices are perceived as more open but less conscientious and emotionally stable [44], that specific intonation patterns increase a speaker's perceived trustworthiness and dominance [81], and that certain prosodic and lexical speech features correlate with observer ratings of charisma [88].

Researchers have also investigated the influence of speech parameters on the perception and treatment of speakers in specific contexts and areas of life. It was found, for instance, that voice cues of elementary school students significantly affect the judgements teachers make about their intelligence and character traits [90]. Similarly, certain speech characteristics of job candidates, including their use of filler words, fluency of speaking, and manner of expression, have been used to predict interviewer ratings for traits such as engagement, excitement, and friendliness [70]. Other studies show that voice plays an important role in the popularity of political candidates as it influences their perceived competence, strength, physical prowess, and integrity [51]. According to [6], voters tend to prefer candidates with a deeper voice and greater pitch variability. The same phenomenon can be observed in the appointment of board members: CEOs with lower-pitched voices tend to manage larger companies, earn more, and enjoy longer tenures. In [65], a voice pitch decrease of 22.1 Hz was associated with $187 thousand more in annual salary and a $440 million increase in the size of the enterprise managed. On top of this, voice parameters also have a measurable influence on perceived attractiveness and mate choice [44].

Based on voice samples, it is possible to predict how strangers judge a speaker along certain personality traits – a technique referred to as "automatic personality perception" [88]. Considering that the impression people make on others often has a tangible impact on their possibilities and success in life [6, 51, 65, 90], it becomes clear how sensitive and revealing such information can be.

2.12 Inference of Socioeconomic Status

Certain speech characteristics may allow insights into a person's socioeconomic status. There is ample evidence, for instance, that language abilities – including vocabulary, grammatical development, complexity of utterances, productive and receptive syntax – vary significantly between different social classes, starting in early childhood [38]. Therefore, people from distinct socioeconomic backgrounds can often be told apart based on their "entirely different modes of speech" [11]. Besides grammar and vocabulary, researchers found striking inter-class differences in the variety of perspectives utilized in communication and in the use of stylistic devices, observing that once the nature of the difference is grasped, it is "astonishing how quickly a characteristic organization of communication [can] be detected." [87].

Not only language skills, but also the sound of a speaker's voice may be used to draw inferences about his or her social standing. The menarcheal status of girls, for example, which can be derived from voice samples, is used by anthropologists to investigate living conditions and social inequalities in populations [15]. In certain contexts, voice cues, such as pitch and loudness, can even reveal a speaker's hierarchical rank [52].

Based on existing research, it is difficult to say how precise speech-based methods for the assessment of socioeconomic status can become. However, differences between social classes certainly appear discriminative enough to allow for some forms of automatic classification.

2.13 Classification of Acoustic Scenes and Events

Aside from human speech, voice recordings often contain some form of ambient noise. By analyzing background sounds, it is possible to recognize the environment in which an audio sequence was recorded, including indoor environments (e.g., library, restaurant, grocery store, home, metro station, office), outdoor environments (e.g., beach, city center, forest, residential area, urban park), and transport modes (e.g., bus, car, train) [43, 97].

It is also possible to automatically detect and classify specific audio events, such as animal sounds (e.g., dog, cat, crow, crickets), natural sounds (e.g., rain, sea waves, wind, thunderstorm), urban sounds (e.g., church bells, fireworks, jackhammer), office sounds (e.g., mouse click, keyboard typing, printer), bathroom sounds (e.g., showering, urination, defecation, brushing teeth), domestic sounds (e.g., clock tick, page turning, creaking door, keys placed on a table), and non-speech human sounds (e.g., crying, sneezing, breathing, coughing) [4, 16, 43, 97].

Algorithms can even recognize drinking and eating moments in audio recordings and the type of food a person is eating (e.g., soup, rice, apple, nectarine, banana, crisps, biscuits, gummi bears) [19, 91]. Commercial applications like *Shazam* further demonstrate that media sounds, such as songs and movie soundtracks, can be automatically identified and classified into their respective genre with high accuracy, even based on short snippets recorded in a noisy environment [49].

Through such inferences, ambient sounds in audio recordings may not only allow insights into a device holder's context and location, but also into his or her preferences and activities. Certain environments, such as places of worship or street protests, could potentially reveal a person's religious and political affiliations.

Sensitive information can even be extracted from ultrasonic audio signals inaudible to the human ear. An example that has received a lot of media attention recently is the use of so-called "ultrasonic beacons", i.e. high-pitched Morse signals which are secretly emitted by speakers installed in businesses and stores, or embedded in TV commercials and other broadcast content, allowing companies to unobtrusively track the location and media consumption habits of consumers. A growing number of mobile apps – several hundred already, some of them very popular – are using their microphone permission to scan ambient sound for such ultrasonic signals, often without properly informing the user about it [59].

3 Discussion and Implications

As illustrated in the previous section, sensitive inferences can be drawn from human speech and other sounds commonly found in recorded audio. Apart from the linguistic content of a voice recording, a speaker's patterns of word use, manner of pronunciation, and voice characteristics can implicitly contain information about his or her biometric identity, body features, gender, age, personality traits, mental and physical health condition, emotions, intention to deceive, degree of intoxication and sleepiness, geographical origin, and socioeconomic status.

While there is a rich and growing body of research to support the above statement, it has to be acknowledged that many of the studies cited in this paper achieved their classification results under ideal laboratory conditions (e.g., scripted speech, high quality microphones, close-capture recordings, no background noise) [10, 20, 30, 36, 55, 60, 70, 82, 94, 107], which may raise doubt about the generalizability of their inference methods. Also, while impressive accuracies have been reached, it should not be neglected that nearly all of the mentioned approaches still exhibit considerable error rates.

On the other hand, since methods for voice and speech analysis are often subject to non-disclosure agreements, the most advanced know-how arguably rests within the industry and is not publicly available. It can be assumed that numerous corporate and governmental actors with access to speech data from consumer devices possess much larger amounts of training data and more advanced technical capabilities than the researchers cited in this paper. Amazon, for example, spent more than $23 billion on research and development in 2017 alone, has sold more than 100 million Alexa-enabled devices and, according to the company's latest annual report, "customers spoke to Alexa tens of billions more times in 2018 compared to 2017" [108]. Moreover, companies can link speech data with auxiliary datasets (e.g., social media data, browsing behavior, purchase histories) to draw other sensitive inferences [47] while the methods considered in this paper exclusively rely on human speech and other sounds commonly found in recorded audio. Looking forward, we expect the risk of unintended information disclosure from speech data to grow further with the continuing proliferation of microphone-equipped devices and the development of more efficient inference algorithms. Deep learning, for instance, still appears to offer significant improvement potential for automated voice analysis [3, 19].

While recognizing the above facts and developments as a substantial privacy threat, it is not our intention to deny the many advantages that speech applications offer in areas like public health, productivity, and convenience. Devices with voice control, for instance, improve the lives of people with physical disabilities and enhance safety in situations where touch-based user interfaces are dangerous to use, e.g., while driving a car. Similarly, the detection of health issues from voice samples (see Sect. 2.9) could help in treating illnesses more effectively and reduce healthcare costs.

But since inferred information can be misused in countless ways [17, 18], robust data protection mechanisms are needed in order to reap the benefits of voice and speech analysis in a socially acceptable manner. At the technical level, many approaches have been developed for privacy protection at different stages of the data life cycle, including operations over encrypted data, differential privacy, data anonymization, secure multi-party computation, and privacy-preserving data processing on edge devices [46, 72, 106]. Various privacy safeguards have been specifically designed or adjusted for audio mining applications. These include voice binarization, hashing techniques for speech data, fully homomorphic inference systems, differential private learning, the computation of audio data in separate entrusted units, and speaker de-identification by voice transformation [72, 73]. A comprehensive review of cryptography-based solutions for speech data is provided in [72]. Privacy risks can also be moderated by storing and processing only the audio data required for an application's functionality. For example, where only the linguistic content is required, voice recordings can be converted to text in order to eliminate all voice-related information and thereby minimize the potential for undesired inferences.

In advocating data collection transparency and informational self-determination, the recent privacy discourse has put a focus on the recording mode of microphone-equipped devices, where a distinction can be made between "manually activated," "speech activated," and "always on" [34]. However, data scandals show that reporting modes cannot always be trusted [105]. And even where audio is only recorded and transmitted with a user's explicit consent, sensitive inferences may unnoticeably be drawn from collected speech data, ultimately leaving the user without control over his or her privacy. Enabling the unrestricted screening of audio data for potentially revealing patterns and correlations, recordings are often available to providers of cloud-based services in unencrypted form – an example being voice-based virtual assistants [1, 22]. With personal data being the foundation for highly profitable business models and strategic surveillance practices, it is certainly not unusual for speech data to be processed in an unauthorized or unexpected manner. This is well illustrated by recently exposed cases where Amazon, Google, and Apple ordered human contractors to listen to private voice recordings of their customers [22].

The findings compiled in this paper reveal a serious threat to consumer privacy and show that more research is needed into the societal implications of voice and speech processing. In addition to investigating the technical feasibility of inferences from speech data in more detail, future research should explore technical and legal countermeasures to the presented problem, including ways to enforce existing data protection laws more effectively. Of course, the problem of undesired inferences goes far beyond microphones and needs to be addressed for other data sources as well. For example, in recent work, we have also investigated the wealth of sensitive information that can be implicitly contained in data from air quality sensors, infrared motion detectors, smart meters [56], accelerometers [57], and eye tracking sensors [58]. It becomes apparent that sensors in many everyday electronic devices can reveal significantly more information than one would assume based on their advertised functionality. The crafting of solutions to either limit the immense amounts of knowledge and power this creates for certain organizations, or to at least avert negative consequences for society, will be an important challenge for privacy and civil rights advocates over the years to come.

4 Conclusion

Microphones are widely used in connected devices, where they have a large variety of possible applications. While recognizing the benefits of voice and speech analysis, this paper highlights the growing privacy threat of unexpected inferences from audio data. Besides the linguistic content, a voice recording can implicitly contain information about a speaker's identity, personality, body shape, mental and physical health, age, gender, emotions, geographical origin, and socioeconomic status – and may thereby potentially reveal much more information than a speaker wishes and expects to communicate.

Further research is required into the privacy implications of microphone-equipped devices, taking into account the evolving state of the art in data mining technology. As it is impossible, however, to meaningfully determine the limits of inference methods developed behind closed doors, voice recordings – even where the linguistic content does not seem rich and revealing – should be regarded and treated as highly sensitive by default. Since existing technical and legal countermeasures are limited and do not yet offer reliable protection against large-scale misuses of audio data and undesired inferences, more effective safeguards and means of enforcement are urgently needed. We hope that the knowledge compiled in this paper can serve as a basis for consumer education and will help lawmakers and fellow researchers in assessing the richness and potential sensitivity of speech data.

References

1. Amazon: Alexa and Alexa Device FAQs. https://www.amazon.com/gp/help/customer/display.html?nodeId=201602230. Accessed Nov 16 2019

2. An, G., et al.: Deep personality recognition for deception detection. In: INTERSPEECH, pp. 421–425 (2018). https://doi.org/10.21437/Interspeech.2018-2269
3. An, G., Levitan, R.: Lexical and acoustic deep learning model for personality recognition. In: INTERSPEECH, pp. 1761–1765 (2018)
4. Aytar, Y., et al.: SoundNet: learning sound representations from unlabeled video. In: Conference on Neural Information Processing Systems (NIPS), pp. 892–900 (2016)
5. Bae, S.-G., et al.: A judgment of intoxication using hybrid analysis with pitch contour compare in speech signal processing. IJAER **12**(10), 2342–2346 (2017)
6. Banai, B., et al.: Candidates' voice in political debates and the outcome of presidential elections. In: Psychology Days in Zadar, pp. 33–39. University of Zadar (2017)
7. Bedi, G., et al.: A Window into the intoxicated mind? Speech as an index of psychoactive drug effects. Neuropsychopharmacology **39**(10), 2340–2348 (2014)
8. Bedi, G., et al.: Automated analysis of free speech predicts psychosis onset in high-risk youths. npj Schizophr. **1**, 15030 (2015)
9. Behravan, H., et al.: i-vector modeling of speech attributes for automatic foreign accent recognition. Trans. Audio Speech Lang. Process. **24**(1), 29–41 (2016)
10. Belin, P., et al.: The sound of trustworthiness: acoustic-based modulation of perceived voice personality. PLoS ONE **12**(10), e0185651 (2017)
11. Bernstein, B.: Language and social class. Br. J. Sociol. **11**(3), 271–276 (1960)
12. Bindahman, S., et al.: 3D body scanning technology: privacy and ethical issues. In: Conference on Cyber Security, Cyber Warfare and Digital Forensic, pp. 150–154 (2012)
13. Bone, D., et al.: Intoxicated speech detection. Comput. Speech Lang. **28**(2), 375–391 (2014). https://doi.org/10.1016/j.csl.2012.09.004
14. Booth, R.: Facebook reveals news feed experiment to control emotions (2014). https://www.theguardian.com/technology/2014/jun/29/facebook-users-emotions-news-feeds
15. Bugdol, M.D., et al.: Prediction of menarcheal status of girls using voice features. Comput. Biol. Med. **100**, 296–304 (2018). https://doi.org/10.1016/j.compbiomed.2017.11.005
16. Chen, J., Kam, A.H., Zhang, J., Liu, N., Shue, L.: Bathroom activity monitoring based on sound. In: Gellersen, H.-W., Want, R., Schmidt, A. (eds.) Pervasive 2005. LNCS, vol. 3468, pp. 47–61. Springer, Heidelberg (2005). https://doi.org/10.1007/11428572_4
17. Christl, W.: How Companies Use Data Against People. Cracked Labs, Vienna (2017)
18. Christl, W., Spiekermann, S.: Networks of Control: A Report on Corporate Surveillance, Digital Tracking, Big Data & Privacy. Facultas, Vienna (2016)
19. Cummins, N., et al.: Speech analysis for health: current state-of-the-art and the increasing impact of deep learning. Methods **151**, 41–54 (2018)
20. Cummins, N., et al.: "You sound ill, take the day off": automatic recognition of speech affected by upper respiratory tract infection. In: IEEE EMBC, pp. 3806–3809 (2017)
21. Desplanques, B., Demuynck, K.: Cross-lingual speech emotion recognition through factor analysis. In: INTERSPEECH, pp. 3648–3652 (2018)
22. Drozdiak, N., Turner, G.: Apple, Google, and Amazon May Have Violated Your Privacy by Reviewing Digital Assistant Commands. https://fortune.com/2019/08/05/google-apple-amazon-digital-assistants/. Accessed 03 Sept 2019
23. Dubey, H., et al.: BigEAR: inferring the ambient and emotional correlates from smartphone-based acoustic big data. In: IEEE CHASE, pp. 78–83 (2016)
24. EEOC: Charge Statistics. https://www.eeoc.gov/eeoc/statistics/enforcement/charges.cfm. Accessed 07 Nov 2019
25. Evans, S., et al.: Relationships between vocal characteristics and body size and shape in human males. Biol. Psychol. **72**(2), 160–163 (2006)
26. Fast, L.A., Funder, D.C.: Personality as manifest in word use: correlations with self-report, acquaintance report, and behavior. J. Pers. Soc. Psychol. **94**(2), 334–346 (2008)

27. Fraser, K.C., et al.: Linguistic features identify Alzheimer's disease in narrative speech. J. Alzheimers Dis. **49**(2), 407–422 (2015). https://doi.org/10.3233/JAD-150520
28. Ghahremani, P., et al.: End-to-end deep neural network age estimation. In: INTERSPEECH, pp. 277–281 (2018). https://doi.org/10.21437/Interspeech.2018-2015
29. González, J.: Correlations between speakers' body size and acoustic parameters of voice. Percept. Mot. Skills **105**(1), 215–220 (2007)
30. Gonzalez, J., Carpi, A.: Early effects of smoking on the voice: a multidimensional study. Med. Sci. Monit. **10**(12), CR649–CR656 (2004)
31. Gosztolya, G., et al.: Identifying schizophrenia based on temporal parameters in spontaneous speech. In: INTERSPEECH, pp. 3408–3412 (2018)
32. Gray, S.: Always On: Privacy Implications of Microphone-Enabled Devices. Future of Privacy Forum, Washington, DC (2016)
33. Grosz, T., et al.: Assessing the degree of nativeness and Parkinson's condition using Gaussian processes and deep rectifier neural networks. In: INTERSPEECH (2015)
34. Grzybowska, J., Ziółko, M.: I-vectors in gender recognition from telephone speech. In: Conference on Applications of Mathematics in Biology and Medicine (2015)
35. Haider, F., et al.: An active feature transformation method for attitude recognition of video bloggers. In: INTERSPEECH, pp. 431–435 (2018)
36. Hashim, N.W., et al.: Evaluation of voice acoustics as predictors of clinical depression scores. J. Voice **31**(2), 256.e1–256.e6 (2017). https://doi.org/10.1016/j.jvoice.2016.06.006
37. Herms, R.: Prediction of deception and sincerity from speech using automatic phone recognition-based features. In: INTERSPEECH, pp. 2036–2040 (2016)
38. Hoff, E.: How social contexts support and shape language development. Dev. Rev. **26**(1), 55–88 (2006). https://doi.org/10.1016/j.dr.2005.11.002
39. Honig, F., et al.: Islands of failure: employing word accent information for pronunciation quality assessment of English L2 learners. In: ISCA SLATE Workshop (2009)
40. HSBC: Welcome to Voice ID. https://www.us.hsbc.com/customer-service/voice/. Accessed 22 Oct 2019
41. Huang, Z., et al.: Depression detection from short utterances via diverse smartphones in natural environmental conditions. In: INTERSPEECH, pp. 3393–3397 (2018)
42. Hughes, S.M., et al.: Sex-specific body configurations can be estimated from voice samples. J. Soc. Evol. Cult. Psychol. **3**(4), 343–355 (2009). https://doi.org/10.1037/h0099311
43. IEEE AASP: Challenge results published. http://www.cs.tut.fi/sgn/arg/dcase2017/articles/challenge-results-published. Accessed 22 Oct 2019
44. Imhof, M.: Listening to voices and judging people. Int. J. List. **24**(1), 19–33 (2010)
45. Jain, A., et al.: Improved accented speech recognition using accent embeddings and multi-task learning. In: INTERSPEECH, pp. 2454–2458 (2018)
46. Jain, P., et al.: Big data privacy: a technological perspective and review. J. Big Data **3**(1), 25 (2016)
47. Jin, H., Wang, S.: Voice-based determination of physical and emotional characteristics of users (2018). https://patents.google.com/patent/US10096319B1/en?oq=10096319
48. Kabil, S.H., et al.: On learning to identify genders from raw speech signal using CNNs. In: INTERSPEECH, pp. 287–291 (2018)
49. Kaneshiro, B., et al.: Characterizing listener engagement with popular songs using large-scale music discovery data. Front. Psychol. **8**, 1–15 (2017)
50. Karpey, D., Pender, M.: Customer Identification Through Voice Biometrics (2016). https://patents.google.com/patent/US9396730
51. Klofstad, C.A., et al.: Perceptions of competence, strength, and age influence voters to select leaders with lower-pitched voices. PLoS ONE **10**(8), e0133779 (2015)
52. Ko, S.J., et al.: The sound of power: conveying and detecting hierarchical rank through voice. Psychol. Sci. **26**(1), 3–14 (2015). https://doi.org/10.1177/0956797614553009

53. Koolagudi, S.G., Maity, S., Kumar, V.A., Chakrabarti, S., Rao, K.S.: IITKGP-SESC: speech database for emotion analysis. In: Ranka, S., et al. (eds.) IC3 2009. CCIS, vol. 40, pp. 485–492. Springer, Heidelberg (2009). https://doi.org/10.1007/978-3-642-03547-0_46
54. Kotenko, J.: To infinity and Beyond Verbal (2013). https://www.digitaltrends.com/social-media/exploring-beyond-verbal-the-technology-of-emotions-analytics/
55. Krauss, R.M., et al.: Inferring speakers' physical attributes from their voices. J. Exp. Soc. Psychol. **38**(6), 618–625 (2002)
56. Kröger, J.: Unexpected inferences from sensor data: a hidden privacy threat in the internet of things. In: Strous, L., Cerf, Vinton G. (eds.) IFIPIoT 2018. IAICT, vol. 548, pp. 147–159. Springer, Cham (2019). https://doi.org/10.1007/978-3-030-15651-0_13
57. Kröger, J.L., et al.: Privacy implications of accelerometer data: a review of possible inferences. In: Proceedings of the 3rd International Conference on Cryptography, Security and Privacy (ICCSP). ACM, New York (2019). https://doi.org/10.1145/3309074.3309076
58. Kröger, J.L., Lutz, O.H.-M., Müller, F.: What does your gaze reveal about you? On the privacy implications of eye tracking. In: Friedewald, M., Önen, M., Lievens, E., Krenn, S., Fricker, S. (eds.) Privacy and Identity 2019. IFIP AICT, vol. 576, pp. 226–241. Springer, Cham (2020). https://doi.org/10.1007/978-3-030-42504-3_15
59. Kröger, J.L., Raschke, P.: Is my phone listening in? On the feasibility and detectability of mobile eavesdropping. In: Foley, S.N. (ed.) DBSec 2019. LNCS, vol. 11559, pp. 102–120. Springer, Cham (2019). https://doi.org/10.1007/978-3-030-22479-0_6
60. Levitan, S.I., et al.: Acoustic-prosodic indicators of deception and trust in interview dialogues. In: INTERSPEECH, pp. 416–420 (2018)
61. Li, J.-L., et al.: Learning conditional acoustic latent representation with gender and age attributes for automatic pain level recognition. In: INTERSPEECH (2018)
62. Lopes, J., et al.: A nativeness classifier for TED Talks. In: ICASSP, pp. 5672–5675 (2011)
63. Magi, T.J.: Fourteen reasons privacy matters: a multidisciplinary review of scholarly literature. Libr. Q. Inf. Community Policy **81**(2), 187–209 (2011)
64. Malhotra, K., Khosla, A.: Automatic identification of gender & accent in spoken Hindi utterances with regional Indian accents. In: IEEE SLT Workshop, pp. 309–312 (2008)
65. Mayew, W.J., et al.: Voice pitch and the labor market success of male chief executive officers. Evol. Hum. Behav. **34**(4), 243–248 (2013)
66. McLaren, M., et al.: The 2016 speakers in the wild speaker recognition evaluation. In: INTERSPEECH, pp. 823–827 (2016). https://doi.org/10.21437/Interspeech.2016-1137
67. Mendels, G., et al.: Hybrid acoustic-lexical deep learning approach for deception detection. In: INTERSPEECH, pp. 1472–1476 (2017)
68. Mohammadi, G., et al.: The voice of personality: mapping nonverbal vocal behavior into trait attributions. In: Workshop on Social Signal Processing (SSPW), pp. 17–20 (2010)
69. Mporas, I., Ganchev, T.: Estimation of unknown speaker's height from speech. Int. J. Speech Technol. **12**(4), 149–160 (2009). https://doi.org/10.1007/s10772-010-9064-2
70. Naim, I., et al.: Automated prediction and analysis of job interview performance. In: IEEE Conference on Automatic Face and Gesture Recognition, pp. 1–6 (2015)
71. Nandwana, M.K., et al.: Robust speaker recognition from distant speech under real reverberant environments using speaker embeddings. In: INTERSPEECH (2018)
72. Nautsch, A., et al.: Preserving privacy in speaker and speech characterisation. Comput. Speech Lang. **58**, 441–480 (2019). https://doi.org/10.1016/j.csl.2019.06.001
73. Nautsch, A., et al.: The GDPR & speech data: reflections of legal and technology communities, first steps towards a common understanding. In: INTERSPEECH, pp. 3695–3699 (2019). https://doi.org/10.21437/Interspeech.2019-2647
74. Norel, R., et al.: Detection of amyotrophic lateral sclerosis (ALS) via acoustic analysis. In: INTERSPEECH, pp. 377–381 (2018). https://doi.org/10.1101/383414

75. Ooi, K.E.B., et al.: Multichannel weighted speech classification system for prediction of major depression in adolescents. IEEE Trans. Biomed. Eng. **60**(2), 497–506 (2013)
76. Perez, M., et al.: Classification of huntington disease using acoustic and lexical features. In: INTERSPEECH, pp. 1898–1902 (2018)
77. Petrushin, V.A.: Detecting emotions using voice signal analysis (2007). https://patents.google.com/patent/US7222075B2/en
78. Pipitone, R.N., Gallup, G.G.: Women's voice attractiveness varies across the menstrual cycle. Evol. Hum. Behav. **29**(4), 268–274 (2008)
79. Polzehl, T., et al.: Automatically assessing personality from speech. In: IEEE Conference on Semantic Computing (ICSC), pp. 134–140 (2010)
80. Polzehl, T.: Personality in Speech. Springer, Cham (2015). https://doi.org/10.1007/978-3-319-09516-5
81. Ponsot, E., et al.: Cracking the social code of speech prosody using reverse correlation. Proc. Natl. Acad. Sci. **115**(15), 3972–3977 (2018)
82. Ranganath, R., et al.: Detecting friendly, flirtatious, awkward, and assertive speech in speed-dates. Comput. Speech Lang. **27**(1), 89–115 (2013)
83. Reavley, N.J., Jorm, A.F.: Experiences of discrimination and positive treatment in people with mental health problems. Aust. N. Z. J. Psychiatry **49**(10), 906–913 (2015)
84. Reubold, U., et al.: Vocal aging effects on F0 and the first formant: a longitudinal analysis in adult speakers. Speech Commun. **52**(7–8), 638–651 (2010)
85. Sadjadi, S.O., et al.: Speaker age estimation on conversational telephone speech using senone posterior based i-vectors. In: ICASSP, pp. 5040–5044 (2016)
86. Sarma, M., et al.: Emotion identification from raw speech signals using DNNs. In: INTERSPEECH, pp. 3097–3101 (2018). https://doi.org/10.21437/Interspeech.2018-1353
87. Schatzman, L., Strauss, A.: Social class and modes of communication. Am. J. Sociol. **60**(4), 329–338 (1955). https://doi.org/10.1086/221564
88. Schuller, B., et al.: A survey on perceived speaker traits: personality, likability, pathology, and the first challenge. Comput. Speech Lang. **29**(1), 100–131 (2015)
89. Schuller, B., et al.: Medium-term speaker states - a review on intoxication, sleepiness and the first challenge. Comput. Speech Lang. **28**(2), 346–374 (2013)
90. Seligman, C.R., et al.: The effects of speech style and other attributes on teachers' attitudes toward pupils. Lang. Soc. **1**(1), 131 (1972)
91. Sim, J.M., et al.: Acoustic sensor based recognition of human activity in everyday life for smart home services. Int. J. Distrib. Sens. Netw. **11**(9), 679123 (2015)
92. Simpson, A.P.: Phonetic differences between male and female speech. Lang. Linguist. Compass **3**(2), 621–640 (2009). https://doi.org/10.1111/j.1749-818X.2009.00125.x
93. Sloboda, J., et al.: Vocal biomarkers for cognitive performance estimation in a working memory task. In: INTERSPEECH, pp. 1756–1760 (2018)
94. Soskin, W.F., Kauffman, P.E.: Judgment of emotion in word-free voice samples. J. Commun. **11**(2), 73–80 (1961). https://doi.org/10.1111/j.1460-2466.1961.tb00331.x
95. Stanek, M., Sigmund, M.: Psychological stress detection in speech using return-to-opening phase ratios in glottis. Elektron Elektrotech. **21**(5), 59–63 (2015)
96. Stanescu, C.G., Ievchuk, N.: Alexa, where is my private data? In: Digitalization in Law, pp. 237–247. Social Science Research Network, Rochester (2018)
97. Stowell, D., et al.: Detection and classification of acoustic scenes and events. IEEE Trans. Multimed. **17**(10), 1733–1746 (2015). https://doi.org/10.1109/TMM.2015.2428998
98. Streeter, L.A., et al.: Pitch changes during attempted deception. J. Pers. Soc. Psychol. **35**(5), 345–350 (1977). https://doi.org/10.1037//0022-3514.35.5.345
99. Swain, M., et al.: Databases, features and classifiers for speech emotion recognition: a review. Int. J. Speech Technol. **21**(1), 93–120 (2018)

100. Trilok, N.P., et al.: Establishing the uniqueness of the human voice for security applications. In: Proceedings of Student-Faculty Research Day, pp. 8.1–8.6. Pace University (2004)
101. Tsai, F.-S., et al.: Automatic assessment of individual culture attribute of power distance using a social context-enhanced prosodic network representation. In: INTERSPEECH, pp. 436–440 (2018). https://doi.org/10.21437/Interspeech.2018-1523
102. Vergyri, D., et al.: Speech-based assessment of PTSD in a military population using diverse feature classes. In: INTERSPEECH, pp. 3729–3733 (2015)
103. Vukovic, J., et al.: Women's voice pitch is negatively correlated with health risk factors. J. Evol. Psychol. **8**(3), 217–225 (2010). https://doi.org/10.1556/JEP.8.2010.3.2
104. Walia, G.S., Sharma, R.K.: Level of asthma: mathematical formulation based on acoustic parameters. In: CASP, pp. 24–27 (2016). https://doi.org/10.1109/CASP.2016.7746131
105. Wolfson, S.: Amazon's Alexa recorded private conversation and sent it to random contact (2018). https://www.theguardian.com/technology/2018/may/24/amazon-alexa-recorded-conversation
106. Zhao, J., et al.: Privacy-preserving machine learning based data analytics on edge devices. In: AAAI/ACM Conference on AI, Ethics, and Society (AIES), pp. 341–346 (2018)
107. Zhou, Y., et al.: Deception detecting from speech signal using relevance vector machine and non-linear dynamics features. Neurocomputing **151**, 1042–1052 (2015)
108. Annual Report 2018. Amazon.com, Inc., Seattle, Washington, USA (2019)

Border Control and Use of Biometrics: Reasons Why the Right to Privacy Can Not Be Absolute

Mohamed Abomhara[1](✉), Sule Yildirim Yayilgan[1], Marina Shalaginova[1], and Zoltán Székely[2]

[1] Department of Information Security and Communication Technology, Norwegian University of Science and Technology, Gjøvik, Norway
{mohamed.abomhara,sule.yildirim,marina.shalaginova}@ntnu.no
[2] Faculty of Law Enforcement, National University of Public Service, Budapest, Hungary
dr.szekely.zoltan@gmail.com

Abstract. This paper discusses concerns pertaining to the absoluteness of the right to privacy regarding the use of biometric data for border control. The discussion explains why privacy cannot be absolute from different points of view, including privacy versus national security, privacy properties conflicting with border risk analysis, and Privacy by Design (PbD) and engineering design challenges.

Keywords: Biometrics · Biometric technology · Border control · Data privacy · Right to privacy

1 Introduction

Biometric technologies are automated methods of recognizing and verifying the identity of individuals based on physiological or behavioral attributes [11]. They are used progressively more and are highly adopted at European borders [20,21]. Strengthening border security, improving border crossing efficiency and facilitating effective migration control and enforcement are among the main grounds for utilizing biometric technologies [26]. Despite the many advantages of biometrics, there are some limitations. The integration of biometric information systems employed for border control leads to increased surveillance that involves collecting and storing personal data such as fingerprints as individuals cross borders, apply for visas or request asylum. According to the General Data Protection Regulation (GDPR) [49], personal information falls in a number of general categories, such as identity number and financial information (Article 4(1) GDPR)

This work is carried out as part of the EU-funded project SMart mobILity at the European land borders (SMILE) (Project ID: 740931), [H2020-DS-2016-2017] SEC-14-BES-2016 towards reducing the cost of technologies in land border security applications.

Published by Springer Nature Switzerland AG 2020
M. Friedewald et al. (Eds.): Privacy and Identity 2019, IFIP AICT 576, pp. 259–271, 2020.
https://doi.org/10.1007/978-3-030-42504-3_17

as well as special categories, for instance, biometric data, sexual orientation, medical information, personal activities, etc. (Article 9(1) GDPR). Such information is a valuable asset because it is crucial for all individuals to be able to keep it to themselves. On one hand, biometric technology has been proven to be cost-effective in enhancing border security, detecting fraud, helping improve border crossing efficiency as well as enabling effective migration control and enforcement [9]. On the other hand, biometric technology has serious impacts on privacy and data protection [33].

Major concerns with the use of biometric technology relate to individuals' privacy reduction and the immutable link between biometric traits and persistent information storage about a person. The tight link between personal information and biometrics can have both positive and negative consequences for individuals' privacy. Recent research [17] explores the possibility of extracting supplementary information from primary biometric traits, e.g. face, fingerprints and iris. These traits denote personal attributes like gender, age, ethnicity, hair color, height, weight and so on. However, a breach (unauthorized acquisition, access, use or disclosure) of such confidential information would violate the principle of the right to privacy for those consenting to cross a border, even if the breach involved innocuous information that would not result in any social, economic, legal, or any other harm.

The right to privacy is described as the right of the individual to be let alone and to decide how, when, with whom and to what degree their personal data should be shared and communicated [30,51]. However, unlike absolute fundamental rights as "the right to life or the prohibition of torture and inhuman or degrading treatment or the right to be free from slavery," which admit of no restriction (judgment of Court of Justice of the European Union (CJEU) of 12 June 2003, Schmidberger, Case C-112/00, paragraph 80 [5]), the right to privacy is not an absolute right and hence can be limited by law [27,35] – for example in time of public emergency that threatens the life of a nation. The arguments made in this paper pertain to why privacy cannot be absolute from different points of view: (1) privacy versus national security; (2) privacy properties conflicting with border risk analysis; and (3) Privacy by Design (PbD) and engineering design challenges.

The remaining part of this paper is organized as follows. Section 2 briefly discusses the use of biometrics for border control and the right to privacy principle. Section 3 investigates and argues why privacy cannot be absolute. Section 4 concludes the paper.

2 Background

2.1 The Use of Biometrics for Border Control

The dramatic advances in biometric technologies have opened doors to unprecedented opportunities in the field of border control. The challenge of border security is to identify with assurance who is crossing the border and decide if the person is authorized to cross or not. Unassisted by technology, a border staff member cannot maintain this degree of swift, assured identification. Border authorities are equipped with biometric technologies to facilitate more efficient checks at borders and contribute to preventing and combating illegal migration, etc. [26]. Biometrics enable accurate identification since each person has their own unique physical characteristics. Moreover, the use of multimodal biometrics [29,37] offers even better results with higher accuracy by combining several biometrics [34]. Multimodal biometrics-integrated border management benefits all stakeholders, including governments concerned with securing national territory, immigration authorities managing controls at ever more crowded borders, and simply travelers who want to enjoy the journey to their destination [21].

European Union (EU)-wide border biometric information management systems [20], including the European Asylum Dactyloscopy Database (EURODAC), Visa Information System (VIS) and Second-generation Schengen Information System (SIS II) have an increasingly important role in the identity establishment process by storing biographic and biometric data of third-country nationals [12,39]. In addition, the centralized Entry/Exit System (EES) of border control is expected to be fully implemented by 2020 in compliance with Regulations (EU) 2017/2225 [2] and (EU) 2017/2226 [3]. Interoperability will allow the EU information systems to complement each other, help facilitate the correct identification of persons, contribute to fighting identity fraud and ease information sharing. Accordingly, the SMart mobILity at the European land borders (SMILE)[1] interoperability with other border information systems is greatly promising in terms of enhancing the speed, efficiency and flow of border crossing mobility as well as border security.

As a border control tool, SMILE encourages the propensity to collect, use and process sensitive biometrics like fingerprints, face and iris data, etc. to improve traveler flow and boost border security. On the one hand, SMILE technologies are intended to enhance security levels and make the traveler identification and authentication procedures easy, fast and convenient. On the other hand, similar to other biometric technologies, SMILE technologies have raised new threats to fundamental rights, data protection and privacy.

2.2 Right to Privacy

Every individual has the right to the privacy protection of their personal information when it is collected and shared. Generally, legal documents on data protection and privacy such as GDPR [49] refer to personal information protection throughout all steps from collection to storage and dissemination. Moreover, provisions of the European Convention of Human Rights (ECHR) constitute the basis for challenging inequitable decisions of public authorities [15,27,50]. Article 8 ECHR ensures everyone's right to have their private and family life, home and correspondence respected without public authority interference. The aforementioned right is mirrored in Article 7 of the Charter of Fundamental Rights (CFR). Moreover, Article 8 of the Charter enshrines the right of everyone to the protection of their personal data. Individuals (data subjects) have the right to

[1] http://smile-h2020.eu/smile/.

exercise control over their personal data (Article 12–23 GDPR [49]). Protection of privacy is frequently seen as a line drawn for how far society can intrude into a person's private affairs [35].

However, these rights can be overruled if a legal basis is laid down for collecting, processing, storing or retaining personal data to achieve a legitimate goal [13]. According to Article 52(1) CFR, "subject to the principle of proportionality, limitations of rights to privacy may be made only if they are necessary and genuinely meet objectives of general interest recognized by the Union or the need to protect the rights and freedoms of others." Moreover, Recital 4 GDPR acknowledges that the right to data protection (as well as the right to privacy) is not an absolute right. Ruling the Eifert case [6], the CJEU holds that "the right to the protection of personal data is not an absolute right, but must be considered in relation to its function in society." Furthermore, the rights to privacy must be proportionately balanced with other fundamental rights. According to the Opinion of Advocate General Jääskinen in the Google Spain Case [7], the fundamental rights to privacy and to the protection of personal data are not absolute and may be limited provided there is a justification acceptable in view of the conditions set out in Article 52(1) of the Charter.

3 Viewpoints of Privacy: Why It Can Not Be Absolute

This section discusses the contradictory interests of privacy versus national security, privacy properties conflicting with border risk analysis and Privacy by Design (PbD) and engineering design challenges.

3.1 Privacy versus National Security

The Schengen Borders Code (SBC) (Regulation (EU) 2016/399) [1] and its amendment (Regulation (EU) 2017/458) [4] set out the rules governing the movement of people across EU's internal and external borders. Internal borders means (a) the common land borders, including river and lake borders, of the Member States; (b) the airports of the Member States for internal flights; (c) sea, river and lake ports of the Member States for regular internal ferry connections. External borders means the Member States' land borders, including river and lake borders, sea borders and their airports, river ports, sea ports and lake ports, provided that they are not internal borders. SBC also defines the rules for the border checks of persons crossing external Schengen borders (border checks on persons). Cross-border movement at external borders shall be subject to minimum and thorough checks by border guards (Article 8 of Regulation (EU) 2016/399). The main objectives of the minimum and thorough checks are to ensure that the persons in question do not represent a threat to public order, internal security or public health, and to improve the security of the EU Member States and their citizens. Therefore, the key issue in question is how to achieve a trade-off between border check requirements (Regulation (EU) 2016/399) to ensure border security and meeting the need for flexible border crossing without compromising individuals' privacy.

According to a survey and interviews carried out for the SMILE project with land border guards (refer to the SMILE public deliverables), the majority of guards claim that the use of biometric authentication and verification at land borders is justified and necessary to improve border security measures and better protect the public interest. Biometrics are believed to help improve the accuracy of traveler identification and verification, meaning the ability to correctly recognize a genuine person and reject an imposter. Moreover, utilizing biometrics promotes the reduction of identity fraud (e.g., fake IDs and passports) as the identification and verification processes do not rely on the human agent. To the best of our knowledge, fraud reduction signifies accuracy increase. Therefore, using biometrics eliminates a considerable integrity threat that border guards face and benefits the authority responsible for border control. As a consequence, biometrics would result in a higher throughput of low-risk travelers without losing accuracy or integrity and allow human resources to focus on potentially higher-risk travelers.

The right of people to be respected for their private and family life, home, etc. (Article 8 of the ECHR) and also protected against unreasonable biometric searches (e.g., unreasonable biometric authentication and verification at the border) shall not be violated. For border authorities to be reasonable with using biometrics, traveler authentication and verification using biometrics must be as limited in its intrusiveness as it is consistent with satisfying the administrative need that justifies it [32]. It is important to argue on the one hand whether the introduction of biometrics for border control can be regarded as being in accordance with the law and if the biometrics satisfy a legitimate aim (public safety, crime prevention, etc.), proportionality and the necessity principles pursuant to Article 8 of ECHR. On the other hand, it is arguable whether breaching the right to privacy is also considered proportionate to the threats that biometric technologies are supposed to prevent. As for necessity and effectiveness, many discussions address whether biometrics actually add value to serving border control interests, including but not limited to, providing the safe, secure, efficient and unobtrusive, on-the-move security control of travelers, fighting terrorism and serious crime, and ensuring high internal security levels.

If national security has greater priority, should individual privacy yield? Besides, would it be ethically justifiable to sacrifice some privacy interests to achieve the highest national security gains possible? Moreover, if a traveler faces the dilemma of either providing biometrics or not being allowed to cross the border (in case no other alternative is provided), the person's right to freedom of movement may be restricted [35]. Our conclusion is not yet sufficiently clear to answer these questions. Even if it is accepted that using biometrics at borders is necessary and proportionate, serious concerns about whether the intrusion is in accordance with the law still remains a question. It may be said that conflicts arising between border security (Regulation (EU) 2016/399) interests and individual privacy cannot be weighed because they are not measurable by the same standards [31]. While acknowledging there could be legitimate aims for the invasion of privacy, the effectiveness of biometrics is still questionable. Therefore, a sensible trade-off is necessary between individual interests (individual privacy) and the legitimate concerns of the Member States such as preventing and investigating crime [40,48]. An appropriate trade-off would involve retaining the benefits of biometric technologies to extend the border control ability to support high border security and reliability while maintaining individual privacy.

3.2 Privacy Properties Conflicting with Border Risk Analysis

Border risk analysis is a governance tool to normalize border and migration risks. It is based on an automated analysis of lager databases (SIS II, etc.) to extract useful information about people and their activities in order to identify behavioral patterns that may point to suspicious activity. Although border risk analysis is useful in decision-making, it can also lead to serious privacy concerns.

The key issue is in the contradictory interests of the principle of data minimization (Article 5 GDPR) that limits personal data collection, use and disclosure and the benefit of the capability to process personal data for performing border risk analysis. The main concerns include the unnecessary and unauthorized collection of biometric data for traveler identification and verification [14,51]. For one, the data minimization principle and strong privacy properties (i.e., unlinkability, anonymity, undetectability) [19] restrict personal data collection and use for further analysis. However, border risk analysis requires the use of personal data for the investigation and/or monitoring of actions/activities of one or more individuals. In other words, the more personal data border authorities, for instance, can obtain about individuals, the better the risk prediction and thus overall risk analysis results will be. There has certainly been considerable progress in privacy preserving techniques for data analytics [38,41]. Nonetheless, even with such privacy-friendly techniques, border risk analysis is prone to privacy violations.

Another major concern is related to the potential for function/purpose creep. Purpose creep occurs when personal data is collected for one specific purpose and subsequently used for another unintended or unauthorized purpose without the user's consent. A famous example of a large-scale biometric function creep is related to the European Dactyloscopy (EURODAC) fingerprint database (Regulation (EC) 2725/2000). The original purpose of the EURODAC database was to compare fingerprints for the effective application of the Dublin convention (Regulation (EU) 603/2013) [51]. EURODAC enables EU countries to identify asylum applicants as well as illegal immigrants within the EU. However, soon after the database was established, other police and law enforcement agencies were also granted access. There are many other large-scale, centralized EU national and international databases, such as SIS II and VIS with the same risks. Similar concerns also arise in the case of border control risk analysis [9]. Hard and soft biometrics [8,18,52] are likely to strengthen the potential for function creep due to the very sensitive nature of the data collected and the possibility to use centrally stored biometric data for purposes other than the original purpose (border crossing). Moreover, such databases offer more attractive targets for outsider attacks and insider misuse.

Therefore, the purpose specification principle (Article 5(1)(b) of GDPR), which is among the main principles of EU data protection legislation, has a key role. According to Article 5(1)(b) of GDPR, personal data must be collected for specified, explicit and legitimate purposes and not be further processed in a way incompatible with those purposes. However, GDPR provides exemptions in Article 23, which stipulates that Member States' laws may restrict the scope of the principles mentioned in Article 5 of GDPR when such restriction constitutes a necessary measure to safeguard national security and public security, to prevent, investigate, detect or prosecute criminal offences or execute criminal penalties, to protect the data subject or the rights and freedoms of others, etc. In our view, the indication of "a necessary measure" means that exemptions are restricted to specific investigations, case-by-case requests, and not to cases where personal data processing is systematic as foreseen by the use of biometrics for border control. As mentioned earlier, the processing of biometric data is questionable even when considering that an exemption might be applicable. Therefore, the problems of function creep and purpose misuse are not to be underestimated. Nonetheless, they can be curbed by stricter laws, particularly by limiting the use of specific biometric data for certain purposes. It can thus be concluded that the clarity of purpose regarding the intention of biometric data collection is paramount. It is important to be clear about the necessity for biometrics and how biometrics will help fulfill specified needs.

3.3 Privacy by Design (PbD) and Engineering Design

Privacy by design (PbD) is a policy measure that guides software developers to apply inherent solutions to achieve better privacy protection [23]. For privacy to be embedded in the system development lifecycle and hence in organizational processes, system developers and policy makers must be ready to embrace and understand the domain [47].

Recent studies [23, 36, 43, 44] reveal that most software developers lack formal knowledge and understanding of the concept of informational privacy. Besides, most have insufficient knowledge of how to develop privacy practices such as PbD [23]. Software developers additionally find it difficult to understand privacy requirements by themselves [44] and require significant effort to estimate privacy risks from a user perspective in order to relate privacy requirements to privacy techniques [36]. Moreover, software developers have trouble evaluating whether they have successfully embedded PbD strategies into the system design.

Privacy design frameworks serve as potential bridges between users, software developers and policy makers. Several studies like [22, 24, 28, 46] discuss privacy design frameworks to assist software developers with addressing privacy during the system development process. However, to the best of our knowledge, it is still unclear how effective these design frameworks are and what the possible limitations for their utilization in everyday privacy engineering practices are. The key elements of PbD are intended to limit the collection, use and disclosure of personal data, to involve individuals in the data lifecycle, and to apply appropriate safeguards in a continuous manner [31]. This means separating personal identifiers, using pseudonyms and anonymization as well as deleting personal data when no longer needed [42]. However, as argued by Leese [31], such practices are undeniably suitable in economic and organizational contexts. But as discussed in Sects. 3.1 and 3.2, border checks and border risk analysis derive decisions exactly through the collection and processing of data, which could ultimately be connected to possible criminal activities, in order to control any risk. On the contrary, essentially PbD principles radically exclude the possibilities that come with advanced data analytics in border control contexts. Thus, the contradictory interests of PbD principles and the benefit of the ability to process biometrics data for border control cannot simply be resolved by technical means.

Thus, if PbD is ever to become a viable practice, a considerable change must be made to prepare the field for the wide implementation of this policy. Privacy implementation guidelines should be provided to help software developers and policy makers embed privacy into the system design. Moreover, an evaluation and demonstration of privacy assurance – as recommended by the ENISA[2] guidelines for privacy and data protection [16] – is required to provide software developers with feedback to verify whether they have successfully followed the guidelines. This would reduce software developers' personal opinions on privacy practices and ease how privacy is embedded into the system. Moreover, the development method (Agile Software Development [10], Security Development Lifecycle [25,45], etc.) used within the organization must be taken into account in order to apply the concepts of privacy throughout the entire system development process. This will enable development teams, policy makers, etc. to take appropriate measures in the relevant phases. Finally, upon design completion, the organization must adopt and monitor it throughout its lifetime.

Alongside the PbD issue is the Privacy by Default obligation. Under this obligation, data controllers must implement appropriate measures on both technical and organization levels to ensure that personal data collected is only used for specific purposes. Essentially, only the minimum amount of personal data required should be collected and stored, while data subjects should be allowed data accessibility and controllability. Ensuring privacy through every phase of the data lifecycle (collection, use, retention, storage, disposal or destruction) has also become crucial to avoiding legal liability, maintaining regulatory compliance, and so on. Therefore, integrating the Data Protection Impact Assessment (DPIA) with the Ethical Impact Assessment (EIA) and Privacy Impact Assessment (PIA) in the earlier stages of any system development would aid with the early identification of ethical and privacy problems and risks. Ideally, a full and detailed description of the processing along with its necessity and proportionality would help manage the risks to the rights and freedoms of natural persons resulting from personal data processing. Furthermore, taking PbD strategies into

[2] European Union Agency for Network and Information Security (www.enisa.europa.eu).

consideration should precede system design to ensure that ethical and privacy principles are taken into account. As a result, data controllers will be more able to comply with the legal requirements of data protection and demonstrate taking appropriate measures where DPIA is used to check compliance against data protection regulations.

4 Conclusions

Border control systems raise the tendency to collect, use and process personal data (e.g. alphanumeric data like names and birth dates; biographic information; biometric data like fingerprints) to optimize and monitor the flow of people at land borders as well as enhance security and detect fraud. However, evidence demonstrates that personal data collection and processing pose several privacy challenges. Privacy is a fundamental human right in EU countries and is controlled by legislation that responds and adapts to data subjects' privacy needs. Moreover, the obligations of the Member States pertaining to personal data collection and processing along with the exchange of personal data among Member States are stated in various EU laws and regulations. It is therefore essential for border authorities to consider the legal consequences of developing and deploying biometric identification and authentication methods.

Personal data flows and ripples are in some ways difficult to predict. Despite all attempts to provide anonymity, biometric data still penetrates a person's physical, psychological and social identity. Biometric technology enables revealing personal information, such as gender, age, ethnicity and even critical health problems like diabetes, vision problems, Alzheimer's disease, etc. Members of particular groups including disabled, transgender and older people, religious groups, and others, can encounter additional negative effects on privacy. Although numerous proposals, recommendations and legal considerations are in place as safeguards, it is unclear how they can ultimately be put in practice. For example, even a well-conceived, general and sustainable data protection and privacy regulation like GDPR is strained by the effort to ensure superior effectiveness with respect to privacy.

It is quite obvious that biometric technology exposes travelers to significant loss of privacy and limitations of other rights and freedoms. This paper discussed concerns with the absoluteness of the right to privacy regarding the use of biometric data for border control. In accordance with Article 8(2) ECHR, "there shall be no interference by a public authority with the exercise of the right to privacy except such as is in accordance with the law." The exemption clause contains two conditions: (1) the necessity for a democratic society that should be proportional to the purposes of the law and (2) exceptions in accordance with the law. These conditions require a specific legal rule that authorizes the interference and sufficient access of individuals to the specific law, and that the law must be precisely formulated in order to ensure that individuals are capable of foreseeing the conditions of its applicability.

In future, the authors plan to investigate the effect of using biometrics for border control on individuals' privacy. Unfortunately, there is still too little knowledge about the real effects on individuals and a system's reputation when privacy breaches occur. This is because on the one hand, very little knowledge exists about the tangible and intangible benefits of personal data collection in EU information systems. On the other hand, it is not clear to what extent people have the right to choose what information about themselves to share and how to engage with border systems and devices such as biometric sensors and readers.

References

1. Regulation (EU) 2016/399 of the European Parliament and of the Council of 9 March 2016 on a Union Code on the rules governing the movement of persons across borders (Schengen Borders Code). Official Journal of the European Union, L 77/1 (2017). https://eur-lex.europa.eu/eli/reg/2016/399/oj
2. Regulation (EU) 2017/2225 of the European Parliament and of the Council of 30 November 2017 amending Regulation (EU) 2016/399 as regards the use of the Entry/Exit System. Official Journal of the European Union, L 327/1 (2017). https://eur-lex.europa.eu/eli/reg/2017/2225/oj
3. Regulation (EU) 2017/2226 of the European Parliament and of the Council of 30 November 2017 establishing an Entry/Exit System (EES) to register entry and exit data and refusal of entry data of third-country nationals crossing the external borders of the Member States and determining the conditions for access to the EES for law enforcement purposes, and amending the Convention implementing the Schengen Agreement and Regulations (EC) No 767/2008 and (EU) No 1077/2011. Official Journal of the European Union, L 327/20 (2017). https://eur-lex.europa.eu/eli/reg/2017/2226/oj
4. Regulation (EU) 2017/458 of the European Parliament and of the Council of 15 March 2017 amending Regulation (EU) 2016/399 as regards the reinforcement of checks against relevant databases at external borders. Official Journal of the European Union, L 74/1 (2017). https://eur-lex.europa.eu/eli/reg/2017/458/oj
5. Court of Justice of the European Union: C-112/00, Eugen Schmidberger, Internationale Transporte und Planzüge v Republik Österreich (2003)
6. Court of Justice of the European Union: Joined Cases C-92/09 and C-93/09, Volker und Markus Schecke GbR and Hartmut Eifert v Land Hessen (2010)
7. Court of Justice of the European Union: Case C-131/12, Google Spain SL and Google Inc. v Agencia Española de Protección de Datos (AEPD) and Mario Costeja González (2014)
8. Abdelwhab, A., Viriri, S.: A survey on soft biometrics for human identification. Mach. Learn. Biom., 37 (2018)
9. Abomhara, M., Yayilgan, S.Y., Nymoen, A.H., Shalaginova, M., Székely, Z., Elezaj, O.: How to do it right: a framework for biometrics supported border control. In: Katsikas, S., Zorkadis, V. (eds.) e-Democracy 2019. CCIS, vol. 1111, pp. 94–109. Springer, Cham (2020). https://doi.org/10.1007/978-3-030-37545-4_7
10. Abrahamsson, P., Salo, O., Ronkainen, J., Warsta, J.: Agile software development methods: review and analysis. VTT Technical Research Centre of Finland, VTT Publications 478 (2017)
11. Bhatia, R.: Biometrics and face recognition techniques. Int. J. Adv. Res. Comput. Sci. Softw. Eng. **3**(5) (2013)

12. Boehm, F.: Information Sharing and Data Protection in the Area of Freedom, Security and Justice: Towards Harmonised Data Protection Principles for Information Exchange at EU-Level. Springer, Heidelberg (2011). https://doi.org/10.1007/978-3-642-22392-1
13. Bonnici, J.P.M.: Exploring the non-absolute nature of the right to data protection. Int. Rev. Law Comput. Technol. **28**(2), 131–143 (2014)
14. Campisi, P.: Security and Privacy in Biometrics, vol. 24. Springer, London (2013). https://doi.org/10.1007/978-1-4471-5230-9
15. Çinar, Ö.H.: The right to privacy in international human rights law. J. İnf. Syst. Oper. Manag. **13**(1), 33–44 (2019)
16. Danezis, G., et al.: Privacy and data protection by design-from policy to engineering (2014). https://www.enisa.europa.eu/publications/privacy-and-data-protection-by-design
17. Dantcheva, A., Elia, P., Ross, A.: What else does your biometric data reveal? A survey on soft biometrics. IEEE Trans. Inf. Forensics Secur. **11**(3), 441–467 (2016)
18. Dantcheva, A., Velardo, C., D'angelo, A., Dugelay, J.L.: Bag of soft biometrics for person identification. Multimedia Tools Appl. **51**(2), 739–777 (2011)
19. Deng, M., Wuyts, K., Scandariato, R., Preneel, B., Joosen, W.: A privacy threat analysis framework: supporting the elicitation and fulfillment of privacy requirements. Requirements Eng. **16**(1), 3–32 (2011)
20. eu-LISA: Biometrics in large-scale it: Recent trends, current performance capabilities, recommendations for the near future. European Agency for the operational management of large-scale IT systems in the area of freedom, security and justice (eu-LISA) (2018). https://www.eulisa.europa.eu/Publications/Reports/Biometrics
21. European Commission: Biometrics technologies: a key enabler for future digital services. Digital Transformation Monitor (2018)
22. Gürses, S., Troncoso, C., Diaz, C.: Engineering privacy by design. Comput. Priv. Data Prot. **14**(3), 25 (2011)
23. Hadar, I., et al.: Privacy by designers: software developers' privacy mindset. Empirical Softw. Eng. **23**(1), 259–289 (2018)
24. Hoepman, J.-H.: Privacy design strategies. In: Cuppens-Boulahia, N., Cuppens, F., Jajodia, S., Abou El Kalam, A., Sans, T. (eds.) SEC 2014. IAICT, vol. 428, pp. 446–459. Springer, Heidelberg (2014). https://doi.org/10.1007/978-3-642-55415-5_38
25. Howard, M., Lipner, S.: The Security Development Lifecycle, vol. 8. Microsoft Press, Redmond (2006)
26. International Civil Aviation Organization (ICAO): Icao tyou rip guide on border control management (2017). https://www.icao.int/Meetings/TRIP-Jamaica-2017/Documents/ICAO
27. Jonsson Cornell, A.: The right to privacy. Oxford University Press (2016). https://oxcon.ouplaw.com/view/10.1093/law:mpeccol/law-mpeccol-e156
28. Kalloniatis, C., Kavakli, E., Gritzalis, S.: Addressing privacy requirements in system design: the PriS method. Requirements Eng. **13**(3), 241–255 (2008)
29. Khoo, Y.H., Goi, B.M., Chai, T.Y., Lai, Y.L., Jin, Z.: Multimodal biometrics system using feature-level fusion of iris and fingerprint. In: Proceedings of the 2nd International Conference on Advances in Image Processing, pp. 6–10. ACM (2018)
30. Kizza, J.M., et al.: Ethical and Social Issues in the Information Age, vol. 999. Springer, Heidelberg (2013)

31. Leese, M.: Privacy and security – on the evolution of a European conflict. In: Gutwirth, S., Leenes, R., de Hert, P. (eds.) Reforming European Data Protection Law. LGTS, vol. 20, pp. 271–289. Springer, Dordrecht (2015). https://doi.org/10.1007/978-94-017-9385-8_11
32. Lind, N.S., Rankin, E.T.: Privacy in the Digital Age: 21st-Century Challenges to the Fourth Amendment: 21st-Century Challenges to the Fourth Amendment, vol. 2. ABC-CLIO, Santa Barbara (2015)
33. Liu, N.Y.: Bio-Privacy: Privacy Regulations and the Challenge of Biometrics. Routledge, Abingdon (2013)
34. Lumini, A., Nanni, L.: Overview of the combination of biometric matchers. Inf. Fusion **33**, 71–85 (2017)
35. Mironenko, O.: Body scanners versus privacy and data protection. Comput. Law Secur. Rev. **27**(3), 232–244 (2011)
36. Oetzel, M.C., Spiekermann, S.: A systematic methodology for privacy impact assessments: a design science approach. Eur. J. Inf. Syst. **23**(2), 126–150 (2014)
37. Parkavi, R., Babu, K.C., Kumar, J.A.: Multimodal biometrics for user authentication. In: 2017 11th International Conference on Intelligent Systems and Control (ISCO), pp. 501–505. IEEE (2017)
38. Rao, P.R.M., Krishna, S.M., Kumar, A.S.: Privacy preservation techniques in big data analytics: a survey. J. Big Data **5**(1), 33 (2018)
39. Robinson, N., Gaspers, J.: Information security and data protection legal and policy frameworks applicable to European Union institutions and agencies (2014)
40. Rung, S., van Lieshout, M., Friedewald, M., Ooms, M., van den Broek, T.: Privacy and security: citizens' desires for an equal footing. In: Surveillance, Privacy and Security, pp. 15–35. Routledge (2017)
41. Saranya, K., Premalatha, K., Rajasekar, S.: A survey on privacy preserving data mining. In: 2015 2nd International Conference on Electronics and Communication Systems (ICECS), pp. 1740–1744. IEEE (2015)
42. Schaar, P.: Privacy by design. Identity Inf. Soc. **3**(2), 267–274 (2010)
43. Senarath, A., Arachchilage, N.A.: Why developers cannot embed privacy into software systems?: An empirical investigation. In: Proceedings of the 22nd International Conference on Evaluation and Assessment in Software Engineering 2018, pp. 211–216. ACM (2018)
44. Sheth, S., Kaiser, G., Maalej, W.: Us and them: a study of privacy requirements across North America, Asia, and Europe. In: Proceedings of the 36th International Conference on Software Engineering, pp. 859–870. ACM (2014)
45. Shostack, A.: Threat Modeling: Designing for Security. Wiley, Hoboken (2014)
46. Spiekermann, S., Cranor, L.F.: Engineering privacy. IEEE Trans. Softw. Eng. **35**(1), 67–82 (2008)
47. Spiekermann-Hoff, S.: The challenges of privacy by design. Commun. ACM (CACM) **55**(7), 34–37 (2012)
48. Valkenburg, G.: Privacy versus security: problems and possibilities for the trade-off model. In: Gutwirth, S., Leenes, R., de Hert, P. (eds.) Reforming European Data Protection Law. LGTS, vol. 20, pp. 253–269. Springer, Dordrecht (2015). https://doi.org/10.1007/978-94-017-9385-8_10
49. Voigt, P., Von dem Bussche, A.: The EU General Data Protection Regulation (GDPR). A Practical Guide, 1st edn. Springer, Cham (2017). https://doi.org/10.1007/978-3-319-57959-7
50. Warren, S., Brandeis, L.: The Right to Privacy. Litres (2019)

51. Zeadally, S., Badra, M.: Privacy in a Digital, Networked World: Technologies, Implications and Solutions. Springer, Heidelberg (2015). https://doi.org/10.1007/978-3-319-08470-1
52. Zewail, R., Elsafi, A., Saeb, M., Hamdy, N.: Soft and hard biometrics fusion for improved identity verification. In: The 2004 47th Midwest Symposium on Circuits and Systems, 2004. MWSCAS 2004, vol. 1, pp. I-225. IEEE (2004)

Tools Supporting Data Protection Compliance

Making GDPR Usable: A Model to Support Usability Evaluations of Privacy

Johanna Johansen[1(✉)] and Simone Fischer-Hübner[2]

[1] Department of Informatics, University of Oslo, Oslo, Norway
johanna@johansenresearch.info

[2] Department of Mathematics and Computer Science, Karlstad University, Karlstad, Sweden
simone.fischer-huebner@kau.se

Abstract. We introduce a new model for evaluating privacy that builds on the criteria proposed by the EuroPriSe certification scheme by adding usability criteria. Our model is visually represented through a cube, called Usable Privacy Cube (or UP Cube), where each of its three axes of variability captures, respectively: rights of the data subjects, privacy principles, and *usable privacy criteria*. We slightly reorganize the criteria of EuroPriSe to fit with the UP Cube model, i.e., we show how EuroPriSe can be viewed as a combination of only *rights* and *principles*, forming the two axes at the basis of our UP Cube. In this way we also want to bring out two perspectives on privacy: that of the data subjects and, respectively, that of the controllers/processors. We define usable privacy criteria based on usability goals that we have extracted from the whole text of the General Data Protection Regulation. The criteria are designed to produce measurements of the level of usability with which the goals are reached. Precisely, we measure effectiveness, efficiency, and satisfaction, considering both the objective and the perceived usability outcomes, producing measures of accuracy and completeness, of resource utilization (e.g., time, effort, financial), and measures resulting from satisfaction scales. In the long run, the UP Cube is meant to be the model behind a new certification methodology capable of evaluating the *usability of privacy*, to the benefit of common users. For industries, considering also the usability of privacy would allow for greater business differentiation, beyond GDPR compliance.

Keywords: Usable privacy · Human-Computer Interaction · Usability goals · Usable privacy criteria · Privacy certification · GDPR

A long version of this paper is available as [17].
We would like to thank the anonymous reviewers for helping improve the paper.
The first author was partially supported by the project IoTSec – Security in IoT for Smart Grids, with nr. 248113. Thanks go also to Josef Noll for introducing me to the topic of privacy evaluations and labeling.

M. Friedewald et al. (Eds.): Privacy and Identity 2019, IFIP AICT 576, pp. 275–291, 2020.
https://doi.org/10.1007/978-3-030-42504-3_18

1 Introduction

The complexity of the privacy concept as such and of digital data and technology, make it difficult for one to evaluate the privacy properties of a specific piece of technology (e.g., web service, Internet of Things (IoT) product, or communication device). The difficulty is not only for average people, but also for regulators to check compliance, and for developers to be able to provide privacy-aware digital services/products/systems.[1] Indeed, there are multiple concepts involved in digital privacy, like data sharing (which for normal business practices nowadays can form a highly intricate network of relationships), ownership and control of data, accountability or transparency (both towards the regulators as well as the users). Many of the privacy concepts are even a challenge by themselves, when it comes to their evaluation, since they are difficult to measure or to present/explain.

For explaining the intricacies of privacy, besides research articles and books [13], there are several legislative texts adopted in different jurisdictions. The General Data Protection Regulation (GDPR)[2] in Europe makes a good effort in clarifying many aspects of data privacy, providing the legislative support to enforce better data protection practices on anyone (within its jurisdiction) collecting and processing personal data. However, these regulations only specify the requirements on the data controllers in the form of basic principles, and the rights of the data subjects, but do not make any strict claims about the extent to which a controller (or processor) should go about implementing these requirements so that they are beneficial for the user, and to what degree.

As such, one motivation for usability evaluations of privacy is the fact that usability goals of GDPR, s.a. "... any information ... and communication ... relating to processing [to be provided] to the data subject in a concise, transparent, intelligible and easily accessible form, using clear and plain language, ..." (Article 12 (1) of GDPR), are left open to the subjective interpretation of both evaluators and controllers. The provisions of GDPR regarding usability are too general and high-level to be suitable for a certification process [18]. To remedy this, we propose a set of criteria thought to produce measurable evaluations of the usability with which privacy goals of data protection are reached.

For evaluating privacy we take as starting point the methodology developed by EuroPriSe [3] that has as purpose to evaluate compliance with GDPR. We are guided by the EuroPriSe criteria when eliciting, what we call, *principles* and *rights*, which form the two variability axes at the basis of our model, i.e., which principles are followed and which rights are respected. However, EuroPriSe does not consider usability, which is the main focus of our work here. As such, one contribution of this paper is to show how to add usability aspects to the existing evaluation criteria of EuroPriSe.

Unlike EuroPriSe (and other existing certification schemes) that provides a seal showing compliance with data protection regulations (or industry stan-

[1] Note that system/product/service are used interchangeably throughout the paper.
[2] GDPR – General Data Protection Regulation from European Union [1].

dards), our evaluation measures on a scale how well data protection obligations are respected and how easy it is for a user to understand that. The measurements can be presented to the user in different ways, e.g., using "traffic light" scales, showing which level of usability has been reached by the privacy of a certain technological product. A "traffic light" presentation of privacy is recommended by [2, Chapter 6(235)] as a way to "foster competition" and "show good practice on privacy policies".

Traditionally, usability is a quality related to the use of a product. In our case, we are not interested in the usability of a product per se, but only in those aspects of a product that concern privacy. Our conceptualization of usable privacy is based on the definition of usability as presented in the ISO 9241-11:2018 [4], which we adapt to include privacy as follows:

> *Usable privacy* refers to the extent to which a product or a service protects the privacy of the users in an efficient, effective and satisfactory way by taking into consideration the particular characteristics of the users, goals, tasks, resources, and the technical, physical, social, cultural, and organizational environments in which the product/service is used.

Our long term goal is to create a methodology to support service providers to make the privacy of their products more usable. The Usable Privacy Cube (UP Cube) described in Sect. 3 and the usable privacy criteria introduced in Sect. 6 are the first building blocks of the methodology we are aiming for. They are meant as tools, for both usability engineering experts and certification bodies, to evaluate if a product was designed to respect and protect the privacy of its users in an usable way. Once privacy measures and privacy enhancing technologies are integrated into the design of a product, it still remains to find out if (and how much or to what extent) those measures empower and respect the rights of their particular user as intended. In Human-Computer Interaction (HCI) this is determined based on user testing and usability evaluations. The criteria we propose presume the use of such established HCI methods for usability evaluations (e.g., [12]).

The legislation does not directly refer to usability goals and context of use as known in the ergonomics/human factors or human-centered design. However, requirements as the one in the Recital (39) of GDPR asking for the information addressed to the data subject to be "easily accessible and easy to understand" are categorized in this paper as usability goals, for which we create usable privacy criteria meant to measure effectiveness, efficiency and satisfaction – as usability outcomes – with regard to privacy aspects (we henceforth call these *Usable Privacy criteria*, and abbreviated it as UP criteria).

After a short digression into Related Work in Sect. 2, we introduce in Sect. 3 the UP Cube model, which is the main contribution of this work. We then continue to detail the UP Cube in the rest of the paper. Section 4 presents the EuroPriSe in the new light of the UP Cube, forming the two axes of criteria at its basis. The third vertical axis of the UP Cube, a genuine contribution of this paper, is formed of the UP criteria detailed in Sect. 6. To the best of our knowledge, there is not other work that extends privacy certification schemes with usability criteria. Section 5 presents usable privacy goals that the criteria are meant to measure. The UP Cube naturally captures Interactions between all the axes, which we talk about in Sect. 7. We conclude in Sect. 8, presenting also some avenues for further work.

2 Putting the Work into Context

Usable Privacy and Security. The present work can be placed in the research field called *usable privacy and security*, with seminal works s.a. [6,10,14,27] and conference series s.a. the Symposium On Usable Privacy and Security (SOUPS). We consider that research on privacy requires, even more than security, an interdisciplinary approach (encompassing the expertise coming from research fields such as Psychology, Law or Human-Computer Interaction). As [5] points out, privacy has its meaning rooted in larger cultural and social practices and has political, ethical as well as personal connotations.

Regarding the relation between security and privacy, in this paper we consider security as one integral aspect of privacy, where privacy implies security but not the other way around. We consider such a clarification necessary, as we have seen a tendency in the general public to equalize the meanings of the two terms in favor of security. In computer science, privacy research has been closely intertwined with security research, reflected e.g. in the contents and the structure of the book [11]. However, in this paper, we favor the term "usable privacy", as it includes by default security, which is in accordance with the data protection legislation, where security (integrity and confidentiality) is specified as one of the several principles to abide by in order to assure the privacy of users' data.

Human-Computer Interaction. Having the goal to evaluate the usability of privacy in technological systems and products, makes our work part of the larger HCI research on privacy [5,19,20,23]. Following the classifications made by Iachello and Hong in their review [16], we approach privacy from a "data protection" perspective by extracting usability related goals from the GDPR. A similar approach is taken in [23], which translates legislative clauses of the Directive 95/46/EC (now replaced by GDPR) into interaction implications and interface specifications.

For evaluating how well a product meets privacy requirements, context of use variables s.a. user capabilities, tasks, the field where the technology is going to be deployed (e.g., healthcare, industrial facilities), should be defined. We thus adopt the ergonomic approach from ISO 9241-11:2018 where *usability is always considered in a specified context of use*, since the usability to be applied to a certain technology can be significantly different for varied combinations of users, goals, tasks and their respective contexts.

3 The Usable Privacy Cube Model

We devise a model for organizing the criteria to use in privacy evaluations and measurements, and represent it as a cube with three axes of variability (see Fig. 1), which we call the Usable Privacy Cube (UP Cube). The two axes found at the base of the UP Cube are composed of the existing EuroPriSe criteria, which we slightly reorganize in the Sect. 4 to fit in one of the two categories: data protection principles or rights of the data subjects.

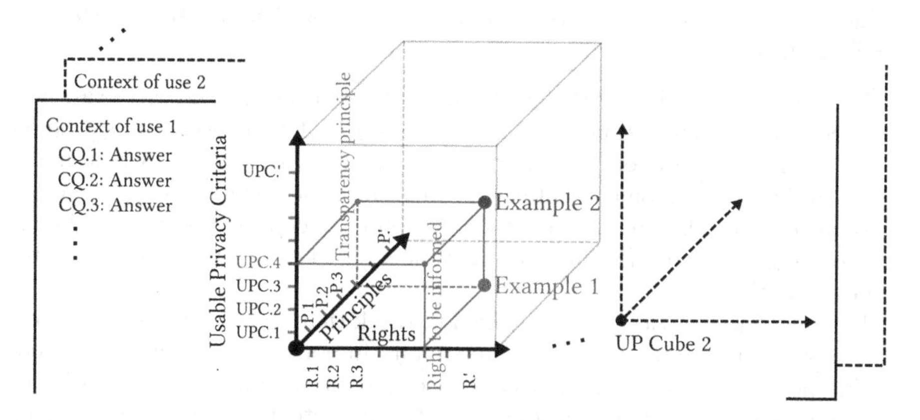

Fig. 1. A generic version of the cube with the three axes of variability: data protection principles, the rights of the data subjects, and usable privacy criteria.

We want to emphasize *two perspectives on privacy* that the UP Cube represents (hence our restructuring of the EuroPriSe criteria): the perspective of the controllers and of the data subjects. The controllers are thus given an overview of the principles that they are obliged to follow, whereas the data subjects are offered an overview of their rights.

The UP Cube allows to visualize interactions between the axes, made easier by our separation of the criteria into the three categories. Each such intersection has its specifics and could be studied in itself; we identify a few exemplary points of intersection between the axes in Sect. 7.

Example 1. The intersection between the transparency principle and the right to be informed is identified in Article 12 of GDPR. The controllers are obliged to provide the data subject information that should be concise, transparent, intelligible and in easily accessible form, using clear and plain language.

The third vertical axis of the cube is composed of our UP criteria, presented in Sect. 6. The UP criteria are determined based on usable privacy goals and are evaluated considering the context of use by following the guidelines in the ISO 9241-11:2018 standard. Interactions exist also with this third axis.

Example 2. For the case presented in Example 1, in order to establish how easily accessible or clear the information is, we must measure the level of efficiency, effectiveness and satisfaction in a specific context of use. Efficiency implies measuring the time and effort spent by a specific user for finding the information needed and for understanding it. Effectiveness measures the completeness with which a goal was achieved. In this case we would like to know how much of the needed information was the specific user able to access and understand. At the same time, what a certain type of user perceives as intelligible information, might be perceived by another as difficult to comprehend. Establishing the perceived characteristics of information is an activity categorized under the satisfaction usability outcome.

The UP Cube also brings the idea of *orderings* on each axis, hence the arrows. Such orderings are important for several reasons, e.g., UP criteria can be ordered based on "how little effort is required to evaluate it compared to how much overall evaluation outcome it entails" or "covers most technologies". Usual for certification methods is to use a decision tree order to capture the impact of each criterion (e.g., choosing the most discriminating first), thus which to prioritize in the evaluation.

Judging from practice, one is inclined to think that an ordering is not always possible to find as some principles are equally important, therefore the orders are not necessarily strict. Moreover, one can even see one principle as more important than another only in some industry or context, whereas in a different industry the same two principles would be ordered the other way, therefore one may think that the orders are only partial (i.e., not total). However, in a specific cube (i.e., used in a specific methodology by a specific authority for privacy usability evaluations in a specific industry and context) there must always be an ordering in which the criteria should be applied. One can always generate a strict and total order from a partial order by just taking a random decision on ordering two criteria when no reasonable order exists. For example, one can any time pick as default order the one arising from the textual placement of the criteria in the data protection legislation texts (maybe considering content from articles as more general than content from recitals), or in the EuroPriSe (or the regulator/company) catalogs. What is certain is that each use case or industry has its specific requirements from which a meaningful ordering would be created.

Forming a specific UP Cube, i.e., deciding on the precise details of each criteria on the three axes and the orderings, is to some degree dependent on the specific context of use for the respective product to be evaluated. Therefore, one can think of *infinitely many cubes*, one for each different context. The criteria will not be different between the cubes, but their scope, depth, and evaluation might be different, depending on the context.

4 EuroPriSe

EuroPriSe originated from the Schleswig-Holstein Data Protection Seal, which was led by the Schleswig-Holstein Data Protection Authority (DPA) from

Table 1. Overview of the the EuroPriSe criteria categorized to fit into our UP Cube model, i.e., as the two axes with Principles and Rights, as well as Context of use.

EuroPriSe Criteria: We list the names of (sub)sections as appearing in the EuroPriSe document [3], which has two parts, the second being subdivided into four *sets* of criteria, whereas the first contains preliminary issues, from where only section C is relevant for us	Principles	Rights	Context
C. Target of Evaluation (ToE)	✓		✓
1.1.1 Processing Operations; Purpose(s)	✓		✓
1.1.2 Processed Personal Data			✓
1.1.3 Controller			✓
1.1.4 Transnational Operations			✓
1.2.1 Data Protection by Design and by Default	✓		
1.2.2 Transparency	✓		
2.1 Legal Basis for the Processing of Personal Data	✓		
2.2 General Requirements	✓		
2.3.1 Data Collection (Information Duties)		✓	
2.3.2 Internal Data Disclosure	✓	✓	
2.3.3 Disclosure of Data to Third Parties	✓	✓	
2.3.4 Erasure of Data after Cessation of Requirement		✓	
2.4.1 Processing of Data by Joint Controllers	✓		
2.4.2 Processing of Data by a Processor	✓		
2.4.3 Transfer to the Third Countries	✓		
2.4.4 Automated Individual Decisions	✓		
2.4.5 Processing of Personal Data Relating to Children			✓
2.5 Compliance with General Data Protection Principles	✓		
Set 3: Technical-Organisational Measures	✓		
Set 4: Data Subjects' Rights		✓	

ca. 2001 until the end of 2013, when it was transferred to a company, EuroPriSe GmbH. The scheme has a history of eighteen years [15] and is one of the oldest privacy and data protection seals based on a law, i.e., the State Data Protection Act of the German federal State Schleswig-Holstein. The role of the seal is to help the vendors of IT products and services to comply with the data protection requirements derived from the applicable law in Europe [7,9,22]. EuroPriSe criteria are already updated to consider the fairly new GDPR.

We have chosen EuroPriSe as the basis for our UP Cube because of its long history, its continuous improvement, strong list of well-developed criteria, being led in the past by a DPA, and being based on the European data protection legislation. EuroPriSe also integrates with widely acknowledged IT security certification methods s.a. ISO 27000 and the The Standard Data Protection Model[3].

The way the criteria are formulated, as questions, also fits with the form of our usable privacy evaluation criteria. In addition, the existing EuroPriSe evaluation, which is at the basis of our model, assures that the GDPR legal grounds are covered, including data protection principles and duties and data subject rights. The UP criteria evaluations come on top, fine-graining the EuroPriSe evaluation with usability measurements, showing how well the legislation is respected.

Another feature that is relevant for our user-centered approach is that the EuroPriSe criteria catalog has been updated to include the data protection by default paradigm, promoting built-in data protection and privacy-friendly default settings. Moreover, EuroPriSe takes into account the technical, organizational and legal framework within which the product or service is operated and asks for considering the requirements of all the parties involved in the system, aiming at strengthening the position of the data subjects. Our work shares with EuroPriSe its high-level goal of making transparent for the general public how companies are managing data protection in their products and services.

In order to build on EuroPriSe, we first look into how its methodology fits with our UP Cube model. We show how EuroPriSe criteria can be redistributed into one of the two axes at the basis, i.e., as either rights of the data subjects or as privacy principles, or otherwise as a context of use criterion. Table 1 gives an overview of this redistribution. The distinction between principles and rights is inspired by the structure in [13], where principles and rights represent the core of this handbook. One purpose of the principles, mentioned in [13], is to serve as the starting point when interpreting the more detailed provisions in the subsequent articles of data protection law. The law also requires that these principles should correspond to the rights presented in the articles 12 to 22. This correspondence can be visualized through the intersection between the respective rights and principles axes of the UP Cube.

5 Usable Privacy Goals

We identify usable privacy goals (henceforth called *Usable Privacy goals*, and abbreviated as UP goals) that appear in the GDPR text. These guide the work

[3] Following the requirement for a consistency mechanism set out in the Article 63 of GDPR, the work of the certifications bodies and DPAs in Germany is coordinated and made consistent through *The Standard Data Protection Model* (https://www.datenschutz-mv.de/datenschutz/datenschutzmodell/), issued by the Conference of the Independent Data Protection Authorities of the Bund and the Länder on 9–10 November 2016. This document is a good reference for methods and guidance for implementing the data protection principles.

in Sect. 6 where we present the UP criteria meant to measure to what extent these goals are being achieved. We give here only some examples of goals, numbered as in the long version [17], where the full list of 30 UP goals can be found. The goals are listed in the order they appear in the legislation. The words ***emphasized*** in each goal relate to usability. The chosen words are those that can be interpreted differently based on the context they are used in, and can result in objective and perceived measurements when evaluated in usability tests. These words also capture goals that can be achieved up to certain degrees, and thus can be translated into a level in a evaluation scale. In addition to the GDPR, there are more specific data protection laws, such as the proposed ePrivacy Regulation that have implication for usability, from where one could eventually extract additional usability goals.

UPG.3 *Consent should be given by a* ***clear*** *affirmative act establishing a* ***freely given***, ***specific***, ***informed*** *and* ***unambiguous*** *indication of the data subject's agreement to the processing of personal data relating to him or her. [Recital (32) of GDPR]*

UPG.8 *Make the natural persons* ***aware*** *of how to exercise their rights in relation to processing of personal data. [Recital (39) of GDPR]*

UPG.18 *Any information addressed to the public or to the data subject to be* ***concise***, ***easily accessible*** *and* ***easy to understand****. [Article 12 (1) and Recital (58) of GDPR]*

UPG.19 *Any information addressed to the public or to the data subject to use* ***clear and plain language****. [Article 12 (1) and Recital (58) of GDPR]*

UPG.21 *Provide information of the intended processing in an* ***easily visible***, ***intelligible*** *and* ***clearly legible*** *manner. [Article 12 (7) and Recital (60) of GDPR]*

UPG.24 *Allow the data subjects to* ***quickly assess*** *the level of data protection of relevant products and services. [Recital (100) linking to Article 42 of GDPR]*

6 Usable Privacy Criteria

The proposed criteria are always measurable, which makes the results of a privacy evaluation easier to present visually through the use of a *privacy labeling* scheme. The use of privacy labels will then fulfill the goal UPG.24. This goal has a special significance from a usability point of view as it reduces considerably the effort spent by the data subject for evaluating privacy, which for most users is not the primary task [5] and it gets in the way of buying or using a product or service.

For generic goals like [17, UPG.1] that regards protection of personal data in general, we formulate a criterion that considers usability as follows:

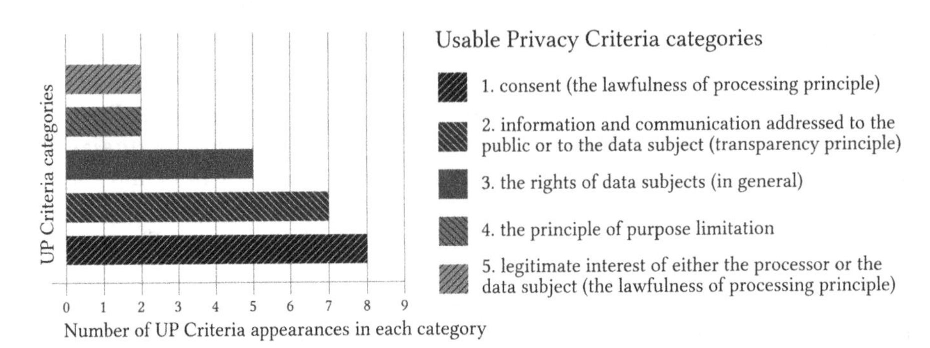

Fig. 2. An overview of the distribution of usable privacy criteria in each category.

What is the level of the *usability* of the personal data protection/privacy that the product or service ensures?

For being able to establish a level of how usable the privacy protection is, the evaluation needs to produce *measurable* outcomes. The structure that we follow is the one proposed in the ISO 9241-11:2018 where the measures consider both the objective and the perceived outcomes of usability (the UP criteria are labeled accordingly). The measurements will produce *counts* or *frequencies* (e.g., how many errors the user does when probed to do certain privacy related tasks) and *continuous data* (e.g., how much time does the user spend on completing a task related to privacy). The evaluation based on the UP criteria proposed below will produce three *main categories of measures*:

1. measures of accuracy and completeness,
2. resource utilization (time, effort, financial, and material resources), and
3. measures resulting from satisfaction scales.

The score for a main UP criterion is established based on evaluations of more specific UP criteria, called subcriteria. In order to reach a *high level* of "control of their own personal data" (Recital (7) of GDPR) the scores from evaluations of the subcriteria should also be high. The resources used to achieve a criterion, i.e., *time, effort, financial, and material* (which we abbreviate TEFM), should be measured to be able to determine the efficiency with which a specific criterion was reached. In addition, the results from the evaluations should show the level of perception that the data subjects have about their data being protected. The data subjects need to be highly satisfied with the offered privacy protection. The "high satisfaction" level is defined based on the user satisfaction evaluation of the respective subcriteria, and is also later important for the adoption of privacy technologies.

The UP criteria are categorized based on their area of application from the GDPR text. Figure 2 gives an overview of the number of criteria in each category.

A high-level UP criterion, like UPC.2, is labeled with the goal that it is related to, UPG.18. We then classify each UP subcriterion (e.g., from UPC.2.1 to

UPC.2.9) into either effectiveness, efficiency, or satisfaction, and label it accordingly. We try to be exhaustive in our UP subcriteria and to give enough questions to cover all major aspects that need to be measured to achieve the respective goal that the high-level UP criterion relates to. The UP subcriteria are labeled with sublabels representing various specific measures of usability for the above three general categories, e.g.: *[Effectiveness:Completeness]*.

6.1 List of up Criteria

We give few examples of the usable privacy criteria, while the full list of all 24 UP criteria can be found in the long version [17]. Since our criteria are modular (i.e., each high-level criterion is thought independent of the other) and can be ordered based on their importance for different application cases, they could be introduced gradually and selectively. It can be that certification bodies (like EuroPriSe) would start to include our UP criteria in their future test catalogs on an article-basis, e.g., a good candidate is Article 12 of GDPR (referring to rights that intersect with the transparency principle) as it contains five UP goals.

UPC.2 Is any information and communication addressed to the public or to the data subjects related to the processing of personal data concise, easily accessible and easy to understand? [UPG.18] *[Type of criteria: Information and communication addressed to the public or to the data subjects]*

> How much [Time/Effort/Financial/Material resources] do the data subjects need to invest in order to [**UPC.2.1** access, **UPC.2.2** read through, **UPC.2.3** understand] the information? *[Efficiency:Time used, Human effort expanded, financial resources expanded, materials expanded] [Measure:Objective]*
> How much of the information were the data subjects able to [**UPC.2.4** access, **UPC.2.5** understand, **UPC.2.6** read through]? *[Measure:Objective][Effectiveness:Completeness]*
>
> **UPC.2.7** To what degree the data subjects perceive the information as concise? *[Satisfaction:Cognitive responses] [Measure:Perceived]*
>
> To what degree the data subjects perceive the information as easy to [**UPC.2.8** access, **UPC.2.9** understand]? *[Satisfaction:Cognitive responses][Measure:Perceived]*

Remark 1. The subcriteria in UPC.2 refer to cognition and understanding, while the subcriteria in UPC.3 refer to visual aspects of the information presented.

Remark 2. In different HCI works one can find different formulations that could seem related to how we formulate the subcriteria, e.g.: "Can the data subjects make sense of the information at all?"; "What is the extent to which the data subjects make sense of the information?". However, we intend to measure the proportion of the information that is made sense of. Therefore we use formulations that give a statistically measurable outcome, such as "How much?", "What is the percentage?", "What is the degree?".

UPC.3 Is the information about the intended processing provided in an easily visible, intelligible and clearly legible manner? [UPG.21]*[Type: Info]*

How much TEFM do the data subjects need to invest in order to [**UPC.3.1** see/locate, and **UPC.3.2** distinguish] the information? *[Ey:Time used, Human effort expanded, Financial resources expanded, Materials expanded]*

How well were the data subjects able to [**UPC.3.3** visually locate and **UPC.3.4** distinguish] the information? *[Es:Accuracy]*

How much of the information were the data subjects able to [**UPC.3.5** visually locate and **UPC.3.6** distinguish]? *[Es:Completeness]*

To what degree the data subjects perceive the information as [**UPC.3.7** easily visible, **UPC.3.8** intelligible, and **UPC.3.9** clearly legible]? *[S:Cognitive responses]*

Remark 3. Poor visibility can affect the perception of trust, as information that has low visibility can appear to be hidden with a purpose. Poor legibility can reflect sloppiness in the way the content is produced, which again can give an impression of lack of professionalism. Poor visibility and legibility affects the satisfaction of the data subjects and it can cause physical discomfort (e.g., to the eyes, by having to read a text written in a very small font).

UPC.4 Is any information and communication addressed to the public or to the data subjects related to the processing of personal data using clear and plain language? [UPG.19] *[Type: Info]*

What is the level of [**UPC.4.1** clearness and **UPC.4.2** plainness] of the language? *[Es:Accuracy]*

UPC.4.3 What is the percentage of the data subjects that understand the language? *[Es:Completeness]*

What is the percentage of the language considered [**UPC.4.4** plain and **UPC.4.5** clear]? *[Es:Completeness]*

How [**UPC.4.6** clear and **UPC.4.7** plain] do the data subjects perceive the language to be? *[S:Cognitive responses]*

Several usability goals are found in the consent related provisions. These provisions are evaluated in detail in the EuroPriSe sections *2.1.1.1 Processing on the Basis of Consent* and *2.1.1.2 Processing on the Basis of a Contract.* The criteria we generate here are meant to complement the ones in the EuroPriSe through bringing in usability concerns. Marc Langheinrich presents several of the problems with how consent can be misused [21]. One of these is the "take it or leave it" dualism where the person does not have a real choice and thus getting consent comes very closed to blackmailing. This problem has been ameliorated in the GDPR law by asking the controllers to allow for separate consent for different data processing operations. A usability evaluation could help further by revealing how the data subjects perceive the consenting act, as well as whether the data subjects consider consent a real choice and if the options to consent to some of the processing operations only, are satisfactory.

UPC.8 Is consent given by a clear affirmative act establishing a freely given, specific, informed and unambiguous indication of the data subjects' agreement to the processing of personal data relating to them? [UPG.3]*[Type: Consent]*

UPC.8.1 How much of the consent text do the data subjects understand? *[Es:Completeness]*

UPC.8.2 How much of the implications of consenting do the data subjects understand? *[Es:Completeness]*

To what degree the data subjects perceive the agreement to be [**UPC.8.3** freely given, **UPC.8.4** informed, and **UPC.8.5** unambiguous]? *[S:Cognitive responses]*

7 Interactions Between the Three Axes

Characteristic to the data legislation text is that it always refers to how principles and rights intersect and depend on each other. In this section, we give examples of such references found in the recitals of GDPR, relevant for some of the identified usability goals. The recitals, though not legally biding, are meant to provide more details to the GDPR's articles. The lawfulness, fairness, and transparency of processing principles, and the right to be informed appear to be closely interrelated, having also the highest occurrence of usability goals.

1. The UP criterion [17, UPC.1] refers to the control the data subjects have over their data. The criterion can be related to the *right to data portability*, through the Recital (68), where due to the aim of strengthening the control of the data subject, the "data subject should also be allowed to receive personal data concerning him or her, which he or she has provided to a controller in a structured, commonly used, machine-readable and interoperable format, and to transmit it to another controller ...". It can also be linked to *data security principle* through the provision in the Recital (75) where the "risk to the rights and freedoms of natural persons" can result in data subjects being deprived of their rights and freedoms or prevented from exercising control over their personal data. The "risk to the rights and freedoms of natural persons" is also mentioned by the [13, pp. 131, 134] in the context of *data security principle.*
2. The UP criteria UPC.2 and UPC.4 are related to the *transparency of processing principle*, which is referred to directly in the Recital (58), where the respective goals are extracted from – "The principle of transparency requires that any information and communication related to the processing of those personal data ..." – and *principles of lawfulness and fairness*, which are also directly referred to in the Recital (39) – "Any processing of the personal data should be lawful and fair".

3. The UPG.8 goal relates to the *fairness and transparency of processing principles*, and is placed under these respective categories, also by the [13, pp. 117, 120].
4. The goal UPG.19 is mentioned in the context of the transparency principle, in the Recital (39), where the information to be given to the data subject relates to the purpose of processing. This connects *the principle of transparency* with *the principle of data minimization.*
5. The UP criterion [17, UPC.22] about "the personal data [being] adequate, relevant and limited to what is necessary for the purposes for which they are processed", is based on the [17, UPG.10], extracted from the Recital (39) of GDPR. This criterion is mentioned in Recital (39) as one of the requirements for complying with *the transparency principle*, while also referring to the purpose of processing. This connects the present criterion also with the *principle of data minimization* and in addition with the *data protection by design principle.* The link between the last two principles can also be seen in EuroPriSe criteria catalog, where data minimization is the focus of the [3, 1.2.1 Data Protection by Design and by Default, p.18] section.

8 Conclusion and Further Work

The benefits of the UP Cube model are multiple: (i) emphasizing both the perspectives of data subjects and of controllers; (ii) representing visually on the three variability axes the existing rights and principles criteria from EuroPriSe, together with our new UP criteria; (iii) visualizing intersections between the three axes; (iv) allowing ordering of the criteria on each axis.

The theory behind our usability evaluation of privacy is based on the well established standards ISO/IEC29100:2011 and ISO 9241-11:2018. We worked directly with the GDPR text, guided by [13], which also inspired our structuring of the EuroPriSe criteria into rights and principles. Our HCI and usability perspective on privacy is influenced by the seminal works [5,6,10,14,19,20,23].

To build the UP Cube we have:

- identified from the GDPR text *30 UP goals*,
- created *24 UP criteria*, each with measurable subcriteria, and
- restructured the criteria of EuroPriSe, laid as the basis of the UP Cube.

Further Work. The UP Cube is meant as the groundwork for building a certification methodology, extending EuroPriSe to evaluate the usability of privacy. The proposed UP criteria are designed to produce measurable evaluations, useful for generating privacy labels in order to guide stakeholders when choosing technological products, by representing and visualizing the different levels of privacy. To achieve this larger goal, one needs to investigate which existing HCI methods for usability testing should be used for each of the UP criteria, and in what way.

One example of such a usability method for measuring the perceived usability of a system is the System Usability Scale (SUS) [8], a ten-item attitude Likert scale questionnaire. The standard [4, Annex B: Usability measurements] also gives examples of methods that produce measurements relevant for our UP criteria, s.a. observing the user behavior to identify the actual usability problems, or asking the users to carry out tasks in a real or simulated context of use and measuring the outcomes. The experts can also run heuristic evaluations following design principles, theories and standards from the design and cognitive fields. More concrete examples of HCI methods and how these could be used for privacy and security solutions can be found in [19].

Which methods are appropriate to use, the number of test persons, and other test related concerns, depend on contextual factors, s.a. the type of technology, users and industry. Defining the required context is what our model offers support for. However, more work (e.g., providing guidelines and examples) is needed on how the context of use can be established.

HCI practices conduct user studies throughout the whole lifecycle of a product. These studies are run by the company itself, with the help of HCI (User Experience or Interaction Design) experts. For certification, the accredited data protection assessors would be using the results provided by the company to answer the UP criteria questions. In the cases of not enough or not reliable results, the assessors can recommend/require further testing. It would be valuable to have guidelines, e.g., in the form of a check-list, to help the assessors with establishing if the results from the company are reliable and sufficient. Recommendations for the businesses are useful as well, to guide how to conduct privacy related user testing, so that the results would be reliable later for certification.

With the same goal of achieving a complete methodology that can be taken in use by the accreditation bodies, building on the present model, one could create a visual representation of the evaluation, i.e., a translation of the measurements of usability of privacy provided by the UP criteria into a visually appealing privacy label. This should serve as a vertically graded scale to differentiate a customer product from another. According to ISO 9241-11:2018, "where usability is higher then expected, the system, product or service can have a competitive advantage (e.g. customer retention, or customers who are willing to pay a premium)". The visuals will be thought to come in addition to the GDPR compliance seal and reflect the usability of the privacy implemented. The purpose will be the same as for the methodology, to help the businesses that have already achieved GDPR compliance to further differentiate themselves on the market. From the point of view of the user of the product, the visual scale would offer support for choosing the service or product that best respects her privacy expectations.

To further validate our UP Cube model and for exemplification, we are applying the UP criteria to *three use cases* taken from pilots done in an ongoing European project called Secure COnnected Trustable Things (SCOTT): *(i)* Assisted Living and Community Care System, *(ii)* Air Quality Monitoring for healthy indoor environments, and *(iii)* Diabetes App. These are examples of IoT systems [24–26] for which our model is especially relevant, as the privacy protection is even more variable and context-dependent. IoT technologies, due to their nature (i.e., ubiquity, invisibility, and continuous sensing) [21], are able to generate granular and intimate data about people and everything or everyone in their surroundings, by that reducing privacy to zero.

References

1. Regulation (EU) 2016/679 of the European Parliament and of the Council of 27 April 2016 on the protection of natural persons with regard to the processing of personal data and on the free movement of such data, and repealing Directive 95/46/EC. Official Journal of the European Union L 119/1 (2016)
2. The House of Lords EU Committee, European Union Committee's report on Online Platforms and the Digital Single Market (2016). https://publications.parliament.uk/pa/ld201516/ldselect/ldeucom/129/12909.htm#_idTextAnchor235
3. EuroPriSe Criteria for the certification of IT products and IT-based services - v201701. Technical report (2017). https://www.european-privacy-seal.eu/AppFile/GetFile/6a29f2ca-f918-4fdf-a1a8-7ec186b2e78a
4. Ergonomics of human-system interaction - Part 11: Usability: Definitions and concepts. Standard ISO 9241–11:2018 (2018)
5. Ackerman, M.S., Mainwaring, S.D.: Privacy issues and human-computer interaction. In: Cranor, L., Garfinkel, S. (eds.) Security and Usability: Designing Secure Systems That People Can Use, pp. 381–399. O'Reilly, Newton (2005)
6. Adams, A., Sasse, M.A.: Users are not the enemy. Commun. ACM **42**(12), 41–46 (1999)
7. Balboni, P., Dragan, T.: Controversies and challenges of trustmarks: lessons for privacy and data protection seals. In: Rodrigues, R., Papakonstantinou, V. (eds.) Privacy and Data Protection Seals. ITLS, vol. 28, pp. 83–111. T.M.C. Asser Press, The Hague (2018). https://doi.org/10.1007/978-94-6265-228-6_6
8. Brooke, J.: SUS - A quick and dirty usability scale. Usability Eval. Ind. **189**(194), 4–7 (1996)
9. Cavoukian, A., Chibba, M.: Privacy seals in the USA, Europe, Japan, Canada, India and Australia. In: Rodrigues, R., Papakonstantinou, V. (eds.) Privacy and Data Protection Seals. ITLS, vol. 28, pp. 59–82. T.M.C. Asser Press, The Hague (2018). https://doi.org/10.1007/978-94-6265-228-6_5
10. Cranor, L.F.: SIGCHI social impact award talk - making privacy and security more usable. In: CHI EA 2018. ACM (2018). https://doi.org/10.1145/3170427.3185061
11. Cranor, L.F., Garfinkel, S.: Security and Usability: Designing Secure Systems That People Can Use. O'Reilly, Newton (2005)
12. Dumas, J.S., Redish, J.C.: A Practical Guide to Usability Testing, Revised edn. Intellect Books (1999)
13. European Union Agency for Fundamental Rights: Handbook on European Data Protection Law - 2018 Edition. Publications Office of the European Union, Luxembourg (2018)
14. Good, N.S., Krekelberg, A.: Usability and privacy: a study of Kazaa P2P file-sharing. In: Proceedings of the SIGCHI Conference on Human Factors in Computing Systems, pp. 137–144. ACM (2003)
15. Hansen, M.: The Schleswig-Holstein data protection seal. In: Rodrigues, R., Papakonstantinou, V. (eds.) Privacy and Data Protection Seals. ITLS, vol. 28, pp. 35–48. T.M.C. Asser Press, The Hague (2018). https://doi.org/10.1007/978-94-6265-228-6_3
16. Iachello, G., Hong, J.: End-user privacy in human-computer interaction. Found. Trends Hum.-Comput. Interact. **1**(1), 1–137 (2007)
17. Johansen, J., Fischer-Hübner, S.: Making GDPR usable: a model to support usability evaluations of privacy. Technical report, arXiv, August 2019. arxiv.org/abs/1908.03503

18. Kamara, I., De Hert, P.: Data protection certification in the EU: possibilities, actors and building blocks in a reformed landscape. In: Rodrigues, R., Papakonstantinou, V. (eds.) Privacy and Data Protection Seals. ITLS, vol. 28, pp. 7–34. T.M.C. Asser Press, The Hague (2018). https://doi.org/10.1007/978-94-6265-228-6_2
19. Karat, C.M., Brodie, C., Karat, J.: Usability design and evaluation for privacy and security solutions. In: Security and Usability, pp. 47–74 (2005)
20. Karat, C.M., Karat, J., Brodie, C.: Privacy security and trust: human-computer interaction challenges and opportunities at their intersection. In: The Human-Computer Interaction Handbook, pp. 669–700 (2012)
21. Langheinrich, M.: Privacy by design—principles of privacy-aware ubiquitous systems. In: Abowd, G.D., Brumitt, B., Shafer, S. (eds.) UbiComp 2001. LNCS, vol. 2201, pp. 273–291. Springer, Heidelberg (2001). https://doi.org/10.1007/3-540-45427-6_23
22. Papakonstantinou, V.: Introduction: privacy and data protection seals. In: Rodrigues, R., Papakonstantinou, V. (eds.) Privacy and Data Protection Seals. ITLS, vol. 28, pp. 1–6. T.M.C. Asser Press, The Hague (2018). https://doi.org/10.1007/978-94-6265-228-6_1
23. Patrick, A.S., Kenny, S., Holmes, C., van Breukelen, M.: Human computer interaction (chap. 12). In: Handbook for Privacy and Privacy-Enhancing Technologies: The case of Intelligent Software Agents, pp. 249–290 (2003)
24. Sicari, S., Rizzardi, A., Grieco, L.A., Coen-Porisini, A.: Security, privacy and trust in Internet of Things: the road ahead. Comput. Netw. **76**, 146–164 (2015)
25. Stankovic, J.A.: Research directions for the internet of things. IEEE Internet Things J. **1**(1), 3–9 (2014)
26. Weiser, M.: Ubiquitous computing. Computer **10**, 71–72 (1993)
27. Whitten, A., Tygar, J.D.: Why Johnny can't encrypt: a usability evaluation of PGP 5.0. In: USENIX Security Symposium, vol. 348 (1999)

Decision Support for Mobile App Selection via Automated Privacy Assessment

Jens Wettlaufer[1(✉)] and Hervais Simo[2(✉)]

[1] Universität Hamburg, Vogt-Kölln-Straße30, 22527 Hamburg, Germany
jens.wettlaufer@uni-hamburg.de
[2] Fraunhofer Institute for Secure Information Technology,
Rheinstraße 75, 64295 Darmstadt, Germany
hervais.simo@sit.fraunhofer.de

Abstract. Mobile apps have entered many areas of our everyday life through smartphones, smart TVs, smart cars, and smart homes. They facilitate daily routines and provide entertainment, while requiring access to sensitive data such as private end user data, e.g., contacts or photo gallery, and various persistent device identifiers, e.g., IMEI. Unfortunately, most mobile users neither pay attention nor fully understand privacy indicating factors that could expose malicious apps. We introduce **APPA** (*Automated aPp Privacy Assessment*), a technical tool to assist mobile users making privacy-enhanced app installation decisions. Given a set of empirically validated and publicly available factors which app users typically consider at install-time, APPA creates an output in form of a personalized privacy score. The score indicates the level of privacy safety of the given app integrating three different privacy perspectives. First, an analysis of app permissions determines the degree of privateness preservation after an installation. Second, user reviews are assessed to inform about the privacy-to-functionality trade-off by comparing the sentiment of privacy and functionality related reviews. Third, app privacy policies are analyzed with respect to their legal compliance with the European General Data Protection Regulation (GDPR). While the permissions based score introduces capabilities to filter over-privileged apps, privacy and functionality related reviews are classified with an average accuracy of 79%. As proof of concept, the APPA framework demonstrates the feasibility of user-centric tools to enhance transparency and informed consent as early as during the app selection phase.

Keywords: Privacy assessment · Mobile apps · Permissions · Privacy policy · User reviews · Privacy perception · Decision support

1 Introduction

While in 2013, only seven percent of companies had provided a corresponding app to their services, in 2017, 67% of small businesses offered mobile apps [40].

Published by Springer Nature Switzerland AG 2020
M. Friedewald et al. (Eds.): Privacy and Identity 2019, IFIP AICT 576, pp. 292–307, 2020.
https://doi.org/10.1007/978-3-030-42504-3_19

This trend, often called *appification*, even goes beyond smartphones. In 2015, Microsoft introduced a unified app marketplace for all Microsoft products. Simultaneously, manufacturers released smart TVs on the basis of apps. Appified systems are also deployed in the context of smart homes and connected vehicles. As such, apps play a major role in our daily routines. They run on devices that surround us all day long, even at night. From the moment they are installed they collect and process personal data, sometimes with little need for human intervention. For example, messenger apps have access to contact details, map apps track our location, and alarm clock apps know at least when users get up. We are motivated by the vision of empowering users with technology based decision support for mobile app selection. This work is a first step in this direction. Specifically, we focus on an automated and user-centric privacy assessment approach that aims at generating privacy recommendations for app users based on machine interpretation of various app attributes that are publicly available on app distribution markets like Google Play. These attributes are supposed to bring transparency and establish competition between apps. However, while users in the online context claim to care about their privacy, studies show that they mostly consider the more functionality informative parameters price, rating, number of installations, and user reviews during their installation decision [5,9,22,26]. This phenomenon is known under the term "privacy paradox" [22,43]. Existing explanations include the view that users often focus on simple parameters [6] because they are limited in time [17] or do not understand certain parameters [6,17]. For example, privacy policies require legal knowledge to fully understand their implications [4,30] and users cannot grasp the impact and consequences of granting certain permissions [7,22,45]. Moreover, observations show that obvious privacy related parameters are placed at disadvantageous positions, e.g., at the bottom of the app page. This is confirmed by research based on user studies revealing that the Google Play permission system is ineffective [7,36,45] mainly because requested permissions need to be granted after the installation decision [14,22], which leads to a desensitization [18].

Efforts to improve this situation by introducing run-time permissions with Android 6.0 have not shown to be effective. For instance, security researchers found that more than 1,000 apps could access permissions that users had explicitly denied before [33,37]. This results in users that can neither judge nor identify privacy-invasive, e.g., over-privileged, apps prior to the installation [19]. Moreover, although studies show that some apps do not comply with given privacy regulations [4,46] and all apps actually need to provide a privacy policy due to the processing of any kind of personal data [10,39], Google Play's privacy policy field is still optional. While the majority of users agree with the terms of services and privacy policies without reading them [3], studies showed that user perception of privacy differs [12,31] and can be categorized into pre-defined privacy profiles [29]. These can help to present users more personalized privacy information.

With the aforementioned vision in mind, in this work, we introduce APPA, a mobile app vetting framework that aims at empowering users towards assessing the level of privacy intrusiveness of apps prior to its installation. APPA takes as input an app's set of attributes from the app market and various natural language processing (NLP) techniques and quantitative models, and computes an overall privacy score for the given app. More specifically, the proposed framework considers three empirically validated app attributes which we hereafter refer to as installation factors: permissions, user reviews and privacy policy. However, note that while this work primarily focuses on these three factors, it is equally applicable to other installation factors deemed relevant to app users (cf. [5,9,22,25,26]).

The rest of this work is structured as follows. Section 2 introduces related work. Section 3 presents an overview of the APPA framework and defines key requirements for the related tool. While Sect. 4 discusses key components of our framework in detail, Sect. 5 presents the evaluation of APPA comparing it to existing work. Finally, Sect. 6 concludes this work and points to future directions.

2 Related Work

The inspection of apps in regard to security is well researched while app privacy analyses first received increased attention in recent years. For example, Kesswani et al. [23] analyze Android app privacy based on requested permissions. They divide permissions into generic and privacy-invasive permissions and classify the app's privacy level respectively. Qu et al. [35] assume descriptions to consistently explain requested permissions and calculate the privacy risk based on the accordance between permissions and descriptions. To identify security and privacy related reviews, Nguyen et al. [34] utilized NLP in combination with machine learning (ML) techniques. They showed that such reviews can have an impact on app related privacy improvements by analyzing and correlating 4.5M historical reviews and app updates over time. Privacy policies are for example examined from Harkous et al. [17] applying self-trained word embeddings and convolutional neural networks in order to generate privacy grading icons. They trained their model on the manually annotated online privacy policy corpus OPP-115 [42]. These works introduce extraction mechanisms from different privacy indicators, but lack of a combined privacy understanding from different perspectives.

Further works intend to compare the described functionality with actual behavior. Therefore, Zimmeck et al. [46] contrast privacy policy statements with the results of a static code analysis. Furthermore, Yu et al. extend the privacy policy findings with description-to-permission fidelity and verify these with a bytecode analysis in order to examine the gap between said and done security and privacy practices. Both approaches take more perspectives into account to quantify the privacy violations of apps and thus the trustworthiness. However, they depend on source or byte code that need to be downloaded and analyzed, which takes more time contrary to a metadata analysis. Particularly, the need for source code to apply static code analysis is not possible in all cases, for example when priced apps are supposed to be investigated prior to the installation. In addition, they both analyze privacy practices on the basis of US regulations, e.g., the Californian Online Privacy Protection Act (CalOPPA), instead of the European GDPR. A GDPR based and multi-source approach is presented by Hatamian et al. [20] suggesting to analyze user reviews, privacy policies, stated permissions and permissions usage. The risk level of ten apps is investigated based on an NLP and ML privacy threat model for user reviews, permission statistics, manual privacy policy analysis as well as a dynamic permission usage analysis over seven days. Thus, their approach lack of real-time app interaction to support users during the decision-making. Additionally, their privacy impact model only provides a subjective risk-perception without considering individual preferences. In contrast, Habib et al. [15] assess trustworthiness by incorporating user sensitivity in privacy issues into their trust score that also comprises the average rating, a sentiment analysis of user reviews, and additional static and dynamic code analysis. Familiarity as well as the posture to desensitization and advertisement frameworks are factors for the personalization. The information is taken by other apps installed on the user device, whether the app is over-privileged or uses an excessive amount of advertisement frameworks. While the first invades user privacy to a certain extent, the other information can only be retrieved using code analysis which introduces a lack of actual on-device real-time assessments. Different personalization techniques in form of privacy profiles are introduced by Liu et al. suggesting the trade-off between functionality and privacy preferences retrieved from app permissions as one measure [28] and learned profiles from a data set of permission settings retrieved by real users with rooted devices [27]. This work retrieves a privacy-to-functionality trade-off from user reviews, but incorporates a similar permission handling as the former. However, the latter is limited to the fact that their training data set was built upon users with rooted devices who are assumed to be more tech-savvy than general users, which affects the quality of the privacy profiles.

3 Approach Overview and Design Goals

3.1 Overview

Properly assessing the level of privacy safety of mobile apps at install-time is laborious and often requires considerable expertise. In many cases, users are not willing to spend substantial amounts of time required for vetting the app prior to its installation, i.e., determining if the later conforms to their personal privacy expectations and preferences. Moreover, due to technical design limitations in today's app ecosystems, mobile users are not always able to make informed privacy decisions on whether a mobile app should be installed on their device [22], nor do they always fully understand implications of particular apps for their privacy [5]. This often results in curious or even malicious apps being installed on users' mobile devices and their overall privacy undermined. Clearly, there is a need for a new privacy-enhancing approach for decision support in the context of app installation. What makes the situation even more challenging is that enhanced transparency and control for app installation decision support should consider the diversity of app users' privacy preferences and the context-dependency of their decisions [24]. Therefore, we propose **APPA**, which is to the best of our knowledge the first decision support tool for personalized app selection through a multi-factor privacy assessment. Our user-centered privacy assessment focuses on a set of empirically validated factors which app users consider before installation [5,9,22,25,26]. More precisely, the proposed solution considers the following three factors: app permissions, user reviews and the app's privacy policy. For each factor, a score is computed. All three scores are subsequently combined into an overall privacy score which, along additional recommendations, is displayed to the app user. For the overall privacy score, optimal combination weights are computed based on empirical insights from [5,9,22,26]. We therefore claim that the weights for the installation factors related scores allow the APPA framework to cover the diversity of app users' privacy preferences. By providing users with the option to explicitly specify weight's values within a pre-defined range, the APPA framework allows minimal user feedback while ensuring that algorithmic generated users' privacy decisions remain context-dependent.

3.2 Requirements

Designing a suitable technology that allows mobile app users to assess the level of privacy safety of any given app and hence answer the question to which extent the app is trustworthy, presents a number of difficult challenges. We argue that APPA regarded as transparency enhancing solution should at least satisfy the following requirements:

R.0 Functionality. The envisaged system shall be able to capture relevant app's metadata from the app market. APPA should be able to autonomously assess the app level of privacy safety, i.e., the app privacy score, in order to link it with privacy recommendations.

R.1 Data Minimization and Privacy-by-Design. The overall APPA framework has to be designed and implemented according to the Privacy-by-Design [16] principle of data minimization. The framework should only store users' digital footprint that is absolutely necessary for analysis and visualization purposes. The framework should not leak any sensitive data to any third party. Access to sensitive data by our framework should require explicit user consent. The confidentiality of the metadata and inferred knowledge, e.g., insights from the associated analysis results, has to be ensured.

R.2 Usability. The APPA framework should not degrade the mobile user's experience. APPA should be implemented as a mobile Android app that does not require root permissions. Users should be able to specify and manage their own privacy and security policies, i.e., rules governing the handling of metadata and functionalities of the framework in a fine-grained manner. The configuration and administration of the app should not require the user to hold specific knowledge about security and privacy. All framework related processes should be mostly automated and require minimal user intervention. Especially the installation process and the specification of privacy preferences and controls should be as simple and unobtrusive as possible. Moreover, the user should have the possibility to realize that she is leaving digital footprints behind while using smartphones as well as the privacy implications of these disclosures. Users should be provided with details about which particular information the APPA framework accesses and which part of these information is stored or further processed. Additionally, the user should be able to modify and delete already collected metadata. To satisfy these usability objectives, three additional challenges have to be met: personalization, minimal overhead and comprehensible visualization. *R.2.1 Personalization.* Given the fact that users perceive privacy differently, personalized recommendations and notices are required. *R.2.2 Minimal Overhead.* The APPA framework should operate with minimal overhead and as efficient as possible. Especially the communication overhead, the battery consumption overhead and the use of computational resources like memory consumption and CPU processing time should be minimal. Unreasonable overheads should be prevented, since they may eventually lead to a decreased user experience and users entirely removing the APPA-app from their devices. *R.2.3 Comprehensible Visualization.* In order for the user to better understand its device's network interactions, APPA should enable a comprehensible visualization of both raw metadata and analysis results. All visualizations should be intuitive to the user, meaning that the user should immediately at first glance be able to grasp important aspects about the data that is presented.

R.3 Extensibility. The APPA framework should be extensible. Components of APPA should be designed and implemented in a way that enables other developers to build upon the framework for future work. For instance, future developers and researchers should be able to leverage our framework to build new, platform-agnostic privacy-enhancing prototypes to be deployed in large scale user behavioral studies.

4 System Design

The APPA framework consists of five main components, as depicted in Fig. 1: App Isolation Module, Metadata Downloader, Parameters Inspection Modules, Personalized Privacy Assessment, Visualization Engine and Graphical User Interface (GUI).

4.1 App Isolation Module

Leveraging the public Virtual Private Network (VPN) API[1] provided by the Android OS (since version 4.0+), we design this component as a firewall that prevent any app under consideration from accessing critical on-device resources or interacting with any remote entities up until the vetting by APPA is completed. Upon analysis by our tool, the user is presented with an overall score and a set of recommendations. Based on this information, the user can either revoke

[1] https://developer.android.com/reference/android/net/VpnService.html.

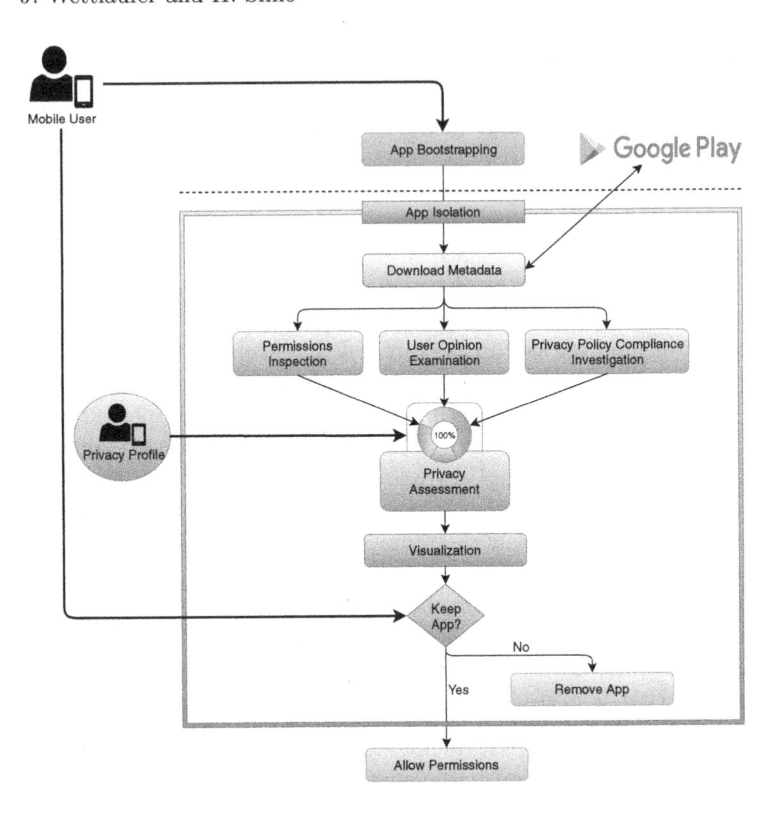

Fig. 1. Privacy-preserving procedure to support users during their app installation decision using an automated and personalized privacy assessment.

the firewall's pre-defined isolation rules and hence allow the installation of the vetted app to be finalized, or reject the app altogether.

4.2 Metadata Downloader

This component handles all tasks related to the interception and aggregation of the app's metadata. In order to successfully complete this task, we designed the Metadata Downloader to be able to query the app market and all related third party domains. As such, it includes three specific sub-components, each focusing on one of the app metadata being considered in this work: A *Permission Collector*, a *User Review Crawler* and a *Privacy Policy Crawler*. Using the unique identifier of the app to be vetted, the Permission Collector fetches all intended permissions as declared by the app developer in the app manifest; the User Review Crawler crawls the top 200 user reviews from the app web page; and the Privacy Policy Crawler searches the app web page for the URL to the privacy policy, follows all subsequent links and extract the policy text. Given the diversity of formats in which privacy policies are displayed, our Privacy Policy Crawler leverages tools such as *textract* [8] and *jusText* [2] to extract text from images to pdf files. The set of metadata capture by all three sub-components of the Downloader is stored on device and made available to other components of APPA for further processing.

4.3 Parameters Inspection Modules

Currently, the APPA framework independently assesses three privacy-indicating app parameters including permissions, user reviews, and the privacy policy in order to inform users about their privateness preservation, i.e., the amount of private mobile information users preserve after the installation, the privacy-to-functionality trade-off, and the app's GDPR compliance, respectively.

Permissions: Privateness Preservation. Modern operating systems such as Android builds upon permission-based security mechanisms to restrict access to sensitive users' data and critical device resources. In Android, app developers must declare all the permissions required by the app to access resources or data outside of the app's own sandbox. Permissions are listed in a so called Manifest file. For a specific subset of the declared access permissions, the normal permissions, the Android OS automatically grants the app access to the related resources or data at install-time. The remaining permissions, the dangerous permissions, are prompted to the user at run-time, requesting her to approve or reject access to sensitive information or resources. However, research has shown that a significant portion of apps overuse access permissions [13] while most people do not fully pay attention nor comprehend permission requests [11]. As component of the APPA framework, the Permission Inspection provides means to quantify possible risks associated with specific permissions. Existing permissions analysis approaches can be compared by three consecutive aspects. First, the set of permissions is determined for the calculation, e.g., all permissions or only a critical subset. Second, the scoring algorithm is defined, e.g., absolute percentage or an ML approach. Finally, each score has a certain interpretation. For example, Kesswani et al. [23] utilize a custom set of privacy invasive permissions, while Hatamian et al. [20] rely on Google's pre-defined permissions with the protection level *dangerous*. These so called *dangerous* permissions have access to personal data such as camera or contacts, according to Google. Finally, they measure the privacy invasiveness and privacy gap, respectively, on the basis of a percentile of app-specific requested permissions.

Our approach extends the use of *dangerous* permissions, based on the fact that Google organizes them in groups and as soon as one permission of a *dangerous* group is granted, all permissions of this group are granted. Therefore, our *advanced dangerous* method counts the number of app specific requested permissions by also taking implicitly granted permissions into account due to the *dangerous* group coherence. Furthermore, we add further permissions to the set that can be hacked as identified by security researchers [44], i.e., `READ_EXTERNAL_STORAGE`. Our score is calculated relatively to the average of *dangerous* permissions. A representative average was calculated by leveraging the permissions of the 40,332 top selling apps of Google Play across 55 categories from March 2019. There are 29 *dangerous* permissions in total. The averages result in $\varnothing_{dangerous} = 5$ and $\varnothing_{adv.dangerous} = 7$, which is used as threshold Q to compute the relative permission score R as follows:

$$R_{\in[0,1]} = \begin{cases} 1 - \frac{\#Permissions}{2*Q} & \text{if } \#Permissions \leq 2*Q \\ 0 & \text{otherwise} \end{cases} \quad (1)$$

Essentially, the permission score is worst in case of at least twice the threshold requested *advanced dangerous* permissions. Furthermore, the score is framed positively, meaning that the score can be interpreted as privateness preservation. Hence, the higher the score the better. The effect of this relative approach (bold line) in contrast to related work (dotted line) can be seen in Fig. 2, which shows the permission scores of 60 randomly chosen apps from a data set of benign and malicious apps [32] separated by their type.

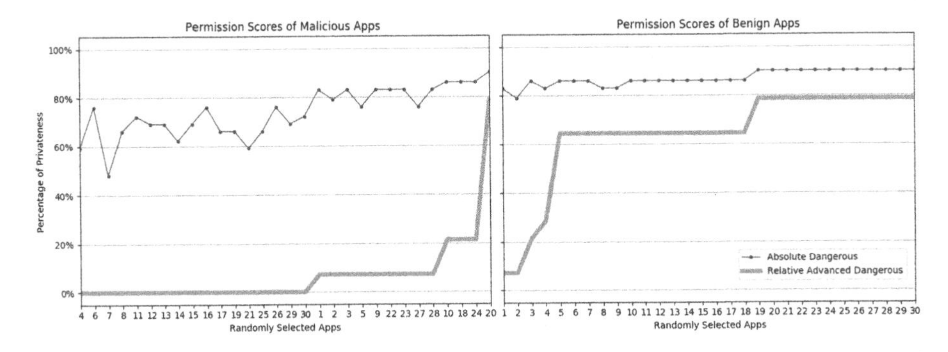

Fig. 2. Comparison of the permission score of related work (dotted) and this work (bold) in regard to malicious and benign apps randomly selected from [32]. The classification of malicious apps can easily be determined with our novel permission score.

User Reviews: Privacy-to-Functionality Trade-Off. User reviews are a means for app users to share their experience and opinion regarding apps in natural language with the community. Previous work shows that reviews contain functionality and privacy related content [34], and can therefore exploited to understand the community opinion. Related work concentrates on identifying security and privacy related comments [20, 34], and consecutively, extracting privacy threat information [20]. These approaches have the disadvantage that users often communicate privacy concerns with emotions instead of technical descriptions. Additionally, user reviews will often contain only such privacy violations that are obvious to app users.

In contrast, our approach bases on the findings of Wottrich et al. [43] that users are confronted with a cost-benefit trade-off, which we interpret as privacy-to-functionality trade-off. Our analysis outweighs community feelings of privacy costs against functionality benefits. For this, we first divide privacy (P) from non-privacy ($nonP$) related reviews using a neural net classifier consisting of two dense layers with 128 and 64 neurons with ReLU and a dropout rate of 0.75. The feature set includes character n-grams to circumvent typos. It was trained on a rather small training set, i.e., 200 privacy and 230 non-privacy reviews, 10-fold cross-validated and scores a mean accuracy of 79%. Second, we use sentiment analysis on each review utilizing SentiStrength, which provides human-level accuracy on short informal texts [41]. This results a sentiment score per review. Third, review related Helpfuls, similar to Likes, are leveraged to calculate a weighted average over all sentiment scores for privacy and non-privacy reviews, respectively. Finally, Eq. 2 computes a trade-off that emphasizes the privacy opinion over functionality.

$$T(P, nonP)_{\in[-1,1]} = \begin{cases} P - nonP & \text{if } P < nonP \\ P & \text{if } P \geq nonP \end{cases} \tag{2}$$

$$S_{\in[0,1]} = \frac{T(P, nonP) + 1}{2} \tag{3}$$

If the privacy sentiment trumps the functionality community opinion, the privacy sentiment value is adopted. However, when functionality exceeds privacy, the distance between both values is negated. To be in line with the other scores, the privacy-to-functionality trade-off is normalized as depicted in Eq. 3. Figure 3 exemplary illustrates the behavior of this score. The graph is sorted by the green line representing the privacy-to-functionality trade-off score. From Pokémon GO to Santander Mobile Banking, the privacy sentiment is lower than the functionality opinion. Therefore, the resulting score is always below 50% because the initially scaled values are below zero. As soon as privacy exceeds functionality, the score makes use of the privacy value and is always greater than 50% due to normalization.

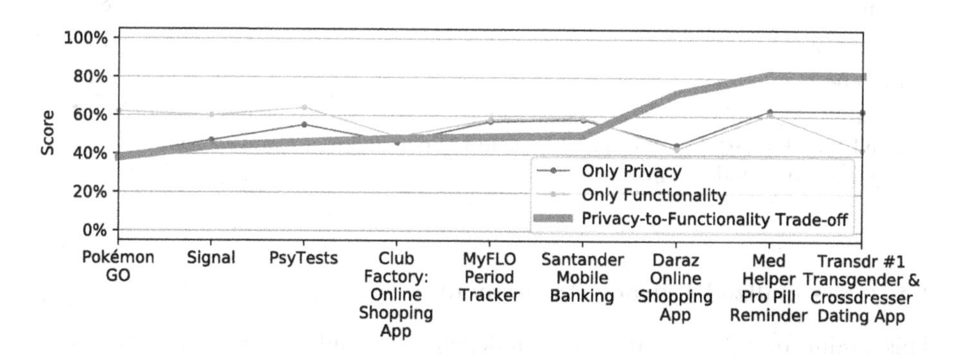

Fig. 3. Comparison of the privacy-to-functionality trade-off score regarding exemplary Android apps. Note that the score differs from the actual privacy values because it is normalized from $[-1, 1]$ to $[0, 1]$.

Privacy Policy: GDPR Compliance. Privacy policies are app corresponding legal documents that are supposed to fully disclose any collection and processing of personal user data. Data protection regulations, such as the CalOPPA in the US and the GDPR in Europe, oblige app developers to have a privacy policy as soon as they process any user data. This work focuses on the European GDPR requiring in Article 5.1 (a) GDPR [10] to fulfill the main principles lawfulness, fairness and transparency in relation to the data subject. As of today, we are only able to assess the transparency of privacy policies. We have already built a fully functional learning based model to cover lawfulness and fairness, but an extensive training set in regard to the GDPR is still work in progress. We measure the readability of policy texts in order to investigate the "concise, transparent, intelligible and easily accessible form, using clear and plain language" as demanded in Article 12.1 GDPR [10]. Consistent to previous work [38], we average the Gunning Fog, Flesch-Kincaid, and SMOG readability metrics to compute a score between 6 and 17 directly mapping to education levels. While 6 corresponds to the sixth grade and 17 to a college graduate, our score defines GDPR compliant readability between 8 and 13. The minimum is set to the age of children where they do not necessarily need a parental guardian anymore, while the maximum corresponds to a college freshman which is still acceptable. Exemplary results of the transparency measurements transformed into GDPR compliance scores can be viewed in Table 1. Privacy policies with scores close to 100% could easily be understood by children in the eighths grade, while scores around 0% indicate that their policies can only be grasped after at least 14 years of education.

Table 1. GDPR compliance scores of 20 popular apps measuring the readability of privacy policies in the above mentioned GDPR compliant range.

Snapchat	86%
Signal	83%
Telegram	70%
Pinterest	54%
WhatsApp, Facebook, Facebook Messenger, Instagram, Google Search, YouTube, Tinder, Twitter, Tumblr, reddit, Pokémon Go, Candy Crush Saga, Spotify, Jodel, Threema	0%

4.4 Personalized Privacy Assessment

This module of APPA aggregates all independent results of the parameter assessments. As shown in Figs. 4 and 5, our visualization repertoire introduces a detailed triangle scheme that represents the scores independently as well as a combined privacy score. While the former on the one hand informs about each score but also reveals the overall privacy impact, the latter enables a personalized privacy score by adopting the weights in accordance to user preferences.

It expresses positively framed privacy safety visualized in five circles with a white plus as studies resulted in it as the most intuitive pattern [6,7].

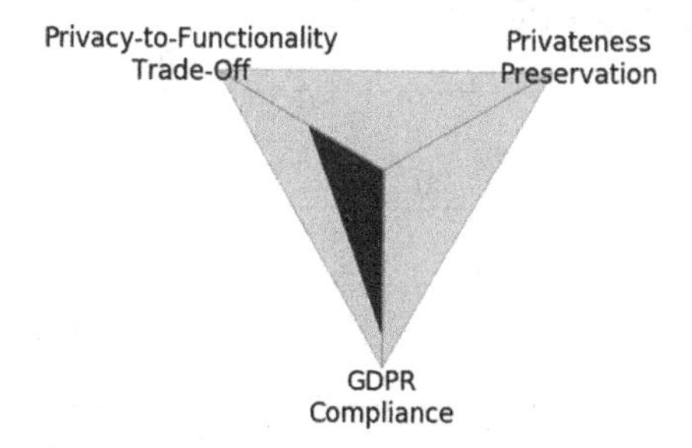

Fig. 4. Visualization of independent results by introducing the overall privacy impact.

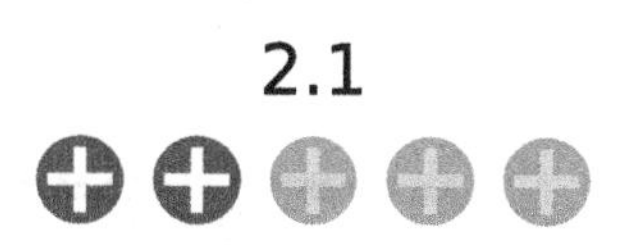

Fig. 5. Intuitive privacy safety score that allows for personalization. This score shows the unpersonalized default option in regard to Fig. 4, i.e., an unweighted average.

4.5 Visualization Engine (VE) and Graphical User Interface (GUI)

The VE is a generic component leveraging free and open source data visualization libraries. Relying on this engine, we implemented a modularized front-end interface, the APPA's GUI. The latter includes a plethora of options for menus, settings and views on details over various app details and the assessment metrics and results. More specifically, the GUI enables visualization of the app's overall privacy score and provides the end user with options to interact with other components of the APPA framework, including activating or deactivating the Isolation Module.

5 Evaluation and Discussion

While the parameter specific assessments were already evaluated in the previous section, the overall outcome of the APPA framework is therefore compared to existing privacy scoring systems as shown in Table 2. *AppCensus*[2] [1] focuses on informing users about which sensitive permissions are accessed and what personal data is shared using a dynamic run-time analysis. *PrivacyGrade*[3] [21] intends to grade the gap between user expectation and the actual app behavior.

Although *AppCensus* states actually used permissions and shared data through a dynamic analysis, users need to have a certain expertise to identify the privacy risks or functionality benefits. The dynamic analysis approaches come with the drawback of high false negative rates, because it might miss certain

[2] https://www.appcensus.mobi/.
[3] http://privacygrade.org/.

Table 2. Exemplary comparison of privacy related scoring schemes in regard to five popular apps. *Triangle scheme labels are identical to Fig. 4.

Pokémon GO	Jodel	Tinder	Telegram	Signal
AppCensus (used sensitive permissions / shared sensitive data)				
1 / 2	0 / 2	1 / 2	3 / 0	1 / 0
PrivacyGrade (#requested sensitive permissions)				
A 9	A 5	A 7	A 20	A 30
Our Score (option: average) *				
0.9	1.2	1.4	1.9	2.1

functionality that uses further permissions or data. In contrast, APPA computes the privacy score by intentionally assuming the most privacy-invasive app state. While *PrivacyGrade*'s grading system might mislead users into thinking that the number of permissions does not affect their privacy, our score precisely indicates that no privateness is preserved when apps use more than 14 sensitive permissions. Similar to *AppCensus*, our score enables users to interpret the independent results based on their personal privacy perception. This can even be automatized in the final privacy safety score.

6 Conclusion and Future Work

The APPA framework achieves an automatic privacy assessment of apps prior to the installation by solely relying on publicly available parameters to lead users to more informed app installation decisions. Therefore, it leverages permissions as indication for the preservation of private information, extracts the user community opinion about the cost-benefit trade-off between privacy and functionality, and investigates the app's legal compliance to the GDPR. The result is visualized in a both detailed and combined privacy safety score that allows for an optional and thus privacy-preserving personalization by weighing the final score in accordance to the user perception. APPA's construction equally brings extensibility of privacy assessment methods as well as transferability to further app based platforms. For example, the GDPR compliance assessment still lacks of justification regarding the main columns fairness and lawfulness. For this, we will extend the assessment with a learning based algorithm that recognizes whether a privacy policy is complete and fair according to the GDPR. On the other hand, future directions of the APPA's privacy safety score include a validating user study and its use as an app wide quality label in regard to privacy due to the fact that only publicly available metadata is needed. In conclusion, the proposed APPA framework enables users to better inform themselves about the privacy safety level of an app without the need to risk their own privacy.

Acknowledgment. This work has been supported in part by the German Federal Ministry of Education and Research (BMBF) within the project "Forum Privatheit und selbstbestimmtes Leben in der Digitalen Welt".

References

1. AppCensus: Appcensus app search (2019). https://search.appcensus.io/. Accessed 20 July 2019
2. Belica, M.: jusText 2.2.0. Python Software Foundation. https://pypi.org/project/jusText/. Accessed 21 Apr 2019
3. Board, T.E.: Opinion: how silicon valley puts the 'con' in consent, February 2019. https://www.nytimes.com/2019/02/02/opinion/internet-facebook-google-consent.html. Accessed 20 July 2019
4. Brandtzaeg, P.B., Pultier, A., Moen, G.M.: Losing control to data-hungry apps - a mixed-methods approach to mobile app privacy. Soc. Sci. Comput. Rev. **37**, 466–488 (2018)
5. Chin, E., Felt, A.P., Sekar, V., Wagner, D.: Measuring user confidence in smartphone security and privacy. In: Proceedings of the Eighth Symposium on Usable Privacy and Security (2012)
6. Choe, E.K., Jung, J., Lee, B., Fisher, K.: Nudging people away from privacy-invasive mobile apps through visual framing. In: Kotzé, P., Marsden, G., Lindgaard, G., Wesson, J., Winckler, M. (eds.) INTERACT 2013. LNCS, vol. 8119, pp. 74–91. Springer, Heidelberg (2013). https://doi.org/10.1007/978-3-642-40477-1_5
7. Chong, I., Ge, H., Li, N., Proctor, R.W.: Influence of privacy priming and security framing on android app selection. In: Proceedings of the Human Factors and Ergonomics Society Annual Meeting (2017)
8. deanmalmgren: textract. GitHub.com (2014). https://textract.readthedocs.io/en/stable/. Accessed 23 Feb 2019
9. Dogruel, L., Joeckel, S., Bowman, N.D.: Choosing the right app: an exploratory perspective on heuristic decision processes for smartphone app selection. Mob. Media Commun. **3**, 125–144 (2014)
10. European Parliament and Council of the European Union: Regulation (EU) 2016/679 (general data protection regulation). Official Journal of the European Union, May 2018. https://eur-lex.europa.eu/eli/reg/2016/679/2016-05-04. Accessed 06 May 2019
11. Felt, A.P., Ha, E., Egelman, S., Haney, A., Chin, E., Wagner, D.: Android permissions: user attention, comprehension, and behavior. In: SOUPS. ACM (2012)
12. Fogg, B.J., Iizawa, D.: Online persuasion in Facebook and Mixi: a cross-cultural comparison. In: Oinas-Kukkonen, H., Hasle, P., Harjumaa, M., Segerståhl, K., Øhrstrøm, P. (eds.) PERSUASIVE 2008. LNCS, vol. 5033, pp. 35–46. Springer, Heidelberg (2008). https://doi.org/10.1007/978-3-540-68504-3_4
13. Gorla, A., Tavecchia, I., Gross, F., Zeller, A.: CHABADA: checking app behavior against app descriptions. In: Proceedings of the 36th International Conference on Software Engineering - ICSE 2014. ACM Press (2014)

14. Gu, J., Xu, Y.C., Xu, H., Zhang, C., Ling, H.: Privacy concerns for mobile app download: an elaboration likelihood model perspective. Decis. Support Syst. **94**, 19–28 (2017)
15. Habib, S.M., Alexopoulos, N., Islam, M.M., Heider, J., Marsh, S., Müehlhäeuser, M.: Trust4App: automating trustworthiness assessment of mobile applications. In: 2018 17th IEEE International Conference on Trust, Security and Privacy in Computing and Communications/12th IEEE International Conference on Big Data Science and Engineering (TrustCom/BigDataSE), pp. 124–135, August 2018
16. Hansen, M.: Data protection by design and by default à la European general data protection regulation. In: Lehmann, A., Whitehouse, D., Fischer-Hübner, S., Fritsch, L., Raab, C. (eds.) Privacy and Identity 2016. IAICT, vol. 498, pp. 27–38. Springer, Cham (2016). https://doi.org/10.1007/978-3-319-55783-0_3
17. Harkous, H., Fawaz, K., Lebret, R., Schaub, F., Shin, K.G., Aberer, K.: Polisis: automated analysis and presentation of privacy policies using deep learning. CoRR (2018)
18. Harris, M., Brookshire, R., Patten, K., Regan, E.: Mobile application installation influences: have mobile device users become desensitized to excessive permission requests? In: Americas Conference on Information Systems (2015)
19. Harris, M.A., Brookshire, R., Chin, A.G.: Identifying factors influencing consumers' intent to install mobile applications. Int. J. Inf. Manag. **36**, 441–450 (2016)
20. Hatamian, M., Momen, N., Fritsch, L., Rannenberg, K.: A multilateral privacy impact analysis method for android apps. In: Naldi, M., Italiano, G.F., Rannenberg, K., Medina, M., Bourka, A. (eds.) APF 2019. LNCS, vol. 11498, pp. 87–106. Springer, Cham (2019). https://doi.org/10.1007/978-3-030-21752-5_7
21. Hong, J.: Privacygrade: grading the privacy of smartphone apps (2014). http://privacygrade.org/home. Accessed 20 July 2019
22. Kelley, P.G., Cranor, L.F., Sadeh, N.: Privacy as part of the app decision-making process. In: Proceedings of the SIGCHI Conference on Human Factors in Computing Systems (2013)
23. Kesswani, N., Lyu, H., Zhang, Z.: Analyzing android app privacy with GP-PP model. IEEE Access **6**, 39541–39546 (2018)
24. Knijnenburg, B.: A user-tailored approach to privacy decision support. Master's thesis, University of California, Irvine, July 2015. http://www.ics.uci.edu/~kobsa/phds/knijnenburg.pdf
25. Kulyk, O., Gerber, P., Marky, K., Beckmann, C., Volkamer, M.: Does this app respect my privacy? Design and evaluation of information materials supporting privacy-related decisions of smartphone users. In: NDSS Symposium 2018 (USEC), San Diego, CA, 18–21 February 2019 (2019)
26. Lim, S.L., Bentley, P.J., Kanakam, N., Ishikawa, F., Honiden, S.: Investigating country differences in mobile app user behavior and challenges for software engineering. IEEE Trans. Softw. Eng. **41**, 40–64 (2015). http://www0.cs.ucl.ac.uk/staff/S.Lim/app_user_survey/
27. Liu, B., et al.: Follow my recommendations: a personalized privacy assistant for mobile app permissions. In: 12th Symposium on Usable Privacy and Security 2016. USENIX Association, Denver (2016)
28. Liu, B., Kong, D., Cen, L., Gong, N.Z., Jin, H., Xiong, H.: Personalized mobile app recommendation: reconciling app functionality and user privacy preference. In: Proceedings of the Eighth ACM International Conference on Web Search and Data Mining, WSDM 2015, ACM, New York (2015)

29. Liu, B., Lin, J., Sadeh, N.: Reconciling mobile app privacy and usability on smartphones: could user privacy profiles help? In: Proceedings of the 23rd International Conference on World Wide Web (2014)
30. Meineck, S.: Komplizierter als Kant: Nerd erstellt Ranking der furchtbarsten AGB (2019). https://www.vice.com/de/article/5974vb/datenschutz-ranking-der-schlimmsten-agb-facebook-airbnb-google-dsgvo. Accessed 28 July 2019
31. Mylonas, A., Theoharidou, M., Gritzalis, D.: Assessing privacy risks in android: a user-centric approach. In: Risk Assessment and Risk-Driven Testing (2014)
32. Urcuqui, C., Navarro, A.: Dataset malware/beningn permissions android (2016). https://doi.org/10.21227/H26P4M
33. Ng, A.: More than 1,000 android apps harvest data even after you deny permissions (2019). https://www.cnet.com/news/more-than-1000-android-apps-harvest-your-data-even-after-you-deny-permissions/. Accessed 20 July 2019
34. Nguyen, D.C., Derr, E., Backes, M., Bugiel, S.: Short text, large effect: measuring the impact of user reviews on android app security & privacy. In: Proceedings of the IEEE Symposium on Security & Privacy. IEEE, May 2019
35. Qu, Z., Rastogi, V., Zhang, X., Chen, Y., Zhu, T., Chen, Z.: AutoCog: measuring the description-to-permission fidelity in android applications. In: Proceedings of the 2014 ACM SIGSAC Conference on Computer and Communications Security - CCS 2014. ACM Press (2014). https://doi.org/10.1145/2660267.2660287
36. Rajivan, P., Camp, J.: Influence of privacy attitude and privacy cue framing on android app choices. In: 12th Symposium on Usable Privacy and Security (2016)
37. Reardon, J., Feal, Á., Wijesekera, P., On, A.E.B., Vallina-Rodriguez, N., Egelman, S.: 50 ways to leak your data: an exploration of apps' circumvention of the android permissions system. In: 28th USENIX Security Symposium (2019)
38. Robillard, J.M., et al.: Availability, readability, and content of privacy policies and terms of agreements of mental health apps. Internet Interv. **17**, 100243 (2019)
39. State of California Department of Justice: Privacy laws. State of California Department of Justice (2003). https://oag.ca.gov/privacy/privacy-laws. Accessed 06 May 2019
40. The Realtime Report: how appification is transforming the internet (2017). https://therealtimereport.com/2017/11/01/appification-transforming-internet/. Accessed 26 July 2019
41. Thelwall, M., Buckley, K., Paltoglou, G., Cai, D., Kappas, A.: Sentiment strength detection in short informal text. J. Am. Soc. Inf. Sci. Technol. **61**, 2544–2558 (2010)
42. Wilson, S., et al.: The creation and analysis of a website privacy policy corpus. In: ACL (2016)
43. Wottrich, V.M., van Reijmersdal, E.A., Smit, E.G.: The privacy trade-off for mobile app downloads: the roles of app value, intrusiveness, and privacy concerns. Decis. Support Syst. **106**, 44–52 (2017)
44. Yin, S.: What can a zero-permissions android app do? April 2012. http://securitywatch.pcmag.com/none/296635-what-can-a-zero-permissions-android-app-do. Accessed 16 June 2019
45. Zhang, B., Xu, H.: Privacy nudges for mobile applications: effects on the creepiness emotion and privacy attitudes. In: Proceedings of the 19th ACM Conference on Computer-Supported Cooperative Work & Social Computing - CSCW 2016 (2016)
46. Zimmeck, S., et al.: Automated analysis of privacy requirements for mobile apps. In: The 2016 AAAI Fall Symposium Series: Privacy and Language Technologies (2016)

Tool-Assisted Risk Analysis for Data Protection Impact Assessment

Salimeh Dashti[1,2(✉)] and Silvio Ranise[1]

[1] Security and Trust - Fondazione Bruno Kessler, Trento, Italy
{sdashti,ranise}@fbk.eu
[2] DIBRIS - University of Genoa, Genoa, Italy

Abstract. Unlike the classical risk analysis that protects the assets of the company in question, the GDPR protects data subject's rights and freedoms, that is, the right to data protection and the right to have full control and knowledge about data processing concerning them. The GDPR articulates Data Protection Impact Assessment (DPIA) in article 35. DPIA is a risk-based process to enhance and demonstrate compliance with these requirements. We propose a methodology to conduct the DPIA in three steps and provide a supporting tool. In this paper, we particularly elaborate on risk analysis as a step of this methodology. The provided tool assists controllers to facilitate data subject's rights and freedoms. The assistance that our tool provides differentiates our work from the existing ones.

Keywords: Data Processing Impact Assessment · Privacy risk analysis · Impact · Rights and freedoms

1 Introduction

Privacy Impact Assessment (PIA) has been gradually developed to assess the impacts on the privacy of a project which involves the processing of personal information [42]. It has been widely adopted and studied, e.g., [5,6,8,9,11,15, 24,28,35,38,40]. The PIA is a tool to *help* [24] controllers who *wishes* [11] to demonstrate their compliance. It has not been an obligation until the General Data Protection Regulation (GDPR) came to enforcement.

The GDPR aims to protect data subjects concerning the processing of personal data. Recital 84 says "in order to enhance compliance with this Regulation where processing operations are likely to result in a high risk to the rights and freedoms of natural persons, the controller should be responsible for the carrying-out of a Data Protection Impact Assessment (DPIA) to evaluate, in particular, the origin, nature, particularity, and severity of that risk". Later in article 35, it provides a minimum guideline for carrying out the DPIA, and do not lay down any further set of requirements. The GDPR is unclear how to implement it [3,7,13,18,38,39]. Different legal bodies and academics,

M. Friedewald et al. (Eds.): Privacy and Identity 2019, IFIP AICT 576, pp. 308–324, 2020.
https://doi.org/10.1007/978-3-030-42504-3_20

e.g., [1,2,4,7,12,13,25,32], started to introduce guidelines and tools to help controllers conduct the DPIA. These works are short in either providing assistance (they work more like a checklist), including all steps of the DPIA (e.g., monitoring), or applying to all domains.

We propose an iterative tool-assisted methodology to assist controllers in different steps of DPIA. We organized our methodology in three steps: Processing Analysis, Risk Analysis, and Run-time Analysis. Each step generates a document. For the DPIA, the asset is the data subject's rights and freedoms. That is to respect the right to data protection, stated in recital 78, and the right of data subjects to fully control their personal data, stated in article 12 to 25.

The focus of this paper is the *Risk Analysis* step. We elaborate on how our methodology assists controllers to facilitate data subject's rights and freedoms. For which, we introduce the main features of the first step, *Processing Analysis*, and its output necessary for the second step. The reader is referred to [17] for further reading on the first step. We leave the description of the third step out, as it is an on-going work.

Plan of the Paper. Section 2 recalls the tools and techniques we have adopted to our methodology. Section 3 discusses the first two steps of our methodology. Section 4 discusses the related work, and Sect. 5 concludes the paper.

2 Background

We briefly recall some notions and techniques that we have used in our work. The descriptions are partial and focus only on the relevant concepts.

The authors in [33] have employed a tool-supported framework for risk modeling and evaluation, called RiskML. The framework is supported by a modeling language and a quantitative reasoning algorithm to analyze models. It allows the user to observe how risks propagate and perform what-if analysis. The RiskML language relies on the following components: a *situation* illustrates the context in which a system is used; an *event* illustrates the situation that has a negative impact on goals; and a *goal* illustrates what the stakeholder aims to protect. These components connect to one another with relations. A situation can connect to an event with the following relations: *expose* when it opens the system to the event, *protect* when it decreases the likelihood of the event, and *increase* when it magnifies impact of the event. Events and goals connect by relation *impact*. All relations can be weighted to determine the significance of their effect. For example, the weight of an *expose* relation determines how likely is that the situation leads to the event.

In [21,29], authors propose a tool-based methodology and a technique to integrate legal compliance and security checks. Their tool checks the compliance of user-specified access control policy p concerning access control related articles in the European Data Protection Directive $p\prime$. Such that a policy p refines a policy $p\prime$ iff every authorization requests permitted or negated by p is also so by $p\prime$.

3 Our Methodology

A (D)PIA must do the following [41]: (1) identify the information flow; (2) identify privacy risks and implemented safeguards, and introduce new safeguards to address the identified risks; and (3) review and update throughout the life of a project. Accordingly, we propose an iterative methodology to conduct DPIA that comprises three steps: *Processing Analysis, Risk Analysis* and *Run-time Analysis*, shown in Fig. 1. The controller needs to conduct the methodology for every data processing they operate in the organization.

Our tool generates a document at the end of each step, namely *Processing specification, Risk summary*, and *Asset and event mapping.* It is difficult to establish a consensus on what needs to be reported in a PIA report [37,42]. In [37], authors have listed the elements that a PIA report needs to contain for different audiences. The list has four main categories: *general system information, assessment information, PIA quality signals*, and *accountability.* The documents that our tool generates cover the four mentioned categories as follows: the document *Processing specification* contains *general system information* and *accountability*; and the *Risk summary* contains *assessment information* and *PIA quality signals.* The document *Asset and event mapping* reports all changes and events that arise during the data processing, as requested in article 33.5.

We do not provide any pre-assessment to check whether conducting the DPIA is necessary, because the minimum requirements that a DPIA shall contain, articulated in article 35.7, has already been stated in different articles of the GDPR. Article 12 requires a written description of the processing operation, as requested by article 35.7.a; article 5 requires purpose limitations and data minimization, as requested by article 35.7.b; article 32 requires to assess risks to rights and freedoms of the data subject, as requested by article 35.7.c,d. Yet, at the end of the first step, our tool informs the controller whether s/he needs to conduct a DPIA for the data processing in question, based on data type and data subject involved, the scale of processing, and usage of new technologies. We recommend the controller to continue, though.

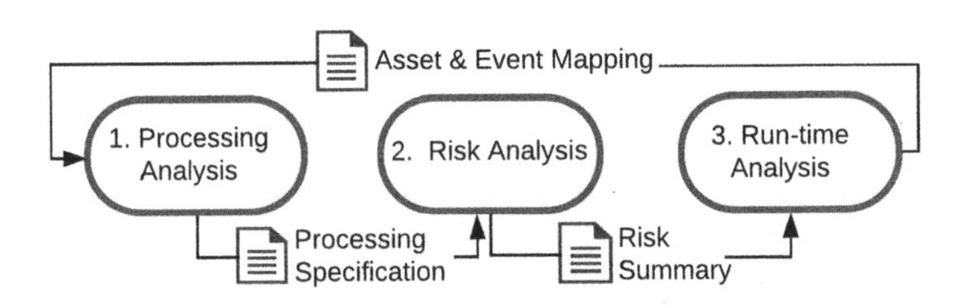

Fig. 1. An overview of our methodology.

Article 35.1 mandates controllers to conduct a DPIA. However, the GDPR obliges processors to comply with many of the requirements which apply to controllers, e.g., implement appropriate technical and organizational measures to ensure the security of processing. For the sake of this paper, we only refer to controllers as the responsible role to follow the steps of our methodology.

Running Example. To illustrate the concepts of our methodology, we consider the following example, taken from [21].

An organization, called ITOrg, needs to compute the salary slips of its employees. Each Employee shall fill in a form, named profile, with some information such as name, surname, address, number of kids, type of contract, etc., and send it to ITOrg. Employees also need to give their consent to use their information to compute the salary slip. ITOrg delegates computing of the salary to its Fin(ancial) Dep(ar)t(ement). The Fin Dept receives selected parts of the profile in a document called fin_profile. In turn, Fin Dept sends it to a company called ACME to compute the salary. Once done, ACME sends the salary slip back to Fin Dept, in a document called salary, and Fin Dept forward it to Employees.

3.1 First Step: Processing Analysis

The first step of our methodology requires a written description of the processing operation. For that, the controller shall provide information, such as: who collects which data (data type), from who (data subject), for how long and why; how the data flow [40,41], etc. Our tool asks the controller to provide this information, by filling a form and drawing a Message Sequence Chart (MSC).

We realized that controllers find it difficult to identify data type and the data subject. To assist them, we have associated the most likely data type and data subject to each economic sector. We have taken the sectors from the Standard Classification of Economic Activities in the European Community.[1] The controller need only to select the sector to which their organization belongs (e.g., education, Manufacturing, construction), and the tool provides a list of related data types and data subjects that the controller can select from. The controller can add others if they do not find what they need in the proposed list. But they need to justify the added items. For more information we refer the reader to [17].

Next, the controller shall draw the MSC of the data processing in question, to capture not only the flow of data but also to identify: (i) the granted permissions to each role who receives or collects the data; (ii) data subject's concerns in each stage of data processing; and (iii) supporting assets used to process data in each stage, for which they need to specify the name of vendor and product, and its version. The processing stages are: *collection, flow* and *processing*; and the permissions are: *r*ead, *w*rite, *u*pdate, and *t*ransfer.

Data subjects have different concerns about their data in different stages of processing. For example, during data collection, they may be concerned about giving too much data; once collected, they worry about the ways their data can be misused. Recital 75 of the GDPR lists some damages caused by data processing: discrimination, identity theft, and damage to the reputation. Other works [12,18,19,27,36] have introduced some further concerns, such as: induced

[1] https://ec.europa.eu/competition/mergers/cases/index/nace_all.html.

disclosure, secondary purpose and identifiability. In this paper, we use the term *concern*, when referring to both damages and concerns. We have used all the concerns mentioned in the literature and the recital, and mapped them to the relevant stage of processing. Table 1 lists an excerpt of the mapping. We have embedded the mapping into the tool. While the controller is drawing the MSC, the tool provides a list of concerns related to that stage.the controller needs to specify the most relevant concerns they believe data subjects might have, according to the type of data and data subject involved, the purpose of data processing (e.g., using new technology), how data are processed, and who else gets to see the data. They need also to specify the supporting assets used in each stage of the processing. That is to be used later in the second step of our methodology.

Once the controller completes the MSC, our tool outputs the user-specified access control, supporting assets, and list of concerns in JSON format. It also generates the *Processing specification* document from information provided by the controller.

Figure 2 illustrates the MSC of our running example, where data subject Employee can read, write and update the document Profile, while controller ITOrg can only read it. Data subject Employee gives profile data using the supporting asset online form (captured by solid rectangles). Data subjects concern (captured by dashed-border rectangles) Unauthorized modification, Exclusion and Economic disadvantage in *processing* stage.

Table 1. An excerpt of the mapping

Privacy concern	Problematic data action	Privacy goal	Stage
Identity theft	Combination Appropriation	Unlinkability	Processing
Induced disclosure	Interrogation	Integrity	Processing
Unauthorized modification	Distortion	Integrity	Collection processing
Economic disadvantage	Distortion poor judgment	Transparency	Collection
Exclusion	Unauthorized purpose	Transparency	Processing

3.2 Second Step: Risk Analysis

Before describing the second step, we introduce the protection goals. We have employed six protection goals to incorporate privacy and data protection [23], that are: confidentiality, integrity, availability (the CIA), unlinkability, transparency, and intervenability. Recent research efforts have come up with these

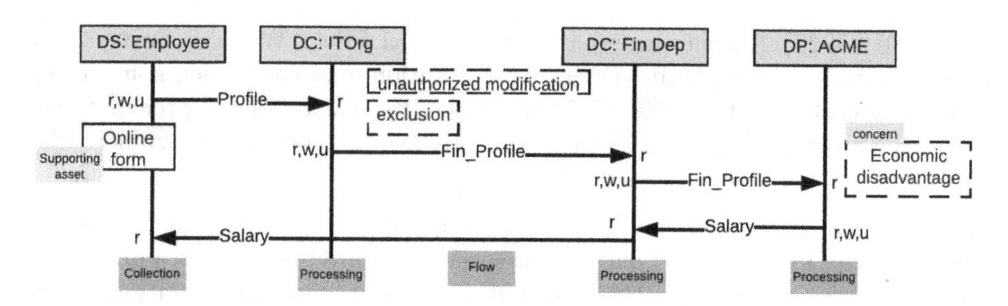

Fig. 2. Annotated Message Sequence Chart of the running example.

goals [16,22,30,31], to provide an interdisciplinary standard model to asses and judge the consequences of utilizing complex IT systems concerning privacy and data protection [23]. As the CIA are well-known, we discuss only the last three. *Unlinkability* relates to the requirements of necessity and data minimization as well as purpose determination, purpose separation, and purpose binding [23], as requested by article 6.4.e and 32.1.a. *Transparency* enables direct controls by entitled entities, such as the data-processing organization itself, a supervisory authority, or the affected human individual whose personal data is processed [23], as requested by article 5.1.a and 12.2. *Intervenability* refers to the requirements that data subjects are effectively granted their rights to notification, information, rectification, blocking and erasure at any time, and that the controller is obliged to implement the appropriate measures [14], as requested by article 12.2.

Classical risk analysis protects assets that organizations value, while for a DPIA assets are data subject's rights and freedoms. Controllers shall respect the data subject's rights and freedoms with regards to (1) data protection, and (2) having full control and knowledge about data processing concerning them. To respect (1), controllers are required to take appropriate technical and organizational measures (recital 78), including the ability to ensure the ongoing confidentiality, integrity, availability, and resilience of processing systems and services (article 32.1.a). To respect (2), controllers shall facilitate the exercise of data subject rights under articles from 15 to 22. Accordingly, we divide the second step of our methodology into two sub-steps: (1) supporting data subject's right to data protection, (2) supporting data subject's right to have full control and knowledge about data processing concerning them.

1. Supporting Data Subject's Right to Data Protection. Controllers are required to take appropriate technical and organizational measures to ensure confidentiality, integrity, availability, and resilience of processing systems and services. Thus, we need to assess whether the implemented security control in the organization in question is appropriate. We need also to identify the vulnerabilities of supporting assets [10–12] and security controls, as they are the source of attacks; that would impact not only the controller but also the involved data subjects. For example, fax-ready HP OfficeJet inkjet printers come out of the box with static buffer overflow vulnerability, which allows remote code execution

(CVE-2018-5925). Another example is Microsoft ADFS 4.0 Windows Server 2016 and previous versions (Active Directory Federation Services) which suffer from SSRF vulnerability (CVE-2018-16794). Below, we discuss our approach to (i) assess the appropriateness of security controls and (ii) identify the vulnerabilities of supporting assets and security controls.

(i) Assess the Appropriateness of Security Controls. Knowing what is an appropriate security level is not trivial. But, we need to ensure that there is a minimum level of security according to the type of data resident on the organization under consideration, as stated in FIPS publication 200 [20]. In this publication, they introduce three sets of security control baselines, named: low-impact, moderate-impact, and high-impact.

Our tool suggests the controller the baseline according to the most sensitive data type that the organization handles. According to the identified baseline, the organization needs to implement one of the appropriately tailored security control baselines from NIST Special Publication 800-53, forth revision [26]. As such, we assign the baseline to data type, as follows: low impact to *personal data*, moderate-impact to *personal data with high-risk*, and the high-impact to a *special category of personal data.* While the first and last data types are taken from the GDPR, we have introduced *personal data with high risk* [17], to make the classification more fine-grained. The data assigned to this category are more sensitive than personal data but less that special category of personal data. For example, geographical location data and financial data are categorized as data with high risk.

To assess whether the baseline controls are implemented, we use a questionnaire, called *Control Questionnaire*, based on the Publication 800-53r4. The *Control Questionnaire* has three sections for the CIA. For example, it asks 'Do you have dual authorization?' (confidentiality), 'Do you review or restrict inputs to trusted sources?' (integrity), 'Do you have a daily backup?' (availability). The controller's response could be: "Yes", "Partially implemented", "No", and "Not applicable". Whenever the answer is one of the first two, the controller shall specify the security mechanism of the control. According to the number of security controls implemented for each goal, we evaluate the likelihood of them to be compromised. The *Control Questionnaire* reports the implemented security controls, what remains to be implemented, and the likelihood of each goal to be compromised. The likelihood—and impact— evaluation is on a scale from 1 (the lowest) to 5 (the highest). It also produces a JSON file, to inputs to RiskML tool (introduced in Sect. 2), that we explain later in this section.

(ii) Identify the Vulnerabilities of Supporting Assets and Security Controls. We use the National Vulnerability Database (NVD).[2] The NVD is the U.S. government repository of standards-based vulnerability management, which includes databases of security checklist references, security related software flaws, misconfigurations, and impact metrics. It specifies the vulnerable object (either an application, an operating system, or a hardware), by name of its vendor and

[2] https://nvd.nist.gov/general.

product, and its version. We have made a list from these objects, to show to the controller when s/he needs to specify the supporting asset (see Sect. 3.1) and the mechanism while answering the *Control Questionnaire*.

Our tool has already generated two JSON files which list the supporting asset, from the MSC chart, drawn in step 1 (see Sect. 3.1), and the security mechanism, from the *Control Questionnaire*. We query the NVD with these two files. If it finds any matches, it retrieves from the database vulnerability type, and the exploitability score.[3] Its output is a JSON file, that is also an input for the RiskML tool. Note that, the *Control Questionnaire* and NVD assist risk analysis and controllers—they do not offer a comprehensive risk analysis.

Controllers may define and enforce access control policies in their organizations, which is enough from a classical risk analysis perspective, but it is not enough to ensure compliance with the GDPR. For example, the classical risk analysis dose not see any risk if a controller uses data subject's personal data for any purpose than the primary purpose for which data are collected, or if they do not give access to data subject to delete or transfer data. However, under the GDPR they put data subject privacy at risks. To avoid such risks, we have used the tool proposed by [21,29], introduced in Sect. 2. We have adapted their tool to the GDPR and extended it to generate its input which are user-specified access control policies, automatically from the MSC (Sect. 3.1). Giving the input, the tool checks it against the access control requirements of the GDPR and reports any non-compliance.

In our methodology, we focus on the processing that involves an individual, not the ones that aggregates data of several individuals by using big data or machine learning techniques. Indeed, there is a lot of data processing which uses neither of them and yet privacy critical. Such as public administrations' process to deliver certificates, e-government processes, or financial transactions that are inherently centered around the personal data of a single individual. For such processing, we consider the necessary access control permissions.

2. Supporting Data Subject's Rights and Freedoms. Article 5.1.a requires controllers to inform the data subject about data operation concerning them and their rights (transparency); articles 15 to 22 require controllers to facilitate the exercise of their rights (intervenability). Controllers need to ensure that their actions do not infringe data subject's rights and freedoms, e.g, to ensure they do not probe for more information [34] or use data for an unauthorized purpose (unlinkability). Some of the concerns could be addressed by security controls, such as insecurity [34], identifiability [19], detectability [19]; but not all. In [14,23], authors have listed some controls to address these protection goals, which are more like actions to take than security mechanisms. Thus, we cannot address them in the *Control Questionnaire*. Sometimes, it is even hard to understand all possible concerns. For example, an organization uses an automated decision-making process to decide on employees' performance. The set of data that has trained the process happened to be biased towards some nationalities.

[3] Please refer to https://www.first.org/cvss/specification-document for more information on how the scores are evaluated. Note that we use the CVSS v3.0.

The chance of capturing such bias is really low. Thus, they keep having *poor judgment* towards some employees, which leads to *discrimination* (recital 75). While drawing the MSC (see Sect. 3.1), our tool provides a list of concerns and asks the controller to specify the most likely ones that data subjects involved in data processing in question have. The list could help controllers to rethink what data subject's concern can be.

In our methodology, we map the data subject's concerns to possible actions that cause them, called *problematic data actions* [27]. We extend the mapping to privacy goal to introduce proper control for counteracting the problematic data action. Table 1 shows the mapping between them. Using a questionnaire, called *Right Questionnaire*, we assess whether the controller takes any of the problematic data actions, and how they facilitate the exercise of data subject's right. For example, it asks 'can data subjects withdraw their consent?', 'May you use the personal data for a purpose other than what you have specified in the document *Processing specification*?', 'do data subjects know how their data are being processed?'. The protection goals mapped to the questions above are interveinability, unlinkability, and transparency, respectively. The questionnaire report which data subject's concerns (from the MSC) and rights are supported, what has remained to act upon, and how likely it is for them to be compromised. It also produces a JSON file from the report to input into the RiskML tool. The *Right Questionnaire* can be seen as a guideline for controllers to ensure they do not miss anything.

To evaluate the impact of data processing on data subjects, we consider the following indicators: (1) type of data; as the impact increases if sensitive data are involved, such as health data, financial data and political opinion; (2) data subject category; as the impact increases, if vulnerable data subjects are involved, such as minors, asylum seekers and employees [4]; (3) the scale; as the impact increases if a large amount of personal data or a large group of data subjects are involved (article 35.3.b, c); (4) usage of new technology (article 35.3.a); as it increases the scale of the impact. Even if only one of these indicators hold, the impact is high on the data subject. In the first step we asked the status of these indicators. Our tool generates a JSON file to report the existing indicators for the data processing in question and the evaluated impact.

Note that, the tool meant to assist the controller and risk analyst, not to be replaced. For example, the control questionnaire is to assess the implemented security mechanism. The controller and risk analyst are responsible to check they are implemented correctly.

Modeling the Risk Using RiskML Tool. In summary, our tool has generated the following JSON files, for the data processing in question: (1) from the MSC, user-specified access control policy, list of concerns and supporting assets; (2) from the *Control Questionnaire*, list of implemented and remained security mechanism and the likelihood of the CIA to be compromised; (3) from the *Right Questionnaire*, list of taken actions to address the concerns/rights, remaining ones, and the likelihood of three protection goals (transparency, intervinability, and unlinkability) to be compromised; (4) from NVD result, list of the vulnerable supporting assets and security mechanisms, and their exploitability score and vulnerability type; (5) list of existing indicators and the impact. The first JSON file is used to check compliance of access control policy. The last four are inputs into the RiskML to generate the risk model.

As mentioned in Sect. 2, RiskML model comprises of situations, events and goals. Situations and events are connected by the *expose* relation, *increase* and *protect*. Events and goals are connected by relation *impact*. We have extended the RiskML tool to automatically generates the risk model from the four JSON files mentioned above. *Situations* will be created from the problematic data actions, vulnerable supporting assets and security mechanism. *Events* will be created from the concerns mapped to the problematic data actions, vulnerability types, or threats raise from missing security mechanism. The weight on *expose* relation, that is the likelihood, comes from the Control Questionnaire, Right Questionnaire and the NVD. The weight on *impact* comes from the last aforementioned JSON file. The ontroller can modify the model, add treatments, and conduct what-if analysis.

Our running example deals with financial data, that belongs to *Personal data with high risk*. Thus, controller ITOrg needs to implement security controls for the *moderate-impact baselines*. The impact of such processing is evaluated 5 because the data type involved is financial data (that is personal data with high risk) and the data subjects involved are among vulnerables [4].

The MSC of this example shown in Fig. 2 depicts that data subjects use an online form. Answering the *Control Questionnaire* we know that there is no control over user input. This may lead to the *event* SQL injection. They also specify in the questionnaire that they use Microsoft ADFS 4.0 Windows Server 2016 to control accesses in the system, depicted as *situation*. Querying the NVD, the tool has retrieved the SSRF vulnerability, depicted as *event*. The vulnerability has an exploitability score of 2.28.[4] Furthermore, the MSC also shows that the data subject Employee concerns that controller ITOrg modifies their data, or excludes them from knowing other purposes for which they may use their data. Employee also concerns that the processor ACME makes them experience economic disadvantage. From the *Right Questionnaire*, we know that there is no action taken to address such concerns. The mapped problematic actions to these concerns are derived from the embedded list into the tool. The controller addresses the event Exclusion by providing a document, namely providing an excerpt of *Processing Specification document* (the output of the first step, see Fig. 1). The document informs the data subject about all the possible purposes, their data may be used for. The controller also addresses the event SQL injection, by using static user query technique. The RiskML tool captures how the risk propagates. Figure 3 shows the propagation by dashed lines. In our running example, the situation Distortion which impacts integrity of profile will also impact the integrity of salary. This propagation is captured through the structure of the documents. The processor ACME computes the salaries based on data he receives

[4] We have normalized the score to 1 to 5.

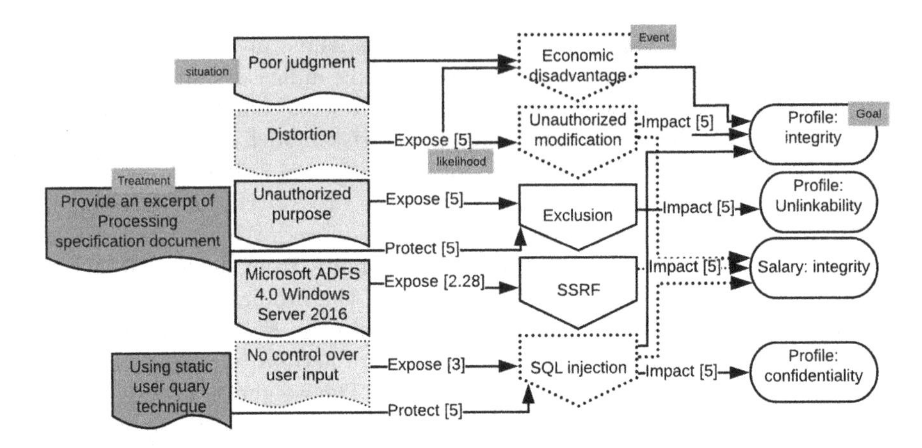

Fig. 3. Annotated Risk model of the Running example.

by document Fin_profile; that is extracted from profile. Thus, any distortion in the profile will affect the Fin_profile, and consequently affect the salary.

We input the user-defined access policy to the tool introduced in Sect. 2 to report any non-compliance in access control policy. Our scenario is not compliance, as the controller does not grant Transfer access to data subject Employee. From the security point of view, this does not make any problem. However, the GDPR grant right to data portability to data subjects (article 20).

4 Related Work

Privacy Impact Assessment (PIA) has gradually developed from the 1990s onward [38] and adopted in Australia, Canada, New Zealand, and the United States. The UK introduced the first PIA methodology in Europe in 2007. The European Data Protection Board (EDPB) set general requirements for PIAs for the RFID [6] in 2009, and smart metering [5] in 2012. Academic works have largely studied it,CNIL e.g., [8,9,15,28,35,40]. Wright defines PIA as "a methodology for assessing the impacts on privacy of a project, policy, program, service, product or other initiative which involves the processing of personal information and, in consultation with stakeholders, for taking remedial actions as necessary in order to avoid or minimize negative impacts" [42]. In his book, he has proposed a sixteen-step optimized PIA process based on a review of various existing PIA methodologies. When these works were developed, conducting a PIA was not a legal obligation. The UK Information Commissioner's Office (ICO) introduced it as "a tool which can help organizations identify the most effective way to comply with their data protection obligations and meet individuals' expectations of privacy" [24]; and the French Commission Nationale de l'Informatique et des Libertés (CNIL) refer to PIA as the process that "[..] controllers who wish to demonstrate their compliance approach and the controls they have selected, as well as for product providers wishing to show that their solutions do not breach privacy" [11].

The General Data Protection Regulation (GDPR) mandates controllers to conduct a Data Protection Impact Assessment (DPIA), stipulated by article 35, when data processing is likely to result in a high risk to the rights and freedoms of natural persons. Article 35(7) provides a minimum standard for carrying out the DPIA and do not lay down any further requirements. It is unclear how to implement it [3,7,13,18,38,39]. National privacy authorities—such as CNIL [12] and ICO [25]—have proposed guidelines to conduct a DPIA. ICO provides a self-assessment toolkit[5], which offers checklists for different roles, namely controller and processor. The toolkit helps to lead a discussion to find out what needs to be done to make to keep people's personal data secure. On the other hand, the CNIL introduces a tool to help controllers build and demonstrate compliance. It requires controllers to respond to some questions in three categories: (1) to describe the context of processing, e.g., applicable standards, data types involved, supporting assets; (2) fundamental principles, e.g., the proportionality of purpose to collected data, accuracy of data, storage duration; (3) risks, e.g., existing controls, possible risks, and the likelihood and impact. Both of these works resemble a check-list that do not provide assistance—which is not an assessment[6]—while our tool assists controllers in different step of the assessment. For instance, our tool assists the user in evaluating risk in two phases by assessing the implemented security countermeasures and the vulnerabilities that may exist in the supporting assets and also implemented security countermeasures. In contrast, the CNIL tool asks the user how they estimate the risk level with no further guidance.

EDPB proposes an iterative process for carrying out a DPIA [4] consisting of the following seven steps organized in a cycle: (1) description of the envisaged processing, (2) assessment of the necessity and proportionality, (3) data protection measures already envisaged, (4) assessment of the risks to the right and freedoms of data subjects, (5) data protection measures envisaged to address the risks, (6) documentation, (7) monitoring and reviewing. We organized our methodology in three steps, namely: Processing Analysis, Risk Analysis, and Run-time Analysis. Each step generates a document. These steps cover those by the EDPB, as follows: Processing Analysis corresponds to steps (1) and (2); Risk Analysis to steps (3), (4), and (5); Run-time Analysis to step (7); and the three documents generated by our tool cover step (6). The authors of [7] use a three-stage process to conduct the DPIA, namely: preparation, evaluation, and report and safeguard. They use the Standard Data Protection Model (SDM) [14] to demonstrate compliance with the requirements of data protection and identify appropriate safeguards. SDM systematizes technical and organizational measures

[5] https://ico.org.uk/for-organisations/data-protection-self-assessment/.

[6] http://www.isaca.org/Knowledge-Center/Blog/Lists/Posts/Post.aspx?ID=864&utm_referrer=direct%2Fnot%20provided&utm_referrer=direct%2Fnot%20provided&utm_referrer=.

to protect the rights of the data subjects based on protection goals. The authors use a catalog of data protection measures developed by the technical working group of the conference of German data protection authorities (AK Technik). These works show the requirements and steps to take to comply, while our work assists the user to meet the DPIA requirements and be compliant.

In the academic world, the authors in [13] use UML class diagrams to specify crucial requirements underlying various aspects of a DPIA, such as consent and necessity. Their focus is to integrate security and privacy requirements engineering processes into a DPIA and understand how a previously developed tool for risk analysis of UML diagrams can be effectively used in this context. The authors of [1] propose UML-based security and privacy analysis. They annotate the UML with the security requirements and check if they are satisfied using UMLsec. To specify the level of privacy, they have extended the Privacy Level Agreement (PLA) to include the following four privacy preferences: purpose, granularity, visibility, and retention. A PLA is an appendix to a service level agreement and provides a structured means to specify privacy preferences and data security requirements. The authors map harmful activities (introduced in [34]), to the privacy threats and appropriate checks. Thus, by identifying the harmful activities in the diagram, they react accordingly. To evaluate impact, they extend the mapping to the privacy target, to identify what has to be protected, and its impact on data subjects and data processor. The authors do not consider to specify likelihood, as they believe that if a privacy threat exists, we control it. They also associate security controls to the mapping. The security controls are based on the NIST 800-53r4, ISO 27001, and German IT baseline. Although we share a similar approach, we acknowledge the fact that privacy and security are two sides of the same coin; thus, we check whether the security controls or supporting assets are vulnerable. Moreover, we check compliance of access control policy, as having them implemented cannot guarantee compliance. Lastly, DPIA is not a one-time activity; instead, it has to continue as long as the life-cycle of the data processing, which the authors of [1] have not considered. In [32], authors have introduced a tool supported by a formal method to verify whether privacy properties are met or not. The authors believe that using such a tool makes the validity of the DPIA more understandable and manageable; also, it verifies whether the DPIA performed by the controller has captured the risks raised from excessive data collection and unauthorized purpose. While using such a tool is useful, they still cannot capture all the DPIA requirements, and instead, focus on purpose and consent. The authors in [2] offer a tool-supported DPIA for Cloud Service Provider (CSP). They have two questionnaires: one to assess whether the DPIA is necessary, and one to establish the effect of interactions among data subjects and CSPs on their rights to data protection. The questions have some pre-defined answers which are associated with privacy indicators and weighted according to the impact they have on them. A global privacy indicator is then calculated based on the indicators. While we share some similarities in the risk analysis phase, our tool is agnostic concerning the technology used to implement the data processing activities. The Risk Analysis step of our methodology is parametric concerning the particular technique used for risk evaluation.

5 Conclusion and Future Work

The DPIA shall assess the risk to data subject's rights and freedoms (article 35.7.c). The GDPR grants to the data subjects the right to data protection and the right to be fully in control of their data and be fully aware of the processing operation concerning them. In this paper, we discussed our three-step DPIA methodology and its supporting tool. In particular, we detailed the second step dedicated to risk analysis. This step assists the controller to exercise data subject's rights and freedoms, in two sub-steps: (1) exercising data subject's right to data protection, and (2) exercising data subject's rights and freedoms. For the former, we ensure that the controller has implemented an appropriate level of security controls and that the supporting assets and the security mechanism are not vulnerable. While for the latter, we ensure that the controller has facilitated the exercise of data subject's rights and has informed them about the data processing that concerns them. To conduct effective DPIA, we need to consider legal, organizational, social, and technical aspects. Applying our methodology on the running example showed that our tool helps to govern the process and tame the complexity.

Working with the public administration of the province of Trento, Italy, we realize that different people are involved in risk analysis in particular in evaluating the impact of data processing. This has led to different impact evaluations for similar processing. To address this issue, we plan to propose a methodology to suggest the most related impact to the data processing according to data type, data subject, and type of processing. For the last step of the methodology, titled *Run-time Analysis* (see Fig. 1), we plan to integrate Inventory Management and Security Information and Event Management to carry out monitoring of compliance as stated in article 40.4.

References

1. Ahmadian, A., Strüber, D., Riediger, V., Jürjens, J.: Supporting privacy impact assessment by model-based privacy analysis. In: Proceedings of the 33rd Annual ACM Symposium on Applied Computing, pp. 1467–1474 (2018)
2. Alnemr, R., et al.: A data protection impact assessment methodology for cloud. In: Berendt, B., Engel, T., Ikonomou, D., Le Métayer, D., Schiffner, S. (eds.) APF 2015. LNCS, vol. 9484, pp. 60–92. Springer, Cham (2016). https://doi.org/10.1007/978-3-319-31456-3_4
3. Alshammari, M., Simpson, A.: Towards an effective PIA-based risk analysis: an approach for analysing potential privacy risks. Technical report CS-RR-18-01, Department of Computer Science, University of Oxford (2017)
4. Article 29 Working Party: Guidelines on data protection impact assessment and determining whether processing is "likely to result in a high risk" for the purposes of regulation 2016/679. https://ec.europa.eu/newsroom/document.cfm?doc_id=47711. Accessed 19 June 2019

5. Article 29 Working Party: Opinion 07/2013 on the data protection impact assessment template for smart grid and smart metering systems ('DPIA template') prepared by expert group 2 of the commission's smart grid task force. https://ec.europa.eu/justice/article-29/documentation/opinion-recommendation/files/2013/wp209_en.pdf. Accessed 19 June 2019
6. Article 29 Working Party: Opinion 5/2010 on the industry proposal for a privacy and data protection impact assessment framework for RFID applications. https://ec.europa.eu/justice/article-29/documentation/opinion-recommendation/files/2010/wp175_en.pdf. Accessed 19 June 2019
7. Bieker, F., Friedewald, M., Hansen, M., Obersteller, H., Rost, M.: A process for data protection impact assessment under the European general data protection regulation. In: Schiffner, S., Serna, J., Ikonomou, D., Rannenberg, K. (eds.) APF 2016. LNCS, vol. 9857, pp. 21–37. Springer, Cham (2016). https://doi.org/10.1007/978-3-319-44760-5_2
8. Clarke, R.: Privacy impact assessments (1999). http://www.xamax.com.au/DV/PIA.html/. Accessed 22 Oct 2019
9. Clarke, R.: Privacy impact assessment: its origins and development. Comput. Law Secur. Rev. **25**(2), 123–135 (2009)
10. CNIL (Commission Nationale de l'Informatique et des Libertés): Methodology for privacy risk management (2012). https://www.cnil.fr/sites/default/files/typo/document/CNIL-ManagingPrivacyRisks-Methodology.pdf
11. CNIL (Commission Nationale de l'Informatique et des Libertés): How to carry out a PIA (2015). https://www.cnil.fr/sites/default/files/typo/document/CNIL-PIA-1-Methodology.pdf
12. CNIL (Commission Nationale de l'Informatique et des Libertés): Privacy risk assessment (PIA) (2018). https://www.cnil.fr/sites/default/files/atoms/files/cnil-pia-1-en-methodology.pdf
13. Coles, J., Faily, S., Ki-Aries, D.: Tool-supporting data protection impact assessments with CAIRIS. In: 5th International Workshop on Evolving Security & Privacy Requirements Engineering, pp. 21–27. IEEE (2018)
14. Conference of the independent data protection authorities of the Federal and State Governments of Germany: The standard data protection model, vol 1.0 EN1 (2017)
15. Cuijpers, C., Koops, B.J.: Smart metering and privacy in Europe: lessons from the Dutch case. In: Gutwirth, S., Leenes, R., de Hert, P., Poullet, Y. (eds.) European Data Protection: Coming of Age, pp. 269–293. Springer, Heidelberg (2013). https://doi.org/10.1007/978-94-007-5170-5_12
16. Danezis, G., et al.: Privacy and data protection by design-from policy to engineering. arXiv preprint arXiv:1501.03726 (2015)
17. Dashti, S., Ranise, S.: A tool-assisted methodology for the data protection impact assessment. In: Proceedings of the International Conference on Security and Cryptography (2019)
18. De, S.J., Le Métayer, D.: PRIAM: a privacy risk analysis methodology. In: Livraga, G., Torra, V., Aldini, A., Martinelli, F., Suri, N. (eds.) DPM/QASA - 2016. LNCS, vol. 9963, pp. 221–229. Springer, Cham (2016). https://doi.org/10.1007/978-3-319-47072-6_15
19. Deng, M., Wuyts, K., Scandariato, R., Preneel, B., Joosen, W.: A privacy threat analysis framework: supporting the elicitation and fulfillment of privacy requirements. Requirements Eng. **16**(1), 3–32 (2011)
20. FIPS (Federal Information Processing Standard Publication) 200: Minimum security requirements for al information and information systems (2006). https://nvlpubs.nist.gov/nistpubs/FIPS/NIST.FIPS.200.pdf

21. Guarda, P., Ranise, S., Siswantoro, H.: Security analysis and legal compliance checking for the design of privacy-friendly information systems. In: Proceedings of the 22nd ACM on Symposium on Access Control Models and Technologies, pp. 247–254 (2017)
22. Hansen, M.: Top 10 mistakes in system design from a privacy perspective and privacy protection goals. In: Camenisch, J., Crispo, B., Fischer-Hübner, S., Leenes, R., Russello, G. (eds.) Privacy and Identity 2011. IAICT, vol. 375, pp. 14–31. Springer, Heidelberg (2012). https://doi.org/10.1007/978-3-642-31668-5_2
23. Hansen, M., Jensen, M., Rost, M.: Protection goals for privacy engineering. In: IEEE Security and Privacy Workshops, pp. 159–166 (2015)
24. ICO (Information Commission's Office): Conducting privacy impact assessments code of practice (2014). https://www.pdpjournals.com/docs/88317.pdf. Accessed 19 June 2019
25. ICO (Information Commission's Office): Data protection impact assessments (2018). https://ico.org.uk/media/for-organisations/guide-to-the-general-data-protection-regulation-gdpr/data-protection-impact-assessments-dpias-1-0.pdf. Accessed 19 June 2019
26. NIST (National Institute of Standard and Technology): Security and privacy controls for federal information systems and organization. NIST special publication 800–53 (2013). https://nvlpubs.nist.gov/nistpubs/SpecialPublications/NIST.SP.800-53r4.pdf
27. NISTIR (National Institute of Standard and Technology Internal Report): Nist privacy risk assessment methodology (PRAM). https://www.nist.gov/itl/applied-cybersecurity/privacy-engineering/resources
28. Oetzel, M.C., Spiekermann, S.: A systematic methodology for privacy impact assessments: a design science approach. Eur. J. Inf. Syst. **23**(2), 126–150 (2014)
29. Ranise, S., Siswantoro, H.: Automated legal compliance checking by security policy analysis. In: Tonetta, S., Schoitsch, E., Bitsch, F. (eds.) SAFECOMP 2017. LNCS, vol. 10489, pp. 361–372. Springer, Heidelberg (2017). https://doi.org/10.1007/978-3-319-66284-8_30
30. Rost, M., Bock, K.: Privacy by design and the new protection goals. Datenschutz und Datensicherheit **35**, 30–35 (2011)
31. Rost, M., Pfitzmann, A.: Datenschutz-schutzziele—revisited. Datenschutz und Datensicherheit **33**(6), 353–358 (2009)
32. Schulz, W., Wittner, F., Bavendiek, K., Schupp, S.: Modeling and verification in GDPR's data protection impact assessment (2019). https://www.cpdpconferences.org/archive
33. Siena, A., Morandini, M., Susi, A.: Modelling risks in open source software component selection. In: Yu, E., Dobbie, G., Jarke, M., Purao, S. (eds.) ER 2014. LNCS, vol. 8824, pp. 335–348. Springer, Cham (2014). https://doi.org/10.1007/978-3-319-12206-9_28
34. Solove, D.J.: A taxonomy of privacy. Univ. Pennsylvania Law Rev. **154**, 477 (2005)
35. Spiekermann, S.: The RFID PIA-developed by industry, endorsed by regulators. In: Wright, D., de Hert, P. (eds.) Privacy Impact Assessment, vol. 6, pp. 323–346. Springer, Heidelberg (2012). https://doi.org/10.1007/978-94-007-2543-0_15
36. Spiekermann, S., Cranor, L.F.: Engineering privacy. Trans. Softw. Eng. **35**(1), 67–82 (2008)
37. Spiekermann, S., Oetzel, M.C.: A systematic methodology for privacy impact assessments: a design science approach. Eur. J. Inf. Syst. **23**(2), 128–150 (2014)
38. Van Dijk, N., Gellert, R., Rommetveit, K.: A risk to a right? Beyond data protection risk assessments. Comput. Law Secur. Rev. **32**(2), 286–306 (2016)

39. Vemou, K., Karyda, M.: An evaluation framework for privacy impact assessment methods. In: 12th Mediterranean Conference on Information Systems (2018)
40. Wright, D.: The state of the art in privacy impact assessment. Comput. Law Secur. Rev. **28**(1), 54–61 (2012)
41. Wright, D., Finn, R., Rodrigues, R.: A comparative analysis of privacy impact assessment in six countries. J. Contemp. Eur. Res. **9**(1), 160–180 (2013)
42. Wright, D., de Hert, P.: Privacy Impact Assessment, vol. 6. Springer, Heidelberg (2012). https://doi.org/10.1007/978-94-007-2543-0

Privacy Classification and Security Assessment

How to Protect My Privacy? - Classifying End-User Information Privacy Protection Behaviors

Frank Ebbers(✉)

Fraunhofer Institute for Systems and Innovation Research ISI,
Breslauer Str. 48, 76139 Karlsruhe, Germany
frank.ebbers@isi.fraunhofer.de

Abstract. The Internet and smart devices pose many risks at users' information privacy. Individuals are aware of that and try to counter tracking activities by applying different privacy protection behaviors. These are manifold and differ in scope, goal and degree of technology utilization. Although there is a lot of literature which investigates protection strategies, it is lacking holistic user-centric classifications.

We review literature and identify 141 privacy protection behaviors end-users show. We map these results to 38 distinct categories and apply hybrid cart sorting to create a taxonomy, which we call the "End-User Information Privacy Protection Behavior Model" (EIPPBM).

Keywords: Privacy protection · Protection behavior · Protection activities · Privacy responses · Taxonomy · Classification · Model · User-Centric

1 Introduction

The Internet and smart devices have become omnipresent. Besides all their advantages, it also poses various information privacy risks. Devices and services in the Internet collect, send and receive personal data of its users. For example data from smartphone sensors, e.g. gyroscopes, can reveal the gender of the user [1]. And even a smart light bulb can reveal a user's location [2]. Different parties collect these personal information, because they are valuable for them [3, 4], for example to offer personalized advertisement or to improve their products and services [5, 6]. However, these parties can misuse personal data for unforeseen and even illegal scenarios [3, 7].

Users in Germany are aware that they are tracked [8]. Although "[i]nvasion of privacy are increasingly regarded as acceptable" [9], recent studies show that users feel that governments and companies do not enough to protect personal data [10, 11]. With the rise of tracking activities in the web and with smart devices [12], users' information privacy concerns rose accordingly in recent years [10]. Thus, individuals try to address their concerns by showing privacy protections behaviors [13] or coping strategies [14]. These can be manifold and vary in scope, goal and technique [15]. Thus they follow a multitude

Published by Springer Nature Switzerland AG 2020
M. Friedewald et al. (Eds.): Privacy and Identity 2019, IFIP AICT 576, pp. 327–342, 2020.
https://doi.org/10.1007/978-3-030-42504-3_21

of specific responses. Scholars have investigated such privacy protection responses and suggested different classifications [e.g. 13, 16]. However, only few researchers visualize their work in a taxonomy or model [e.g. 17]. A structured visualization, could support users to identify behaviors that fit their needs or coincide their skills. For research a state-of-the-art classification can be a useful tool for investigating user behavior [14, 18].

There is an ongoing debate about if individuals should be responsible for their data protection and if they could protect effectively [19]. However, our work concentrates on creating a model of end-user privacy protection without considering this debate. Accordingly our research question is: *How can end-user privacy protection behaviors be represented in a comprehensive model while incorporating prior classification approaches.* To address this research question we conducted a literature review and identified 141 protection behaviors. In a next step, we mapped these behaviors to 38 distinct categories. Based on this we finally apply card sorting to created a hierarchical model.

The remainder of this article is structured as follows: In the second section, an overview of users' protection behaviors and classification approaches is presented. In the methodology (Sect. 3), we describe the iterative literature review and model creation. Section 4 presents the model details. The article concludes with a discussion of the findings (Sect. 5) and an outlook (Sect. 6).

2 Related Work on End-User Privacy Protection Classification and Modeling

Researchers have proposed different classification approaches that vary in scope, criteria and the way of representation. Further they differ in denominations so that a plurality of terms is found: practices [16, 19], activities [20], strategies [14, 19, 21, 22], responses [17], mechanisms [23, 24] and reactions [14]. Other classifications involve coping strategies [14] by which users asses "the expectancy that one's response can reduce the actual danger" [14]. For simplicity reasons we subordinate coping to behavior and use this as an umbrella term in this paper. Behavior refers to "a particular way of acting" [25] and therefore includes all previously mentioned terms. Further we use the term privacy but refer to information privacy as the "access to individually identifiable personal information" as defined by [26].

Privacy protection behavior of end-users is manifold and depends on diverse characteristics, such as user concerns and willingness to take risks, digital literacy and experience [13, 14, 27, 28]. Thus, authors identify myriad protection behaviors. Some categories are simple, whereas others are more advanced and complex.

2.1 High-Level Classifications

Several authors suggest high-level classifications. For example Buchanan et al. [29] identifies two factors based on an analysis of users' privacy concerns: "general caution" (e.g. search for privacy certifications) and "technical protection of privacy", which needs a certain level of computer literacy to e.g. delete cookies. A similar differentiation is found in Lwin et al. [30]. They differentiate between deflective behaviors, which means to avoid data collection, and defensive behaviors, which focusses on removing personal information from a vendor's database. Others emphasize limiting information sharing [13, 20, 28]. Xu et al. [15] classify protection behaviors by emphasizing control of information flows between data subjects and service providers. They distinguish personal control agency, which "empowers individuals with direct control over how their personal information may be gathered by service providers" [15]. Whereas proxy control agency concerns industry self-regulation and government legislation.

2.2 Active and Passive

Other authors differentiate between active and passive behaviors, [e.g. 22, 31, 32 or 33]. Active behaviors means that users engage to utilize countermeasures directly. Passive behavior primarily involves the "general decision to share or not to share personal information" [19]. Other passive behaviors involves relying on external entities, such as data protection authorities [34] or to ask other individuals, e.g. parents [35].

2.3 Chronological Classifications

Further, one can distinguish countermeasures adopted before or after a data disclosure. Several authors write about preventive (ex ante) and protective or reactive strategies (ex post) [22, 36]. Lampinen et al. [21] call these preventive and corrective strategies. Lwin et al. [30] differentiates between deflective (prevention-focused) and protective measures. A similar distinction is introduced by [14]. They distinguish threat appraisal (ex ante), the analyzation of "probabilistic characteristics of a potential threat" [14] and coping appraisal, where users evaluate options to diminish the threat ex post. Moshki and Barki [14] highlight three categories of coping based on a temporal sequence: before an event (anticipation period), during (impact period) and after (post-impact period). Another classification mentions prevention, avoidance and detection [24]. Prevention mechanisms try to ensure that "undesirable use of private data will not occur" (ex ante) [24]. Avoidance mechanism minimize risks associated with data exchanges "by carefully considering the context in which they take place" (ex ante) [24]. Detection mechanisms seek to find privacy incidents (ex post).

2.4 Classification from a Technological Perspective

Protection can be supported by technology or not [15, 29]. Several authors concentrate purely on a technological perspective. Jiang et al. [24] identifies "mechanisms" that are supported by tools. Other authors mention to use ad blockers, cookie management or enabling Do-Not-Track functions in web browsers [13]. Further authors, such as [37], focus on countermeasures utilizing privacy enhancing/preserving technologies and distinguish them on a technological basis [18].

2.5 Fine-Grained Approaches

As an overview Yap et al. [16] identify seven categories of what they call "privacy management practices", namely: (1) withdraw, (2) defend, (3) neutralize, (4) feint, (5) (counter) attack, (6) perception management, and (7) reconcile. Lwin et al. [38] define three defensive measures, namely: fabricate (use of false information), protect (utilization of technologies), and withhold (refusal to provide data). A quite similar wording is found in Metzger [39]. She states three behavior types, what she calls "rules": withholding information, falsifying information and information seeking (informing oneself about a company before disclosing data).

2.6 Taxonomies and Models

Taxonomies and models aim to make difficult relationships easy-understandable or to visualize hierarchical structures. However only very few authors created such to visualize the relationship of their privacy protection categories. In the next sections two models are presented, which gained wide recognition.

One prominent example is introduced by Son and Kim [17]. They define a set of responses to privacy threats, calling their model "Information Privacy-Protective Responses" (IPPR) (Fig. 1). Son and Kim [17] focus on three main behavioral responses:

1. *Information provision* affects users when they are asked to disclose personal information in registration forms on websites. The authors suggest two possible responses: refuse to disclose information or to falsify them (misrepresentation). Whereas refusal often goes along with a loss of functionality, misrepresentation is considered as "a less costly and more convenient option" [17].
2. *Private action* represents the fightback once users lost control over their data, e.g. when receiving unwanted marketing emails. The authors divide this complaining behavior into removal (e.g. by opting-out) and negative word-of-mouth recommendation to damage a company's reputation.
3. *Public action* means complaining directly to the company or complaining indirectly via independent third-party privacy groups, such as data protection authorities. In contrast to private actions, these complaints are broadcasted to public.

Another model is suggested by Ochs et al. [40]. They classify protection practices in a two-by-two grid, which distinguishes between users' digital literacy and data intensity

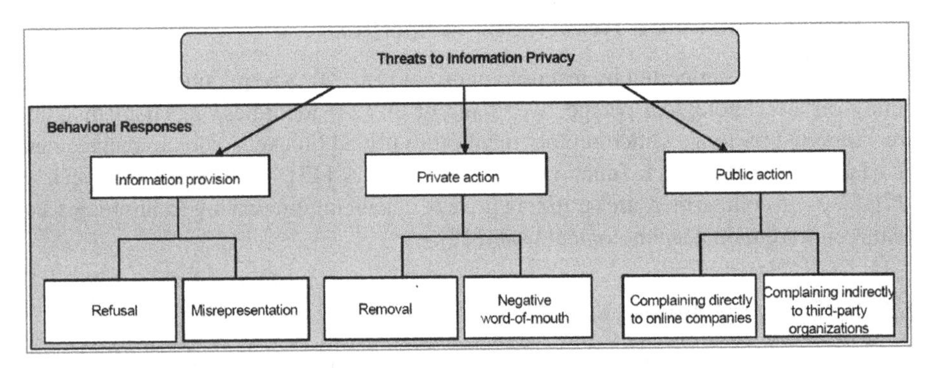

Fig. 1. "Information Privacy-Protective Responses" (IPPR) by [17]

(Fig. 2). Digital literacy means the ability to evaluate and manage information and to facilitate computer technologies. Data intensity concerns the amount of information on which the protection is based on. Data intensity is low for defensive practices. For example, an individual who uses the internet only temporarily create only few information and thus aim for data minimization. Whereas offensive practices are used if individuals don't want to minimize their data disclosure but instead try to obfuscate or falsify information. Accordingly there is a high data intensity.

"*Exclusion practices*" (category A) are defensive approaches that require less digital competencies in order to be successfully applied. Users do not disclose any personal information permanently. "*Controlling practices*" (category B) are defensive approaches that require a certain level of digital competency. The general idea is to control information flows of personal data individually. Whereas category A and B emphasize on individual approaches, "*Exoteric practices*" (category C) rely also on external involvement. These require less digital competency, however the amount of data that is disclosed is much higher than using exclusion practices. With "*Networking practices*" (category D) users try to codify their information exchange. For example users agree upon several abbreviations or code words for communication. Other users obfuscate their data or create fake profiles (avatars) to participate in networks but not reveal their real identity. This requires a certain level of digital literacy.

Digital literacy requirements ↑ / Data intensity →	Defensive	Offensive
Privacy practices with high requirements	B: Controlling practices – "Shut down" – Audience segregation – Selective data disclosure – Encryption	D: Networking practices – Obfuscation – Avatar generation – Social steganography
Privacy practices with low requirements	A: Exclusion practices – Self-censorship – Selective usage – Temporary offline – Non-usage	C: Exoteric practices – Trust – External control – Delegation – Assertiveness

Fig. 2. Privacy practices adopted from [40] and translated into English by the author

The literature research has shown that there is a myriad of different protection behaviors, which users can apply before and after a data disclosure. These can be of active or passive manner. A majority of behaviors aim for limiting information disclosure. Furthermore protection can be achieved with and even without technical literacy. Lastly the chapter shows two classification models, which represent two very different perceptions of classifying privacy protection behaviors.

3 Methodology

Our research approach was two-fold. First, we conducted an iterative literature review for end-user privacy protection and user-centric classifications utilizing the publication database Scopus[1]. Second, we compared and mapped the findings to create a model.

3.1 Literature Review

We conducted an iterative literature search in March and April 2019 using the database Scopus to cover top peer-reviewed journals in fields like computer science, communication science, psychology and law. These seemed most promising to find suitable literature, as privacy is a multi-disciplinary concept [41]. We applied an iterative review approach, following [42] to refine and extend our initial search.

As a first step, we aimed at consolidating privacy protection behaviors of users and came up with 16 keyword strings for our search. After several search iterations (including synonyms and limited to years 2000 until 2019), we reviewed protection behaviors and created a list of their denominations and meanings - resulting in 141 entries. As a last step, we scanned the resulting literature for the keywords "classification", "categorization", "framework" or "model" to find existing approaches. The results show that only two papers [17, 40] deal with visualizing user privacy protection. We conducted a forward and backward search for both papers.

3.2 Classification of Behaviors and Model Generation

We manually examined the list to compare the meanings. In cases where wordings were identical, we merged them immediately and selected the original denomination (e.g. refusal [17]). In other cases, a deeper look at the meaning was needed (e.g. feint [16]). Once there was a match with regard to wording, we set the denominations in relationship to each other and chose a coherent name. For example preventive measures [36] and threat appraisal [14] are shown before data is disclosed and therefore belong to the pre-disclosure category[2]. In some cases, the denominations in the literature could be mapped to several others. Finally, we identified 38 distinct categories.

Next, we started creating the model by applying hybrid card sorting. Originally card sorting was used in usability engineering to create menu structures [43]. However it has gained popularity in creating structures and hierarchies as well [44]. Hybrid card sorting means that users can add new categories to some pre-defined ones. To do so we wrote each category on a note. We used the model proposed by Son and Kim [17] as our starting point for the hybrid card sorting, as it has been widely used in literature. However we allowed to change the nominations of their categories. After nine rounds of sorting, a final structure was created. As a last step, we indicated if there is a need for technological or non- technological means to fulfill the protection practice in each category on the third and fourth level.

[1] https://www.scopus.com.

[2] A figure of the complete taxonomy can be found at: linksplit.io/EIPPBM.

4 Results and Explanation of the Model

Our results consist of two parts:

1. a list of privacy protection behaviors derived from a literature review and its mapping to distinct categories (see footnote 2), and
2. a classification model (Fig. 3), which we call "End-User Information Privacy Protection Behavior Model" (EIPPBM).

The literature review and the classification show that protective behaviors are multi-faceted. Our literature review ends up with a list of 141 behaviors, which are assigned to 38 distinct categories. Authors name behaviors very differently. This might be because privacy means different things to different people [26, 45]. To make up the EIPPBM, the categories are arranged on four hierarchical levels. In the name of model we use the term "information privacy" to make absolutely clear that we do not refer to physical privacy, even if someone watches the model without reading this paper. Same applies to the term "end-user" to make clear that the model is not addressed to developers or other high-level groups. The model indicates which behaviors are supported by technological means and which do not need any technology or digital literacy to be applied. Our results show that users can do a majority of behaviors (79 percent) with little or no digital literacy. In the following paragraph the four levels of the model are explained in detail. We want to note that the explanatory names of the levels do not necessarily correspond with the categories identified in the related work section.

Level 1 - Chronological Distinction
In contrast to [17] the EIPPBM introduces a chronological distinction on the first level. These pre- and post-disclosure behaviors can be interpreted as preventive or defensive actions and are made by several authors [such as 14, 18, 22, 30, 46]. This distinction on the very first level seems plausible, because all subordinated behaviors can be allocated clearly to a temporal order.

Level 2 - Active vs. Passive
The second level is characterized by the distinction between active and passive behaviors. Whereas passive behaviors involve the "general decision to share or not to share personal information" [19], active behaviors "serve to build a protected sphere" [19]. The latter need a direct involvement of the user. This distinction is widely used in literature [such as 22, 31–33].

Level 3 - Superordinate Behaviors
Groups of behaviors are introduced on the third level. These categories are superordinate and base on behaviors found in literature. Hence, these categories do not necessarily represent a specific or direct behavior and instead signify general goals a user tries to achieve. These are manifold and thus can be split on the subjacent (fourth) level. Further, a distinction between behaviors that are supported by technological or non- technological means is introduced at this level and indicated by the black (non-technological) and grey (technological) boxes. There are the following categories:

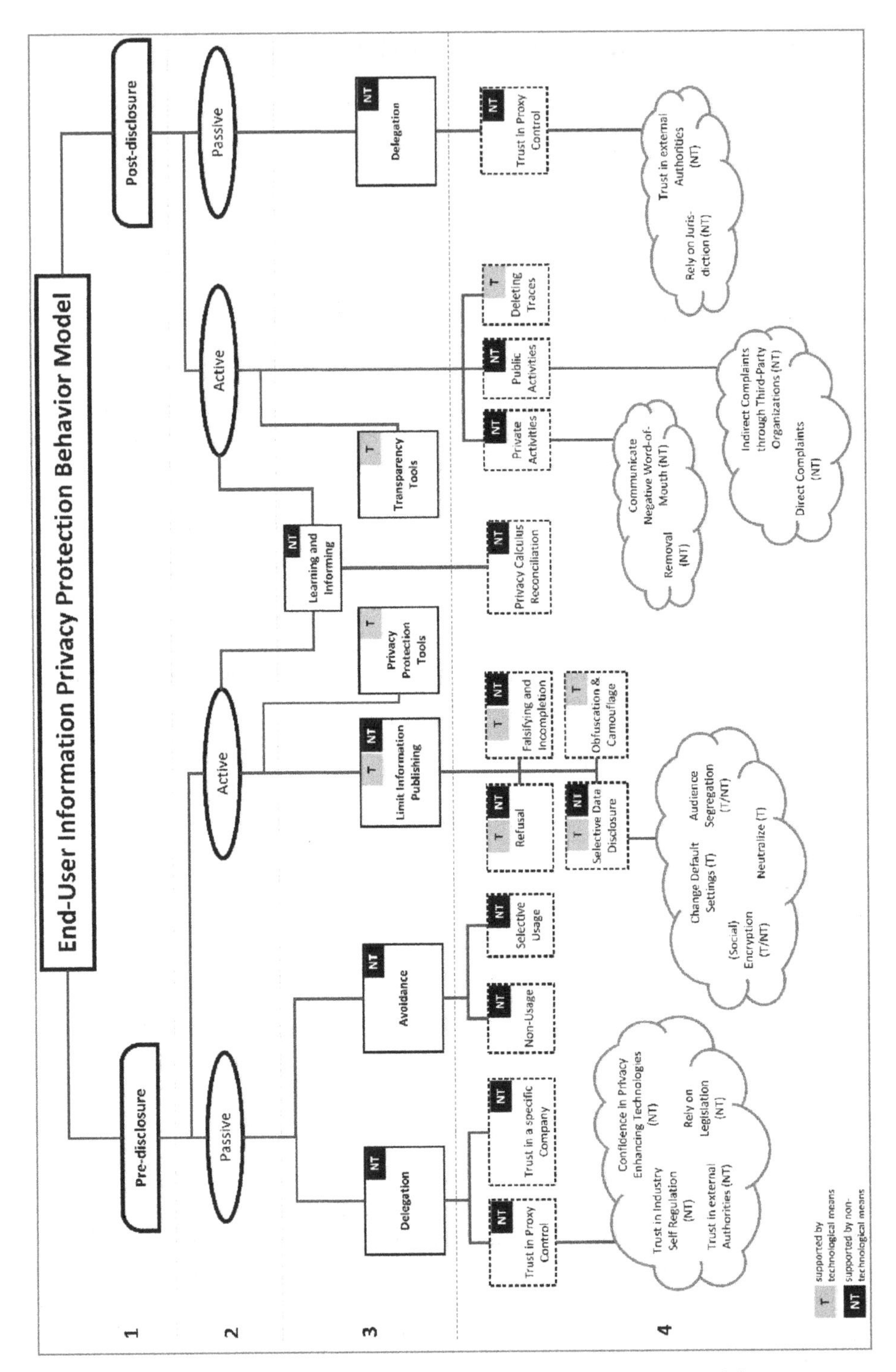

Fig. 3. End-User Information Privacy Protection Behavior Model

- *Delegation* appears before and after a disclosure. Accordingly, this category occurs twice in the model. It is applied by users when they feel overstrained with internet usage and seek for external support [40].
- *Avoidance* refers to the concept of users to "strategically removing themselves from potential privacy-related situations" [16], for example by simply not using an Internet service [40]. Other authors name these approaches escape [14] and deflective behavior [30].
- *Limit information publishing* deals with all behaviors that aim at limiting information disclosure [13, 28] and consolidates many specific approaches. Son and Kim [17] identify this as an important "privacy response".
- The category *privacy protection tools* refers to the countermeasure by applying tools or apps, such as ad blockers, to prevent tracking [28]. Different authors highlight the preventive character [24, 30, 36].
- After a data disclosure users can consult *transparency enhancing tools* to get "insight about what data have been processed about them and what possible consequences might arise" [47].
- *Learning and informing* occupies a special position as users can apply it before and after a data disclosure. This can be either self-learning to avoid privacy pitfalls or as a "lessons learnt" after a privacy incident [35]. Informing refers to the idea of detection, which "assumes that some undesirable use will occur, and seeks to find such incidents" [24]. This category links both pre- and post-disclosure branches. Therefore, it can be seen as a central point of privacy protection behaviors.

Level 4 - Specific Behaviors

This last level represents specific practices and are subjacent to the categories named in the third level. These represent behaviors, which users show when having a specific goal or are more aware of their protective options. In sum, there are thirteen categories. However, each category itself can contain a plurality of subjected behaviors (cloud shape). The cloud shape symbolizes that these behaviors are not collectively exhaustive and may contain many more.

Pre-disclosure behaviors

- Users can *trust in proxy control* before and after a data disclosure. This can be considered as coping strategies. Users "attempt to align themselves such that they are able to gain control through powerful others" [15]. Powerful others can be *external authorities* such as data protection authorities. Further individuals can *trust in industry self-regulation* or in *general legislation* [15]. Moreover users can have *confidence in privacy enhancing technologies* [48]. As there might be several more, these practices are put in the cloud shape.
- Users can *trust in a specific company* and thus do not feel concerned with privacy issues [49, 50] (coping).
- *Non-Usage* is the most radical way of protection [40].
- With *selective usage* individuals choose to be temporarily offline or to use only services that seem trustworthy to them [40].

- Users can *refuse* to publish information [16, 17, 38], if they are not mandatory.
- If data fields are mandatory, users can provide incorrect or incomplete data [17, 20, 35], which refers to *falsifying and incompletion.*
- *Selective data disclosure* contains different approaches. Users can *select the audience* or the type of information they want to disclose [40]. With *(social) encryption* two individuals communicate in a language that an analytical program cannot encode [40]. *Changing default privacy settings* can limit unintended data disclosure [13, 20]. Also blacken personal documents before sending them to third-party [51], called *neutralize* by [16], is a countermeasure.
- *Obfuscation* "is the production of noise modeled on an existing signal" [52]. For example consumers use multiple email addresses [16]. *Camouflage* use a similar approach, however "try[ing] to vanish from view entirely" [52].
- *Privacy calculus reconciliation* is the only practice within the model that is intrapersonal. Thus it rather represents a coping strategy than an actual behavior. Users "engage in an intrapersonal dialogue [...] to rationalize away their desires and concerns for privacy [...] to convince themselves that they remain in control" [16].

Post-disclosure behaviors

- Even after a data disclosure, users can *trust in proxy control* such as *external authorities* [34]. Depending on the legal circumstances in the respective country, users can *rely on jurisdiction* [53].
- Making *public activities* users complain publicly and manifestly about a company. This behavior fits into two types: "*direct complaints* to sellers and *indirect complaints* made to third-party organizations" [17].
- Users make *private activities* when they boycott a particular company, e.g. by asking to be *removed* from a mailing list. Alternatively customers can *communicate* their negative experience to relatives or friends [17]. However, these behaviors are not done publicly. In contrast to [17], we name it "activities" instated of "actions" to emphasize that some behaviors can be permanent.
- In contrast to removal, *deleting traces* refers to data that is stored at the user's devices, such as cookies [16, 19, 54].

Summarizing, 141 protection behaviors were found and mapped to 38 distinct categories. The EIPPBM represents the relationships between these distinct categories. The model introduces several categories on four hierarchical levels. Each category is derived from a literature review based on a mapping of nomination or meanings. On the next page in Fig. 3 the End-User Information Privacy Protection Behavior Model is presented.

5 Discussion

Our models aims to create an overview of user-centric privacy protection behaviors.

Although we also allude the term coping, we use "behavior", as it refers to a way of acting [25]. Whereas classifications in literature have only few categories, our model provides more categories covering a broader field of applicability, because privacy concerns "the Internet as a whole" [13]. In contrast, [17] mainly concentrated on privacy responses in electronic commerce or [29] distinguishes between "general caution" and "technical protection.

Although there is an ongoing debate about if individuals should be responsible for their data protection [19], our work purely concentrates on identifying protection behavior, without taking a position in this debate.

Nonetheless, it is not trivial or unambiguous to organize protection behaviors in a taxonomy or model, because categories could mean different things to people [26]. Whereas some authors label behaviors literally equal (e.g. fabricate [35, 38]), other scholars rely on own nominations (e.g. perturbation [55]). We tried to stick to generally recognized wordings. But sometimes there was a need to create own nominations which literature do not discuss throughout to cover the different authors' aspects (e.g. trust in proxy control). There are also some categories in literature, which did not fit into our model. For example Moshki and Barki say that coping can happen during "a stressful encounter" [14] and call this the "impact period". However, we argue that it is difficult to identify a privacy breach on the fly. Finally, sometimes authors' classifications are contradictory. For example Yap et al. [16] call withdraw as an active behavior, whereas Gurău and Ranchhod [22] consider it as passive. Furthermore we argue that the model can never be considered as complete. As technology is changing with high pace, new protective technologies and behaviors are likely to arise. Adding insights from further disciplines, such as sociology, might bring up other or new categories to the model.

5.1 Implications

Our work offers implications for theory, as well as for practice. For scholars this work offers an overview of current privacy protection behaviors of end-users and helps to understand the range of protection mechanisms [13, 16, 41]. Additionally, it shows which denominations authors use. Further it shows how behaviors could be linked to each other. As the model concentrates on a user-centric approach, it answers the call from [18]. The model is not fixed. Instead, it is highly adaptive. Researchers can edit, add or delete categories without losing the general meaning. In addition, the model is not fixed to a special domain, thus scholars can adopt the classifications to special fields of research.

For practice, our model helps end-users to identify tools and techniques to protect themselves, as users are often not aware how to safeguard their data [13, 41]. Further, data protection officers could use the insights for digital literacy programs. The fact that many protective behaviors can be applied without any technological literacy could encourage novice. However, at the current stage of research, this statement is theoretical and will need an evaluation. Additionally it could help freshmen to choose behaviors that fit to their skills or specific situation. For example, they can distinguish on the very first level if they need advice before or after a data disclosure. Thus it serves as a guideline or manual. However, we have to mention that the usefulness and applicability of the model might depend on the individual user, its awareness and its threat perception. Moreover, users can show different behaviors simultaneously.

5.2 Limitations

Although we try to ensure a high quality of our work, there are some limitations. First to mention, our literature review has no claim to be collectively exhaustive. Some classification could be overlooked, others might be misinterpreted due to the bilingual keyword search. Interestingly very tech-savvy protective countermeasures, such as running own mail servers, did not come up in the literature review. One reason could be that it is considered more a security issues. Further, academics should not consider our classifications as mutually exclusive. Due to complexity reasons, we did not include the motivational factors or influences (such as emotions, cognitions and the characteristics of the environment [14]) on privacy protection. We excluded a consideration of the effectiveness of each behavior for two reasons. First, effectiveness depends on a specific situation and on technological advancements [18]. Second, behaviors appear together and their effects cannot be considered solely [15]. As we aim at an overview without implications or biases, we excluded an analysis of frequency of usage. Lastly we argue that some of our identified behaviors could be considered as coping strategies. E.g. rationalizing away privacy concerns trough delegation is rather a way of coping.

6 Conclusion and Future Work

Our work tried to create a comprehensive view on end-user privacy protection behaviors. To do so we conducted an iterative literature review and identified 141 protection behaviors. We mapped the results and came up with 38 distinct categories. By means of card sorting, we created the "End-User Information Privacy Protection Behavior Model", which represents a taxonomy of behaviors. The work by [17] served as a basis for our model. Our four-level model distinguishes behaviors which users do before or after information disclosure. Further, it differentiates between active and passive behaviors. One central point is learning and information, which links pre- and post-disclosure behaviors. Furthermore, the EIPPBM indicates which behavior need technological or non- technological means. Our literature research has shown that a majority of behaviors can be done without any digital literacy.

Further work could evaluate the model in a representative user study, for example by using cart sorting with a large sample of users. In addition, users can be asked to assign specific protection practices to the proposed categories. It could be interesting to classify behaviors based on groups of privacy threats, as identified by [45] and [56]. Future work could consider involving a typology of users and show which users' personality traits lead to a specific protection behavior. This could be accompanied by a consideration of the effectiveness and applicability of each behavior. Lastly, we encourage scholars from different disciplines to edit our model by adding, deleting or rearranging categories.

Acknowledgement. This work is partially funded by the German Ministry of Education and Research within the project 'Forum Privacy and Self-determined Life in the Digital World', https://www.forum-privatheit.de.

References

1. Malekzadeh, M., Clegg, R.G., Cavallaro, A., et al.: Protecting sensory data against sensitive inferences. In: Maia, F., Mercier, H., Brito, A. (eds.) Thirteenth EuroSys Conference 2018, pp. 1–6. ACM, New York (2018)
2. Crisler, V., Richardson, B., DiGerolamo, J.: The state of IoT security: it is time for action (2018). https://darkcubed.com/iot-security-technical
3. Acquisti, A., Taylor, C., Wagman, L.: The economics of privacy. J. Econ. Lit. **54**(2), 442–492 (2016). https://doi.org/10.1257/jel.54.2.442
4. Tucker, C.E.: The economics of advertising and privacy. Int. J. Ind. Organ. **30**(3), 326–329 (2012). https://doi.org/10.1016/j.ijindorg.2011.11.004
5. Barathi, J.J., Kavitha, G., Imran, M.M.: Building a Mobile Personalized Marketing system using multidimensional data. In: Proceedings of the 2015 International Conference on Smart Technologies and Management for Computing, Communication, Controls, Energy and Materials: ICSTM, 6th–8th May 2015, pp. 133–137. IEEE, Piscataway (2015)
6. Junglas, I.A., Johnson, N.A., Spitzmüller, C.: Personality traits and concern for privacy: an empirical study in the context of location-based services. Eur. J. Inf. Syst. **17**(4), 387–402 (2008). https://doi.org/10.1057/ejis.2008.29
7. Belanger, F., Xu, H.: The role of information systems research in shaping the future of information privacy. Inf. Syst. J. **25**(6), 573–578 (2015). https://doi.org/10.1111/isj.12092
8. DIVSI: Internetnutzung - Risikowahrnehmung unter Jugendlichen und jungen Erwachsenen in Deutschland 2018 (2018). https://de.statista.com/statistik/daten/studie/943840/umfrage/befuerchtete-risiken-der-internetnutzung-unter-jungen-menschen-in-deutschland/. Accessed 17 July 2019
9. Bennett, C.J., Raab, C.D.: The Governance of Privacy: Policy Instruments in Global Perspective, 1st edn. Routledge, Abingdon (2017)
10. CIGI-Ipsos: 2019 CIGI-Ipsos Global Survey on Internet Security and Trust (2019). https://www.cigionline.org/internet-survey-2019. Accessed 23 Apr 2019
11. GPRA: Datenschutz - Vertrauen in Internetunternehmen in Deutschland 2017 (2018). https://de.statista.com/statistik/daten/studie/790373/umfrage/vertrauen-in-den-datenschutz-von-internetunternehmen-in-deutschland/. Accessed 23 Sept 2019
12. Wambach, T., Bräunlich, K.: The evolution of third-party web tracking. In: Camp, O., Furnell, S., Mori, P. (eds.) ICISSP 2016. CCIS, vol. 691, pp. 130–147. Springer, Cham (2017). https://doi.org/10.1007/978-3-319-54433-5_8
13. Boerman, S.C., Kruikemeier, S., Zuiderveen Borgesius, F.J.: Exploring motivations for online privacy protection behavior: insights from panel data. Commun. Res. **25**, 1–25 (2018). https://doi.org/10.1177/0093650218800915
14. Moshki, H., Barki, H.: Coping with information privacy breaches: an exploratory framework. In: Ågerfalk, P.J., Levina, N., Kien, S.S. (eds.) Proceedings of the International Conference on Information Systems - Digital Innovation at the Crossroads, ICIS 2016, Dublin, Ireland, 11–14 December 2016. Association for Information Systems (2016)
15. Xu, H., Teo, H.-H., Tan, B.C.Y., et al.: Effects of individual self-protection, industry self-regulation, and government regulation on privacy concerns: a study of location-based services. Inf. Syst. Res. **23**(4), 1342–1363 (2012). https://doi.org/10.1287/isre.1120.0416
16. Yap, J.E., Beverland, M.B., Bove, L.L.: "Doing Privacy": consumers search for sovereignty through privacy management practices. In: Scott, L.M., Belk, R.W., Askegaard, S. (eds.) Research in Consumer Behavior. Emerald, Bingley (2012)
17. Son, J.-Y., Kim, S.S.: Internet users' information privacy-protective responses: a taxonomy and a nomological model. MIS Q. **32**(3), 503–529 (2008). https://doi.org/10.2307/25148854

18. London Economics: Study on the Economic Benefits of Privacy-enhancing Technologies (PETs): Final Report to The European Commission, DG Justice, Freedom and Security. London Economics (2010)
19. Matzner, T., Masur, P.K., Ochs, C., von Pape, T.: Do-it-yourself data protection—empowerment or burden? In: Gutwirth, S., Leenes, R., De Hert, P. (eds.) Data Protection on the Move. LGTS, vol. 24, pp. 277–305. Springer, Dordrecht (2016). https://doi.org/10.1007/978-94-017-7376-8_11
20. Büchi, M., Just, N., Latzer, M.: Caring is not enough: the importance of Internet skills for online privacy protection. Inf. Commun. Soc. **20**(8), 1261–1278 (2016). https://doi.org/10.1080/1369118X.2016.1229001
21. Lampinen, A., Lehtinen, V., Lehmuskallio, A., et al.: We're in it together: interpersonal management of disclosure in social network services. In: Tan, D., Fitzpatrick, G., Gutwin, C., et al. (eds.) The 29th Annual CHI Conference on Human Factors in Computing Systems: Conference Proceedings and Extended Abstracts, pp. 3217–3226. ACM, New York (2011)
22. Gurău, C., Ranchhod, A.: Consumer privacy issues in mobile commerce: a comparative study of British, French and Romanian consumers. J. Consum. Mark. **26**(7), 496–507 (2009). https://doi.org/10.1108/07363760911001556
23. Singh, N., Singh, A.K.: Data privacy protection mechanisms in cloud. Data Sci. Eng. **3**(1), 24–39 (2018). https://doi.org/10.1007/s41019-017-0046-0
24. Jiang, X., Hong, J.I., Landay, J.A.: Approximate information flows: socially-based modeling of privacy in ubiquitous computing. In: Borriello, G., Holmquist, L.E. (eds.) UbiComp 2002. LNCS, vol. 2498, pp. 176–193. Springer, Heidelberg (2002). https://doi.org/10.1007/3-540-45809-3_14
25. Heacock, P. (ed.): Cambridge Academic Content Dictionary, 1st edn. Cambridge University Press, Cambridge (2009)
26. Smith, H., Dinev, T., Xu, H.: Information privacy research: an interdisciplinary review. MIS Q. **35**(4), 989–1015 (2011). https://doi.org/10.2307/41409970
27. Acquisti, A., Brandimarte, L., Loewenstein, G.: Privacy and human behavior in the age of information. Science **347**(6221), 509–514 (2015). https://doi.org/10.1126/science.aaa1465
28. Baruh, L., Secinti, E., Cemalcilar, Z.: Online privacy concerns and privacy management: a meta-analytical review. J. Commun. **67**(1), 26–53 (2017). https://doi.org/10.1111/jcom.12276
29. Buchanan, T., Paine, C., Joinson, A.N., et al.: Development of measures of online privacy concern and protection for use on the Internet. J. Am. Soc. Inf. Sci. Technol. **58**(2), 157–165 (2007). https://doi.org/10.1002/asi.20459
30. Lwin, M.O., Wirtz, J., Stanaland, A.J.S.: The privacy dyad: antecedents of promotion- and prevention-focused online privacy behaviors and the mediating role of trust and privacy concern. Internet Res. **26**(4), 919–941 (2016). https://doi.org/10.1108/IntR-05-2014-0134
31. Gurung, A., Jain, A.: Antecedents of online privacy protection behavior: towards an integrative model. In: Chen, K. (ed.) Cyber Crime: Concepts, Methodologies, Tools and Applications, pp. 69–82. IGI Global, Hershey (2012)
32. Dolnicar, S., Jordaan, Y.: A market-oriented approach to responsibly managing information privacy concerns in direct marketing. J. Advertising **36**(2), 123–149 (2007). https://doi.org/10.2753/JOA0091-3367360209
33. Li, Y., Dai, W., Ming, Z., et al.: Privacy protection for preventing data over-collection in smart city. IEEE Trans. Comput. **65**(5), 1339–1350 (2016). https://doi.org/10.1109/TC.2015.2470247
34. Xu, H., Dinev, T., Smith, J., et al.: Information privacy concerns: linking individual perceptions with institutional privacy assurances. J. Assoc. Inf. Syst. **12**(12), 1 (2011)
35. Youn, S.: Determinants of online privacy concern and its influence on privacy protection behaviors among young adolescents. J. Consum. Affairs **43**(3), 389–418 (2009). https://doi.org/10.1111/j.1745-6606.2009.01146.x

36. Anderson, C.L., Agarwal, R.: Practicing safe computing: a multimedia empirical examination of home computer user security behavioral intentions. MIS Q. **34**(3), 613–643 (2010). https://doi.org/10.2307/25750694
37. Fritsch, L.: State of the art of privacy-enhancing technology (PET): deliverable D2.1 of the PETweb project (2007)
38. Lwin, M., Wirtz, J., Williams, J.D.: Consumer online privacy concerns and responses: a power–responsibility equilibrium perspective. J. Acad. Mark. Sci. **35**(4), 572–585 (2007). https://doi.org/10.1007/s11747-006-0003-3
39. Metzger, M.J.: Communication privacy management in electronic commerce. J. Comput.-Mediated Commun. **12**(2), 335–361 (2007). https://doi.org/10.1111/j.1083-6101.2007.00328.x
40. Ochs, C., Büttner, B., Hörster, E.: Das Internet als »Sauerstoff« und »Bedrohung«. In: Friedewald, M. (ed.) Privatheit und selbstbestimmtes Leben in der digitalen Welt. D, pp. 33–80. Springer, Wiesbaden (2018). https://doi.org/10.1007/978-3-658-21384-8_3
41. Bélanger, F., Crossler, R.E.: Privacy in the digital age: a review of information privacy research in information systems. MIS Q. **35**(4), 1017–1042 (2011)
42. Vom Brocke, J., Simons, A., Riemer, K., et al.: Standing on the shoulders of giants: challenges and recommendations of literature search in information systems research. CAIS **37** (2015). https://doi.org/10.17705/1cais.03709
43. Benyon, D.: Designing Interactive Systems: A Comprehensive Guide to HCI, UX and Interaction Design, 3rd edn. Pearson, Harlow (2014)
44. Spencer, D.: Card Sorting: Designing Usable Categories. Rosenfeld Media, New York (2009)
45. Solove, D.J.: A Taxonomy of Privacy. Univ. Pa. Law Rev. **154**(3), 477 (2006). https://doi.org/10.2307/40041279
46. Folkman, S., Moskowitz, J.T.: Coping: pitfalls and promise. Annu. Rev. Psychol. **55**(1), 745–774 (2004). https://doi.org/10.1146/annurev.psych.55.090902.141456
47. Murmann, P., Fischer-Hubner, S.: Tools for achieving usable ex post transparency: a survey. IEEE Access **5**, 22965–22991 (2017). https://doi.org/10.1109/ACCESS.2017.2765539
48. ENISA: Readiness Analysis for the Adoption and Evolution of Privacy Enhancing Technologies (2016). https://www.enisa.europa.eu/publications/pets. Accessed 14 June 2017
49. Teo, H.H., Wan, W., Li, L.: Volunteering personal information on the Internet: effects of reputation, privacy initiatives, and reward on online consumer behavior. In: Sprague, R.H. (ed.) Proceedings of the 37th Annual Hawaii International Conference on System Sciences, pp. 1–10. IEEE Computer Society Press, Los Alamitos (2004)
50. Bitkom: Datenschutz - Vertrauen in Organisationen im Umgang mit persönlichen Daten in Deutschland (2018). https://de.statista.com/statistik/daten/studie/936247/umfrage/vertrauen-in-organisationen-im-umgang-mit-persoenlichen-daten-in-deutschland/. Accessed 09 July 2019
51. Kung, A., et al.: A privacy engineering framework for the internet of things. In: Leenes, R., van Brakel, R., Gutwirth, S., De Hert, P. (eds.) Data Protection and Privacy: (In)visibilities and Infrastructures. LGTS, vol. 36, pp. 163–202. Springer, Cham (2017). https://doi.org/10.1007/978-3-319-50796-5_7
52. Brunton, F., Nissenbaum, H.: Obfuscation: A User's Guide for Privacy and Protest. MIT Press, Cambridge (2015)
53. ROLAND-Gruppe: Roland Rechtsreport 2019 (2019). https://www.roland-rechtsschutz.de/unternehmen/presse_2/publikationen/publikationen.html. Accessed 16 July 2019
54. Statista: Datenschutz - Maßnahmen in Deutschland 2017 (2017). https://de.statista.com/statistik/daten/studie/712775/umfrage/massnahmen-zum-datenschutz-in-deutschland/. Accessed 10 July 2019

55. Shin, K.G., Ju, X., Chen, Z., et al.: Privacy protection for users of location-based services. IEEE Wireless Commun. **19**(1), 30–39 (2012). https://doi.org/10.1109/MWC.2012.6155874
56. Kasper, D.V.S.: The evolution (or devolution) of privacy. Sociol. Forum **20**(1), 69–92 (2005). https://doi.org/10.1007/s11206-005-1898-z

Annotation-Based Static Analysis for Personal Data Protection

Kalle Hjerppe[1(✉)], Jukka Ruohonen[2], and Ville Leppänen[2]

[1] Geniem Oy, Tampere, Finland
kalle.hjerppe@geniem.com
[2] University of Turku, Turku, Finland
{juanruo,ville.leppanen}@utu.fi

Abstract. This paper elaborates the use of static source code analysis in the context of data protection. The topic is important for software engineering in order for software developers to improve the protection of personal data during software development. To this end, the paper proposes a design of annotating classes and functions that process personal data. The design serves two primary purposes: on one hand, it provides means for software developers to *document* their intent; on the other hand, it furnishes tools for automatic *detection* of potential violations. This dual rationale facilitates compliance with the General Data Protection Regulation (GDPR) and other emerging data protection and privacy regulations. In addition to a brief review of the state-of-the-art of static analysis in the data protection context and the design of the proposed analysis method, a concrete tool is presented to demonstrate a practical implementation for the Java programming language.

Keywords: Data protection · Privacy · Static analysis · Java · GDPR

1 Introduction

The famous General Data Protection Regulation (GDPR) in the European Union [37] places various new requirements for software architectures as well as their design, development, and maintenance. Thus, this paper builds on previous work on eliciting requirements from the GDPR in the context of software architectures and their design [22]. While the previous work concentrated on high-level design, the present work takes a step down to the actual implementation of data protection solutions during software development. The approach presented belongs to the domain of static analysis of software source code.

Data protection is relevant for both organisational and software security. Numerous recent data breaches, such as the high-profile Equifax breach of 2017 [4], blatantly demonstrate how negligence has real consequences. The GDPR mandates organisations to invest in data protection to be able to process personal data securely and legally. For the present purposes, the juridical

Published by Springer Nature Switzerland AG 2020
M. Friedewald et al. (Eds.): Privacy and Identity 2019, IFIP AICT 576, pp. 343–358, 2020.
https://doi.org/10.1007/978-3-030-42504-3_22

aspects can be also used to distinguish the concept of *data protection* from the technical concept of (information) *security*, which does not cover the legality of data processing. Furthermore, the concepts of *privacy* and data protection are often used synonymously, but the latter is important without the presence of the former. That is, data should be protected even in a context that does not respect privacy.

The GDPR imposes various implicit requirements for software engineering, software development, and software architectures. Among these are general requirements related to concepts such as confidentiality, integrity, and availability. In several occasions, the GDPR also mentions *appropriate* data protection measures (see Articles 5, 25, and 32). These measures are not binary-valued "on/off requirements" in their nature, however. The scale is continuous: the requirements can be improved with design patterns and different software development technologies. Static analysis is a good example about such technologies.

In essence, static analysis is a method of analysing a program (*analysis*) without running it (*static*). Static source code analysis can be applied to any software project with appropriate tools. It is also generally accepted as a "*security-by-design*" best practice [9]. Using tools to improve software quality during software development is also very much what the GDPR asks for. Therefore, static analysis naturally extends also to the present "*data-protection-by-design*" context.

Given this background, this paper answers to two research questions (RQs):

- RQ1: *What parallels there are between those static analysis solutions designed for security and those that seek to improve privacy and data protection?*
- RQ2: *How the practical state-of-the-art of static analysis solutions can be applied in the context of the GDPR's data protection requirements?*

The answer to the first research question is sought by analysing and categorising a few recent static analysis tools. There is a well-established literature on static analysis in general and using static analysis for improving security in particular (see, for instance, [9] and [25]). There exists also some previous work on using static analysis for the GDPR's requirements [17,18]. However, a synthesis is still lacking—in fact, only little has been written about static analysis in terms of the intersection between security and data protection.

Building upon the categorisation presented (RQ1), the second research question is answered by presenting a design of a concrete tool for using static analysis to improve data protection. To briefly outline the background of this tool and its design, it suffices to note that common static source code analysis tools used to prevent (security) bugs analyse the abstract syntax tree (AST) of a software source code under inspection. The so-called `FindBugs` [3] static code analyser is a good example in this regard. This inspection allows the tools to make judgements of the logical content of a program. By implication, however, the tools are also limited to the logical content; they cannot analyse design, developer intent, or data semantics. For these reasons, the tool presented augments AST inspection with personal data annotations.

The structure of the paper's remainder follows the two research questions examined. In other words, Sect. 2 evaluates the state of the art and categorises static analysis solutions for data protection (RQ1). The subsequent Sect. 3 elaborates the tool implemented for improving personal data processing and its protection (RQ2). Section 4 discusses the implications and future directions.

2 Background and Related Work

A sensible starting point is that security in general is a *prerequisite* for data protection, which, in turn, is a partial legal prerequisite for privacy. If data is not secured, it is also meaningless to discuss data protection and privacy. In other words, compromising security allows to compromise the other two concepts.

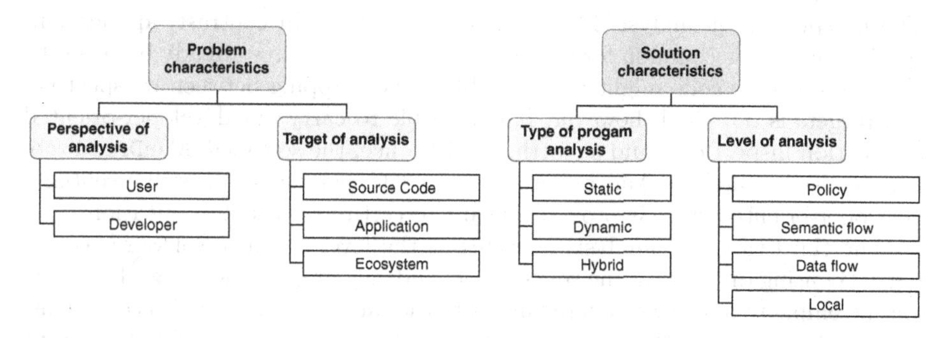

Fig. 1. Characteristics of existing approaches of analysis tools for data protection

Given this assumption, a sensible next step for categorising the extensive literature on static analysis is the work of Sadeghi et al. who consider both security and privacy (data protection) in the context of the Android operating system and its software ecosystem [35]. They categorise different research approaches by considering the problems addressed ("what") and the solutions to the problems ("how"). This fundamental taxonomy works also in the present context. Building upon their categories, different dimensions of data protection are thus illustrated in Fig. 1. By analysing these characteristics from the dual perspectives of data protection and security, it is possible to reason about differences as well as parallels. When these characteristics are combined, it is further possible to demonstrate a research gap. The following will briefly elaborate this gap.

2.1 Categorisation by Problem Characteristics

Threat modelling is a distinct characteristic of program analysis tools used for security assessments [35]. In other words, an analysis tool seeks to address a particular threat or a set of threats. In terms of data protection, unintended information disclosure is the obvious threat—and a data breach a manifestation of the threat's realisation. For the present purposes, however, threat models are not sufficient for a categorisation of existing static analysis tools.

There appears to be a fundamental difference that separates static analysis approaches for data protection and privacy: whether an analysis is done from a *perspective* of a user or a provider of the software the user is using. The user-oriented approaches typically focus on analysing software for potentially malicious or erroneous activity (see, for instance, [24] and [39]). In contrast, the developer-oriented approaches typically focus on verifying and validating software. For example, Calcagno et al. [6] incorporate developer-oriented static analysis to software development for verifying memory safety. Obviously, it is possible to use a single approach from both perspectives; therefore, it is important to also consider the target of a static analysis tool.

In the present work, *the target of analysis* refers to a particular software artifact under inspection. Traditional static analysis tools focus on source code. The perspective of analysis is on the developer side. In contrast, application-specific analysis is common for the user-oriented perspective already because the associated source code might not be public. The mapping between perspectives and targets is not rigid, however. It is possible to carry out developer-oriented application inspections, and so forth. For instance, the so-called `SCanDroid` tool may be used to analyse Android applications [20], yet the tool is still essentially developer-oriented since it provides means to certify security specifications.

The third target of analysis operates at the level of whole software ecosystems. By using different frameworks for dependency graphs, there have been various attempts to analyse software vulnerabilities at the level of whole ecosystems. A typical example is the *npm* ecosystem and its dependencies for JavaScript libraries [30,40]. Another example would be the software vulnerabilities in the packages stored to the Python's PyPI repository [34]. However, according to the literature review conducted, no notable previous work has been done to extend these approaches toward data protection.

2.2 Categorisation by Solution Characteristics

Turning to the how-question, a reasonable separation can be done in terms of the type and level of a software analysis solution. In terms of the *type*, there are three common cases: static, dynamic, and hybrid. Each of these three categories have representatives also in the data protection and privacy contexts [2,15,38]. While each type has its merits [2], the paper's focus is strictly on static analysis.

The *level* of analysis is the last but not the least dimension considered. The dimension builds on the noted premise that both data protection and privacy build upon security. This dimension is also the one through which many notable differences between different security and data protection (privacy) solutions differ. Four levels are considered: local, data flow, semantic flow, and policy.

In terms of the local level, static analysis tools typically consider the syntax of the source code without much semantics and context. A good example would be so-called linters, which are reasonably well-suited for detecting simple syntax errors but also some more serious issues, such as Null pointer accesses. In contrast, the data flow, semantic flow, and policy levels seek to follow data through a software and make judgements based on this following. For instance, basic tools for finding structured query language (SQL) injections operate at the data flow level. Semantic flows add meaning to the data and its flows. For instance, the so-called `TaintDroid` tool [15] labels sensitive data according to the source of it, such as sensors. Another recent approach is the so-called `DroidRista` tool [1], which achieves high privacy leak detection rates also at the semantic flow level.

A policy-level analysis adds security and data protection (privacy) policies to the semantic flow of data in order to verify that a policy is followed. An example of a policy-level approach is the `MorphDroid` tool [19]. However, there are also more comprehensive languages for formally specifying data protection and privacy policies. These levels of analysis are also an important aspect when considering differences between security-focused approaches and those designed for data protection and privacy. While there is a long history of using policies for security, many of the classical solutions do not consider semantics well. Whether it is mandatory access control mechanisms, group-based solutions, or capabilities, data is often seen uniformly as data worth protecting regardless whether the data is sensitive personal data or not. In contrast, recent formal languages specific to data protection and privacy, such as the so-called Layered Privacy Language (LPL) introduced by Gerl et al. [21], have been strongly motivated by the semantics of personal data. This language provides different constructs to reason about privacy policies. For the purposes of the tool soon elaborated, the LPL's *Policy*, *Purpose*, and *Data* constructs are particularly useful.

The categories briefly described are useful for distinguishing different approaches and their underlining perspectives. For instance, an interesting comparison could be done regarding the different levels of analysis when the user-oriented and developer-focused approaches are used. In fact, it seems that a lot of the literature about the policy-level is tied to the user-oriented perspective; a typical goal is to verify that an application follows a privacy policy. If this generalisation is accepted, it could be also stated that a goal of a developer-oriented policy-level tool, in turn, would be to *generate* the given policy from the source code. On the developer side, however, many current tools specific to data protection and privacy operate on the data flow and local levels of analysis. This observation provides the rationale for the tool presented; the goal is to move from the data flow level to the arguably more important policy level.

2.3 Related Work

The high-level categorisation presented can be used to group many (if not most) static analysis approaches. To augment the categorisation, the approach proposed can be further explicitly compared with the following four previous works:

1. Taint analysis can be done both statically and dynamically. Static taint analysis has the potential to solve the same issues as the approach of this paper. For instance, the work of Arzt et al. [2] finds privacy leaks in Android applications. Their approach is limited to programs for which data sensitivity can be deduced as coming from user input. Server-side taint analysis, in turn, is limited to the data flow level of analysis—unless an appropriate context for

data sensitivity can be provided (such as the annotations presented). As a consequence, taint analysis tools are generally either for client-side analysis or they focus on security vulnerabilities in server-side applications.
2. Formal languages, coupled with a compiler, such as the one developed by Ramezanifarkhani et al. [31], possess the potential to make privacy policies statically verifiable. These languages have the potential to statically enforce data protection stronger than the lightweight analysis proposed. Although these would limit the usefulness of static analysis approaches, it should be emphasised that the approaches are not yet ready for production use.
3. The work of Myers et al. [27] shares a similar goal than the approach proposed: information flow control. They use labels in source code (like annotations, but as an extension of a language) to protect variables from improper reading or modification. While their decentralised end-to-end approach is valid, it is user-focused in contrast to the developer-oriented perspective pursued in this paper. Their design is also more pervasive, requiring a specific execution platform and an extension for the programming language.
4. A lightweight static analysis design similar to the approach proposed has been presented by Evans et al. [16]. They leverage comments (acting as annotations) in C code in order to detect security vulnerabilities. While their approach is also developer-oriented, the low-level focus differs from the present work. Furthermore, the approach neither recognises personal data nor focuses on data protection as such. That is, the principle of annotating relevant parts of code is the same, but the approach proposed considers documentation as a supplement to technical verification of code.

3 Tool

The tool proposed seeks to improve the analysis level of source code analysis tools by focusing on the developer-oriented perspective. The underlying rationale is to analyse personal data in order to prevent accidental leaks of this data. In essence, the tool's goal is to help software developers by warning them about possible privacy leaks in source code during software development or immediately afterwards. The focus is thus in the implementation phase in a typical software development life cycle. This focus allows to catch design and implementation mistakes early on. In this developer-oriented context, *personal data* simply means sensitive values in run-time memory. A typical example is a typed object of a class. Traditional source code analysis cannot know which objects contain personal data without some outside information. This constraint places them to the data flow level and limits their usefulness for data protection.

The goal is also easy to justify with respect to regulations. Personal data could be processed in an unintended way as a result of a bug. Saving values of personal data into a log file would be a good example. With regards to the GDPR, such a bug may violate confidentiality and the need to know where all personal data is stored. Warning a developer from possible unintended processing with a short feedback loop should improve these and other related risks.

A brief further point can be made about performance requirements. Recently, Distefano et al. have discussed the issue of scaling static analysis solutions for large code bases [14]. Their main scaling properties are *composionality* [7,12] and *abstraction* [11]. They define composionality so that a program analysis is composional when "*the analysis result of a composite program is defined in terms of the analysis results of its parts and a means of combining them*" [14, p. 70]. Abstraction, on the other hand, refers to considering only parts of the procedures that are relevant to the analysis, and discarding the rest. The gain in having these properties is the ability to parallelise the analysis, or doing incremental analysis only on parts of a code base. These properties are taken into account in the approach proposed—after all, scaling and performance are important aspects for any software development tool.

3.1 Design of the Tool

The design of the tool has two layers. The first layer refers to the annotating of personal data objects in source code. This annotating is useful for raising the analysis level for source code analysis solutions in general. The second layer builds on a light source code analysis on the semantic flow level of analysis. This layer is constructed with two additional annotations to further specify a data processing context. The method demonstrates how a higher analysis level is reachable with relatively simple rules without a heavy analysis process. As will be discussed, the method scales also to larger problems.

The solution proposed uses the following three annotations:

- A1: `@PersonalData`
- A2: `@PersonalDataHandler`
- A3: `@PersonalDataEndpoint`

In essence, the `@PersonalData` annotation (A1) should be used to document all classes containing personal data. This annotation gives the tool the necessary context to separate personal data types from other types. It also serves as a documentation of the source code. The other two annotations, `@PersonalDataHandler` (A2) and `@PersonalDataEndpoint` (A3), are used to further document the context within which personal data is processed. There are three specific rules (R) for this processing of personal data:

- R1: *Data classes storing personal data are annotated with A1.*
- R2: *Any instance of an A1-annotated class used in a context not annotated with A2 is a violation.*
- R3: *In A2-annotated contexts, calling a non-A3 function outside the A2-context is a violation if an argument is A1-annotated.*

Rule R1 is the basis of the analysis, enabling the other two. R2 covers most data processing by requiring explicit handling of personal data to be documented. R3 allows establishing boundaries for personal data processing contexts within an application. A good example would be a generic database method for fetching a single entity. To further clarify, an A1-annotated class includes type variables for instances of generic classes and classes inheriting actual A1-annotated classes. Contexts annotated with A2 and A3, in turn, refer to functions or classes within which these are defined. The use of these three annotations is illustrated in Fig. 2. The illustration demonstrate a part of a class structure of an imaginary web application following the classical *model-view-controller* design pattern. The illustration also demonstrate basic violations of the R2 and R3 rules.

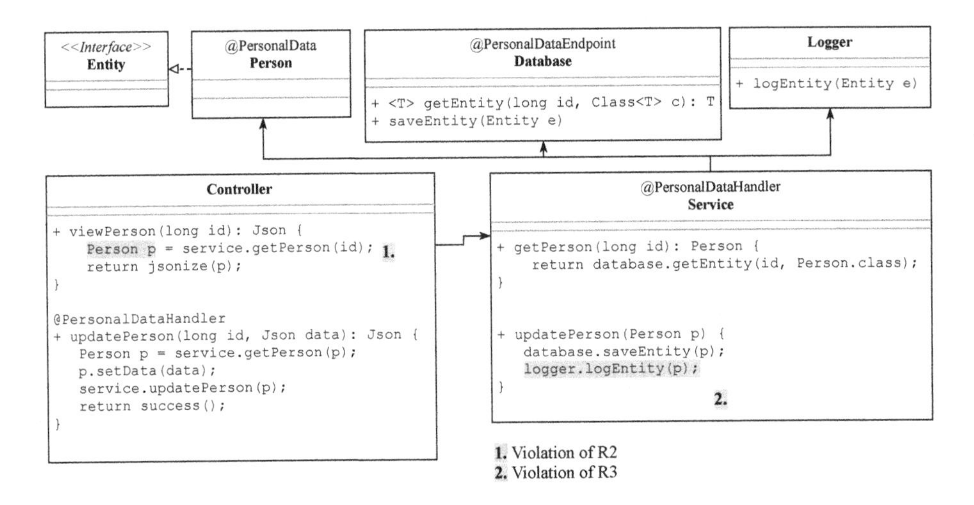

Fig. 2. Diagram and pseudo-code example of R2 and R3 violations.

Three additional points are warranted about the design. First, in LPL's terms, A1-annotated classes correspond with (groups of) *Data* elements. Such elements can be further mapped to distinct entry points, such as function-to-URL mappings of a web service. With such mappings and further annotations, it is possible to construct distinct execution trees that mimic LPL's *Purpose*.

Second, the design adheres to the noted composionality requirement: a method under analysis only needs to know the annotations present in referred methods and classes, not whether they themselves would pass the analysis. By implication, the analysis can focus on units of code rather than the whole system—though, of course, to reach full coverage, eventually all of the code has to be analysed. When compared to more comprehensive designs, the design is also lightweight. Given that using static analysis for software development is perceived to have a significant cost [8], the lightweight design can ease adoption and shorten the feedback cycle for software developers. This practical point should not be undermined.

Third, in the terms of data flow analysis, the design can be briefly evaluated also in terms of so-called sensitivity properties [28]. The design discards path sensitivity and instead considers all possible branches of execution. Flow and context sensitivity are both needed to validate the rules described. Here, flow sensitive analysis implies being dependent on the order of instructions in the code, while context sensitivity comes from taking into account the calling context of method calls instead of analysing functions in a vacuum.

3.2 Implementation

The tool was implemented for the Java programming language using the Annotation Processing Tool (APT) functionality [32]. The implementation is also packaged and can be thus attached as a dependency for further projects. Command line build tools such as *Maven*[1] and integrated developer environments like *IntelliJ IDEA*[2] allow adding annotation processors to projects' build processes. Therefore, the implementation can be also embedded to continuous integration environments. Extending IDEs to integrate further tools to enhance developers is a common practice (see `DebtFlag`, for instance [23]). The source code of the implementation is published[3] under an open source license.

The implemented tool consists four classes: the three annotations discussed and a `PersonalDataAnnotationProcessor` (PDAP) class. The annotation processor inherits the APT's `AbstractProcessor` and implements `TaskListener` to hook into the Java compile process. This hooking allows to carry out the static analysis each time a software is compiled. A more heavyweight solution would not allow such a fast feedback loop. When compared to using only the `AbstractProsessor` model, which is limited to method signatures, making use of `JavacTask` allows traversing the entire AST of a program. Analysing the AST instead of the source code eliminates unnecessary noise from the analysis.

After compilation of the target program, the PDAP class is notified by a `finished(TaskEvent task)` method call. This call provides the access to the compilation unit, and allows to traverse the AST with the *visitor* [29] pattern. The result is a warning for each violation of R2 and R3, with a marker to the specific line of code. An example of presenting the violations is shown in Fig. 3.

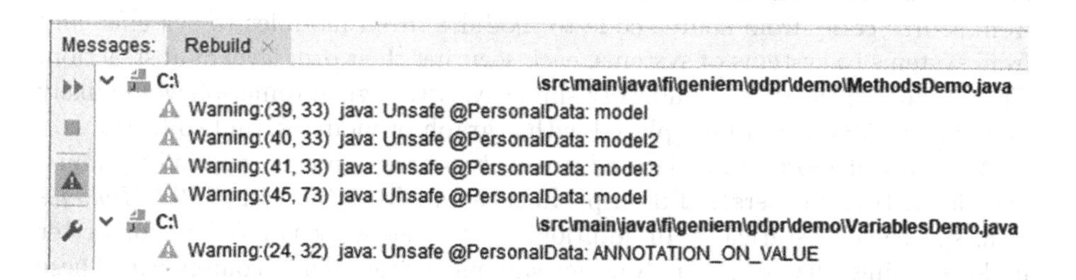

Fig. 3. Example of warnings produced by the analysis in *IntelliJ IDEA*.

[1] https://maven.apache.org/.
[2] https://www.jetbrains.com/idea/.
[3] https://github.com/devgeniem/personaldataflow.

The rules defined cover typical cases of processing personal data. The Java implementation passes conventional tests. However, there is one known issue: the tool does not cover representation exposure. For instance, a warning is not produced when extracting a primitive value from an A1-annotated object and then logging it. Improving the coverage of the tool for these situations should be investigated in further work. The current advise for this problem is to use specific classes for values of personal data rather than primitives. In the future, the coverage of the tool also improves as missing cases are added when discovered.

4 Discussion

The approach presented and the tool implemented improves data protection qualities of software systems processing personal data. As discussed by Chess and McGraw [9], the goal of static analysis for software security should be "good enough" coverage rather than guarantees. This objective underlines the developer-focused viewpoint to static analysis—to find as many bugs as possible. Privacy and data protection bugs do not mark an exception from this objective.

While the static analysis tool presented finds possible development errors, the `@PersonalData` annotations also document the nature of personal data in source code. If even more lightweight approach is desirable, a project might use only the `@PersonalData` annotations. When all three are used, processing of personal data is documented in source code rather extensively. Obviously, the downside is the increased effort for developers to annotate the code and maintain the annotations. While there is no single universal answer to a question about whether the quality increases are worth the effort, the annotations provide opportunities to develop further tools that increase the return on investment.

In other words, the annotations described enable the development of further tools, whether for code analysis, visualisation, or software analytics. Annotation-based reasoning is not strictly limited to source code analysis. It is possible to imagine an analysis of higher-level modules or even ecosystems in a similar fashion. In fact, a distant possibility would be to annotate all abstraction levels from source code; from source code to modules, from modules to systems, and from systems to systems of systems, each securing their own level of abstraction. The design and the rules would work similarly. Although automation would likely be difficult, AST could be replaced with a graph of modules, and so forth.

In terms of more immediate advances, two areas appear especially promising: (a) metrics for personal data prevalence and (b) generating LPL's *Policies* from source code statically. In addition, (c) further work is required for better understanding what static analysis actually means for GDPR compliance. These three topics are discussed in what follows.

Before continuing, however, three limitations can be briefly noted. First, the approach described is not granular and thus does not support separating singular paths in source code, unlike for example annotating distinct allowed execution paths in the code. This is a trade-off between developer effort and sensitivity of results. Second, the Java implementation is unable to find processing that goes beyond standard language usage, such as using reflection or type casts extensively. A developer must take this into account when relying on the results of the analysis. Last but not least: when compared to the whole scope of the GDPR, the results of this paper are only a small part of building systems with better data protection. Needless to say, the entire issue goes beyond technical measures altogether.

4.1 Metrics for Personal Data Prevalence

Software metrics are commonly used for evaluating source code. The use cases range from quality control (since Boehm et al. [5]) to predicting change prone areas [33], for example. Metrics for software security can also be derived from source code [10]. However, to the best of our knowledge, mature metrics for personal data processing in source code are lacking. The approach of using annotations naturally extends to metrics for personal data and its processing.

A good topic would be the distribution of personal data across a system. While centralisation may not be optimal for security, reducing the distribution of personal data is a worthwhile goal in data protection terms [22]. In fact, the centralisation of personal data to as few occurrences as possible is in spirit with the data minification principle of the GDPR. It can be applied from the architectural level also to the source code level. To facilitate this goal, developing metrics about personal data in code bases is relevant.

Even simple metrics (such as the share of `@PersonalData` classes to all classes) might be useful for better understanding the data protection aspects of a particularly large software architecture. Cohesion and coupling of different personal data processing call trees might also be valuable. In this respect, there is also a good literature base to build upon [26]. With associated metrics, refactoring could be guided for more robust personal data processing.

Regardless of particular metrics, the annotations described enable automatic calculation of the values. This aspect also signifies the usefulness of the semantic flow analysis level: without knowledge of the personal data semantics, calculating the values would need to be done manually or supplied in an *ad hoc* fashion.

Any metrics from source code are naturally not metrics of *privacy* itself, as argued by Danezis et al. [13]. However, focusing on detecting *code smells* in this way is much easier to reason about than handling the abstract privacy concept as a whole. Source code metrics about data protection could point out especially sensitive areas of software architectures to focus efforts on, for example.

4.2 Finding LPL Purposes

Another promising next step would be to generate LPL's *Policies* (or templates for the policies) from `@PersonalData` annotations coupled with web framework annotations. Common use cases for Java applications include server-side application programming interfaces (APIs) that essentially listen to different HTTP requests' URLs and execute a part of an application code for each. Using these in combination with the annotations, it should be possible to generate also policies.

Consider, for instance, the Spring framework[4] `@RequestMethod` annotation. This annotation is used to map service URLs to application *entry points*. In essence, this mapping means that the application code itself does not handle web requests; the framework controls which method is called by which HTTP request. Therefore, the application source code can be viewed as distinct directed graphs of (possible) method calls, starting from the endpoint method. Each graph can be understood to cover one possible way to call the application. In addition, a conventional use case is to do an Object Relational Mapping (ORM) to map Java classes to database relations using another library. From the Java Persistence API specification, `@Entity` is a common way to implement this mapping.

By having an application built as described, the relevant entry points can be considered as LPL's *Purposes*. This part is enabled trivially by the annotations: if an entry point method has a `@PersonalDataHandler` annotation, it is a *Purpose*. Following from that, the source code and the execution paths can be traversed for finding uses of classes annotated with both `@Entity` and `@PersonalData`. These classes would then form LPL's *Data* elements. The accessed *Data* elements and the different *Purposes* using them provide a good starting point for generating a *Policy*. The same idea could be extended to analyse communication between modules through HTTP, for instance. This would require mapping the *Purpose* of the first module to the *Purpose** of another module, and a way to analyse the APIs statically in both ends. They would then combine into a higher abstraction level *Purpose*. An example would be to use Web Service definitions with corresponding annotations on both the service provider and the client.

A weakness is that a certain architecture is required for an application in order to derive the policies. The idea also breaks down in case all functionality is coupled to a single entry point. It should be also noted that the framework code, being imported from a library, would be outside of the scope of the compile time analysis presented in this paper, which is an advantage in this case. The analysis would then start from the entry point and cover the execution flow.

4.3 Regarding the GDPR and Static Analysis

It is necessary to point out the obvious: it is not possible to achieve compliance with the GDPR merely with just static analysis. While technical requirements can be derived from the regulation [22], the GDPR itself does not lay any exact technical requirements for compliance. Against this backdrop, Schneider [36] argues that *a posteriori* methods, such as static analysis conducted after software is designed, are not useful for achieving "*privacy-by-construction*". That would be the ultimate goal of privacy engineering research. As he concludes, however, achieving "privacy-by-construction" is extremely difficult and expensive—if not impossible. Thus, in practice, other methods must suffice.

What the GDPR actually requires is ensuring appropriate personal data security with appropriate technical and organisational measures. Integrity and

[4] https://spring.io/.

confidentiality are the underlying concepts (Article 5). Article 25 further qualifies these with the following remark:

> *"Taking into account the state of the art, the cost of implementation and the nature, scope, context and purposes of processing as well as the risks of varying likelihood and severity for rights and freedoms of natural persons posed by the processing [...]"*

This quote would allow to argue that "privacy-by-construction" is a laudable goal but not a requirement. If the state of the art progresses to fully robust solutions, the goal may turn into a requirement, however.

In the meantime, the usefulness of quality control methods such as static analysis seems apparent. Software quality is arguably also the essential aspect when contemplating about compliance. Many pragmatic industry practitioners fulfil legal requirements with minimum viable products and concentrate their efforts to business concerns. Such pragmatic reasoning tends to downplay the fact that quality control is a requirement in most software projects. Although not defined in exact terms, quality is also present in the GDPR. Therefore, the question of using static analysis is similar to a question of whether to use unit testing for improving software quality. Both questions are often cultural issues in software development organisations; some tools and approaches are preferred over others, and any method that increases costs is often debated.

It is impossible to say whether static analysis would have prevented any given data breach. It is also hard to evaluate whether a potential leak caught in analysis would not have been fixed otherwise. Since the GDPR does require appropriate data protection measures, investments to data protection tools may signal that an organisation was not *negligent* even if a breach was to happen. While this ought to provide a motivation for engineers in itself, having lightweight tools well-integrated to the software development process lowers the threshold of adoption.

5 Conclusion

This paper studied static analysis for data protection. In addition to categorising state of the art tools, the paper provided a novel approach to improve software data protection qualities. The implementation of the presented tool is available to the public. The conclusions to the two research questions can be summarised as follows:

Regarding RQ1, three parallels were found between security and data protection focused static analysis tools: although (a) data protection analysis tools build upon security analysis tools, (b) threat models are less useful to categorise data protection tools. That said, the most important difference is that (c) data protection is more concerned about data semantics. These semantics imply a need for a higher *level of analysis* for static analysis tools, from local level to the data flow level and from there to the semantic flow and policy levels.

Regarding RQ2, to demonstrate the applicability of static analysis for meeting the GDPR's data protection requirements, a novel design of a rule-based source code annotation method was presented alongside a concrete light-weight implementation. Although further work is required, the tool alone is sufficient for concluding that static analysis is applicable also in the GDPR context.

References

1. Alzaidi, A., Alshehri, S., Buhari, S.M.: DroidRista: a highly precise static data flow analysis framework for android applications. Int. J. Inf. Secur. **1**, 1–14 (2019)
2. Arzt, S., et al.: Flowdroid: precise context, flow, field, object-sensitive and lifecycle-aware taint analysis for android apps. ACM SIGPLAN Not. **49**(6), 259–269 (2014)
3. Ayewah, N., Pugh, W., Hovemeyer, D., Morgenthaler, J.D., Penix, J.: Using static analysis to find bugs. IEEE Softw. **25**(5), 22–29 (2008). https://doi.org/10.1109/MS.2008.130
4. Berghel, H.: Equifax and the latest round of identity theft roulette. Computer **50**(12), 72–76 (2017). https://doi.org/10.1109/MC.2017.4451227
5. Boehm, B.W., Brown, J.R., Lipow, M.: Quantitative evaluation of software quality. In: Proceedings of the 2nd International Conference on Software Engineering, pp. 592–605. IEEE Computer Society Press (1976)
6. Calcagno, C., et al.: Moving fast with software verification. In: Havelund, K., Holzmann, G., Joshi, R. (eds.) NFM 2015. LNCS, vol. 9058, pp. 3–11. Springer, Cham (2015). https://doi.org/10.1007/978-3-319-17524-9_1
7. Calcagno, C., Distefano, D., O'Hearn, P., Yang, H.: Compositional shape analysis by means of bi-abduction. In: ACM SIGPLAN Notices, vol. 44, pp. 289–300. ACM (2009)
8. Chatzieleftheriou, G., Katsaros, P.: Test-driving static analysis tools in search of C code vulnerabilities. In: 2011 IEEE 35th Annual Computer Software and Applications Conference Workshops, pp. 96–103. IEEE (2011)
9. Chess, B., McGraw, G.: Static analysis for security. IEEE Secur. Priv. **2**(6), 76–79 (2004). https://doi.org/10.1109/MSP.2004.111
10. Chowdhury, I., Chan, B., Zulkernine, M.: Security metrics for source code structures. In: Proceedings of the Fourth International Workshop on Software Engineering for Secure Systems, pp. 57–64. ACM (2008)
11. Cousot, P., Cousot, R.: Abstract interpretation: a unified lattice model for static analysis of programs by construction or approximation of fixpoints. In: Proceedings of the 4th ACM SIGACT-SIGPLAN Symposium on Principles of Programming Languages, pp. 238–252. ACM (1977)
12. Cousot, P., Cousot, R.: Modular static program analysis. In: Horspool, R.N. (ed.) CC 2002. LNCS, vol. 2304, pp. 159–179. Springer, Heidelberg (2002). https://doi.org/10.1007/3-540-45937-5_13
13. Danezis, G., et al.: Privacy and data protection by design-from policy to engineering. arXiv preprint arXiv:1501.03726 (2015)
14. Distefano, D., Fähndrich, M., Logozzo, F., O'Hearn, P.W.: Scaling static analyses at Facebook. Commun. ACM **62**(8), 62–70 (2019)
15. Enck, W., et al.: TaintDroid: an information-flow tracking system for realtime privacy monitoring on smartphones. ACM Trans. Comput. Syst. (TOCS) **32**(2), 5 (2014)
16. Evans, D., Larochelle, D.: Improving security using extensible lightweight static analysis. IEEE Softw. **19**(1), 42–51 (2002)

17. Ferrara, P., Olivieri, L., Spoto, F.: Tailoring taint analysis to GDPR. In: Medina, M., Mitrakas, A., Rannenberg, K., Schweighofer, E., Tsouroulas, N. (eds.) APF 2018. LNCS, vol. 11079, pp. 63–76. Springer, Cham (2018). https://doi.org/10.1007/978-3-030-02547-2_4
18. Ferrara, P., Spoto, F.: Static analysis for GDPR compliance. In: ITASEC 2018 (2018). http://ceur-ws.org/Vol-2058
19. Ferrara, P., Tripp, O., Pistoia, M.: MorphDroid: fine-grained privacy verification. In: Proceedings of the 31st Annual Computer Security Applications Conference, pp. 371–380. ACM (2015)
20. Fuchs, A.P., Chaudhuri, A., Foster, J.S.: Scandroid: automated security certification of android. ACM, Technical report (2009)
21. Gerl, A., Bennani, N., Kosch, H., Brunie, L.: LPL, towards a GDPR-compliant privacy language: formal definition and usage. In: Hameurlain, A., Wagner, R. (eds.) Transactions on Large-Scale Data- and Knowledge-Centered Systems XXXVII. LNCS, vol. 10940, pp. 41–80. Springer, Heidelberg (2018). https://doi.org/10.1007/978-3-662-57932-9_2
22. Hjerppe, K., Ruohonen, J., Leppänen, V.: The general data protection regulation: requirements, architectures, and constraints. In: 2019 IEEE 27th International Requirements Engineering Conference (RE), p. (to appear). IEEE (2019). https://arxiv.org/abs/1907.07498
23. Holvitie, J., Leppänen, V.: DebtFlag: technical debt management with a development environment integrated tool. In: Proceedings of the 4th International Workshop on Managing Technical Debt, pp. 20–27. IEEE Press (2013)
24. Lin, J., Liu, B., Sadeh, N., Hong, J.I.: Modeling users' mobile app privacy preferences: restoring usability in a sea of permission settings. In: 10th Symposium On Usable Privacy and Security (SOUPS 2014), pp. 199–212. USENIX Association, Menlo Park (2014)
25. Louridas, P.: Static code analysis. IEEE Softw. **23**(4), 58–61 (2006). https://doi.org/10.1109/MS.2006.114
26. Mäkelä, S., Leppänen, V.: Client-based cohesion metrics for Java programs. Sci. Comput. Program. **74**(5–6), 355–378 (2009)
27. Myers, A.C., Liskov, B.: Protecting privacy using the decentralized label model. ACM Trans. Softw. Eng. Methodol. (TOSEM) **9**(4), 410–442 (2000)
28. Nielson, F., Nielson, H.R., Hankin, C.: Principles of program analysis. Springer, Heidelberg (2015). https://doi.org/10.1007/978-3-662-03811-6
29. Palsberg, J., Jay, C.B.: The essence of the visitor pattern. In: Proceedings of The Twenty-Second Annual International Computer Software and Applications Conference (Compsac 1998) (Cat. No.98CB 36241), pp. 9–15, August 1998
30. Pfretzschner, B., ben Othmane, L.: Identification of dependency-based attacks on Node.js. In: Proceedings of the 12th International Conference on Availability, Reliability and Security, ARES 2017, pp. 68:1–68:6. ACM (2017)
31. Ramezanifarkhani, T., Owe, O., Tokas, S.: A secrecy-preserving language for distributed and object-oriented systems. J. Log. Algebr. Methods Program. **99**, 1–25 (2018)
32. Rocha, H., Valente, M.T.: How annotations are used in Java: an empirical study. In: SEKE, pp. 426–431 (2011)
33. Romano, D., Pinzger, M.: Using source code metrics to predict change-prone Java interfaces. In: 2011 27th IEEE International Conference on Software Maintenance (ICSM), pp. 303–312. IEEE (2011)

34. Ruohonen, J.: An empirical analysis of vulnerabilities in Python packages for web applications. In: Proceedings of the 9th International Workshop on Empirical Software Engineering in Practice (IWESEP 2018), pp. 25–30. IEEE, Nara (2018)
35. Sadeghi, A., Bagheri, H., Garcia, J., Malek, S.: A taxonomy and qualitative comparison of program analysis techniques for security assessment of android software. IEEE Trans. Softw. Eng. **43**(6), 492–530 (2016)
36. Schneider, G.: Is privacy by construction possible? In: Margaria, T., Steffen, B. (eds.) ISoLA 2018. LNCS, vol. 11244, pp. 471–485. Springer, Cham (2018). https://doi.org/10.1007/978-3-030-03418-4_28
37. The European Union: Regulation (EU) 2016/679 of the European Parliament and of the Council of 27 April 2016 on the Protection of Natural Persons with Regard to the Processing of Personal Data and on the Free Movement of Such Data, and Repealing Directive 95/46/EC (General Data Protection Regulation) (2016)
38. Wang, X., Continella, A., Yang, Y., He, Y., Zhu, S.: LeakDoctor: toward automatically diagnosing privacy leaks in mobile applications. Proc. ACM Interact. Mob. Wearable Ubiquit. Technol. **3**(1), 28 (2019)
39. Yang, Z., Yang, M., Zhang, Y., Gu, G., Ning, P., Wang, X.S.: AppIntent: analyzing sensitive data transmission in android for privacy leakage detection. In: Proceedings of the 2013 ACM SIGSAC Conference on Computer and Communications Security, pp. 1043–1054. ACM (2013)
40. Zimmermann, M., Staicu, C., Tenny, C., Pradel, M.: Small world with high risks: a study of security threats in the NPM ecosystem. In: Proceedings of the 28th USENIX Security Symposium. USENIX, Santa Clara (2019)

Order of Control and Perceived Control over Personal Information

Yefim Shulman[1(✉)], Thao Ngo[2], and Joachim Meyer[1]

[1] Tel Aviv University, Tel Aviv, Israel
efimshulman@mail.tau.ac.il, jmeyer@tau.ac.il
[2] University of Duisburg-Essen, Duisburg, Germany
thao.ngo@uni-due.de

Abstract. Focusing on personal information disclosure, we apply control theory and the notion of the *Order of Control* to study people's understanding of the implications of information disclosure and their tendency to consent to disclosure. We analyzed the relevant literature and conducted a preliminary online study ($N = 220$) to explore the relationship between the Order of Control and perceived control over personal information. Our analysis of existing research suggests that the notion of the Order of Control can help us understand people's decisions regarding the control over their personal information. We discuss limitations and future directions for research regarding the application of the idea of the Order of Control to online privacy.

Keywords: Personal information disclosure · Perceived information control · Order of Control · Privacy

1 Introduction

The desire to control the important aspects of our lives is common, and it may be rooted in nature [16]. In this paper, we address one aspect that is becoming increasingly important, namely online privacy regarding the disclosure of personal information.

People's ability to decide which information concerning themselves others should have, and which they should not, is key to the major approach in privacy research that defines "privacy" as control over information (e.g., [20,23,30,32]). A major challenge in this approach is the question whether and how control can be achieved. This is a crucial point in the critique raised by alternative approaches to privacy, such as the theory of "contextual integrity", originating in Nissenbaum [22].

"Privacy as control" arises in jurisprudence and ethical philosophy, and, with the technological developments, is more relevant than ever. To overcome the issues related to controllability of information, new tools defining and operationalizing control have been developed, making their way into the legislation. For instance, the traditional triad of information security goals (confidentiality,

Published by Springer Nature Switzerland AG 2020
M. Friedewald et al. (Eds.): Privacy and Identity 2019, IFIP AICT 576, pp. 359–375, 2020.
https://doi.org/10.1007/978-3-030-42504-3_23

integrity and availability) was expanded to include three additional goals, specific to privacy protection: unlinkability, transparency and intervenability [9]. This privacy protection triad was intended to address the needs of individuals and society at large. It was originally proposed to form the paradigm of "linkage control" [10], where each of the goals describes an information control relationship between the people (users, data subjects) and data controllers (entities deciding on the purposes and means of personal data processing [7]). Particularly relevant to personal information control are transparency and intervenability. The former states the need for clear descriptions of the intended processing, so people can understand what will be done with the data. The latter goal calls for the ability to intervene in the data processing to allow erasure of data, revision of processing, etc., and to physically exercise control over what may be, can be, and is happening with the data. These two goals are relevant for both the data subjects and the data controllers/processors. Privacy protection through control over personal information is now implemented in legislation, such as the EU's General Data Protection Regulation [7], which is also applied extraterritorially[1].

One of the cornerstones of the discussions of the personal information control in law is the data subject's ability to give and revoke their consent to the data processing. A person's decision to grant consent to access or distribute certain information can be considered a control action, performed by the person to achieve a desired level of exposure of the information. These decisions are informed by the data subjects' perceptions regarding the possible implications from revealing this information. In particular, a person needs to consider whether having knowledge of some information (perhaps in combination with additional data) allows others to infer other information that the person did not explicitly reveal. Users' perceptions of control are a major determinant of their behavior. Perceived control is linked to privacy concerns (e.g., [19]), and it can elicit risk-compensation in privacy-related scenarios [1,4,14,15].

Control theory deals with the formal study of phenomena related to the control of systems and processes. In the previous article, we argued that the framework of control theory can be useful for the analysis of personal information control [24]. The closed-loop (feedback) control model can serve to describe and achieve privacy protection goals. Control actions, informed by the feedback mechanism, would represent the data processing interventions, informed through ex-ante and ex-post transparency. The informative feedback can provide ex-post transparency to the data subjects, while predictive feedback (or somewhat more peculiar, a feedforward loop) can deliver ex-ante transparency (in line with the discussions on transparency-enhancing technologies [21]). In our previous work (Shulman and Meyer [24]), we presented the conceptual control theoretic analysis of privacy, expressed through personal information disclosure. The conceptual controlled system included: (1) a person performing actions; (2) a process depending, in part, on these actions; (3) disclosure of personal information as a controlled output; and (4) an evaluation of disclosure. Our analysis emphasized the importance of the control properties of the systems and processes, and their

[1] According to the American Bar Association: http://tiny.cc/sf74gz.

effects on user behavior. One of the relevant properties is the *Order of Control*. This paper presents an analysis of phenomena related to the Order of Control, aiming to assess how it might affect the users' perceptions of control over the disclosure of personal information.

2 Order of Control for Personal Information

In this work, we rely on, and further develop the ideas first presented in Shulman and Meyer [24]. We focus on the person (i.e., the data subject), interacting with a process (e.g., using a mobile app, browsing for information, setting up a device, etc.) through some control actions (e.g., pressing buttons) and receiving some feedback on the outcomes of the process (Fig. 1). The process holder can be another individual or a legal entity who is responsible for the development and/or support of the process or may have administrative access to the process. Personal information is transferred from the person to the process during the interaction. Subsequently, this personal information may be known to the process holder, who assumes the role of the data controller/processor. Ideally, people should have control over their personal information, up to the point where they should be able to have information erased, should they so desire. The GDPR, for instance, mandates the ability of a data subject to give and revoke their consent and fulfil the deletion requests, partly or fully, regarding their personal information [7].

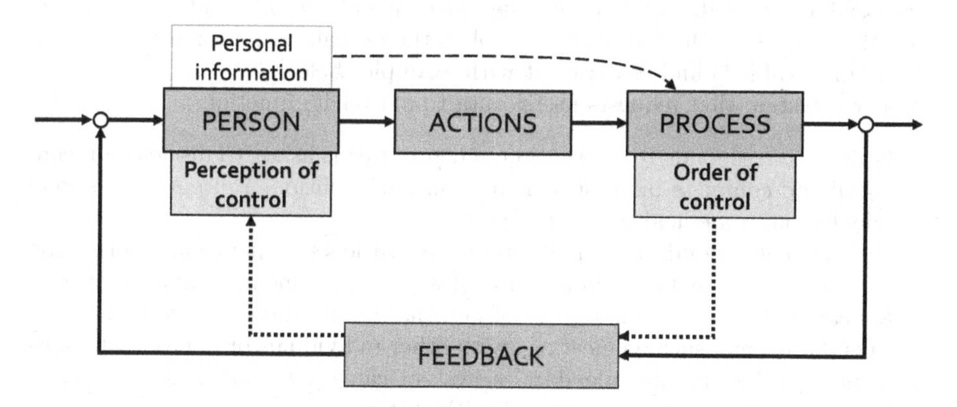

Fig. 1. Order of Control as a characteristic of the process affecting the individual's perception of control

In our model, the acts of divulging and deleting personal information constitute control of disclosure, relating to the mechanisms of granular consent, specified by the data protection legislation (i.e., the GDPR). The normative properties of the process (interacting with technology) may affect the person's attitudes and behavior. Specifically, the Order of Control may affect the individual's perception of control, whereby the information about the order itself can be part of the informative feedback the person receives. The Order of Control may serve as a measure of transparency and as a description of the intervenability for the data subjects, affecting the individuals' perception of their ability or agency over their personal information. Additionally, the Order of Control may be applied to analyze privacy protection risks from the standpoint of data controllers or regulators, but in this paper we focus on the individuals', i.e. data subjects', point of view.

In control theory, the Order of Control is usually defined as the highest order of a differential equation that describes a controlled element. Thus, the Order of Control refers to a dynamic relation between the behavior of the controlled element and the control action, and describes how the input generates the output throughout the control system. This relation may be linear or non-linear, depending on whether the output can be expressed as a linear transformation of the input. The higher the Order of Control, the more difficult it is to control a process, and the more skill is required from the human controller. Thus, the Order of Control of a system or a process defines how much learning will be required from an individual to control it efficiently.

The Order of Control is a fundamental property of the control of a process. We assume that people's own perception of their control over the process should be affected by it. Clearly, the meaning of the Order of Control in the context of privacy differs from the meaning in the context of actual dynamic systems. As we operationalize privacy as personal information disclosure, the most prominent difference is that the Order of Control is a property of the information being disclosed or revealed, together with the processing of this information. We translate the concept of the Order of Control to the domain of personal information disclosure (Table 1) and illustrate it with examples below.

For a system that requests users' data to deliver its functions:

- Order 0 corresponds to a situation, when a user is asked to disclose a meaningful and complete piece of their personal information, an item (e.g., sexual orientation, some health fact, etc.).
- Order 1 corresponds to a request for constituents of personal information (e.g., calendar events), accumulation of which over time may allow to extract a meaningful and complete piece of personal information. For example, from calendar events, such as meetings with other individuals or doctor's appointments, marked in the calendars, a system can draw conclusions about an individual's sexual orientation or health status.
- Order 2 corresponds to allowing access to data (e.g., third-party databases, location tracking, etc.) that, when aggregated, may allow to infer a meaningful and complete piece of personal information. For example, a record of locations with timestamps juxtaposed with the data and records of other residents of the area, customers of local businesses, doctors, etc., allows to infer the meetings and appointments, and reveal the sexual orientation and medical diagnoses.
- Order 3 and higher may describe, for example, access to the metadata and data of an even higher granularity and lower abstraction level.

Table 1. Order of Control for personal information disclosure.

Order of Control	Example from manual control	Example for personal information disclosure
0	Position (displacement) Control of position	Disclosing a meaningful and complete item of personal information – a fact
1	Position over time Control of speed – 1st derivative of position	Disclosing {pieces of personal information, accumulation of which over time, may allow to learn} the fact
2	Speed over time Control of acceleration – 2nd derivative of position	Divulging [data, aggregation of which over time, may allow to infer {the pieces of personal information, accumulation of which over time, may allow to learn}] the fact
3+	Jerk, jounce, "crackle and pop", etc	Divulging ⟨higher granularity data, collection of which over time, may allow to construct [data, aggregation of which over time, may allow to infer {pieces of personal information, accumulation of which over time, may allow to learn}]⟩ the fact

An important question that follows is how the Order of Control relates to the users' perceived control over their personal information disclosure in interactions with online systems. We conjecture that users' perceptions of control over the process (personal information disclosure) and over their personal information are inversely related to the Order of Control of the controlled process (personal information disclosure) and that of the information. In other words, the higher the Order of Control, the lower the perceived control over personal information.

3 Analysis of Literature on Perceptions of Control

It is possible to analyze the existing literature to gain some insights into the relation between the Order of Control and perceived control. Relying on the methodological recommendations by Webster and Watson [31], we looked for papers reporting empirical studies[2] dealing with perceptions of control and personal information disclosure (Table 2). The goal of the literature review was to investigate how the information regarding disclosure and sharing was communicated to the participants in terms of our analysis of the Order of Control of the information. Therefore, we restricted our review to the papers reporting studies, where participants were able to interact with privacy-related interfaces (i.e., privacy policies, permission managers, privacy warnings).

[2] Note, we refer to "a paper" and "a study" not as interchangeable terms. Papers study a subject, but report one or more studies. A study is a single complete item of empirical work.

Table 2. Literature survey methodology

Selection criteria	Selection values
Publication language	English
Venue type	Peer-reviewed journal or conference
Publication period	2004–2019
Publication type	1. Research article (paper) containing research results 2. Workshop article containing relevant research results
Search queries	1. 'perceived control' & 'personal information disclosure' 2. 'perceived control' & 'personal information sharing'
Search query target	Titles, abstracts, full texts (when available)
Search engines and databases	Google Scholar; ACM Digital Library, IEEE Xplore, Scopus (see *Search strategy* below)
Initial relevance	Screening of the titles and abstracts
Search & initial selection strategy	1. Google Scholar search to acquire relevant initial results 2. Identifying databases containing relevant initial results 3. Database search to check for further relevant initial results 4. Lists of references in the initially relevant items 5. Lists of mentions of the initially relevant items
Inclusion requirements	1. Empirical research on perceived control and personal information disclosure; and 2. Experiment with human subjects as participants; and 3. Privacy-related user interface (UI) as stimulus material (communicating information to the participants); and 4. The communicated information must contain details related to data collection and/or processing; and 5. Participants ought to interact with the privacy-related UIs
Exclusion bases	Upon further screening: not satisfying inclusion requirements

The search retrieved 57 initially relevant entries. Overall, after the quality assessment of the search results, we included 16 papers in our analysis (Table 3) that satisfied all the aforementioned criteria (Table 2).

The reviewed papers cover various application domains: social networks, permission management in mobile apps, location-based services, privacy policies and privacy settings. Four papers report usability studies of specific tools or designs: [5,17,26,28], while the rest investigate people's attitudes and behaviors in a wider sense: [1,2,4,8,11–13,25,27,29,33,34].

We identified the levels of the Order of Control used in the studies (see Sect. 2), based on the stimulus material and the design descriptions provided in the papers. Participants could only receive information regarding the Order of Control in their interaction with the stimulus materials and experimental platforms. All studies communicated disclosure-related information at either Order

0: [2,4,8,13,17,33], or at an Order higher than 2: [1,2,5,8,11,12,25–29,33,34], with no studies having intermediate values[3]. The studies that aimed for ecological validity presented the information similarly to its appearance in the "real world", tending to present information of the higher Order. The papers that studied particular factors in models of disclosure presented the information as simple and meaningful as possible, therefore naturally presenting information of the lower Order.

Drawing more general conclusions from the literature proved to be problematic, due to incomparable research questions and hypotheses, differences in the research designs and methods, and unreported effect sizes. Another complication arises, because "control" in different papers was used as a dependent or independent variable, moderator or mediator, or as a combination thereof. We can highlight some particular issues in the literature in more detail.

Hoadley et al. [11] studied people's reaction to Facebook changing the design of its user pages. As Facebook introduced a news feed, which was solely a design change, the Order of Control of information disclosed on the social network did not change. However, the users may have perceived that the Order of Control increased (as Facebook, effectively, started to take the sharing decisions upon itself), and that after the change in the interface design, they had less control over their personal information. This discrepancy in perception has been attributed simply to the way the information regarding disclosure was communicated to the users. The "Order-of-Control" interpretation of this observation supports our initial conjecture on the relation of the Order and perception of control (Sect. 2).

Gerlach et al. [8] studied how a social network's data-handling practices (as reflected in their privacy policy) relate to the users' willingness to disclose personal information. In a sense, the paper reports a comparison between the information of Order 0 and Orders higher than 2. The authors tested 8 permutations of simplified privacy policy designs, where one of the permutations communicated information with Order 0 and the rest – with Orders higher than 2. For our purposes, there is a clear asymmetry in the experimental design. However, the results indicate that the effect of the Order of Control might be mediated by privacy risks perceptions, lowering the tendency to disclose personal information.

Brandimarte et al. [4] looked into the relations between actual and perceived control, disclosure, and privacy concerns. Their experiments communicated 0 Order of Control to the participants. Their results indicated that when people were unable to estimate the risks of access to, and usage of their personal information (or perceived the risks to be high), and the actual control was perceived to be low, the tendency to disclose decreased. Overall, the higher the perceived control within the same Order of Control, the higher the tendency to disclose.

[3] It must be noted that in Xu (2007) [33] the stimuli were designed ad hoc and the description is limited. Therefore, it is not clear whether it corresponds to Order 0 (due to the "ad hoc" nature of the experiment) or 1 (as it likely would be in real life circumstances).

Table 3. Publications included in the analysis.

Paper	Year	Domain	Method	Control as variable	Order communication
Arcand et al. [2]	2007	E-Commerce	2 studies: experiments	Dependent variable (DV)	Privacy policies
Xu [33]	2007	Location-based service	Experiment	Multiple independent variables (IVs)	On-screen instructions
Hoadley et al. [11]	2010	Social networks	Survey	Self-reported measure	Facebook feed design change
Lipford et al. [17]	2010	Privacy settings	Usability study	n/a	Privacy policy
Wang et al. [28]	2011	Social networks	Qualitative research	n/a	Permissions interface
Brandimarte et al. [4]	2013	Social networks	3 studies: experiments	IV	Experimental description of processing/publishing
Christin et al. [5]	2013	Permissions management	Usability study	IV	Data access settings
Knijnenburg et al. [13]	2013	Websites, privacy setup	Experiment	IV	Experimental on-screen information
Keith et al. [12]	2014	Social networks, mobile gaming	Field study	IV/DV, mediator	Permission requests, privacy settings
Gerlach et al. [8]	2015	Social networks	Experiment	IV	Privacy policy
Tschersich [27]	2015	Social networks	Experiment	IV	Privacy settings
Wang et al. [29]	2015	Mobile apps, permissions	Experiment	IV	Permissions dialogue
Aïmeur et al. [1]	2016	E-Commerce (allegedly)	Experiment	IV/DV, mediator	Privacy policies
Steinfeld et al. [25]	2016	Privacy policies	Experiment	n/a	Privacy policy interaction
Zhang & Xu [34]	2016	Permissions management	Experiment	IV/DV, mediator	Experimental description of access to data
Tsai et al. [26]	2017	Permissions management	Usability study	n/a	Experimental interface

Arcand et al. [2] investigated the impact of reading privacy policies on the perceptions of control over privacy and trust towards an e-commerce website (through standardized inventories). From our standpoint, the authors tested policies with Order of Control 0 and Order 2 or higher. The policy with "opt-in" options represented the lower Order of Control. Compared to the higher Order "opt-out" policy, the former one increased perceived control over personal information, supporting our aforementioned conjecture (Sect. 2).

Studying the effects of privacy control complexity on consumer self-disclosure behavior, Keith et al. [12] argued that perceived ease-of-use determined actual control usage, noting that "...the more usable the controls, the more likely participants were to adjust the privacy control settings 'downward' (make them more restrictive) from the default setting that allowed sharing with all players". Since the Order of Control in their experiment was higher than 2, the participants might have felt less control, hence trying to restrict personal information disclosure with available actual control. Additionally, Zhang and Xu [34] showed that feedback on how permissions were used (with Order higher than 2) might decrease perceived control over personal information.

Overall, the reviewed body of research indicates that our conjecture regarding the relation between the Order of Control and perceptions of control (Sect. 2) is justified. One challenge in postulating an effect of the Order of Control would be to show that the control over personal information has levels between what we described as Order 0 and Orders higher than 2. To the best of our knowledge, no systematic study so far looked at the Order of Control in empirical privacy research.

4 Preliminary Study on the Order of Control and Perceptions of Control

We conducted a first empirical study to investigate whether there is a relationship between the Order of Control, as a property of information, and the user's perception of control over their personal information. We hypothesized that perceived user control over personal information is inversely related to the Order of Control of the personal information disclosure. In the study, we specifically examine Orders 0, 1 and 2, omitting Orders 3 and higher. This allowed us to simplify the experiment.

In the instructions, we described a fictional home sharing website "The Platform". It matches travellers with hosts who are offering a place to stay at their homes. We told the participants that to register for the service, they would need to provide some personal information. Three experimental conditions, representing Orders of Control 0, 1 and 2, differed in the descriptions of the ways in which the information people provide would be processed. In Order 0, the information would be treated as is, without further processing. In Order 1, the information was presented as something that would be accumulated at the request of other users and shared through the Platform. No suggestions were given on what the other users may have wanted to do with the user's personal information. In Order 2, the site stated explicitly that the information could be combined with external or third party databases. We predicted that the participants would consider the possibility that the data they could provide may be used to infer additional information when assessing the sensitivity of providing different types of data. The participants were randomly assigned to one of the three levels of the Order of Control.

4.1 Method

Participants. All in all, 220 participants (age range: 18–69, 39.1% female, 0.9% undisclosed gender) were recruited through the crowd-sourcing platform Amazon Mechanical Turk and completed the task. The self-reported educational levels varied from no degree to Master's and professional degrees, with the majority holding a Bachelor's degree or its equivalent. The distribution of the participants between the three Order of Control groups was comparable (75, 76 and 69 people in each group, respectively).

Study Design. We applied a between-subject design, using the *Order of Control* as an independent variable with three conditions related to the different Orders of Control described in Table 1:

1. Order of Control 0, referring to personal information disclosure as revealing of a fact;
2. Order of Control 1, referring to disclosure of the elements of personal information over time – constituting elements of a new fact;
3. Order of Control 2, referring to disclosure of the data over time, combination of which with other data constitutes the elements, from which a new fact can be inferred.

As a dependent variable, we measured perceived controllability over personal information, perceived information control, as well as online and physical privacy concerns through standardized questionnaires.

Materials. Our experiment contained several measures, some of which were based on tools we adapted from the literature.

Perceived Controllability. To measure the perceived controllability of personal information, we created a list of 15 personal information items. This list was also used in the scenarios, presented at the beginning of the experiment. After the presentation of the scenarios, the participants had to respond to the question of "How easy will it be for you to control the access and use of the personal information that you may disclose through the platform?" on a 9-point Likert-type scale including the "*I do not want to answer*" option. The 15 items included:

- First name;
- Last name;
- Email address;
- Phone number;
- A personal photo;
- National ID, residence permit or equivalent;
- Address;
- Links to social media accounts (e.g., Facebook, Twitter);
- Hobbies;
- Countries visited so far;
- Photos of the apartment;
- The amount of rent paid per month;
- Sleeping time;
- The time at home;
- Favorite places in the city.

Perceived Information Control. Four items were used to measure perceived control over personal information, adapted from Dinev et al. [6], e.g., "I think I have control over what personal information is released by the home sharing platform." The items were measured on a 7-point Likert-type scale, ranging from "1 = I do not agree at all" to "7 = I fully agree". The internal consistency of this measure in the original paper was reported as good (Cronbach's alpha = .89).

Online Privacy Concerns. Four items were used to measure online privacy concerns, adapted from Lutz et al. [18]. The participants were asked to indicate their level of concern about potential online privacy risks that could arise from personal information disclosure on the home sharing platform (the risk of identity theft, hacking, cyberstalking and publishing personal information without consent). The items were measured on a 5-point Likert-type scale, ranging from "1 = No concern at all" to "5 = Very high concern". The internal consistency of the measurement in the original paper was reported as good (Cronbach's alpha = .92).

Physical Privacy Concerns. To measure physical privacy concerns, we used four items adapted from Lutz et al. [18]. The participants had to indicate their level of concern about potential privacy risks that could arise when hosting somebody via a home sharing platform. These risks included damage to personal belongings, the guest snooping through personal belonging, the guest entering personal areas without permission, and the guest using items that should not be used. The items ranged from "1 = No concern at all" to "5 = Very high concern". The internal consistency of the measurement in the original paper was reported as good (Cronbach's alpha = .89).

4.2 Procedure

The participants were randomly assigned to one of the three between-subject conditions, corresponding to the three Orders of Control (Table 1). In each condition, a scenario about a registration on a novel (fictive) home sharing platform was presented in text. The scenarios were introduced as follows:

Imagine you are about to register on and start using a home sharing website, called "The Platform". You can use it, when travelling yourself, to stay at other users' homes, or you can host other travellers in your own home, or both. If you decide to use the service, a profile will be created for you on The Platform. Your profile will only be accessible by you and by the members of this respective home sharing community, who registered the same way as you did. Your profile information will be disclosed to the home sharing platform itself. In order to use the home sharing platform, some information has to be shared with other users of The Platform.

The list of requested information consisted of the 15 personal information items. We manipulated the Order of Control through the description of the way the information would be used (Fig. 2).

Condition: 0 Order of Control	Condition: 1 Order of Control
Therefore, your profile must disclose several types of information upon registration: • [List of requested information]	The Platform can ask you to disclose one or more types of personal information to enable a certain function or action. The requests are usually made on behalf of other users, but information is shared through The Platform itself. These requests may include the following types of information: • [List of requested information]

Condition: 2 Order of Control

The Platform can ask you to disclose one or more types of personal information to enable a certain function or action. The requests are usually made on behalf of other users, but information is shared through The Platform itself. These requests may include the following types of information:

- [List of requested information]

The Platform will be able to fetch some of your personal information via your connected accounts on your social networks, or make requests to the national and municipal registries and databases regarding your insurance coverage, property ownership or credit history, and to other third-party databases. This does not mean that The Platform will definitely access these third-party databases, but this is possible while you are a user (were active within 365 days).

Fig. 2. Manipulation of the Order of Control for each scenario

The experiment flow was as follows:

1. Greetings and informed consent form.
2. Presentation of a scenario per randomly assigned condition.
3. Items measuring the dependent variable.
4. Measurement of information control perceptions and privacy concerns.
5. General demographics questions and attention check.
6. Debriefing. The participants received monetary compensation for participation through Amazon Mechanical Turk.

4.3 Results

We performed an exploratory factor analysis (EFA) on the 15 items measuring perceived information control. The EFA (with primary axis factoring and oblimin rotation) revealed that 13 of our 15 personal information items reliably reflected 2 factors (Cronbach's alpha of .93 and .90, respectively): *"control over disclosing identity"* (first and last name, email, phone, address, personal photo, national ID, and photos of the apartment); and *"control over disclosing preferences"* (sleeping schedule, favorite places in the city, visited countries, home presence time, and hobbies). The KMO measure was not significant >.8, and Bartlett's test of sphericity was significant <.001. The links to social media and the amount of monthly rent did not load sufficiently and were excluded from further analyses.

We analyzed the relation of the *Order of Control* and *perceived information control* with several configurations of analysis of variance (ANOVA, MANOVA, mixed ANOVA, testing our hypothesis from different angles), finding no significant effect of the former at $p < .05$.

4.4 Discussion

In our preliminary study, we aimed to investigate the relationship between the Order of Control and perceived control over personal information. The results revealed no significant effect of the Order of Control on the perceptions of control over personal information. However, the exploratory factor analysis, which we performed to explore latent variables, revealed that perceived control over personal information items can be meaningfully split into two categories, which we termed: *control over disclosing identity* and *control over disclosing preferences*. This finding might be helpful for conceptualizing Order of Control in the context of privacy. For instance, we can now hypothesize that *control over disclosing identity* might differ in execution and perception from *control over disclosing preferences*, as the latter occurs as unintended information disclosure (e.g., online search, or clicking behavior). Therefore, the relation between the Order of Control and perceptions of control may differ for different types of personal information. This may be a topic for future research.

In our experiment, we found no significant effect of the Order of Control on perceived control over personal information. This null-effect could be due to one of two causes. First, the Order of Control may have little value in the context of privacy and control over personal information. Second, our study had methodological shortcomings that caused our results to be non-significant. In particular, our manipulation of the Order of Control may have been inadequate. We did not conduct a systematic manipulation check, so we do not know if the participants indeed perceived the different conditions as different Orders of Control. Future research needs to address this shortcoming.

Specifically, in future studies, qualitative pre-tests could be run before the main study to ensure the comprehension and readability of the scenarios. The manipulation check should also help to understand the potential outcomes of the experiment. The conceptual conjecture regarding the Order of Control may be split into separate hypotheses to explicitly test its underlying components. At a technical level, the difference between the Orders of Control in the experiment may not have been clear enough to obtain a significant effect.

5 General Discussion and Conclusion

In this paper, we extend our previously proposed approach to study privacy-related phenomena through the lens of the control theoretic framework [24]. We added an analysis of the effects of intrinsic properties of controlled systems on users' attitudes towards control. We focus on and conceptualize the Order of Control as one such property of information disclosure in online systems.

We looked at how the Order of Control relates to the users' perceptions of control over their personal information. We conjectured an inverse relation between the Order of Control and the users' perception of control, and described the necessary steps to investigate the proposed relation. The analysis of the relevant literature supported our idea regarding the Order of Control and highlighted the challenges arising from the delicate interplay of actual control, perceived control, risk perception, feedback presentation and personal information disclosure.

We also attempted to study this relation with an online experiment. Our analysis revealed two categories of perceived control: *control over disclosing identity*, and *control over disclosing preferences*. This distinction may have profound implications for the meaningful communication of privacy-related information via notices and indications, for individuals' privacy risk perceptions, and for the efficient application of transparency-enhancing technologies. These potential implications may warrant future research.

We found no significant effect of the Order of Control on perceptions of control. This, in our opinion, highlights some methodological challenges in the study of the effects of the Order of Control and related issues in the context of privacy. Further research is needed to establish a better understanding of the determinants of the controllability and perceptions of control over personal information.

Finally, having focused on the individuals' perceptions in this paper, we can see another promising direction for future research. The Order of Control can also provide a measure of granularity of the personal information disclosed to the online systems, and, potentially, to the data controllers/processors. It can characterize data processing breadth and depth, serving as the basis for the evaluation of the information inference. As a property of information, the Order of Control can be used to evaluate transparency and to describe how information may affect the individuals' perceptions of intervenability. Furthermore, the Order of Control may help system designers and auditors estimate transparency as a whole and quantify intervenability of data processing for a given information system interaction. Thus, the Order of Control may not only be a useful concept for studying the individuals' attitudes and behaviors, but it may also be relevant for data controllers and regulators, performing privacy risks assessment ("data protection impact assessment" under the GDPR [3]) procedures.

Funding. This research is partially funded by the EU Horizon 2020 research and innovation programme under the Marie Skłodowska-Curie grant agreement No 675730 "Privacy and Us", and by the Deutsche Forschungsgemeinschaft (DFG) under Grant No. GRK 2167, Research Training Group "User-Centered Social Media".

References

1. Aïmeur, E., Lawani, O., Dalkir, K.: When changing the look of privacy policies affects user trust: an experimental study. Comput. Hum. Behav. **58**, 368–379 (2016). https://doi.org/10.1016/j.chb.2015.11.014

2. Arcand, M., Nantel, J., Arles-Dufour, M., Vincent, A.: The impact of reading a web site's privacy statement on perceived control over privacy and perceived trust. Online Inf. Rev. **31**(5), 661–681 (2007)
3. Bieker, F., Friedewald, M., Hansen, M., Obersteller, H., Rost, M.: A process for data protection impact assessment under the European general data protection regulation. In: Schiffner, S., Serna, J., Ikonomou, D., Rannenberg, K. (eds.) APF 2016. LNCS, vol. 9857, pp. 21–37. Springer, Cham (2016). https://doi.org/10.1007/978-3-319-44760-5_2
4. Brandimarte, L., Acquisti, A., Loewenstein, G.: Misplaced confidences: privacy and the control paradox. Soc. Psychol. Pers. Sci. **4**(3), 340–347 (2013). https://doi.org/10.1177/1948550612455931
5. Christin, D., Michalak, M., Hollick, M.: Raising user awareness about privacy threats in participatory sensing applications through graphical warnings. In: Proceedings of International Conference on Advances in Mobile Computing and Multimedia, MoMM 2013, pp. 445:445–445:454. ACM, New York (2013). https://doi.org/10.1145/2536853.2536861
6. Dinev, T., Xu, H., Smith, J.H., Hart, P.: Information privacy and correlates: an empirical attempt to bridge and distinguish privacy-related concepts. Eur. J. Inf. Syst. **22**(3), 295–316 (2013). https://doi.org/10.1057/ejis.2012.23
7. EU 2016/679: Regulation (EU) 2016/679 of the European parliament and of the council of 27 April 2016 on the protection of natural persons with regard to the processing of personal data and on the free movement of such data, and repealing directive 95/46/EC (general data protection regulation). Official Journal of the European Union L119, 1–88 May 2016. http://eur-lex.europa.eu/legal-content/EN/TXT/?uri=OJ:L:2016:119:TOC
8. Gerlach, J., Widjaja, T., Buxmann, P.: Handle with care: how online social network providers' privacy policies impact users' information sharing behavior. J. Strateg. Inf. Syst. **24**(1), 33–43 (2015). https://doi.org/10.1016/j.jsis.2014.09.001
9. Hansen, M., Jensen, M., Rost, M.: Protection goals for privacy engineering. In: 2015 IEEE Security and Privacy Workshops, pp. 159–166, May 2015. https://doi.org/10.1109/SPW.2015.13
10. Hansen, M.: Top 10 mistakes in system design from a privacy perspective and privacy protection goals. In: Camenisch, J., Crispo, B., Fischer-Hübner, S., Leenes, R., Russello, G. (eds.) Privacy and Identity 2011. IAICT, vol. 375, pp. 14–31. Springer, Heidelberg (2012). https://doi.org/10.1007/978-3-642-31668-5_2
11. Hoadley, C.M., Xu, H., Lee, J.J., Rosson, M.B.: Privacy as information access and illusory control: the case of the Facebook news feed privacy outcry. Electron. Commer. Res. Appl. **9**(1), 50–60 (2010). https://doi.org/10.1016/j.elerap.2009.05.001. special Issue: Social Networks and Web 2.0
12. Keith, M., Maynes, C., Lowry, P., Babb, J.: Privacy fatigue: the effect of privacy control complexity on consumer electronic information disclosure, December 2014. https://doi.org/10.13140/2.1.3164.6403
13. Knijnenburg, B.P., Kobsa, A., Jin, H.: Counteracting the negative effect of form auto-completion on the privacy calculus. In: 34th International Conference on Information Systems, Milan, Italy 15–18 December 2013
14. Kowalewski, S., Ziefle, M., Ziegeldorf, H., Wehrle, K.: Like us on Facebook! - analyzing user preferences regarding privacy settings in Germany. Procedia Manuf. **3**, 815–822 (2015). https://doi.org/10.1016/j.promfg.2015.07.336, http://www.sciencedirect.com/science/article/pii/S2351978915003376, 6th International Conference on Applied Human Factors and Ergonomics (AHFE 2015) and the Affiliated Conferences, AHFE (2015)

15. Krol, K., Preibusch, S.: Control versus effort in privacy warnings for webforms. In: Proceedings of the 2016 ACM on Workshop on Privacy in the Electronic Society, WPES 2016, pp. 13–23. ACM, New York (2016). https://doi.org/10.1145/2994620.2994640
16. Kunkel, J., Luo, X., Capaldi, A.P.: Integrated TORC1 and PKA signaling control the temporal activation of glucose-induced gene expression in yeast. Nat. commun. **10**(1), 1–11 (2019)
17. Lipford, H.R., Watson, J., Whitney, M., Froiland, K., Reeder, R.W.: Visual vs. compact: a comparison of privacy policy interfaces. In: Proceedings of the SIGCHI Conference on Human Factors in Computing Systems, CHI 2010, pp. 1111–1114. ACM, New York (2010). https://doi.org/10.1145/1753326.1753492
18. Lutz, C., Hoffmann, C.P., Bucher, E., Fieseler, C.: The role of privacy concerns in the sharing economy. Inform. Commun. Soc. **21**(10), 1472–1492 (2018). https://doi.org/10.1080/1369118X.2017.1339726
19. Malhotra, N.K., Kim, S.S., Agarwal, J.: Internet users' information privacy concerns (IUIPC): the construct, the scale, and a causal model. Inf. Sys. Res. **15**(4), 336–355 (2004). https://doi.org/10.1287/isre.1040.0032
20. Moore, A.: Defining privacy. J. Soc. Philos. **39**(3), 411–428 (2008)
21. Murmann, P., Fischer-Hübner, S.: Tools for achieving usable ex post transparency: a survey. IEEE Access **5**, 22965–22991 (2017). https://doi.org/10.1109/ACCESS.2017.2765539
22. Nissenbaum, H.: Privacy as contextual integrity. Wash. Law Rev. **79**(1), 119–157 (2004). www.scopus.com/inward/record.uri?eid=2-s2.0-1842538795&partnerID=40&md5=377fc1b3e8b0a416836505aaea590b01
23. Parent, W.A.: Privacy, morality, and the law. Philos. Public Affairs **12**(4), 269–288 (1983). www.jstor.org/stable/2265374
24. Shulman, Y., Meyer, J.: Is privacy controllable? In: Kosta, E., Pierson, J., Slamanig, D., Fischer-Hübner, S., Krenn, S. (eds.) Privacy and Identity Management. Fairness, Accountability, and Transparency in the Age of Big Data: 13th IFIP WG 9.2, 9.6/11.7, 11.6/SIG 9.2.2 International Summer School, Vienna, Austria, 20–24 August 2018, Revised Selected Papers, pp. 222–238. IFIP Advances in Information and Communication Technology, Springer International Publishing, Cham (2019). https://doi.org/10.1007/978-3-030-16744-8_15, the authors version is available in open access via. http://arxiv.org/abs/1901.09804
25. Steinfeld, N.: "I agree to the terms and conditions": (how) do users read privacy policies online? an eye-tracking experiment. Comput. Hum. Behav. **55**, 992–1000 (2016). https://doi.org/10.1016/j.chb.2015.09.038
26. Tsai, L., et al.: Turtle guard: helping android users apply contextual privacy preferences. In: Thirteenth Symposium on Usable Privacy and Security (SOUPS 2017), pp. 145–162. USENIX Association, Santa Clara (2017). https://www.usenix.org/conference/soups2017/technical-sessions/presentation/tsai
27. Tschersich, M.: Comparing the configuration of privacy settings on social network sites based on different default options. In: 2015 48th Hawaii International Conference on System Sciences, pp. 3453–3462, January 2015. https://doi.org/10.1109/HICSS.2015.416
28. Wang, N., Xu, H., Grossklags, J.: Third-party apps on Facebook: privacy and the illusion of control. In: Proceedings of the 5th ACM Symposium on Computer Human Interaction for Management of Information Technology, CHIMIT 2011, pp. 4:1–4:10. ACM, New York (2011). https://doi.org/10.1145/2076444.2076448

29. Wang, N., Zhang, B., Liu, B., Jin, H.: Investigating effects of control and ads awareness on android users' privacy behaviors and perceptions. In: Proceedings of the 17th International Conference on Human-Computer Interaction with Mobile Devices and Services, MobileHCI 2015, pp. 373–382. ACM, New York (2015). https://doi.org/10.1145/2785830.2785845
30. Warren, S.D., Brandeis, L.D.: The Right to Privacy, pp. 193–220, Harvard Law Review (1890)
31. Webster, J., Watson, R.T.: Analyzing the past to prepare for the future: writing a literature review. MIS Q. **26**(2), 5–7 (2002). www.jstor.org/stable/4132319
32. Westin, A.: Privacy and freedom. Atheneum, New York (1967–1970)
33. Xu, H.: The effects of self-construal and perceived control on privacy concerns. In: ICIS 2007 Proceedings - Twenty-Eighth International Conference on Information Systems (2007)
34. Zhang, B., Xu, H.: Privacy nudges for mobile applications: effects on the creepiness emotion and privacy attitudes. In: Proceedings of the 19th ACM Conference on Computer-Supported Cooperative Work and Social Computing. CSCW 2016, pp. 1676–1690. ACM, New York (2016). https://doi.org/10.1145/2818048.2820073

Aggregating Corporate Information Security Maturity Levels of Different Assets

Michael Schmid[1,2](✉) and Sebastian Pape[1](✉)

[1] Chair of Mobile Business & Multilateral Security, Goethe University Frankfurt, Frankfurt, Germany
{michael.schmid,sebastian.pape}@m-chair.de
[2] Hubert Burda Media Holding KG, Munich, Germany

Abstract. General Data Protection Regulation (GDPR) has not only a great influence on data protection but also on the area of information security especially with regard to Article 32. This article emphasizes the importance of having a process to regularly test, assess and evaluate the security. The measuring of information security however, involves overcoming many obstacles. The quality of information security can only be measured indirectly using metrics and Key Performance Indicators (KPIs), as no gold standard exist. Many studies are concerned with using metrics to get as close as possible to the status of information security but only a few focus on the comparison of information security metrics. This paper deals with aggregation types of corporate information security maturity levels from different assets in order to find out how the different aggregation functions effect the results and which conclusions can be drawn from them. The required model has already been developed by the authors and tested for applicability by means of case studies. In order to investigate the significance of the ranking from the comparison of the aggregation in more detail, this paper will try to work out in which way a maturity control should be aggregated in order to serve the company best in improving its security. This result will be helpful for all companies aiming to regularly assess and improve their security as requested by the GDPR. To verify the significance of the results with different sets, real information security data from a large international media and technology company has been used.

Keywords: Information security · Information security management · ISO 27001 · Aggregation functions · Information security controls · Capability maturity model · Security maturity model · Security metrics framework

1 Introduction

Approximately 18 months ago the General Data Protection Regulation (GDPR) containing requirements regarding the processing of personal data of individuals

Published by Springer Nature Switzerland AG 2020
M. Friedewald et al. (Eds.): Privacy and Identity 2019, IFIP AICT 576, pp. 376–392, 2020.
https://doi.org/10.1007/978-3-030-42504-3_24

became operative. The GDPR states that organizations must adopt appropriate policies, procedures and processes to protect the personal data they hold. Article 32 of the GDPR specifically requires organizations to ensure confidentiality, integrity, availability and resilience (core principles of the information security) of processing systems and services, and to implement a process for regularly testing, assessing and evaluating the effectiveness (e.g. with KPIs) of technical and organizational measures for ensuring secure processing [27]. Thus, in addition to presenting a state of the art security level, this article emphasizes the importance of a process for regularly testing, assessing and evaluating the security. However, it does not provide detailed guidance on how to achieve these goals.

It is difficult to judge whether the security level is sufficient from a management perspective. Managers often act according to the maxim 'minimal effort maximum success', since the budget is usually limited. Of course, this also applies to the area of information security and varies depending on the industry and the self-perception of IT security within it. This is justifiable from an economic point of view, but it has an influence on how information security is dealt with in the company. In this situation, it is important to create transparency regarding the state of information security, within an organization to determine how good the process is, as well as in comparison to other companies operating in the same environment. This transparency can be used to demonstrate/ensure that (information) security does not suffer from budget constraints.

An established way to monitor and steer the information security is the implementation of an information security management system (ISMS). With the most popular standard in this field, ISO/IEC 27001 [14], it is possible to manage the information security in a company through the ISO-controls. An effective ISMS that conforms to ISO/IEC 27001 meets all requirements of GDPR's article 32.

The information security status of an environment like a company is a very individual observation [1]. To estimate the actual status of information security normally metrics or key performance indicators (KPI) are taken into account [21]. The information gathering of these KPIs is usually done through different technical or organizational metrics of a company. Using KPI/Metric/-Maturity for the status of information security is only an indicator of improvement or deterioration since there is unfortunately no gold standard for this [4]. It would be very complex and expensive to first collect or generate these KPIs for this evaluation. It is important therefore, to work with the data/metrics already available and no need for further data collection. In this context, it should not go unmentioned that another standard exists in this environment, the ISO/IEC 27701 [15]. This standard deals with how to establish and run a Privacy Information Management System (PIMS) that adds Personally Identifiable Information (PII) security protection to an existing ISMS. In order to assess the status of information security as well as the quality of the process, mostly a maturity model is used. A common method for the assessment of the maturity is the COBIT control maturity model from the ISACA framework [13]. With the help of this model it is possible to assess the goodness of the ISO-controls on a 0 to 5 scale. The assessment supports the improvement of the organization's security and delivers the management perspective in the fulfillment of regulatory requirements.

With the maturity level, the manager has a relatively good overall view of the status of information security. However, this is usually a very aggregated view of the status, as a company will operate different types of IT systems/applications to support its business process. The information assets worth protecting (e.g. customer data, trade secrets, source code, etc.) are not only processed or stored on one IT system, but on several. As a consequence, the maturity level may differ between systems. Therefore, many companies not only collect a maturity level for the whole company, but also a maturity level per system for each control [11]. An ISO control such as A.12.6.1 (Vulnerability Management) will only be able to reflect a combined value from several IT systems/applications. That's why, different values exist for different assets per ISO control (see Fig. 1).

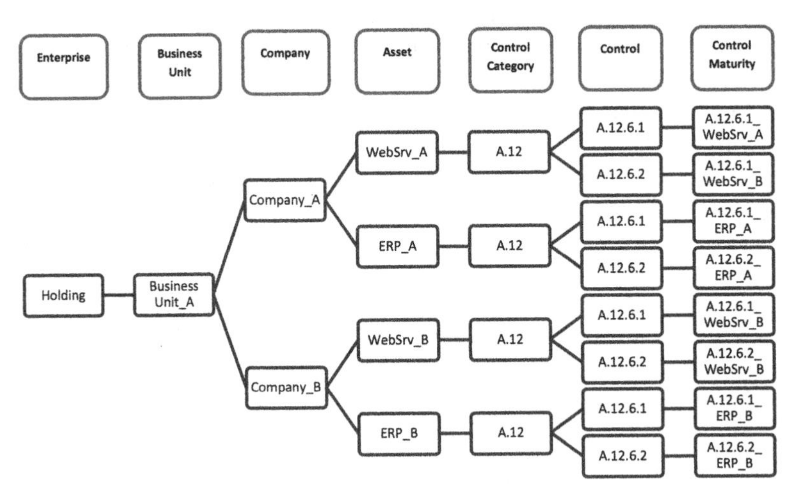

Fig. 1. Exemplary holding structure with different assets and control maturity for ISO-controls

In order to derive a KPI from the assets' control maturity level or use them as input for existing approaches [24,25], the questions arises how they can be meaningfully aggregated.

RQ1: How can maturity levels for one control be meaningfully aggregated across different assets?

Different aggregation types can not only influence the outcome of the approach, but also influence the managers which security controls should be improved.

RQ2: How would a manager's optimization strategy depend on the different aggregation methods?

And finally, it's equally important to consider the aggregation's influence on the final result of the algorithm.

RQ3: How much does the outcome of a holistic approach actually change depending on different aggregation types?

To examine this research question, we first discuss different types of aggregation for maturity levels. In the next step, for each of the aggregations we derive possible security managers' optimization strategies in order to establish which control to improve next. For a reality check, we examine asset's maturity levels from real company data to check if our assumptions are realistic. As a final step, we also use real companies' maturity levels to examine how much the outcome of [24] would be changed by applying a different aggregation.

The remainder of this work is structured as follows: In Sect. 2 we give a brief overview of related work. Section 3 describes our methodology how we developed our approach for each research question shown in Sect. 4. Our results are shown in Sect. 5 and discussed in Sect. 6, respectively Sect. 7.

2 Background and Related Work

In addition to the differences in the assessment of information security, all assessment procedures also have in common that the ratings of the maturity level and the weighting of weights are not allocated to a common overall value in the sense of an 'information security score'. It is, therefore up to the evaluator to carry out the respective evaluation, as he or she is forced to choose between these two quantitative aspects of the evaluation, e.g. the ratings on the one hand and the weighting on the other [17]. Savola [23] discussed a broader approach to finding a metrics which can be used in the field of different security disciplines like management and engineering practices. In contrast to this, the works of Böhme [8] and Anderson [4] deal more with the economic impact of investments in information security. There are also other models that deal with the measurement of information security using maturity levels e.g. the Information security maturity model (ISMM) [22] and the Open Information Security Maturity Model (O-ISM3) [22]. ISMM is intended as a tool to evaluate the ability of organizations to meet the objectives of security and O-ISM3 aims to ensure that security processes operate at a level consistent with business requirements. However, both models refer more to the process level than to the asset level. The focus of this work is to compare the different aggregation types of maturity within an industry. This could later lead to a monetary assessment of information security or maturity.

2.1 Aggregation Types

Unfortunately, the precise process of how to aggregate maturity levels is neither well documented nor comprehensively studied or understood (from a psychological perspective), so most of this labor is done by rule-of-thumb [26]. As mentioned, our approach varies between four aggregation types - namely the minimum, maximum, average and median - to compare their different potential impacts on decision making. Regarding the two measures of central tendency (average, median), strengths and weaknesses have been discussed in scientific literature. Averages are strongly influenced by extreme values. In our context, this could lead to an over- or underestimate of control maturity. In contrast, the median is not skewed by extreme values, consequently running the risk of overestimating control maturity [10]. The opposite can be the case when there are multiple non-values (e.g. zeros) in a data sample, as laid out by Anderson et al. [5]. The relative position of average and median differs in skewed distributions. A distribution skewed to the left will lead to a smaller median compared to the average, while a right-skewed distribution reverses the relation [18]. Overall, it makes sense to include both measures of central tendencies in our analysis to compensate for weaknesses and bias. The minimum and maximum further alleviate potential misrepresentations of control maturity, as they provide the numerical range of scores and expose potential outliers [7]. Logically, both measures are most sensitive to outliers in a data set but are nevertheless useful in our analysis when used in combination with the measures of central tendency.

2.2 Aggregation of Security Metrics

Although the domain of security metrics has been covered by a number of authors [3], only limited work on the area of metrics aggregation has been carried out. Ramos et al. [20] provided a detailed survey on models for quantifying networks resilience to attacks. The authors used stochastic techniques and attack graphs to map the possible routes an attacker could take to compromise a system. Abraham et al. [2] discussed the challenges faced by practitioners in the field of security measurements and highlighted the need to develop a mechanism for quantifying the overall security of all the systems on the network. The authors proposed a predictive framework that uses stochastic techniques based on attack graphs and incorporated temporal factors relating to the vulnerabilities such as availability of patch and exploits predicting the future state of the system. Cheng et al. [9] proposed a model for aggregating security metrics using Common Vulnerability Scoring System (CVSS) base metrics to estimate the exploitability of the vulnerabilities. Homer et al. [12] and Beck et al. [6] proposed a mathematical security model for aggregating vulnerabilities in risks in enterprise networks based on attack graphs. An aggregated numeric value was assigned to show the likelihood of a vulnerability being exploited by an attacker.

3 Research Methodology

The general aim of our approach is to determine which effect the different aggregation types of the maturity control of assets have on the information security of the companies. In order to do this it is important to create transparency around the state of information security. The method should take into account the different requirements of the different research questions set out in Sect. 1.

We derive the different aggregation methods in the next subsection for our approach, then determine the proper algorithm and finally describe the data collection of our approach.

3.1 Different Aggregation Functions

First, we examine which functions are suitable to verify the approach described above. As shown in Sects. 2.1 and 2.2 with the different aggregation functions e.g. average, median, minimum and maximum it is possible to form a single summary value from a group of data. The challenge now is to find the right aggregation functions to support the approach provided. These aggregation functions have in common that they can represent the impact of decisions by information security managers, each type in its own way. The hypotheses provide an outlook how information security managers might behave in terms of aggregation.

3.2 Data Collection

It would be very complex and expensive to first collect or generate these KPIs for this evaluation. It is important to use data/metrics already available (e.g. information security maturity level). To test the above approach it is necessary to set up the model and verify it with real data. We need a maturity assessment of the ISO/IEC controls and to weight and aggregate them according to the specific industry. We focused on the eCommerce industry for the following reasons:

- Available data from a large range of companies
- Excellent data quality and validity
- High actuality of the existing data
- Very good know-how available in the expert assessment of the industry

We collected data from Hubert Burda Media (HBM), an international media and technology company (over 12,500 employees, more than 2.5 billion annual sales, represented in over 20 countries). This group is divided into several business units that serve various business areas (including print magazines, online portals, eCommerce etc.). The business units consist of over 250 individual companies with about 30 of them being in the eCommerce industry. Each subsidiary operates independently of the parent corporation. There is a profit center structure, so the group acts as a company for entrepreneurs and the managing directors have the freedom to invest money in information security and to choose the appropriate level of security. We will briefly describe how this data is collected before going into more detail on the data used for the comparison. Each individual company in the group operates its own Information Security Management System (ISMS) in accordance with ISO/IEC 27001, which is managed by an Information Security Officer (ISO) on site and managed by a central unit in the holding company. As part of the evaluation of the ISMS, the maturity level of the respective ISO 27001 controls is ascertained - very granularly at the asset level (application, web-server, CRM etc.). The maturity level is collected/updated regularly once a year as part of a follow-up procedure.

3.3 Algorithm Method Selection

Taking all requirements of the method into account, a previously developed approach from Schmid and Pape [24] is applicable. The primary objective of this approach was to show how to use the analytic hierarchy process (AHP) to compare the information security controls of a level of maturity within an industry in order to rank different companies. The AHP is one of the most commonly used Multiple Criteria Decision Methods (MCDM), combining subjective and personal preferences in the information security assessment process [19]. It allows a structured comparison of the information security maturity level of companies with respect to an industry [26] and to obtain a ranking [16]. This allows the definition of a separate weighting of information security metrics for each industry with respect to their specifics while using a standardized approach based on the maturity levels of the ISO/IEC 27001 controls.

To achieve the aim of this paper it is necessary to calculate the control maturity of the assets with different aggregation types such as: minimum, maximum, average or median. This shows how strong the characteristics of the individual aggregation types are in comparison to the real data. Out of this, the first indicators can then be derived to clarify which effect the aggregation types have on the information security for individual companies. The following chapter describes the implementation of the approach for each of the 3 research questions.

4 Discussion of Different Aggregations

As outlined in the previous chapter the different aggregation functions have a very likely a different outcome when it comes down comparing them with each other. Among other things, this chapter will describe the different characteristics of the aggregation functions as well as the effects of the various IT assets of a company and how they affect the results. A vivid example with real world data illustrates how the various aggregations affect the final result and ultimately the behaviour of those responsible for information security.

4.1 General Aggregation Functions

The great advantage of the aggregation functions average, median, minimum and maximum is that by aggregating (key) figures differences can be identified in the results and thus comparisons can be made. These could be a strength or weakness per each aggregation type. In contrast to this, there is no difference in the comparison of the results for the aggregation functions sum, range and count, for example. A further advantage of the four aggregation functions mentioned above is the adaptability of these types to a different number of values. They work nicely even if each company has a different number of assets considered. This makes it possible to derive different scenarios for the comparison.

4.2 Derived Optimization Strategies

If the results of the different aggregation functions are compared with each other, different optimization strategies can be derived in the end. This is particularly

Table 1. Maturity levels of different collective assets for the ISO-control A.12.6.1 from five companies

Asset	Company1	Company2	Company3	Company4	Company5
1	4	0	3	3	4
2	4		2	2	4
3	4		2	3	
4	1			1	
5	0				

Table 2. Maturity level results from different aggregation functions

Aggregation	Company1	Company2	Company3	Company4	Company5
average	2.6	0	2.3	2.25	4
median	4	0	2	2.5	4
minimum	0	0	2	1	4
maximum	4	0	3	3	4

important for those who are responsible for information security. Due to the different aggregations, it is possible that different optimization possibilities can be shown in the evaluation of information security. The information security manager can then decide which optimization strategy/aggregation function brings him the most benefit. If we take a closer look at the 4 aggregation functions mentioned above and examine them for the possible outcome, we obtain the following hypotheses:

- minimum → improve only the worst value (weakest chain, can make sense),
- maximum → improve only the best value (is this desirable?),
- average → improve any value (probably the easiest ones first) and
- median → may lead to a really two-fold security level with $\frac{n-1}{2}$ insecure services and $\frac{n+1}{2}$ secure services.

As next step we validate these hypotheses using an example with real world data.

4.3 Example with Real World Data

In order to compare the results of the different aggregation functions we need real data. Section 3.2 describes how these real data, in this case the COBIT maturity, are collected. For a concrete example we use the maturity level for a specific ISO-Control (here A.12.6.1 'Management of Technical Vulnerabilities') because this control focuses on an IT asset. As an example, we use data from five companies and their various IT assets (see Table 1).

Based on this data, the calculations of the four different aggregation functions are now performed (see Table 2) for the five companies. The colored cells highlight the aggregation functions and the maturity levels used. These exemplary calculations are based on the maturity levels of companies with different IT assets. A company uses many different IT assets to support its core and support processes. The next chapter examines these different types of IT assets in more detail.

4.4 More Complex Aggregations

In order to steer manager's optimization strategy one needs to integrate weightings for the different assets. This leads to the problem that many approaches, e.g. AHP [24] only work with a fixed number of assets. Considering only a fixed set of assets for each domain would narrow the defined scope, thus it should be possible to still evaluate a different number of assets. Conclusion: Define most important assets and their weighting and build an asset class for all remaining assets. This way, at least the impact of the manager's optimization strategies is more limited and only usable among the assets within the 'special class'. Arising Question: How to derive the priorities for all the classes?

When considering the core business processes for an eCommerce company, the web presence, a merchandise management system and a customer management system are normally expected. For this stage, we examined the prevailing situation of the IT assets used by 25 eCommerce companies from HBM and evaluated them. Almost all eCommerce companies had a web sever (24), a database server (24), an ERP system (22) and a CRM system (20). Further IT assets, which did not have such a high frequency were mail servers (14), file servers (14), dev servers (12), git (9), ftp servers (7), etc. This also coincides with the assumption resulting from the core business processes. Resulting from this the core IT assets of an eCommerce company, a web sever, a database server, an ERP system and a CRM system were selected.

Only considering these core IT assets would not reflect the overall picture of an eCommerce company. In order to have a comprehensive picture we also need the assets that are used in the IT department (e.g. file server, dev server, ftp server etc.). We have combined these IT assets into one collective asset for the comprehensive picture. In a further step, this collective asset, or better the maturity level, is calculated or evaluated using various aggregation types (minimum, maximum, average, median). In combination with the 4 core assets, aggregated values of the collective assets are included in the calculation as 5th assets (with 20%). This can provide the first insights as to whether a certain aggregation method might influence the units or sub-companies decision, hence which control should be improved next.

4.5 Priorization of Asset Classes

The core IT assets are equally important (e.g. 25% for each) at the moment. An interesting question would be e.g. how much more important is the web server of an eCommerce company compared to the ERP system? It would be necessary to add an additional layer of prioritization in order to differentiate between the differing control requirements. In order to implement this we could use the CIA triad model which encompasses a triangle of tension between the three principles Confidentiality, Integrity and Availability. When applied to our use case, the principles of importance vary between control objectives and is represented by a score for the CIA principles according to their importance for these control objectives. This would provide for an extension of the approach by the CIA values of the individual assets. In order to do this, we need the CIA evaluation per IT asset. The information (e.g. customer data, contracts etc.) is stored or processed on an IT asset. It allows conclusions to be drawn as to how this asset should be treated in terms of confidentiality, integrity and availability. This means that there is at least one information asset per asset, but usually several information assets per asset, which are evaluated according to the CIA criteria with a 3-step classification (normal, advanced and high). A web server will, for example, process or even store information assets such as customer data, bank details, etc. If the information values 'customer data' and 'bank details' for a web server are uniformly evaluated for confidentiality, integrity and availability according to a given system, this can be set in relation to an ERP system with the information values 'purchasing conditions' and 'master data'. A further step was needed to convert our CIA data to pairwise comparisons on our AHP score, as depicted in Table 3a. We define a factor of equal importance regarding the CIA triad of all four core assets as a proportion percentage of 25% each. Consequently, we can conduct pairwise comparisons related to the proportion gaps in our data, which are then normalized based on the AHP preference score i.e. equal importance (AHP score: 1) is expressed by tiny differences in proportion to percentage of smaller than 2.77%, while the highest order of relative importance (AHP score: 9) means a difference of 25% in proportion to percentage (see Table 3b).

Table 3. Combined GAP of core assets and AHP Score

AHP Score	Verbal description
9 8	Extreme preference
7 6	Very strong preference
5 4	Strong preference
3 2	Moderate preference
1	Equal preference

(a) Fundamental AHP Score

AHP Score	Proportional CIA differences	Verbal description
9 8	22.22 - 25.00 19.45 - 22.21	Extreme preference
7 6	16.67 - 19.44 13.89 - 16.66	Very strong preference
5 4	11.12 - 13.88 08.34 - 11.11	Strong preference
3 2	05.56 - 08.33 02.78 - 05.55	Moderate preference
1	00.00 - 02.77	Equal preference

(b) AHP Score vs. GAP of the CIA differences

5 Results of the Holistic Approach Considering Different Aggregation Types

The aim of this paper is to find out which effects the different aggregation functions have on the results and which conclusions can be drawn from them. The different aggregation functions can not only influence the outcome of the approach, but also influence the manager's decision as to the order in which control's maturity levels should be increased. They can influence the manager's optimization strategy depending on the different aggregation functions. At present, the maturity levels have not yet been examined with a view to optimization.

Table 4. Comparison of different aggregation types from 5 companies only for control A.12.6.1

Aggregation/proportion	Company1	Company2	Company3	Company4	Company5
Average	15.4%	7.7%	30.8%	30.8%	15.4%
Median	12.6%	12.6%	27.4%	34.9%	32.0%
Minimum	10.0%	10.0%	40.0%	20.0%	20.0%
Maximum	22.2%	11.1%	22.2%	22.2%	22.2%

Table 5. Comparison (proportion) of different aggregation types from 5 companies for control category A.12

Aggregation	Company1	Company2	Company3	Company4	Company5
Average	1.7% *(17.9%)*	1.2% *(12.6%)*	2.3% *(24.2%)*	2.1% *(22.1%)*	2.2% *(23.1%)*
Median	1.6% *(16.8%)*	1.7% *(17.9%)*	2.4% *(25.3%)*	1.9% *(20.0%)*	1.9% *(20.0%)*
Minimum	1.4% *(14.7%)*	1.2% *(12.6%)*	2.8% *(29.5%)*	2.1% *(22.1%)*	2.0% *(21.0%)*
Maximum	1.8% *(18.9%)*	1.3% *(13.7%)*	1.7% *(17.9%)*	1.6% *(16.8%)*	3.1% *(32.6%)*

5.1 Results of Aggregated Maturity Levels

The AHP was used to compare the maturity levels in order to work out how a maturity control should be determined to best serve the company in improving its security with reference to the first research question [24]. Table 4 shows a comparison of results with different aggregation types from five companies only for control A.12.6.1 'Management of Technical Vulnerabilities'. Because this control is asset-based, this value is composed of different IT assets that were calculated with each of the 4 different aggregation types.

As expected, Company 2 is very weakly developed if the raw data in Table 1 is considered. Company 1 is also quite clearly recognizable with regard to the minimum and maximum. Company 3 has the highest proportion concerning the minimum (40.0%). The results show that a detailed look at Company 5 would be worthwhile, as the largest fluctuations between average and median (15.4%–32.0%) can be observed here.

If we now abstract this comparison to a higher level, e.g. no longer to the control level but to control category level, the results should no longer fluctuate greatly. In the case of control categories, we are concentrating only on the most important ones for the eCommerce industry. The weighting of the respective control categories can be seen from the results of the AHP [24]. 'A.14' (System Acquisition, Development and Maintenance) is the most important for the eCommerce industry with 16.5%, followed by 'A.17' (Information Security Aspects of Business Continuity Management) with 14.7% and then 'A.12' (Operations security) with 9.5%. Table 5 shows how the individual eCommerce companies weighting is compared with each other and the four different aggregation types for 'A.12' Operations security are compared in detail.

Table 6. Comparison of different aggregation types from 5 companies for the complete ISO/IEC 27001

Aggregation/proportion	Company1	Company2	Company3	Company4	Company5
Average	16.7% *(4.)*	15.4% *(5.)*	19.8% *(1.)*	18.3% *(3.)*	19.5% *(2.)*
Median	16.7% *(4.)*	16.3% *(5.)*	19.8% *(1.)*	18.8% *(2.)*	18.1% *(3.)*
Minimum	16.6% *(4.)*	14.6% *(5.)*	21.3% *(1.)*	18.7% *(2.)*	18.5% *(3.)*
Maximum	17.5% *(2.)*	15.6% *(5.)*	16.1% *(4.)*	16.2% *(3.)*	24.2% *(1.)*

The rows total up to 9.5% because it is the ratio of 'A.12' weighting in contrast to the overall control categories. The distribution of values within an aggregation type per company is specified in brackets. The differences are marginal but a closer inspection more pronounced differences can be observed at the control level and therefore tendencies are recognizable. Company 3 has again the highest proportion concerning the minimum (29.5%)

The last comparison in this environment is the application of the four different aggregation types to the complete controls of Annex A of ISO/IEC 27001. This is ultimately the highest expected level of aggregation of this approach. It is to be expected that the results will no longer differ so much from each other. Table 6 shows the results of the comparison.

The rows total up only to 89.9% because 11.1% is a 'measure of the error due to inconsistency' which is provided by the AHP. The ranking within all companies is specified in brackets. Concerning the outcome of the comparison, Company 5 stands out with a high value for maximum aggregation (24.2%) and Company 1 looks very stable concerning the different aggregation types. Generally, the minimum does not fluctuate as much as the maximum. Company 1 to 3 have no high fluctuation in common and concerning Company 3 there is not a lot of variance can be observed.

5.2 Results of Priorization the Asset

The descriptive statistic of HBMs information asset presence is used to begin with the set of four core assets, namely web server (24), database server (24), ERP system (22) and CRM system (20). Besides, computing our input scores as well as defining our priorities for sub criteria level requires the processing of the CIA inputs. The summarizing statistic is presented in Table 7 below.

All CIA scores are summed up for each asset and divided by the total number (see Table 8). The lowest sum resulted from the CRM asset with 100, and is hence our base value.

Concerning the priorization of asset classes Table 9 shows a pairwise comparison of the core assets from one eCommerce company. The deviation is then transformed into the AHP scores with the help of the intervals from the GAP of core assets (see Table 3b). It is clear that the biggest difference lies between the web server and the CRM system (11.7%) and the smallest difference between

Table 7. CIA of information assets from different IT assets of one company

Company	Information asset for	Confidentiality	Integrity	Availability	Sum of CIA
Company_1	Web-Server	2	2	3	**7**
	Web server	3	3	3	**9**
	Web server	3	3	2	**8**
	Web server	2	3	2	**7**
	Web server	3	3	2	**8**
	Database server	2	2	2	**6**
	Database server	2	2	2	**6**
	ERP system	2	2	2	**6**
	ERP system	2	2	2	**6**
	ERP system	2	2	2	**6**
	ERP system	2	2	2	**6**
	CRM system	2	2	2	**6**
	CRM system	2	2	2	**6**
	CRM system	1	2	2	**5**
	CRM system	1	2	2	**5**
Company_2	...	...	...	...	...

Table 8. Distribution of assets

Asset	CIA sum	Distribution
WEB	156	32.5%
ERP	104	25.0%
DB	120	21.7%
CRM	100	20.8%

the CRM system and the database server (0.7%). With the help of this score it is possible to weight the core assets based on their CIA assessment and process them with the AHP.

6 Discussion

Based on these results, we discuss the main findings as follows. The results show that it is possible to elaborate differences in the assessment and comparison of IT assets with the help of different aggregation types. The main goal of this paper, to assist managers in how they can improve their information security by comparing different aggregated information security maturity levels on asset level has shown several outcomes. The results show that a certain type of aggregation affects a company when trying to improve its maturity levels (see Table 4). Company 1 and 2 would improve first the collective assets with a low control maturity if a minimum aggregation is used. If the aggregation function maximum is used Company 3 would try to improve one collective asset in order to maximize only one control maturity (see Table 5). Concerning the big picture in Table 6 the ranking of the companies differs only for Company 1 and 3. Company 1 has already very high control maturities, so it is not as easy for them to improve. Company 3 almost a very homogenous control maturity thats why the would probably improve only one collective assets if the maximum aggregation is chosen. The other companies are more or less stable concerning the ranking, e.g. Company 2 does not changes at all.

Table 9. AHP Comparison with core assets

Sub criteria A	Sub criteria B	A/B	Deviation	Score
WEB	ERP	A	+7.25%	3
WEB	DB	A	+10.8%	4
WEB	CRM	A	+11.7%	5
ERP	DB	A	+2.3%	1
ERP	CRM	A	+4.1%	2
DB	CRM	A	+0.7%	1

With the help of the CIA prioritization is possible to first weight and then aggregated the different IT systems and applications with each other (see Table 9). The results show hat for an eCommerce company it is obvious that the web server is more important than the ERP-System in supporting the business processes.

6.1 Limitations

Maturity levels are not assessed automatically but by each of the individual companies' information security officer (ISO). Therefore, there may be discrepancies in the way the maturity levels are understood and assessed. This is clearly a limitation of any approach based on security maturity levels, but it might limit the informative value of the collected maturity levels. Moreover, the maturity levels are reported to the management and they result in a key performance indicator (KPI) for security for that specific unit. Thus, it can be assumed that each ISO has an interest in having a good evaluation. Therefore, ISOs might be tempted to assess the maturity levels more optimistically or to limit the scope of the information security management system in order to achieve better evaluations more easily. A common understanding of the different maturity levels is already established by guidelines and manuals provided to the ISOs (of HBM). This could be expanded further in order to reach a better understanding for the assessment of control maturity levels. Furthermore, deviations can be addressed if the companies are (externally) audited from time to time to double check the maturity levels.

7 Conclusion and Future Work

The discussion of how an overall score for a maturity level for security controls across different assets shows that the aggregation is an important tool needed to distinguish how the information security managers would optimize information security. In practice it makes a big difference which aggregation is used because it could lead to optimizing only the control maturity levels which are easily reachable. The defined priorization is necessary in order not to depend too much on the different kind of optimization strategies of the managers. This way, it can be steered more directly where the security should be enhanced and it probably also reflects better the current security level of companies. This approach is a helpful result for all companies aiming to regularly assess and improve their security as requested by the GDPR in order to ensure the confidentiality, integrity, availability and resilience of IT assets and evaluating the effectiveness of the technical and organizational measures for ensuring the security process.

As future work the outcome with other approaches could be compared to sen how the aggregation has changes the influence. Additionally, one might need to find other ways to prioritize the different controls, since in this case it was easy since it's one of the AHPs natural properties. Further investigations have to been carried out in order to clarify the validity of the control maturity levels because of the containing bias. Additional work could also be carried out to check validity of scope in order to measure any changes in the results after the metrics have been introduced.

References

1. Abbas Ahmed, R.K.: Security metrics and the risks: an overview. Int. J. Comput. Trends Technol. **41**(2), 106–112 (2016)

2. Abraham, S., Nair, S.: A predictive framework for cyber security analytics using attack graphs. Int. J. Comput. Netw. Commun. **7**(1), 1–17 (2015)
3. Ahmed, Y., Naqvi, S., Josephs, M.: Aggregation of security metrics for decision making: a reference architecture. In: ACM International Conference Proceeding Series (2018)
4. Anderson, R., et al.: Measuring the cost of cybercrime. In: Böhme, R. (ed.) The Economics of Information Security and Privacy, pp. 265–300. Springer, Heidelberg (2013). https://doi.org/10.1007/978-3-642-39498-0_12
5. Anderson, R., et al.: Measuring the changing cost of cybercrime our framework for analysing the costs of cybercrime. In: Workshop on the Economics of Information Security (WEIS), pp. 1–32 (2019)
6. Beck, A., Rass, S.: Using neural networks to aid CVSS risk aggregation - an empirically validated approach. J. Innov. Digit. Ecosyst. **3**(2), 148–154 (2016)
7. Bland, M.: Estimating mean and standard deviation from the sample size, three quartiles, minimum, and maximum. Int. J. Stat. Med. Res. **4**(1), 57–64 (2015)
8. Böhme, R.: Security metrics and security investment models. In: Echizen, I., Kunihiro, N., Sasaki, R. (eds.) IWSEC 2010. LNCS, vol. 6434, pp. 10–24. Springer, Heidelberg (2010). https://doi.org/10.1007/978-3-642-16825-3_2
9. Cheng, P., Wang, L., Jajodia, S., Singhal, A.: Aggregating CVSS base scores for semantics-rich network security metrics. In: Proceedings of the IEEE Symposium on Reliable Distributed Systems (2012)
10. Doane, D.P., Seward, L.E.: Applied Statistics in Business and Economics. McGraw-Hill Higher Education, New York (2016)
11. Gordon, L.A., Loeb, M.P.: The economics of information security investment. ACM Trans. Inf. Syst. Secur. **5**(4), 438–457 (2002)
12. Homer, J., et al.: Aggregating vulnerability metrics in enterprise networks using attack graphs. J. Comput. Secur. **21**(4), 561–597 (2013)
13. ISACA: COBIT 5: A business framework for governance and management of enterprise IT (2012)
14. ISO/IEC 27001: Information technology - security techniques - information security management systems - requirements. International Organization for Standardization (2013)
15. ISO/IEC 27701: Security techniques - extension to ISO/IEC 27001 and ISO/IEC 27002 for privacy information management - requirements and guidelines. International Organization for Standardization (2019)
16. Khajouei, H., Kazemi, M., Moosavirad, S.H.: Ranking information security controls by using fuzzy analytic hierarchy process. Inf. Syst. e-Bus. Manag. **15**(1), 1–19 (2017)
17. Lee, M.C.: Information security risk analysis methods and research trends: AHP and fuzzy comprehensive method. Int. J. Comput. Sci. Inf. Technol. (IJCSIT) **6**, 29–45 (2014)
18. Manikandan, S.: Measures of central tendency: median and mode. J. Pharmacol. Pharmacother. **2**(3), 214–215 (2011)
19. Nasser, A.A.: Measuring the information security maturity of enterprises under uncertainty using fuzzy AHP. I.J. Inf. Technol. Comput. Sci. **4**, 10–25 (2018)
20. Ramos, A., Lazar, M., Filho, R.H., Rodrigues, J.J.: Model-based quantitative network security metrics: a survey. IEEE Commun. Surv. Tutor. **19**(4), 2704–2734 (2017)
21. Rudolph, M., Schwarz, R.: Security indicators - a state of the art survey public report. FhG IESE VII(043) (2012)

22. Saleh, M.: Information security maturity model. Int. J. Comput. Sci. Secur. (IJCSS) **5**, 21 (2011)
23. Savola, R.M.: Towards a taxonomy for information security metrics. In: Proceedings of the ACM Conference on Computer and Communications Security, pp. 28–30 (2007)
24. Schmid, M., Pape, S.: A structured comparison of the corporate information security maturity level. In: Dhillon, G., Karlsson, F., Hedström, K., Zúquete, A. (eds.) SEC 2019. IAICT, vol. 562, pp. 223–237. Springer, Cham (2019). https://doi.org/10.1007/978-3-030-22312-0_16
25. Schmitz, C., Pape, S.: LiSRA: lightweight security risk assessment for decision support in information security. Comput. Secur. **90** (2020)
26. Syamsuddin, I., Hwang, J.: The application of AHP to evaluate information security policy decision making. Int. J. Simul. Syst. Sci. Technol. **10**(4), 46–50 (2009)
27. Vinet, L., Zhedanov, A.: A 'missing' family of classical orthogonal polynomials. J. Phys. A Math. Theor. **44**(8), 16 (2011)

Privacy Enhancing Technologies in Specific Contexts

Differential Privacy in Online Dating Recommendation Systems

Teresa Anna Steiner(✉)

DTU Compute, Technical University of Denmark, 2800 Kongens Lyngby, Denmark
terst@dtu.dk

Abstract. By their very nature, recommendation systems that are based on the analysis of personal data are prone to leak information about personal preferences. In online dating, that data might be highly personal. The goal of this work is to analyse, for different online dating recommendation systems from the literature, if differential privacy can be used to hide *individual* connections (for example, an expression of interest) in the data set from any other user on the platform - or an adversary that has access to the information of one or multiple users. We investigate two recommendation systems from the literature on their potential to be modified to satisfy differential privacy, in the sense that individual connections are hidden from anyone else on the platform. For Social Collab by Cai et al. we show that this is impossible, while for RECON by Pizzato et al. we give an algorithm that theoretically promises a good trade-off between accuracy and privacy. Further, we consider the problem of *stochastic matching*, which is used as the basis for some other recommendation systems. Here we show the possibility of a good accuracy and privacy trade-off under edge-differential privacy.

1 Introduction

By their very nature, recommendation systems that are based on the analysis of personal data are prone to leak information about personal preferences. Calandrino et al. [3] showed that given little auxiliary information even a passive adversary can infer other user's individual transactions for many user-to-item recommendation systems from the recommendations given to them by the system. In online dating that data might be highly personal. The goal of this work is to analyse, for different online dating recommendation systems from the literature, if differential privacy can be used to hide *individual* connections (for example, the existence of a dialogue, or an expression of interest) in the data set from any other user on the platform - or an adversary that has access to the information of one or multiple users.

Our contribution is summarized as follows. We investigate two recommendation systems from the literature on their potential to be modified to satisfy differential privacy, in the sense that individual connections are hidden from anyone else on the platform. For Social Collab [2] we show that this is impossible, while for RECON [21] we give an algorithm that theoretically promises a good accuracy

Published by Springer Nature Switzerland AG 2020
M. Friedewald et al. (Eds.): Privacy and Identity 2019, IFIP AICT 576, pp. 395–410, 2020.
https://doi.org/10.1007/978-3-030-42504-3_25

and privacy trade-off. Further, we consider the problem of *stochastic matching*, which is used as the basis for some other recommendation systems [4,22]. Here we show the possibility of a good accuracy and privacy trade-off under edge-differential privacy, though the running time of the algorithm is exponential in the size of the input graph.

Related Work. The definition of differential privacy is due to Dwork et al. [6]. It gives strong privacy guarantees and has been broadly researched in the contexts of data analysis and machine learning. For surveys on differential privacy in general and its applications in machine learning specifically see for example [7,14,25].

For a survey on privacy aspects of recommender systems see [9]. Differential privacy in recommendation systems has been widely researched in applications where items (goods at an auction, movies, etc.) are recommended to users [8,10,12,16,18,23,26]. A common feature of all differentially private methods for preserving privacy in recommendation systems is that the goal is to hide individual connections (purchases, ratings, etc.), which is our goal as well. There are two main differences in this work:

1. The recommendation systems themselves are different: all recommendation systems considered in this work are *reciprocal*, which means they take the preferences of both parties into account, that is, the taste of the recommended user and the user that is recommended to are considered. For online dating reciprocal recommendation systems have been shown to widely outperform non-reciprocal ones in practice [20].
2. The data that needs to be protected is different. While in most other applications, the privacy of the items is unimportant, here, everyone's privacy matters equally.

Further, Machanavajjhala et al. [17] studied social recommendation systems under differential privacy. The recommendation systems in their work are based on the assumption that it is much more likely that a user will form a connection if any of their friends formed the same connection. They concluded that a good accuracy and privacy trade-off is not possible for those systems. None of the recommendation systems considered in this work use direct mutual connections.

The stochastic matching problem is equivalent to the maximum matching problem. Matching and allocation problems have only been studied more recently in the differential privacy literature [1,13,15]. Notably, Hsu et al. [13] give an infeasibility result for differentially privately matching goods to people. Their counter-example heavily relies on having multiple copies of the same good (and as such, the same weight from a person to copies of the same good). Consequently, Hsu et al. [13] and Kannan et al. [15] use relaxed versions of differential privacy. In contrast, we assume that the weights are independent probabilities. Anandan and Clifton [1] focus on constructing a data oblivious algorithm and then show how to output the value of the minimum matching in a differentially private way.

Outline. This paper is structured as follows. In the preliminaries (Sect. 2) we define differential privacy, collect some basic results, and introduce the recommendation systems we analyse in this work. In Sect. 3.1 we show by a counter-example that a sensible privacy and accuracy trade-off is impossible for the Social Collab recommendation system by Cai et al. [2]. In Sect. 3.2 we give a differentially private algorithm for RECON based on the Laplace mechanism and give arguments as to why this seems to promise a good trade-off. In Sect. 3.3 we consider the stochastic matching problem, and give a differentially private algorithm with a good accuracy trade-off, but which is inefficient. The algorithm achieves essentially the same as the exponential mechanism by McSherry and Talwar [19] on the set of all matchings, but we provide a direct analysis for our specific application.

2 Preliminiaries

In this section we define differential privacy, collect some basic results, and introduce the recommendation systems we adapt in this work.

2.1 Differential Privacy

The definition of differential privacy is due to Dwork and McSherry [6]. The basic idea is to find a randomized algorithm with the property that the output distributions are similar if the input data sets differ in a single entry. This is generally achieved by adding noise at the cost of accuracy.

For the formal definition, we need the notion of *neighbouring data sets.*

Definition 1. *Let $\mathcal{D}$ be a universe of data sets. Two data sets $x \in \mathcal{D}$ and $y \in \mathcal{D}$ are called* neighbouring *if they differ in at most one entry (what this means exactly will depend on $\mathcal{D}$). We will write $x \sim y$.*

Now, the differential privacy property says that the output distributions of a randomized mechanism have to be close for neighbouring data sets.

Definition 2. *A mechanism $\mathcal{M}$ is called* (ϵ, δ)-differentially private *if for every S in the range of $\mathcal{M}$ and two neighbouring data sets x and y it holds that*

$$P(\mathcal{M}(x) \in S) \leq \exp(\epsilon) P(\mathcal{M}(y) \in S) + \delta. \tag{1}$$

For small values of ϵ and δ this means that the output distributions are similar if the data sets differ in a single entry. If $\delta = 0$, the function is called ϵ-differentially private, and the smallest parameter ϵ for which (1) is true is referred to as the "privacy loss".

The results presented in this work use the Laplace mechanism by Dwork et al. [6] as a basic tool. We need the notion of *global sensitivity*, which is a measure for the maximum output difference of a function evaluated on two neighbouring data sets.

Definition 3. *The* global sensitivity *of a function $f : \mathcal{D} \to \mathbb{R}^d$ is defined as*

$$\Delta f = \max_{x \sim y} \|f(x) - f(y)\|_1. \tag{2}$$

Next, we define the Laplace distribution.

Definition 4 (Laplace Distribution). *The Laplace distribution $Lap(\mu, b)$, where μ is the mean and $b > 0$ is the scale parameter, is defined by the density function $f(x|\mu, b) = \frac{1}{2b} \exp\left(-\frac{|x-\mu|}{b}\right)$. If we omit the first parameter we assume $\mu = 0$.*

The Laplace mechanism adds Laplace noise scaled with the global sensitivity to the output of a function.

Theorem 1 (Dwork et al. [6]). *For a function $f : \mathcal{D} \to \mathbb{R}^d$ let $(Y_1, \ldots, Y_d)$ be independent random variables drawn from the Laplace distribution with scale parameter $b = \frac{\Delta f}{\epsilon}$. Then the mechanism $\mathcal{M}(x) := f(x) + (Y_1, \ldots, Y_d)$ is ϵ-differentially private. This mechanism is called the* Laplace mechanism.

Further, we need two basic results about differential privacy. The first one is the post-processing rule, which states that any transformation of a differentially private output preserves the same (or better) privacy guarantees.

Lemma 1. *If $\mathcal{M}$ is differentially private, any mechanism $g \circ \mathcal{M} : x \mapsto g(\mathcal{M}(x))$ is also differentially private, for any function g defined on the range of $\mathcal{M}$.*

Secondly, a basic composition theorem by Dwork et al. [5] states that the privacy loss of a composition of several differentially private functions is no more than the sum of the privacy losses of each individual function.

Theorem 2 (Dwork et al. [5]). *The composition of k differentially private mechanisms with parameters $(\epsilon_1, \delta_1), \ldots, (\epsilon_k, \delta_k)$ is (ϵ, δ)-differentially private with $\epsilon = \sum_{i=1}^k \epsilon_i$ and $\delta = \sum_{i=1}^k \delta_i$.*

When working with graphs, there are mulitple ways of defining neighbouring data sets [11], depending on whether the goal is to hide a person's presence in the data set (resulting in the definition of *node differential privacy*), or individual connections. The latter is formalized by the definition of *edge-differential privacy*, which is due to Hay et al. [11].

Definition 5 (Edge-Differential Privacy). *Let $\mathcal{D}$ be a set of graphs on a vertex set V with $|V| = n$. A mechanism is* edge-differentially private, *if it is differentially private over $\mathcal{D}$, when for $x = G_x = (V, E_x) \in \mathcal{D}$ and $y = G_y = (V, E_y) \in \mathcal{D}$ the neighbouring relation $x \sim y$ means there exists an edge e such that $E_x = E_y \backslash \{e\}$ or $E_y = E_x \backslash \{e\}$.*

In this work we will only consider variations of edge-differential privacy.

2.2 Online Dating Recommendation Systems

By a *recommendation system*, in general form, we mean any mechanism that uses data from an online dating platform, and gives an individual, ranked list of users as recommended connections to every user on the platform. When we talk about the recommendations given to one specific user, we call this user the *active user*. Often, the space of potential connections for one active user is not the entire data base, but some restricted subset based on for example location, gender, age, etc. The choices for recommendation systems considered in this work are mostly influenced by an extensive review by Pizzato et al. [20]. Additionally to summarizing and categorizing existing strategies, they point out two specific characteristics which show improved success rate of recommendations:

- Recommendation systems which are based on *implicit* preferences (i.e. based on existing matches and past behaviour on the platform) are more effective than *explicit* preferences (i.e. preferences the user states).
- Recommendation systems which take preferences of both the active user and the user we potentially want to recommend into account are more effective than considering preferences of the active user only.

Social Collab. In the model by Cai et al. [2] the data set is modelled as a directed graph, where nodes represent users and edges represent some form of "like", for example liking a profile, or sending a message, or the positive reply to a message. They define two users to be *similar in attractiveness* if they were liked by at least one person in common, and *similar in taste* if they liked at least one person in common; see Fig. 1(a). A user r is a *predicted match* for the active user a if there exists

1. at least one user similar in taste to r that liked a and
2. at least one user similar in attractiveness to r that was liked by a.

See Fig. 1(b).

All predicted matches for user a are ranked by the number of total users for which either of the two conditions above is true. The output is, for some N, the sorted list of the top N predicted matches.

RECON. The RECON system by Pizzato et al. [21] considers a model where users can like other users' profiles. Each profile consists of several attributes. Based on the attributes of users that a liked before and b's profile, they predict a score $C^+(a, b)$ of how likely it is that user a will like user b. The data set is modelled as a directed graph where users are nodes and an edge (u, v) means that user u liked user v. Let $\mathcal{N}(a)$ be the set of users that user a liked, and $d(a) = |\mathcal{N}(a)|$ the outdegree of a. Denote $\text{Att}(b)$ the set of all attributes that user b possesses.

Definition 6. *The* positive compatibility $C^+(a, b)$ *of a candidate b for user a is defined as follows:*

$$C^+(a,b) = \frac{\sum_{u\in\mathcal{N}(a)} \sum_{t\in Att(b)} \mathbb{1}\,(t \in Att(u))}{d(a)\,|Att(b)|}.$$

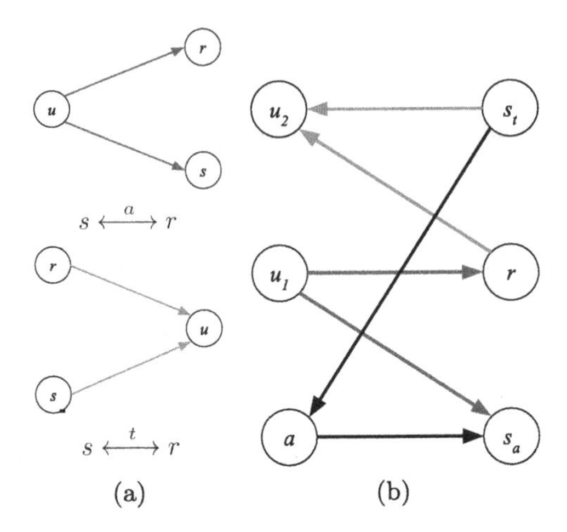

Fig. 1. Social Collab. (a) The users s and r are similar in attractiveness or taste, respectively. (b) Since s_t is similar in taste to r and s_t likes a, and s_a is similiar in attractiveness to r and is liked by a, user r is a predicted match for active user a.

As mentioned before, an important characteristic that has been shown to improve the success rate of recommendations is reciprocability. In their paper, Pizatto et al. [21] define a reciprocal version of RECON by using the harmonic mean between $C^+(a, b)$ and $C^+(b, a)$. The *positive compatibility between a and b* is defined as

$$C^+_{rec}(a, b) = C^+_{rec}(b, a) = \frac{2}{\frac{1}{C^+(a,b)} + \frac{1}{C^+(b,a)}}.$$

As such, a low compatibility score for one of either $C^+(a, b)$ or $C^+(b, a)$ will result in a low combined compatibility score. This means we will only recommend b to a if it is considered likely that b will like a back. Pizatto et al. [21] showed that the reciprocal version of RECON has a significantly higher success rate than the non-reciprocal one.

Stochastic Matching. Next, we will define the stochastic matching problem, which is not a recommendation system itself, but is used as a model for some of them. Instead of singling out one active user and ranking recommendations for this user, the idea here is to always recommend user b to user a exactly when we also recommend user a to user b. Instead of outputting a sorted list of recommendations for one user, we will output a list of total recommendations, and the goal is to have the same number of recommendations for each user. The motive of this strategy is, as argued by Pizzato and Silvestrini [20], to ensure that there is no difference made between popular and unpopular users. Receiving either too many or too few messages can cause frustration in a user.

Both the recommendation systems by Pizzato and Silvestrini [20] and Chen et al. [4] model the problem as a weighted, undirected graph $G = (V, E)$.

The vertices in V represent the users, and for each edge $e = (u, v)$, the edge weight $0 \leq w(e) \leq 1$ is an estimate for the probability of a *successful match* between u and v. A successful match is defined differently in the two papers, but the common ground is that the recommendation leads to some sort of positive interaction between the users. For simplicity we assume $E = V \times V$ and set the weight of non-existing edges to zero. The methods used to estimate the edge weights differ for each system and will not be further discussed here.

The *weighted maximum matching problem* is defined as follows: given a weighted, undirected graph $G = (V, E)$ with a non-negative weight function w on E and a given number $N \geq 1$, the goal is to find a subset M of edges such that

1. any vertex is adjacent to at most N edges in M and
2. the value $w(M) := \sum_{e \in M} w(e)$ is maximized with respect to all subsets M satisfying condition 1.

We will call any subset of edges satisfying condition 1 a *matching* on G. Furthermore, we will restrict ourselves to the simplified case where $N = 1$.

In Pizzato and Silvestrini [20] and Chen et al. [4] the *stochastic matching problem* is formulated slightly differently: given the probabilities of any successful match, the goal is to maximize the expected number of successful matches, under the condition that each user receives at most N recommendations. Note that since the edge weights represent the probabilities of any match being successful, the problem is equivalent to the one formulated above.

3 Results

First, we have to define what privacy notion we aim to achieve. As mentioned in the preliminaries, one could aim to hide either the presence of a person in the data set, or individual connections. We focus on hiding individual connections. This has two reasons: it makes more practical sense, because the presence of someone on a dating platform can usually not be entirely hidden. Also, it is much easier to achieve, since an individual connection influences a recommendation less than the full data of a person.

Note that when recommending to a, we do not need to protect outgoing edges from a, since they only encode information about a's own preferences. Our aim is to hide a user's connections from anyone who has access to the recommendations of any other user on the platform. An adversary in this setting could be someone who creates fake profiles and gathers information from the recommendations given to those fake users. Formally, for a fixed active user a, we define two data sets $x = (V, E_x)$ and $y = (V, E_y)$ with $a \in V$ as neighbouring, $x \sim y$, if there exists an edge $e \neq (a, v)$ for any $v \in V$ such that $E_x \backslash \{e\} = E_y \backslash \{e\}$. This means two neighbouring data sets differ in an edge that is not an outgoing edge of a, because those are the edges we want to keep private when recommending to a. We will use differential privacy based on this definition of neighbouring data sets for Social Collab and RECON.

3.1 Social Collab

We give a counter-example to show that it is not possible to directly modify Social Collab to satisfy differential privacy under the definition of neighbouring data sets described above.

Consider the (sub-)graph given by Fig. 2. Assume we have an active user a and a subset of k possible candidates $r_1, \ldots, r_k$. Everyone of the r_i, for $i = 1, \ldots, k$, likes a user u_2, who is also liked by a set of users $s_{t1}, \ldots, s_{tl}$. As such, all the users r_i and s_{tj} are similar in taste. Additionally, all users s_{tj}, for $j = 1, \ldots, l$ like user a. Then, there is a user s_a who is liked by a, and a user u_1, which likes every r_i for $i = 1, \ldots, k$. Now, if u_1 also likes s_a, all users r_i are predicted matches for a with weight $l + 1$, since they are similar in taste to l users that liked a and similar in attractiveness to a user s_a who was liked by a. On the other hand, if the edge (u_1, s_a) does not exist, there is no user similar in attractiveness to any of the r_i which was liked by a, which means none of them is a predicted match.

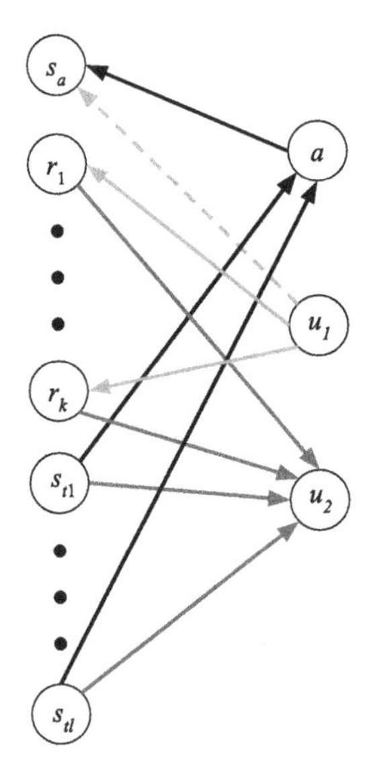

Fig. 2. This figure shows the counter-example for Social Collab. The dashed edge is present in a data set x, but not in its neighbouring data set y. Each s_{tj} is similar in taste to each of the r_i. In x, each r_i is similar in attractiveness to s_a.

Note that l and k can be arbitrarily large. This means that one edge can influence the rating of arbitrarily many users from very high, to not even being recommended at all. Thus, we cannot hope to preserve differential privacy here without destroying all information.

The example also demonstrates which kind of potential attacks we are trying to prevent: An adversary aware of the subgraph in Fig. 2 can find out if the edge (u_1, s_a) exists by looking at recommendations for user a. As such, an adversary can try to build specific subgraphs using fake profiles to uncover additional information about the network.

3.2 RECON

The goal of this section is to find a way to privately output the vector consisting of $C^+_{rec}(a,u)$, for every u which is a candidate for a, to a under the privacy model defined in the beginning of Sect. 3. This means that two neighbouring data sets differ in an edge that is not an outgoing edge of the active user a.

First, notice that because differential privacy is immune to post-processing (Lemma 1), if we show how to make both $C^+(a,u)$ and $C^+(u,a)$ differentially private, their private versions can be used to compute $C^+_{rec}(a,u)$ while still preserving differential privacy.

Deleting any (u,v) for $u \neq a$ can not change $C^+(a,b)$ because that quantity depends only on outgoing edges of a and attributes of b. Thus, no noise has to be added to privately output $C^+(a,b)$. Consider now a as the active user. Denote U_a the set of candidate matches for a (e.g. users living close to a). The following Lemma shows that deleting or adding an edge (b,v) where $b \neq a$ can change only $C^+(b,a)$ by at most $\frac{1}{d(b)}$.

Lemma 2. *For any $b \in U_a$ with $d(b) > 1$, deleting or adding an outgoing edge of b can change $C^+(b,a)$ by at most $\frac{1}{d(b)}$.*

Proof. Fix a data set $x = G_x = (V, E_x)$. We consider neighbouring data sets y which differ from x in an outgoing edge from b. We will use x and y as subscripts for d, $\mathcal{N}(b)$ and C^+ to differentiate the value of the respective functions in the different data sets.

First, assume x and y differ in an edge (b,v) that is present in x but not in y. We have

$$\begin{aligned}
&\left|C^+_x(b,a) - C^+_y(b,a)\right| \\
&= \left|\frac{\sum_{u\in\mathcal{N}_x(b)}\sum_{t\in\mathrm{Att}(a)} \mathbb{1}\left(t \in \mathrm{Att}(u)\right)}{d_x(b)\left|\mathrm{Att}(a)\right|} - \frac{\sum_{u\in\mathcal{N}_x(b)\setminus\{v\}}\sum_{t\in\mathrm{Att}(a)} \mathbb{1}\left(t \in \mathrm{Att}(u)\right)}{\left(d_x(b)-1\right)\left|\mathrm{Att}(a)\right|}\right| \\
&= \left|\frac{\sum_{u\in\mathcal{N}_x(b)\setminus\{v\}}\sum_{t\in\mathrm{Att}(a)} \mathbb{1}\left(t \in \mathrm{Att}(u)\right)}{\left|\mathrm{Att}(a)\right|}\left(\frac{1}{d_x(b)} - \frac{1}{d_x(b)-1}\right)\right. \\
&\quad \left. + \frac{\sum_{t\in\mathrm{Att}(a)} \mathbb{1}\left(t \in \mathrm{Att}(v)\right)}{d_x(b)\left|\mathrm{Att}(a)\right|}\right|.
\end{aligned}$$

Note that since

$$\left(\frac{1}{d_x(b)} - \frac{1}{d_x(b)-1}\right) = -\frac{1}{d_x(b)\left(d_x(b)-1\right)} < 0,$$

we have that the difference between $C_x^+(b,a)$ and $C_y^+(b,a)$ is bounded by

$$\max\left(\left|\frac{\sum_{u\in\mathcal{N}_x(b)\setminus\{v\}}\sum_{t\in\mathrm{Att}(a)}\mathbb{1}\left(t\in\mathrm{Att}(u)\right)}{|\mathrm{Att}(a)|\,d_x(b)\,(d_x(b)-1)}\right|,\left|\frac{\sum_{t\in\mathrm{Att}(a)}\mathbb{1}\left(t\in\mathrm{Att}(v)\right)}{d_x(b)\,|\mathrm{Att}(a)|}\right|\right).$$

Bounding each $\mathbb{1}\left(t\in\mathrm{Att}(u)\right)$ by 1 we get

$$\left|C_x^+(b,a)-C_y^+(b,a)\right|\leq\frac{1}{d_x(b)}.$$

In the other case where y has one edge (b,v) more than x, the analysis from above goes through with reversed roles of x and y and we get

$$\left|C_x^+(b,a)-C_y^+(b,a)\right|\leq\frac{1}{d_y(b)}<\frac{1}{d_x(b)}.$$

Assuming our data universe is such that there is a lower bound $T\leq d(u)$ for all potential matches u of a, this means that the global sensitivity of $(C^+(u,a))_{u\in U_a}$ is bounded by $1/T$ - since one edge can influence one entry in the vector by at most $1/T$. As such, we can use the Laplace mechanism with scale $\frac{1}{T\epsilon}$ to privately output $(C^+(u,a))_{u\in U_a}$. Then we can compute $(C_{rec}^+(a,u))_{u\in U_a}$ privately by the post processing rule.

In the general case, we will not have a good lower bound on the outdegree of all potential candidates. In fact, in the worst case there might exist a user with outdegree zero, for example, a new user on the platform. Intuitively though, if a user u does not have a lot of outgoing edges, the information gained by considering $C^+(u,a)$ is not reliable anyway - there is not enough data to make conclusions about the user's taste.

We propose the following idea: we use the noisy $C_{rec}^+(a,u)$ only for elements $u\in U_a$ which have an outdegree above a certain threshold, and $C^+(a,u)$ for all others. This means that whenever u has sufficiently high degree, that is, sufficient activity on the platform, we use both u's and a's preferences in the recommendation. Otherwise, we will only consider a's preferences.

The difficulty with this approach is that the decision whether or not the degree is above a certain threshold can leak information. This means we first have to find a differentially private approximation of the outdegrees of all candidates u. The vector containing all outdegrees has a sensitivity of 1, since adding or removing an edge can change the outdegree of one node. In Algorithm 1, we use the Laplace mechanism for finding a differentially private degree sequence. Alternatively, one could use Hay et al.'s [11] modification for finding a differentially private degree sequence. We stick to the Laplace mechanism for simplicity.

Our approach is now summarized as follows: We first add noise to make the vector of outdegrees differentially private. Then we use this vector to decide if the outdegree for any u is above a certain threshold or below. Then, depending on which case we are in, we either use the idea described above to output a noisy $C_{rec}^+(a,u)$, or output $C^+(a,u)$ directly. The details are shown in Algorithm 1.

The $(T-\alpha)$ in Algorithm 1 originates from the fact that if we know that the noisy degree is above a certain threshold, we do not have a guarantee that the *actual* degree is above that threshold; but we will show that with appropriate choice of α, with high probability, the noisy degree will be above $T-\alpha$. If we are in that case, adding noise scaled with $\frac{1}{T-\alpha}$ is sufficient for preserving privacy. With non-zero probability, though, the noisy outdegree will be above T, while the actual one is smaller than $T-\alpha$. In that case, we cannot guarantee that the privacy loss in our algorithm is less than ϵ. We prove $(\epsilon,\delta)-$differential privacy for $\delta > 0$. The details follow in the proof of Theorem 3.

Algorithm 1. Private RECON $(a, U_a, \epsilon, T, \alpha)$

for every $u \in U_a$ **do**
  Let $Y_1 \sim \text{Lap}\left(\frac{2}{\epsilon}\right)$ and $Y_2 \sim \text{Lap}\left(\frac{2}{(T-\alpha)\epsilon}\right)$
  if $d(u) + Y_1 > T$ **then**
    Compute $\tilde{C}^+(u,a) := C^+(u,a) + Y_2$ and use it to compute $c_u := \frac{2}{\frac{1}{C^+(a,u)} + \frac{1}{\tilde{C}^+(u,a)}}$
  else
    set $c_u := C^+(a,u)$
return $(c_u)_{u \in U_a}$

Theorem 3. *Algorithm 1 is (ϵ,δ)-differentially private for $\alpha > \frac{2(\log(|U_a|)-\log(\delta))}{\epsilon}$.*

Proof. First, note that providing (ϵ,δ)-differential privacy is equivalent to guaranteeing a privacy loss of at most ϵ with probability at least $1-\delta$. Since the vector $(d(u))_{u \in U_a}$ has sensitivity 1, adding independent noise with scale $2/\epsilon$ makes the output of the noisy vector $(\epsilon/2)$-differentially private. This means outputting which users are in the first or second case of the algorithm preserves $(\epsilon/2)$-differential privacy.

As argued before, outputting $C^+(a,u)$ for each user u in the second case can be done without further privacy loss. The interesting case to consider is first case, when the noisy degree is above threshold T.

Claim. With probability at least $1-\delta$, all u from the first case satisfy $d(u) > T-\alpha$.

If the claim is true, then with probability $1-\delta$, adding independent Laplace noise scaled with $\frac{2}{(T-\alpha)\epsilon}$ to $C^+(u,a)$ for each u in the first case will preserve $(\epsilon/2)$-differential privacy - since by Lemma 2, changing one edge can change $C^+(u,a)$ for at most one u by at most $1/(T-\alpha)$, thus the L_1 norm of the vector by at most $1/(T-\alpha)$. By the post-processing rule and the composition theorem (Lemma 1 and Theorem 2) we are done.

Proof (of Claim). Let u be any node with $d(u) \leq T - \alpha$. We then have that

$$P\left(d(u) + Y_1 > T\right) \leq P(|Y_1| > \alpha) = \exp\left(-\frac{\alpha\epsilon}{2}\right). \quad (3)$$

By the union bound, the probability that any u with $d(u) \leq T - \alpha$ gets classified into the first case is at most $|U_a| \exp\left(-\frac{\alpha\epsilon}{2}\right) < \delta$ by choice of α.

This concludes the proof of the theorem.

To preserve privacy in practice, the parameter δ is usually recommended to be $o(1/n)$, where n is the data set size. Since our algorithm only operates on the set U_a, if we choose e.g. $\delta = \frac{1}{|U_a|^2}$, we satisfy this condition and can choose $\alpha = \frac{6\log(|U_a|)}{\epsilon}$. Clearly, if we choose α as above, we have to choose $T = \Omega\left(\frac{\log(|U_a|)}{\epsilon}\right)$. Often, we require δ to be smaller than the inverse of any polynomial of the data set size [7]. Note that if we choose e.g. $\delta = \frac{1}{|U_a|^{\log|U_a|}}$, we can choose $T = \Theta\left(\frac{\log^2(|U_a|)}{\epsilon}\right)$. Now, the higher we choose T, the less noise we have to add to high degree nodes. On the other hand, the higher we choose T, the fewer nodes will be classified as high degree nodes, which means we take the preferences of fewer candidates into account. As such, a good threshold T can only be empirically optimized given data - assuming that we cannot estimate mathematically how many profiles a user would have to like in order for us to make sensible predictions about their taste.

3.3 Stochastic Matching

For the stochastic matching problem, we give an exponential running time algorithm which is both accurate and private, showing the theoretical feasibility of a good trade-off. The definition of privacy we use is edge-differential privacy from Definition 5. Note that for the privacy definition to make sense in this application, we implicitly assume that the edge weight estimations are independent for each edge, which might not be true for all applications. In [24] we show that the simple idea of adding Laplace noise to the weight of each edge fails to give a good privacy and accuracy trade-off. The algorithm we present here is similar to the algorithm Report Noisy Max found in [7]. The output distribution of this algorithm is almost equivalent to the *exponential mechanism* by McSherry and Talwar [19], as shown in [7]. In our algorithm, we compute all feasible matchings, and add Laplace noise to the value of each matching. Then we choose the matching with maximum noisy value.

Theorem 4. *Algorithm 2 is ϵ-differentially private.*

Proof. We will sketch the proof by showing the properties used in Claim 3.9 in [7], which proves ϵ-differential privacy for Report Noisy Max, and then follow the steps of their proof.

Fix two neighbouring data bases x and y such that there exists an edge e with $w_x(e) \geq w_y(e)$. For each possible matching M of G, denote its weight in x

Algorithm 2. Noisy Max Matching (G, w, ϵ)

for every possible matching M in G **do**
  compute its weight $w(M)$
  draw $Y \sim \text{Lap}(1/\epsilon)$
  set $\tilde{w}(M) = w(M) + Y$
return $\text{argmax}(\tilde{w})$

by $w_x(M) = \sum_{e \in M} w_x(e)$. Note that, since the graph without weights is the same in both data sets, the set of possible matchings is independent of the data set.

Now, it is easy to see that the two properties used in the proof in [7] hold:

1. Monotonicity: For each possible matching M we have $w_x(M) \geq w_y(M)$.
2. Lipschitz Property: For each possible matching M we have $1 + w_y(M) \geq w_x(M)$.

We will use these properties later in the proof. Denote by M_G the set of all possible matchings on G and fix one matching $M_0 \in M_G$. Define the vector of independent Laplace variables $Y = (Y_M)_{M \in M_G}$, where $Y_M \sim \text{Lap}(1/\epsilon)$. That is, the algorithm chooses $\text{argmax}_{M \in M_G}(w(M) + Y_M)$. Further, denote Y_{-M_0} the random vector of Y without the entry corresponding to M_0.

Now, fix a realization z_{-M_0} of Y_{-M_0}, that is, a vector where each coordinate is drawn from $\text{Lap}(1/\epsilon)$. We will show the property from Definition 2 for $\delta = 0$ separately for each condition $(Y_{-M_0} = z_{-M_0})$. For simplicity, we write z_M for the coordinate corresponding to matching M in z_{-M_0}.

Denote $r := \max_{M \in M_G}(w_x(M) + z_M - w_x(M_0))$. Thus, we output M_0 on database x if and only if $Y_{M_0} > r$. By the Monotonicity and Lipschitz property above, we have, for all $M \neq M_0$,

$$1 + w_y(M_0) + r \geq w_x(M_0) + r \geq w_x(M) + z_M \geq w_y(M) + z_M.$$

This means that if $Y_{M_0} > 1 + r$, then M_0 is the output on data base y. We have

$$\begin{aligned} P(\text{Noisy Max Matching}(y, \epsilon) = M_0 | Y_{-M_0} = z_{-M_0}) &\geq P(Y_{M_0} > 1 + r) \\ &= \frac{1}{2}\exp(-\epsilon(1 + r)) \\ &= \exp(-\epsilon)P(Y_{M_0} > r) \end{aligned}$$

$$= \exp(-\epsilon)P(\text{Noisy Max Matching}(x, \epsilon) = M_0 | Y_{-M_0} = z_{-M_0}),$$

where for the first and second equality we use Theorem 1 and the symmetry of the Laplace distribution.

Similarly, one can prove the other inequality, that is,

$$\begin{aligned} &P(\text{Noisy Max Matching}(y, \epsilon) = M_0 | Y_{-M_0} = z_{-M_0}) \\ &\leq \exp(\epsilon)P(\text{Noisy Max Matching}(x, \epsilon) = M_0 | Y_{-M_0} = z_{-M_0}). \end{aligned}$$

When we have that, we are done, since if the inequalities hold under each realization of Y_{-M_0}, they also hold for the marginal distributions $P\,(\text{Noisy Max Matching }(x, \epsilon) = M_0)$ and $P\,(\text{Noisy Max Matching }(y, \epsilon) = M_0)$.

Next, we will show that this algorithm actually provides a good trade-off. The result is comparable to the accuracy achieved by outputting only the value of the maximum matching using the Laplace mechanism: Since the global sensitivity of the value is at most 1 and by the properties of the Laplace distribution, the expected error for this is at most $\frac{1}{\epsilon}$. Even though outputting an actual matching instead of only the value seems to provide much more information, we achieve almost the same error guarantees.

Lemma 3. *The value of the matching output by Algorithm 2 differs from the optimal by at most α with probability at least $1 - 2\exp\left(-\frac{\epsilon\alpha}{2}\right)$. The expected difference is at most $\frac{2}{\epsilon}$.*

Proof. Let M denote the matching output by Algorithm 2 and M^* any maximum matching. Further, denote Y_M the random variable added to $w(M)$ and Y_{M^*} the random variable added to $w(M^*)$ in the algorithm. We have

$$w(M) + Y_M \geq w(M^*) + Y_{M^*}.$$

It follows

$$|w(M) - w(M^*)| = w(M^*) - w(M) \leq Y_M - Y_{M^*} \leq |Y_M - Y_{M^*}|.$$

Using this, we get

$$\begin{aligned} P\left(|w(M) - w(M^*)| \geq \alpha\right) &\leq P(|Y_{M^*} - Y_M| \geq \alpha) \\ &\leq P(2\max(|Y_{M^*}|, |Y_M|) \geq \alpha) \\ &\leq 2P(2|Y_M| \geq \alpha) \\ &= 2P\left(|Y_M| \geq \frac{\alpha}{2}\right) = 2\exp\left(-\frac{\epsilon\alpha}{2}\right) \end{aligned}$$

where the second inequality follows from triangle inequality, and the third from the union bound together with the fact that Y_{M^*} and Y_M are identically distributed. The last equality holds because the absolute value of a Laplace distribution follows an exponential distribution. We conclude that our solution is within an error α with probability at least $1 - 2\exp\left(-\frac{\epsilon\alpha}{2}\right)$, or equivalently, with probability at least $1 - \beta$ our error is within $\frac{2}{\epsilon}\log\left(\frac{2}{\beta}\right)$.

By the same argument, the expected error will be at most

$$\mathbb{E}\left(|Y_{M^*} - Y_M|\right) \leq \mathbb{E}\left(|Y_{M^*}| + |Y_M|\right) = 2\mathbb{E}\left(|Y_M|\right) = \frac{2}{\epsilon}.$$

The last equality follows again from the exponential distribution of $|Y_M|$.

4 Conclusion and Open Problems

This work shows that it is certainly feasible to design recommendation systems for online dating which satisfy differential privacy, and opens many directions for future research. First, the theoretical results of this paper should be tested on real data to verify their practicality. Secondly, finding a more efficient solution to the differentially private maximum matching problem is an interesting open question. Potentially strategies from approximation algorithms could be interesting, since they trade accuracy for efficiency and, in a differentially private setting, we are not looking for a perfectly accurate solution. Lastly, all results presented in this work model the problem to be static: we have a static data set and give a set of recommendation once. In practice, both users and edges will appear on or leave the platform, and we will give recommendations to users over a longer period of time. Specifically, this will make the assumption that the probabilities are independent for stochastic matching invalid. To capture these properties it would be necessary to consider dynamic models.

Acknowledgments. I want to thank Inge Li Gørtz, Philip Bille and Sune Lehmann for helpful suggestions and discussions.

References

1. Anandan, B., Clifton, C.: Secure minimum weighted bipartite matching. Proc. DSC **2017**, 60–67 (2017)
2. Cai, X., et al.: Collaborative filtering for people to people recommendation in social networks. In: Proceedings of 23rd AI, pp. 476–485 (2010)
3. Calandrino, J.A., Kilzer, A., Narayanan, A., Felten, E.W., Shmatikov, V.: You might also like: privacy risks of collaborative filtering. In: Proceedings of 32nd IEEE Symposium on Security & Privacy, pp. 231–246 (2011)
4. Chen, N., Immorlica, N., Karlin, A.R., Mahdian, M., Rudra, A.: Approximating matches made in heaven. In: Albers, S., Marchetti-Spaccamela, A., Matias, Y., Nikoletseas, S., Thomas, W. (eds.) ICALP 2009. LNCS, vol. 5555, pp. 266–278. Springer, Heidelberg (2009). https://doi.org/10.1007/978-3-642-02927-1_23
5. Dwork, C., Kenthapadi, K., McSherry, F., Mironov, I., Naor, M.: Our data, ourselves: privacy via distributed noise generation. In: Vaudenay, S. (ed.) EUROCRYPT 2006. LNCS, vol. 4004, pp. 486–503. Springer, Heidelberg (2006). https://doi.org/10.1007/11761679_29
6. Dwork, C., McSherry, F., Nissim, K., Smith, A.: Calibrating noise to sensitivity in private data analysis. In: Halevi, S., Rabin, T. (eds.) TCC 2006. LNCS, vol. 3876, pp. 265–284. Springer, Heidelberg (2006). https://doi.org/10.1007/11681878_14
7. Dwork, C., Roth, A.: The algorithmic foundations of differential privacy. Found. Trends Theor. Comput. Sci. **9**(3–4), 211–407 (2014)
8. Friedman, A., Berkovsky, S., Kaafar, M.A.: A differential privacy framework for matrix factorization recommender systems. User Model. User-adap. Inter. **26**(5), 425–458 (2016)
9. Friedman, A., Knijnenburg, B.P., Vanhecke, K., Martens, L., Berkovsky, S.: Privacy aspects of recommender systems. In: Ricci, F., Rokach, L., Shapira, B. (eds.) Recommender Systems Handbook, pp. 649–688. Springer, Boston (2015). https://doi.org/10.1007/978-1-4899-7637-6_19

10. Guerraoui, R., Kermarrec, A.-M., Patra, R., Taziki, M.: D2P: distance-based differential privacy in recommenders. Proc. VLDB Endowment **8**(8), 862–873 (2015)
11. Hay, M., Li, C., Miklau, G., Jensen, D.: Accurate estimation of the degree distribution of private networks. Proc. ICDM **2009**, 169–178 (2009)
12. He, K., Mu, X.: Differentially private and incentive compatible recommendation system for the adoption of network goods. In: Proceedings of 15th ACM EC, pp. 949–966 (2014)
13. Hsu, J., Huang, Z., Roth, A., Roughgarden, T., Wu, Z.S.: Private matchings and allocations. SIAM J. Comput. **45**(6), 1953–1984 (2016)
14. Ji, Z., Lipton, Z.C., Elkan, C.: Differential privacy and machine learning: a survey and review. arXiv preprint arXiv:1412.7584 (2014)
15. Kannan, S., Morgenstern, J., Rogers, R., Roth, A.: Private pareto optimal exchange. ACM Trans. Econ. Comput. **6**(3–4), 12 (2018)
16. Liu, X., et al.: When differential privacy meets randomized perturbation: a hybrid approach for privacy-preserving recommender system. In: Candan, S., Chen, L., Pedersen, T.B., Chang, L., Hua, W. (eds.) DASFAA 2017. LNCS, vol. 10177, pp. 576–591. Springer, Cham (2017). https://doi.org/10.1007/978-3-319-55753-3_36
17. Machanavajjhala, A., Korolova, A., Sarma, A.D.: Personalized social recommendations: accurate or private. Proc. VLDB Endowment **4**(7), 440–450 (2011)
18. McSherry, F., Mironov, I.: Differentially private recommender systems: building privacy into the Netflix prize contenders. In: Proceedings of 15th ACM SIGKDD, pp. 627–636 (2009)
19. McSherry, F., Talwar, K.: Mechanism design via differential privacy. In: Proceedings of 48th FOCS, vol. 7, pp. 94–103 (2007)
20. Pizzato, L., Rej, T., Akehurst, J., Koprinska, I., Yacef, K., Kay, J.: Recommending people to people: the nature of reciprocal recommenders with a case study in online dating. User Model. User-adapt. Inter. **23**(5), 447–488 (2013)
21. Pizzato, L.A., Rej, T., Yacef, K., Koprinska, I., Kay, J.: Finding someone you will like and who won't reject you. In: Konstan, J.A., Conejo, R., Marzo, J.L., Oliver, N. (eds.) UMAP 2011. LNCS, vol. 6787, pp. 269–280. Springer, Heidelberg (2011). https://doi.org/10.1007/978-3-642-22362-4_23
22. Pizzato, L.A., Silvestrini, C.: Stochastic matching and collaborative filtering to recommend people to people. In: Proceedings of 5th RecSys, pp. 341–344 (2011)
23. Shin, H., Kim, S., Shin, J., Xiao, X.: Privacy enhanced matrix factorization for recommendation with local differential privacy. IEEE Trans. Knowl. Data Eng. **30**(9), 1770–1782 (2018)
24. Steiner, T.A.: Differential privacy in graphs. Master's thesis, Technical University of Denmark (2019)
25. Zhu, T., Li, G., Zhou, W., Philip, S.Y.: Differentially private data publishing and analysis: a survey. IEEE Trans. Knowl. Data Eng. **29**(8), 1619–1638 (2017)
26. Zhu, T., Ren, Y., Zhou, W., Rong, J., Xiong, P.: An effective privacy preserving algorithm for neighborhood-based collaborative filtering. Future Gener. Comput. Syst. **36**, 142–155 (2014)

Distributed Ledger for Provenance Tracking of Artificial Intelligence Assets

Philipp Lüthi, Thibault Gagnaux, and Marcel Gygli(✉)

Fachhochschule Nordwestschweiz FHNW, Institut für Interaktive Technologien (IIT), Windisch, Switzerland
{philipp.luethi,thibault.gagnaux,marcel.gygli}@fhnw.ch

Abstract. High availability of data is responsible for the current trends in Artificial Intelligence (AI) and Machine Learning (ML). However, high-grade datasets are reluctantly shared between actors because of lacking trust and fear of losing control. Provenance tracing systems are a possible measure to build trust by improving transparency. Especially the tracing of AI assets along complete AI value chains bears various challenges such as trust, privacy, confidentiality, traceability, and fair remuneration. In this paper we design a graph-based provenance model for AI assets and their relations within an AI value chain. Moreover, we propose a protocol to exchange AI assets securely to selected parties. The provenance model and exchange protocol are then combined and implemented as a smart contract on a permission-less blockchain. We show how the smart contract enables the tracing of AI assets in an existing industry use case while solving all challenges. Consequently, our smart contract helps to increase traceability and transparency, encourages trust between actors and thus fosters collaboration between them.

Keywords: Artificial intelligence · Blockchain · Transparency · Provenance

1 Introduction

Artificial intelligence (AI) is continuously becoming more critical for businesses. As reported by the *Big Data and AI Executive Survey 2019* 92% of the firms are increasing their pace of investing in Artificial Intelligence and Big Data [21]. The availability of data drives AI development. Google, for example, surpassed 3 billion searches worldwide per day in 2012 [8]. Moreover, data has replaced oil as the most valuable resource and is becoming the new currency of the digital era [7].

Creating value from data consists of multiple steps. For Machine Learning (ML), Baylor et al. identified eight steps shown in Fig. 1. Between each of these phases, value is passed in the form of assets. Value chains are created by linking

P. Lüthi and T. Gagnaux—Equal contribution.

Published by Springer Nature Switzerland AG 2020
M. Friedewald et al. (Eds.): Privacy and Identity 2019, IFIP AICT 576, pp. 411–426, 2020.
https://doi.org/10.1007/978-3-030-42504-3_26

these assets together, named AI value chain if all involved assets are relevant for the creation of an AI solution. Machine or deep-learning that are used to solve a defined task such as medical image analysis [12] are examples of AI value chains.

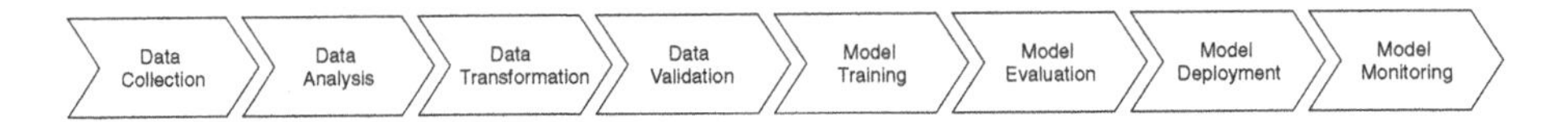

Fig. 1. The eight phases in typical machine learning workflows. Each phase creates or adds value to assets that are transferred between different phases.

Today, an AI value chain often involves experts from different organizations. It is also possible that multiple value chains coexist or interact with each other as shown in Fig. 2. As a result, AI assets need to be exchanged between actors possibly merging or splitting value chains. The figure shows how three different companies exchange AI assets to create value. Additionally, points of friction are shown where challenges arise. These are the challenges we address within this work.

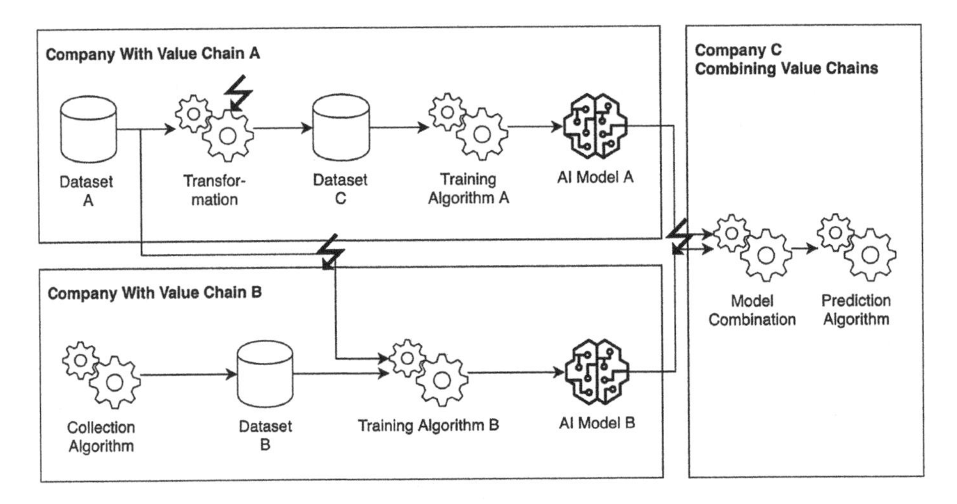

Fig. 2. Coexisting and interacting AI value chains involving different companies. Datasets and Models are shared between the three companies. Friction points where challenges arise are marked with the lightning symbol. Trust issues occur when handing over an AI asset from one company to another. Privacy and confidentiality concerns arise whenever data is processed within a value chain. Traceability and fair remuneration need to be addressed, as otherwise, companies will often not collaborate.

Participants of such AI value chains face several challenges when they try to exchange AI assets:

Trust is one of the core issues that need to be addressed. Currently, when handing over an AI asset, the provider needs to trust the receiver as well as the medium of transfer because they lose sovereignty and control over the asset. This setting discourages the sharing of AI assets because there are no good systems that allow creating such trust between two parties that do not know each other.

Privacy: From a provider's perspective, it is vital to know how their asset is used to prevent abuse. Especially if personal information is in the data, data regulations such as GDPR[1] might apply. We do not aim at providing a solution towards data privacy within this work, but are interested in providing solutions such that data providers can keep track where data is being used. In regards to GDPR, this would potentially enable the retraction of data items from multiple AI value chains at once.

Confidentiality: A provider has a key interest in keeping their assets either private or only visible to a selected number of other actors. This allows providers to protect their business interests and for example share sensitive data only with certified companies.

Traceability, Transparency and Auditability: If AI assets are exchanged without their provenance, the receiver may not have any possibility to verify their correctness. By verifying the asset's correctness, errors that otherwise might propagate through the value chain unnoticed can be detected early.

Fair Remuneration: A provider wants to know who is making use of their assets to claim their reward. From a receiver's perspective, it might be of interest to identify the involved actors to fairly compensate them.

Distributed Ledger Technology (DLT) [15] allows creating a distributed database, spread across many nodes. Blockchain [5] is one of the most prominent concepts for the actual implementation of DLT. In a blockchain all blocks are cryptographically linked with one another, depending on the data that is stored within them. This ensures that data cannot be modified once it has been stored.

In this paper, we introduce a smart contract [4] concept that allows tracking the provenance of AI assets along their value chains. The smart contract can be deployed on the Ethereum blockchain [22]. Data providers will be able to register their data using the smart contract and data consumers can register how they used this data and what kind of operations they applied on it. This smart contract creates trust for all involved parties, as it allows for independent auditing of all registered transactions to verify its integrity. We will show how this approach solves the privacy, auditability, and fair remuneration challenges.

The rest of the paper is structured as follows: Sect. 2 describes the approaches and solutions of other provenance systems and identifies a gap that we will address. Section 3 illustrates how we define our provenance model and how it solves the first three challenges privacy, auditability and fair remuneration.

[1] See https://eugdpr.org.

Section 4 shows the functionality of the smart contract and the protocol specification that allow us to solve the remaining trust and confidentiality challenges. We validate our solution using a real-world medical use case in Sect. 5. The paper closes with a discussion and pointers towards future work.

2 Related Work

In the following, we introduce existing approaches for the purpose of tracing the provenance of data in AI, and then existing provenance tracing solutions that are using blockchain technology. We discuss these approaches, introduce the limitations of existing approaches and identify a gap that we address within this work.

2.1 Provenance Tracing for Data in AI Development

With the rise of Big Data, traditional data processing and provenance tracing [2,23] became inapplicable. However, Provenance tracking has been identified as a key requirement for Big Data applications [10]. MapReduce [6] is a specialized framework enabling parallel processing of high volume data. Provenance capturing for MapReduce workflows is possible. Park et al. developed RAMP [16], a provenance capturing system that extends *Hadoop*[2].

In the area of machine learning, tracking the provenance of data points and training algorithms can be automated. The importance of single data points can be calculated. Ma et al. designed LAMP [14] to automate the partial derivative calculation of each data point evaluating its importance on the machine learning algorithm's result. Schelter et al. designed a system [19] that automatically extracts and stores provenance information of common artifacts in machine learning experiments. The system can be integrated into many popular machine learning frameworks to improve the reproducibility and comparability of machine learning experiments.

2.2 Provenance Tracing Using Blockchain

Blockchain technology is well suited in environments where trust between actors is needed as it makes a middleman obsolete. Additionally, storing provenance information on a blockchain is beneficial due to its immutable nature. Liang et al. [11] introduce ProvChain, a cloud architecture that gathers and validates provenance data by inserting them into blockchain transactions. Also, ProvChain provides security features such as tamper-proof provenance and user privacy.

Ramachandran and Kantarcioglu [17] use Ethereum to develop a secure and immutable scientific data provenance management framework called SmartProvenance that validates the provenance data using a use case-tailored verification script. During the verification, involved actors approve or reject proposed changes in a voting process.

[2] See https://hadoop.apache.org.

ProvChain and SmartProvenance focus on tracking provenance on a single file and solve auditability by storing each file change on the blockchain. As they have only one value chain they do not need to address trust or confidentiality.

Sarpatwar et al. [18] combined blockchain and AI and propose a concept for trusted AI. They illustrate the needed requirements and key blockchain constructs for trusted AI and demonstrate how these can be used to represent provenance using a federated learning [9] use case. For their specific use case, they did not need to address the trust and confidentiality challenges. They also do not address the issue of auditability of data.

2.3 Limitations of State of the Art

Existing solutions allow storing provenance of digital assets on central or distributed databases [2,23]. These databases are controlled by a single authority, making them vulnerable to untruthful modification. This requires the trust of all involved parties into this central authority, which hinders collaboration.

For Big Data and machine learning workflows, provenance tracking of individual data points with traditional databases is not feasible. Therefore, specialized frameworks have been developed that allow tracing the provenance using MapReduce.

Blockchain technology allows storing provenance without the need for a centralized authority. Consequently, all actors can participate equally and validate the transactions of other actors. Furthermore, provenance information stored on the blockchain is immutable and therefore false modifications are impossible.

Existing blockchain solutions can track the changes applied to single digital assets and AI models trained in a federated learning scenario. They, however, cannot trace all phases of a typical AI development workflow. In particular, the exchange and transformation of AI assets between interacting AI value chains are not supported, as can be seen in Fig. 2. For example, it is not possible to track how data is collected or transformed before it is used for AI model training. Furthermore, current blockchain solutions do not address how assets can be shared confidentially and selectively. The models trained in federated learning are directly stored on the blockchain and thus publicly accessible.

None of the abovementioned works cover all outlined challenges in a sufficient manner. This calls for further research on how to design a system that addresses all of them. In this work, we generalize the provenance model of Sarpatwar et al. [18] to be able to represent interacting AI value chains of any sort. Additionally, we build a system that stores this provenance model and supports the exchange of confidential assets without the need for a centralized authority.

3 Provenance Model for AI Assets on a Public Permission-Less Blockchain

Our provenance model extends the model of Sarpatwar et al. [18], which differentiates between datasets, operations, and models. This existing concept solves the privacy, auditability, and fair remuneration within the federated learning [9] context. Thus, it has several limitations: Datasets and models can exist only with a corresponding operation and the types of operations are finite. Operations always result in a model. Trust is generated by making generated models as well as their coefficients public. Confidentiality is addressed only for private datasets. Besides, their provenance model does not allow to track transformations on datasets and thus does not solve any challenges for general interacting value chains where an exchange of datasets is needed as illustrated in Fig. 2.

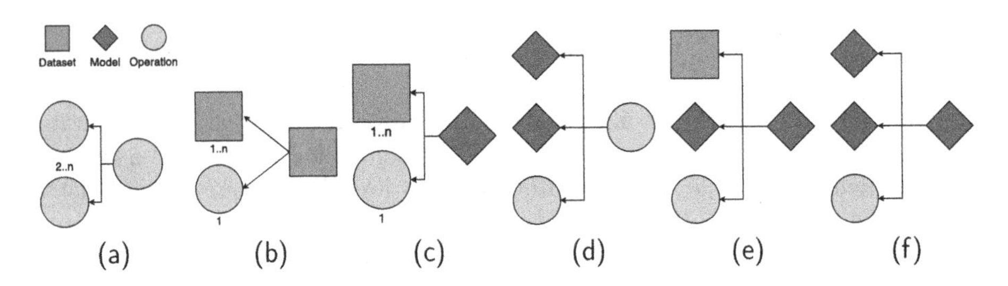

Fig. 3. The basic building blocks can be used to combine operations (a), transform datasets (b), or create models (c). By combining these types we are able to model concrete examples such as combining models to bring them into application (d), transfer learning (e), or the fusion part of federated learning (f).

Our goal is to generalize this provenance model and support interacting AI value chains and add the ability to track datasets and models without the need for the corresponding operation. Furthermore, the provenance model will allow participants to define their operations. To achieve this goal we redefine the three types of AI assets and their relations as follows:

Operation: An operation may represent any executable algorithm. In an AI value chain operations might be used for data collection, transformation, combination, reduction, analysis, training, etc. Multiple operations can be combined into one single operation as shown in Fig. 3a.

Dataset: A dataset represents any composition of digital data. A dataset might be transformed (e.g. an anonymization algorithm) resulting in a new dataset as shown in Fig. 3b or multiple datasets might be reduced (e.g. a filtering algorithm) to one dataset as shown in Fig. 3c.

Model: Combining an operation with a dataset (e.g. a classification algorithm) can result in a model as shown in Fig. 3d. Additionally, combining a model, a dataset and an operation might result in a new model (e.g. transfer learning) as shown in Fig. 3e. Lastly, multiple models can be combined into one model (e.g., through federated learning) as shown in Fig. 3f.

We represent the provenance model as a directed acyclic graph (DAG) [1] with nodes representing the AI assets. Edges in this graph either represent a **Parent Of** or **Child Of** relationship between two assets. Figure 4 shows how the interacting value chains of Fig. 4 are transformed into this graph. The "Collection Algorithm" is a parent of "Dataset B", and similarly "AI Model B" is a child of "Dataset B".

Using this graph representation, the traceability, fair remuneration, and privacy challenges can be solved using graph traversal algorithms. This will be shown in detail in Sect. 4.3.

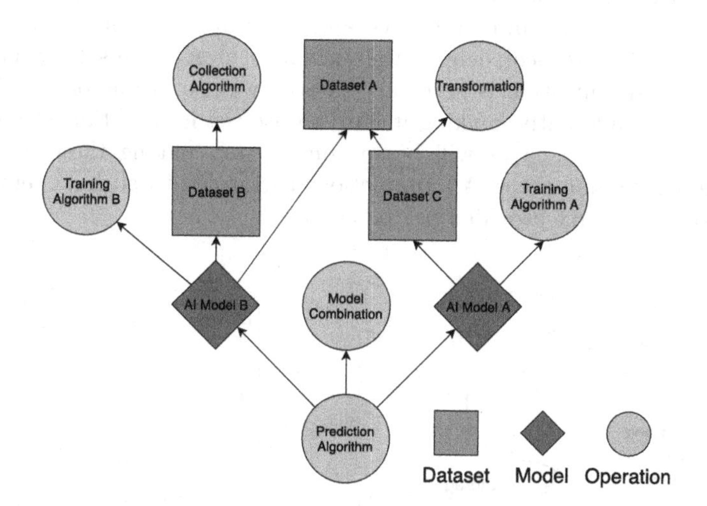

Fig. 4. The provenance model created from the interacting value chains shown in Fig. 2. The "Collection Algorithm" has a parent of relation with "Dataset B", and similarly the "AI Model B" has a child of relation with "Dataset B".

4 Implementation

In this section, we propose the implementation of the provenance model outlined in Sect. 3 on a public blockchain. Our implementation introduces the possibility to exchange confidential AI assets between different actors, therefore addressing the confidentiality challenges. We use the Ethereum blockchain [22] as a deployment and execution framework. In this framework, code is executed through so-called *smart contracts* [3]. These smart contracts can write data in two different ways *State Storage* or *Logs*. *State Storage* stores information directly in the state of the smart contract and can be modified by it. *Logs* are a cheaper form of data storage that can not be read or modified by the smart contract. When Logs are written to the blockchain they emit *events* on the smart contract, for which client applications are able to listen to.

Additionally, we define a protocol to interact with our implementation. Based on this protocol, we will show how we provide solutions for all challenges introduced in Sect. 1. Every client that implements the protocol accordingly, helps to build and enforce provenance, therefore increasing trust into the registered AI assets.

4.1 System Overview

Our system is comprised of a smart contract and a protocol specification to interact with it as shown in Fig. 5. The smart contract writes the results of all actions performed on it into the blockchain. By specifying what information is stored the contract enforces the provenance model from Sect. 3. The protocol specification defines how to work with the smart contract in such a way that the stored information can be retrieved and the provenance model can be built. Furthermore, the protocol defines the exchange of AI assets between actors. This protocol specification is necessary as otherwise, every actor could use the smart contract differently, making it hard to use the information stored on the blockchain. Finally, a client will be responsible for making these interactions accessible to a human actor. All interactions performed by such a client will use the Ethereum account provided by the actor.

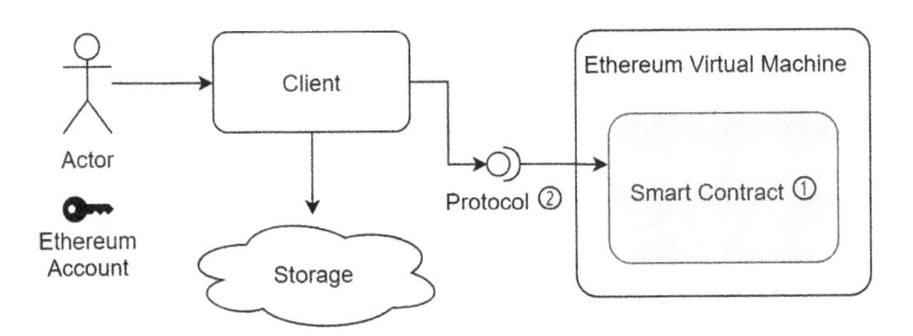

Fig. 5. System context diagram: The complete system context with the described smart contract ① (see Sect. 4.2) on the Ethereum Network and the specified protocol ② (see Sect. 4.3).

4.2 Smart Contract

The smart contract exposes the functionality that Clients will interact with. For our use, the smart contract will only store information in immutable *Logs* by executing events. These logs are searchable and remain retrievable forever. All information stored in logs this way is publicly accessible. Additionally, blockchain technology does not allow to store large amounts of data directly on the blockchain. Confidential data should, therefore, be encrypted and stored on an external storage solution as shown in Fig. 5. Our protocol will introduce a way for exchanging such encrypted data that is provided through such storage solutions.

Table 1 shows all functions exposed by the smart contract.

A new AI asset is registered using the *addAsset* function providing the following information: An *asset identifier* which is generated by computing a hash of the data item that is provided with the asset; The *URL* through which the data item of the asset can be retrieved; Additional *meta-information* that describes the contents of the asset; This information needs to be provided as a JSON

Table 1. Functions exposed on the smart contract, their description and required parameters.

Function	Description
addAsset	Register a new AI asset. The caller of the function will automatically be assigned as maintainer
transfer	Transfer ownership of an AI asset from one blockchain user to another. This operation can only be performed by the maintainer of the asset
addUrl	Add a download URL to an asset. This operation can only be performed by the maintainer of the asset
requestAccess	Request access to an AI asset. This operation can be performed by any blockchain user
grantAccess	Grant access to an AI asset. This operation can only be performed by the maintainer of the asset
getMaintainer	Retrieve maintainer of an AI asset. This operation can be performed by any blockchain user

object. The set of *parents* that are in a relationship with this asset. When the smart contract is executed, the *maintainer* of the asset is automatically set to the user that called the function.

The execution of this function will trigger several events, which in turn will store the information as logs on the blockchain. Listing 4.1 shows the events related to the provenance tracking of AI assets. When an AI asset is registered, the following events are emitted: A *Register* event that writes a log with the metadata of the asset, as well as a *URL* event that stores the URL. Lastly, all parents are written using *ParentOf* events and the inverse *ChildOf* events to ensure that the provenance graph remains in order and without cycles.

The *FormerMaintainer* event is only used when the ownership of an AI asset changes. Using these events allows us to store the complete provenance model to the blockchain, and therefore solve the privacy, auditability, and fair remuneration challenges.

```
event Register(asset_id, metadata);
event URL(asset_id, url);
event FormerMaintainer(asset_id, previous_maintainer);
event ParentOf(asset_id, parent_id);
event ChildOf(asset_id, child_id);
```

Listing 4.1. Provenance related events of the smart contract

To address the trust and confidentiality challenges, the smart contract needs to provide the functionality to exchange AI assets between actors in a confidential manner. This works under the assumption that the data item, that can be downloaded from an AI asset URL, is encrypted. The smart contract provides the two functions *requestAccess* and *grantAccess* that facilitate the exchange of the cryptographic information needed to decrypt the asset. Listing 4.2 shows the additional events that store this information on the blockchain in the form of logs. In the following section, we will introduce how the protocol specification facilitates this exchange.

```
event RequestAccess(asset_id, accessor, encryption_algorithm,
    public_key);
event GrantAccess(asset_id, accessor, encrypted_AEK);
```

Listing 4.2. Event logs on the smart contract for the exchange of AI assets

4.3 Protocol Specification

The smart contract specification from above defines a set of functionality that can be used by any application. To ensure that all applications use the smart contract in the same fashion, we design an interaction protocol. This will allow all systems implementing this protocol to rebuild the complete provenance of all AI assets stored on the blockchain. Interactions with the smart contract are enabled through the JSON-RPC interface provided by the Ethereum network[3].

Registration: Registering an AI asset is the key step to enable provenance for any asset. Figure 6a shows a sequence diagram of the protocol. It will insert a new node into the provenance model and make it visible to others. As already outlined, our system only supports assets that are made available through an external file storage provider. Should the asset contain multiple files, it can be compressed into a ZIP file. First, the Client computes a hash of the file to generate the asset identifier, and then encrypts it. We call the encryption key used in this process the Asset Encryption Key (AEK) and it will be later used when access to an asset is granted. Using the *addAsset* function of the smart contract, the asset is then registered on the blockchain, and the account making this request will automatically become the maintainer of the asset.

Accessing AI Assets: To solve the confidentiality challenge the data files representing an asset are encrypted as shown previously. The smart contract and the protocol need to provide ways for exchanging the cryptographic material in a secure way that is registered on the blockchain. Providing access to an AI asset needs an action of two actors. First, the accessor creates an encryption key pair consisting of a private (PrK) and public key (PuK). Then, they start the access process using the *requestAccess* method of the smart contract, indicating the asset with its identifier. A Client on the premises of the maintainer will react to the emitted *RequestAccess* event. If access should be granted the maintainer encrypts the AEK with the provided PuK, and invokes the *grantAccess* method on the smart contract. The Client of the accessor listens for the emitted *GrantAccess* event containing the encrypted AEK that they will be able to decrypt using their PrK. This then will allow them to download the asset from the external storage and decrypt it using the AEK. The sequence diagram of this protocol is provided in Fig. 6b. Our system does not prevent misuse of

[3] https://github.com/ethereum/wiki/wiki/JSON-RPC.

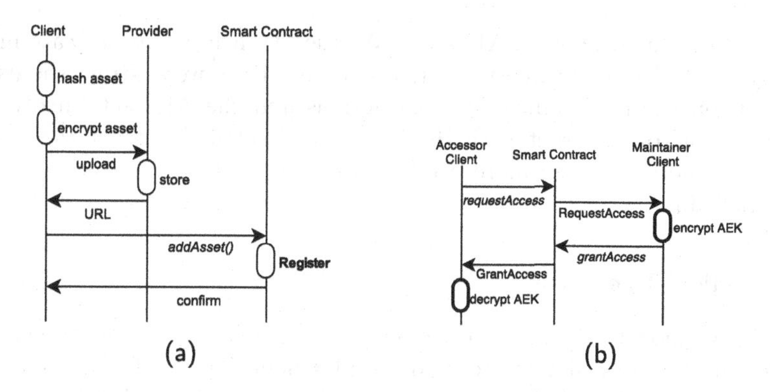

Fig. 6. Sequence diagrams of the registration (a) and the access retrieval (b) process. Function calls on the smart contract are in italics and emitted events in bold.

the keys by the involved parties, e.g. the receiver simply providing the AEK to third-parties. But the maintainer at least knows exactly whom they provided access to, limiting the number of potential bad actors. It would also allow for the creation of systems that would automatically take the actual assets offline.

Retrieve Accessors: We solve the outlined privacy challenge by making it possible for a maintainer to retrieve a list of actors who possibly accessed their AI assets. As the smart contract emits the *GrantAccess* event for all granted access requests, this can be solved easily. A client can retrieve and filter all past events using the Ethereum JSON-RPC API. Using this information, a client can then visualize for each AI asset, which Ethereum accounts currently have access. In the case of Data Erasure requests through the GDPR, the maintainer can then notify all users of such a data item to remove it as well.

Retrieve Usages and Build Provenance: Finally, we need to provide a solution for the traceability and fair remuneration challenges. We represent the potential usage of an AI asset if it is linked in a *ParentOf* relationship to any other AI asset. This also means that we can solve both challenges by building the provenance graph. As our provenance model is represented by an acyclic graph, we are able to compute it using existing graph traveling algorithms. A Client can build the provenance graph by retrieving the *ParentOf* events recursively, starting with the AI asset they are interested in.

This shows, that the protocol, combined with our generalized provenance model, addresses all challenges in Sect. 1.

5 Validation

We validate our implementation by modeling the provenance of the *Surgical Workflow Recognition for Collaborative Operation Theatre* [20] use case. The use case has been documented within the Bonseyes [13] project and reflects existing real-world use cases. In this section, we first introduce the use case, its actors, actions, and involved AI assets. We then translate this use case into the provenance model introduced in Sect. 3. For simplicity, we abstract the use case and focus on the provenance affecting actions and the AI assets involved. We show which actors interact with the smart contract and present the resulting provenance model. Furthermore, we look into the costs of registering assets on the blockchain.

5.1 Medical Use Case

In Fig. 7a we show the complete use case. It covers all steps of a machine learning workflow from data collection to model deployment. The goal of this use case is to develop AI models that support surgery inside the operation theatre. Data is acquired from different sources in an operation theatre (e.g. cameras, and sensors). The data collected during this phase is processed and managed by the data management action. Different filter and anonymization algorithms support the data management action, all represented in a single action. The result of this action is heterogeneous but anonymized RAW data. Before labeling the data needs to be pre-processed. This is performed by an algorithm that transforms and possibly aggregates the RAW data into a new dataset containing unlabeled data. This data is then fed into a labeling tool where expert labelers annotate the data.

As a result, we end up with a dataset viable for model development. Models are generated within the university hospital (e.g. by students) and externally by partners. In order to be able to compare the generated models, the data is split into three parts (training, validation, and testing). Only the training and validation datasets are made available to the development teams such that the testing dataset can be used for final evaluation. For the external partners, the data leaves the network of the hospital for the first time. Each development party uses its training algorithms that use the training data to generate an ML model. These algorithms, as well as the resulting models, need to be registered using the smart contract. Once the external parties provide their models back to the hospital, they again cross network boundaries. For an audit of the training method, also the training algorithm might be exchanged. The developed models will go through the model evaluation activity before they might get integrated into smart services.

5.2 Smart Contract Interactions and Costs for TUM Use Case

Along the outlined value chain, various AI assets are created and need to be registered on the smart contract. As data acquisition and data management are performed in sequence by the same actor, we abstract these two actions into one. In Fig. 7b, we show all AI assets identified in the value chain and the corresponding provenance graph. Each of these assets is registered according to our protocol definition. As in the real world, we use separate Ethereum accounts for every involved actor.

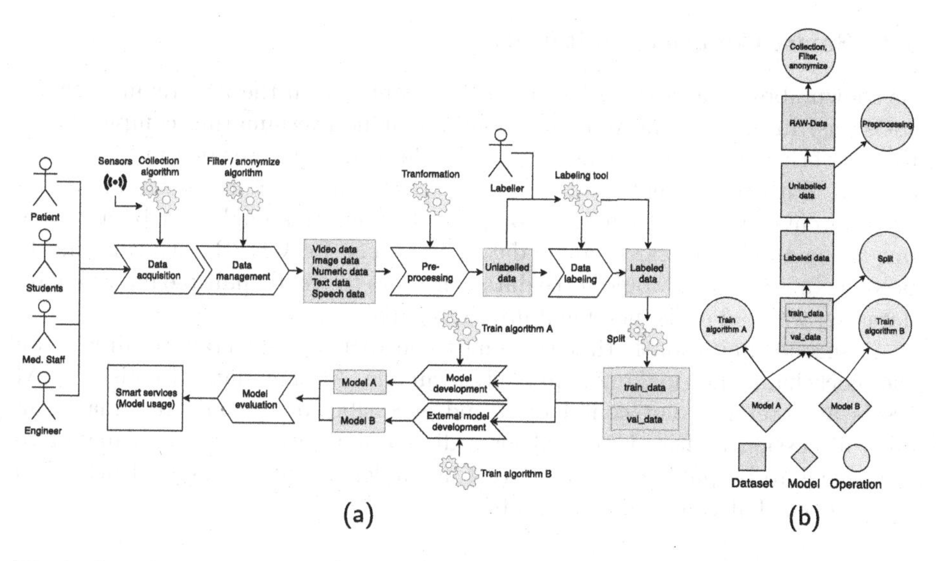

Fig. 7. Simplified use case: Surgical Workflow Recognition for collaborative operation theatre from Technical University Munich

In Table 2, we list the costs for each action that occurs inside the presented use case. These actions correspond with the smart contract interactions presented in Table 1 using the protocol introduced in Sect. 4.3. Ether is the currency on the Ethereum blockchain and gas is the execution fee for an operation. Each interaction with the smart contract costs around 0.15 USD. The variations in costs are explained by the different amount of metadata that is provided for each AI asset. Therefore, the complete provenance model of the outlined use case can be stored on the blockchain for less than 3 USD.

Table 2. Gas cost in USD-Cents of registering and exchanging AI assets for each action of the TUM use case (Ether price: 210 USD, Gas price: 1E-9 Ether)

Action	Gas	Cost(¢)
TUM registers data management algorithm	74'669	15,7
TUM registers RAW data	77'868	16,4
TUM registers unlabeled data and preprocessing algorithm	150'769	31,7
TUM registers now labeled data	80'321	16,9
TUM registers split algorithm and train/val archive	156'525	32,9
ExternalDataScientist requests access for archive	72'573	15,2
TUM encrypts AEK for archive to grant access	69'056	14,5
TUM registers own model and algorithm	149'296	31,4
ExternalDataScientist registers their model and algorithm	149'552	31,4
TUM requests access for Model B	72'573	15,2

5.3 Smart Contract Limitations

As shown above, each execution of a smart contract on the Ethereum Network costs gas. In order to prevent over-complex or non-terminating computations, the network limits the amount of gas a single transaction (or one function call) can use. With the current gas limit, an AI asset can have a maximum of around 1200 parents. In case of tracking complete datasets this might not be an issue, but if one aims at tracking single data points, e.g. for GDPR, this can quickly become an issue. This would force data collectors to create multiple intermediate datasets, where data is incrementally aggregated.

A second limitation is that the smart contract stores metadata directly on the blockchain, thus making it publicly available. All metadata of registered AI assets are therefore not confidential and accessible by anyone. It remains an open discussion if there is a need for a distinction between required metadata that needs to be publicly available (e.g. for independent audits) and metadata information that can be made private.

6 Discussion

Multiple experts from different companies collaborate in interacting AI value chains by exchanging AI assets. Tracing the provenance of these AI assets from start to finish and especially across different AI value chains bears many challenges including trust, privacy, confidentiality, traceability and fair remuneration as defined in Sect. 1.

We introduce a graph-based provenance model generalizing the federated learning provenance model from Sarpatwar et al. [18] to support interacting value chains by solving the traceability, fair remuneration, and privacy challenges. Furthermore, we provide a smart contract for the Ethereum blockchain implementing the provenance model and removing the need for a middleman, thereby solving the trust challenge. Additionally, the smart contract offers the ability to exchange assets in a confidential manner specified in a protocol fulfilling the confidentiality challenges.

Combining the provenance model and our protocol definition in a smart contract allowed us to track interacting value chains from start to finish. We validated our smart contract with an industry use case from the Technical University of Munich, covering all phases of a typical machine learning workflow except the model monitoring phase. Our results show that the smart contract is able to sufficiently trace the produced and exchanged AI assets at a low cost, as shown in Table 2.

Comparing our concept to traditional centralized provenance tracing systems, we found that trust is a significant issue when collaborating. Removing the centralized party with a decentralized blockchain increases trust among the participants and encourages collaboration. Other blockchain-based provenance work, such as ProvChain [11] and SmartProvenance [17], exceed at tracking single file changes with sophisticated privacy features but cannot trace AI assets along all phases of value chains and are therefore not addressing the abovementioned challenges. The work of Sarpatwar et al. [18] focuses on enabling trusted AI for interacting value chains performing federated learning. However, as the exchange of datasets is by design not supported, it is not possible to track interacting value chains outside of the federated learning context. This gap is where our contribution steps in providing a solution for all challenges mentioned in Sect. 1.

As every transaction on a blockchain has its cost, our solution has its limitations. We found that, with the current gas limit of ca. 7×10^6 on Ethereum, we can insert assets referencing up to 1200 other AI assets as parents. We consider this sufficient for most use cases. Our smart contract was only tested on a local network discarding the waiting time that every transaction generates during block mining. Furthermore, non-deterministic operations such as many AI training algorithms do not allow to reproduce the output asset completely and thus hinder auditability. Finally, every client using our smart contract must implement our protocol to support the confidential exchange of assets as the smart contract is not able to enforce confidentiality without it.

For future work, it would be beneficial to test our smart contract with a sophisticated web client on the Ropsten or Ethereum network to include usability related aspects, such as the above mentioned waiting time when registering or exchanging assets. Future extensions of the provenance model's type definitions, e.g. data streams, might allow increased provenance tracking coverage. Furthermore, a comparison of our solution to a smart contract on a permissioned blockchain such as Hyperledger, which does not need any exchange management in the smart contract would help to decide if the blockchain or the smart contract should be responsible for the management of the AI asset exchange. Finally, zero-knowledge proofs may provide more secure auditability capabilities because access to assets would not be required.

Acknowledgements. The project leading to this application has received funding from the European Union's Horizon 2020 research and innovation programme under grant agreement No 732204 (Bonseyes). This work is supported by the Swiss State Secretariat for Education, Research and Innovation (SERI) under contract numbers 16.0159. The opinions expressed and arguments employed herein do not necessarily reflect the official views of these funding bodies.

References

1. Bondy, J.A., Murty, U.S.R., et al.: Graph Theory with Applications, vol. 290. Macmillan, London (1976)
2. Buneman, P., Khanna, S., Wang-Chiew, T.: Why and where: a characterization of data provenance. In: Van den Bussche, J., Vianu, V. (eds.) ICDT 2001. LNCS, vol. 1973, pp. 316–330. Springer, Heidelberg (2001). https://doi.org/10.1007/3-540-44503-X_20
3. Buterin, V., et al.: A next-generation smart contract and decentralized application platform. White Paper **3**, 37 (2014)

4. Clack, C.D., Bakshi, V.A., Braine, L.: Smart contract templates: foundations, design landscape and research directions. arXiv:1608.00771 [cs], August 2016
5. Crosby, M., Pattanayak, P., Verma, S., Kalyanaraman, V., et al.: Blockchain technology: beyond bitcoin. Appl. Innov. **2**(6–10), 71 (2016)
6. Dean, J., Ghemawat, S.: MapReduce: simplified data processing on largeclusters. Commun. ACM **51**, 107–113 (2008)
7. Economist, T.: The World's Most Valuable Resource is No Longer Oil, but Data. The Economist, New York (2017)
8. Google LLC: Zeitgeist (2012)
9. Konečný, J., McMahan, H.B., Yu, F.X., Richtárik, P., Suresh, A.T., Bacon, D.: Federated learning: strategies for improving communication efficiency. arXiv:1610.05492 [cs], October 2016
10. Labrinidis, A., Jagadish, H.V.: Challenges and opportunities with big data. Proc. VLDB Endow. **5**, 2032–2033 (2012)
11. Liang, X., Shetty, S., Tosh, D., Kamhoua, C., Kwiat, K., Njilla, L.: Provchain: a blockchain-based data provenance architecture in cloud environment with enhanced privacy and availability. In: Proceedings of the 17th IEEE/ACM International Symposium on Cluster, Cloud and Grid Computing (2017)
12. Litjens, G., et al.: A survey on deep learning in medical image analysis. Med. Image Anal. **42**, 60–88 (2017)
13. Llewellyn, T., et al.: BONSEYES: platform for open development of systems of artificial intelligence. In: ACM International Conference on Computing Frontiers 2017. ACM Digital Library (2017)
14. Ma, S., et al.: LAMP: data provenance for graph based machine learning algorithms through derivative computation. In: Proceedings of the 2017 11th Joint Meeting on Foundations of Software Engineering (2017)
15. Maull, R., Godsiff, P., Mulligan, C., Brown, A., Kewell, B.: Distributed ledger technology: applications and implications. Strateg. Change **26**(5), 481–489 (2017). https://doi.org/10.1002/jsc.2148
16. Park, H., Ikeda, R., Widom, J.: Ramp: a system for capturing and tracing provenance in mapreduce workflows (2011)
17. Ramachandran, A., Kantarcioglu, M.: Smartprovenance: a distributed, blockchain based dataprovenance system. In: Proceedings of the Eighth ACM Conference on Data and Application Security and Privacy (2018)
18. Sarpatwar, K., et al.: Towards enabling trusted artificial intelligence via blockchain. In: Calo, S., Bertino, E., Verma, D. (eds.) Policy-Based Autonomic Data Governance. LNCS, vol. 11550, pp. 137–153. Springer, Cham (2019). https://doi.org/10.1007/978-3-030-17277-0_8
19. Schelter, S., Boese, J.H., Kirschnick, J., Klein, T., Seufert, S.: Automatically tracking metadata and provenance of machine learning experiments. In: Machine Learning Systems Workshop at NIPS (2017)
20. Stauder, R., et al.: Surgical data processing for smart intraoperative assistance systems. Innov. Surg. Sci. **2**(3), 145–152 (2017). https://doi.org/10.1515/iss-2017-0035
21. Davenport, T.H., Bean, R.: Big data and AI executive survey (2019). Technical report, NewVantage Partners (NVP) (2019)
22. Wood, G., et al.: Ethereum: a secure decentralised generalised transaction ledger. Ethereum Project Yellow Paper **151**, 1–32 (2014)
23. Woodruff, A., Stonebraker, M.: Supporting fine-grained data lineage in a database visualization environment. In: Proceedings 13th International Conference on Data Engineering (1997)

A Survey-Based Exploration of Users' Awareness and Their Willingness to Protect Their Data with Smart Objects

Chathurangi Ishara Wickramasinghe(✉) and Delphine Reinhardt(✉)

Georg-August-Universität Göttingen, 37073 Göttingen, Germany
c.wickramasinghe@stud.uni-goettingen.de, reinhardt@cs.uni-goettingen.de

Abstract. In the last years, the Internet of Things (IoT) and smart objects have become more and more popular in our everyday lives. While IoT contributes in making our everyday life more comfortable and easier, it also increases the threats to our privacy, as embedded sensors collect data about us and our environment. To foster the acceptance of IoT, privacy-preserving solutions are therefore necessary. While such solutions have already been proposed, most of them do not involve the users in their design. In this paper, we therefore adopt a user-centric approach and lay the ground for the future design of user-centric privacy-preserving solutions dedicated to smart home environments. To this end, we have designed and distributed a questionnaire fulfilled by 229 anonymous participants. Our objectives are two-fold: We aim at investigating (1) requirements for end user-involved privacy-preserving solutions and (2) users' readiness to be involved in their own privacy protection. Our results show that the majority of our participants are aware of the data collection happening as well as the associated privacy risks and would be ready to control and audit the collected data.

Keywords: Internet of Things · IoT · Social IoT · Privacy · Data protection · Data collection · Smart objects · Smart home · Smart environments

1 Introduction

In the last decade, the interest in IoT has tremendously increased, resulting in different products now available for and usable by the general public [7]. IoT is based on a network, where the physical objects of our environment, such as homes and workplaces, gain the ability to provide services and simultaneously play an active role in our environment via embedded systems [7]. The IoT is composed of different smart objects, which adapts to both, users' behavior and the environment. For example, smart objects include smart lamps, smart fridges, smart door locks, and smart parking systems [7]. This rapid technological development is foreseen to continue in the coming years, reaching billions

Published by Springer Nature Switzerland AG 2020
M. Friedewald et al. (Eds.): Privacy and Identity 2019, IFIP AICT 576, pp. 427–446, 2020.
https://doi.org/10.1007/978-3-030-42504-3_27

of smart objects. These objects further contribute in improving our lives in different areas, including our homes and workplaces [7]. Note that smart objects do not only present advantages in households, such as helping in managing our energy consumption, but also in companies, which can benefit from automated context-aware processes [7,12,16].

To provide these services, smart objects with embedded sensors continuously collect a vast amount of data about their environments and potential users, thus potentially endangering the privacy of their owners as well as of potential bystanders [36,38]. Privacy issues especially arise when sensitive personal data are collected and disclosed to third parties without the users' consent by smart object providers [18,27,38]. Cyber attacks caused by security vulnerabilities [28, 38], which among others are enabled by the use of low-power hardware in smart objects, can also result in information leaks and endanger users' privacy [28,38]. Recently, Bloomberg reported that thousands of the Amazon workers listen, how the users interact with Alexa, the virtual assistant in Amazon Echo devices [11]. Despite the phenomenon of the privacy paradox[1] [10,20], laws [1] still call for more user involvement in their own privacy protection process, because (1) users have the fundamental right of protecting their personal data [1,27] and (2) users' privacy behavior highly depends on the context [10].

Furthermore, the European General Data Protection Regulation (GDPR) with different rights, such as "Right for Access" and "Right to be Forgotten", calls for giving the users more transparency regarding the personal data processing, empowering users to be responsible and to have more control for the protection of the personal data processing [1]. Therefore, it is important to put the user in the center while designing usable privacy-preserving solutions for smart home environments.

Within the scope of this paper, we primarily focus on (1) the exploration of user's willingness to control the disclosure of their data and their need for transparency regarding the data collection. Based upon the results, we further focus on (2) identifying requirements in the form of user centric control mechanisms for privacy-preserving solutions for smart home environments.

The remaining paper is structured as follows. We first discuss related work in Sect. 2. We next detail the methodology of our empirical study in Sect. 3. In Sect. 4, we present the demographics and the results of our survey. In Sect. 5, we formulate design requirements based on the survey results for end user-centric-privacy-preserving solutions. Discussions and closing remarks conclude this paper in Sects. 6 and 7, respectively.

2 Related Work

Existing works can be classified as follows: (1) user surveys regarding privacy issues in IoT and (2) technical approaches allowing users to apply control mechanisms for their privacy protection.

[1] Privacy paradox explains the discrepancy between the users' stated preferences with regard to privacy protection and their actual behavior.

The first category includes surveys, which are carried out with smart objects' consumers in order to find out users' perception and opinions regarding privacy issues in IoT. Based on interviews with eleven smart home owners, Zheng et al. outline that the users' primary motivation of using smart objects lies on the convenience and connectedness [37]. They recommend developers to focus on designing (mobile) applications, allowing the users to access and control the collected data [37]. In [36], Zeng et al. also encourage developers to design smart objects considering users' privacy needs. Additionally, the user study by Martin and Nissenbaum [23] outlines that users find that the usage of their data is more relevant to users' privacy opinion than the sensitivity level of the collected data. Moreover, few large-scale surveys [3,21,25] were also carried out in order to find out users' privacy preferences while using smart objects. The results of these studies confirm that privacy issues regarding IoT objects highly depend on the context [3,25]. Some user studies also focus on privacy issues regarding smart watches and toys connected to the Internet [24,30]. These studies investigate users' awareness of privacy issues while using such smart objects. They give hints for the designers and smart object providers how to deal with users' needs regarding such smart objects in order to increase the acceptance of smart objects. In comparison to the previous works, our questionnaire-based approach focuses on identifying control mechanisms that users want to have in the data collection and disclosure process of smart home environments. These control mechanisms should empower users to protect their own privacy in their smart home environment. Additionally, our study helps to understand, whether the users want to have the empowerment to control their personal data protection while living in smart home environments.

In the second category, we consider technical solutions that allow users to apply control mechanisms for their privacy protection. Solutions such as [16,17,28,35] aim at avoiding the misuse of IoT objects and collected data by attackers for burglaries. While [16] implements a strong password authentication policy in their smart home automation system, the approach in [35] includes a set of new security policies for detecting abnormal behavior of each device. In addition, the solution presented in [17] introduces a new context-based permission system, which allows the user to decide based on collected context information, whether an abnormal action will be performed. Perera et al. propose in [28] a Privacy-by-Design framework, allowing the evaluation of IoT applications and middleware platforms based on a set of guidelines. These guidelines can be categorized in four elements: (a) Minimizing data collection, storage and disclosure without users' consent; (b) reducing the data granularity and controlling data; (c) anonymizing data and encrypting data communication and processing; (d) publishing source code, data flow diagrams of IoT applications, certifications and fulfilled compliance. Few technical frameworks, such as [4,8,14,15,26], present Role Based Access Control (RBAC) including k-anonymity mechanisms and privacy preserved access control protocols for IoT environments. These frameworks include authentication protocols to identify the user and to allow users the event-based data sharing for user-defined roles, such as doctor, partner, etc. The functionality of the frameworks is mostly explained with the help of the collected sensor data based on smart healthcare systems and other devices, such as wearables as well as few home and hotel automation devices [4,8,14,15,26]. Further approaches introduce a privacy preserving policy, authentication protocols and data encryption methods in order to protect the collected sensor data and thus users' privacy [2,5,6,9,13,22,29,31–34]. However, most of these solutions reduce the availability of original data with time delay [36]. Finally, in [19], Khan et. al. present a solution to improve the privacy concerns in case of ownership change of the smart objects. These considerations show us that the proposed technical solutions include less user involvement. In comparison to previous works, our survey thus focuses on deriving control mechanisms from the end user perspectives. The proposed technical solutions in this category can be considered in the technical implementation of the derived requirements of this paper.

To the best of our knowledge, the contribution of our research work to this body of literature is two-fold: (1) We show users' readiness to be involved in their own privacy protection, (2) we derive requirements for end user-centric-privacy-preserving solutions. This lays the ground for our future work.

3 Methodology

In order to gather insights regarding our main objectives, we carried out an online questionnaire based survey[2]. Our questionnaire including 22 questions is in English and available in Appendix A.

It is structured as follows. It gathers insights in participants' knowledge and experience with smart objects. Next, it addresses the potential participants' awareness of data collected and disclosed by smart objects and their related privacy risks. It then focuses on the participants' potential willingness to inform themselves and control the data collected and shared by smart objects, before analyzing their requirements and motivation to use privacy-preserving solutions. We distributed our questionnaire on online social network platforms, such as Xing, LinkedIn, SurveyCircle, IoT Subreddit and the community platforms of several companies in order to reach frequent Internet users. No incentives were given to the participants. It required approximately ten to fifteen minutes to be answered and consisted of multiple choice and open-ended questions. Main goal of the survey is to conduct a preliminary study as a basis for future studies rather than collecting representative insights, which are valid for the whole population.

In total, 229 participants completed the questionnaire. We have discarded invalid data sets and this resulted in 209 valid data sets. Moreover, during our

[2] At the beginning of the survey, we informed the participants that both data collection and processing take place anonymously. Note that the survey was carried out at the University of Bonn, which did not have an ethical board for reviewing user studies in our field at the time of the study. We have, however, limited the data collection to the minimum and conducted it anonymously. The participants were informed that they could opt out at any time and that their data would be removed. After agreeing to participate, each participant has been assigned a pseudonym and asked to answer a questionnaire to gather his/her demographics.

analysis we derived and tested five hypotheses based on Q_{16}, applied statistical tests, such as Mann-Whitney, multiple linear regression and correlation tests and carried out comparisons of different participant groups in order to get more insight regarding user-centric control mechanisms for privacy-preserving solutions.

4 Results

4.1 Demographics

Our respondents are predominantly male (69%). Most of them are between 26 and 50 years old (58%). 16% are under 26 and 25% over 51. The majority are German citizens (74%) followed by US Americans (7%), Sri Lankans (5%), and British citizens (5%). The remaining citizenships are distributed among 15 other nationalities from all over the world. Among the 209 participants, 166 indicated their annual income range, which ranges between "less than 25.000 Euro" and "more than 100.000 Euro". However, most of these participants (34%) annually earn "between 40.000 and 75.000 Euro".

4.2 Knowledge and Experience

In our sample, about 93% of our participants indicated that they have already heard about IoT ($Q_1, n_{Q1} = 209$). In order to get more insight, we asked our participants, in which context they have heard about IoT (Q_2). In Q_2, we also specified what we meant by IoT, by giving some examples for orientation, such as smart home, smart factory, smart city, etc. The mentioned answers were smart home (ca. 27%), Industry 4.0 (ca. 20%), smart/intelligent things (ca. 19%), smart city (ca. 19%), and smart factory (ca. 13%) ($Q_2, n_{Q2} = 209$).

In the free text box further answers were given such as smart vehicles, smart clothes, wearables, smart meters, smart grids, smart supply chain, smart campus, smart agriculture, robotic machines, smart logistics, smart health devices, and predictive maintenance. Additionally, 89% of the participants mentioned that they know or use smart objects ($Q_3, n_{Q3} = 209$). The most cited answers were: Controlling home technology apps (12%), smart voice control objects, like Amazon Echo (10%), smart health devices (8%), smart door/window locks (8%), smart bulbs (7.5%), smart fridge (7.2%), augmented /virtual reality glasses (7%), smart washing machine (6%), smart alarm clock (5.7%), smart toothbrush (4%), smart grid apps (3%) and smart scale (2.7%). The majority uses the specific smartphone apps for this purpose (71%), while 11% uses the associated web interface (Q_4). Regarding Q_5 with "How frequently do you use a device connected to the Internet, such as smart scale, fridge, wearables, watch, etc.? (smartphone, computer, smart TV does not count as smart devices in this question)", about 70% of the participants indicated that they use devices connected to the Internet frequently ($n_{Q_5} = 206$). Among the participants frequently using smart objects, 76% use them at least once per day, while the remaining 24%

only use them occasionally. The cross tables grouped by gender and age groups show that male participants and participants in general aged between 26 and 50 years significantly use connected devices more frequently than others. Based on the answers to Q_5, we derived three user categories that we use in our further analysis.

1. Frequent users: They use connected devices several times a day,
2. Average users: They use connected devices at most once a day or less,
3. Non users: They do not use any connected devices.

One of the questions we asked the participants using a 5-point Likert scale[3] was to indicate their degree of agreement regarding the statement: "In a few years, I believe that it will be difficult to live without using smart objects" (Q_{11}, $n_{Q_{11}} = 208$). About 87% of the participants agreed that it will be difficult to live without smart objects. A majority appreciated the potential advantages offered by smart objects (Q_{12}), while only 20% of the sample stated that there are no advantages offered by smart objects. The seven most frequently mentioned advantages can be summarized as follows: Facilitating to fulfill routine tasks, high comfort and convenience, low error rates, setting adjustments according to lifestyle, recording interesting personal information, outline the optimization potentials and specific things are automatically done.

4.3 Collection, Disclosure, and Privacy

A large majority (93%) of our participants believe that smart objects collect information about themselves and their environments (Q_6, $n_{Q_6} = 200$). However, only 58% agreed with the statement: "I believe that I know the information collected by smart objects." (Q_7, $n_{Q_7} = 193$). Most cited answers were location (29%) and health (25%) followed by browsing (24%) and personal data (19%), like bank details etc. (Q_8). With regard to the derived user profiles in Sect. 4.2, frequent and average users appear to be more aware of the data collection than the non users (Mann-Whitney test frequent users vs. non users: $p - value = 0.003 < 0.05$, r = 0.248, average users vs. non users: $p - value = 0.048 < 0.05$, r = 0.195). The boxplots in Fig. 1 confirm the above mentioned results. The outliers present the divergent answers from the frequently mentioned answers by most users and each number presents the data set of the correspondent anonymous respondent.

Additionally, only 24% of the sample believe in knowing the third parties, who receive the data collected by smart objects (Q_9, $n_{Q_9} = 209$: "I believe that I know the parties who have access to collected data and receive the collected data from my smart objects (Parties can be: hospital, doctors, insurance companies, institutes using data for statistics, etc.)"). As expected, the majority (72%) indicated that they do not know the third parties who get access to their collected data. Note that few participants mentioned some of the third parties.

[3] A score of 1 corresponds to a strong disagreement, while a score of 5 to a strong agreement.

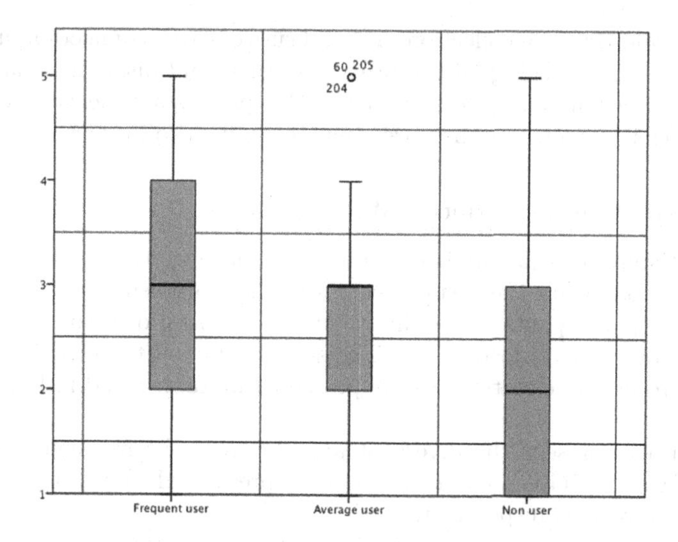

Fig. 1. Boxplots regarding "I believe that I know the information collected by smart objects." (Q_7) clustered by user profiles, derived based on the results of Q_5 (1 = strong disagreement, while 5 = strong agreement)

The mentioned parties can be clustered as follows: retail companies (like Amazon, Apple, etc.), service providers (like Google), cyber security firms, social media companies (such as Facebook, etc.), several smart object/telco providers, institutes/companies using data for statistics and analyses, (health) insurance companies, hospitals, doctors, manufacturers of the heating systems/cars, banks and government departments.

We further asked the participants to mention potential privacy issues and privacy risks in the context of IoT in a free text box (Q_{14}, $n_{Q14} = 209$). About 55% of the participants filled it. To sum up their statements, they mentioned that the smart objects on the one hand make their lives and every day routines easier, but on the other hand that those objects collect a vast amount of data and transfer those data to third parties, which are used for personalized services or offers, to create (more or less) detailed personal profiles of the users and to manipulate the smart object owners. Additionally, the participants also indicated that today they actually do not have any means to protect those data, before sharing it with third parties and that there is a lack of strict regulations regarding privacy aspects in smart environments. Moreover, approx. 93% of the participants agreed to the statement: "I believe that smart objects can endanger my privacy." (Q_{15}, $n_{Q_{15}} = 205$). The results of the Fisher's Exact test outline that there is no dependency between the participants' gender and their answers regarding Q_{15}. While 38% indicated that they take special measures to protect their privacy when using smart objects, 48% denied to do so (Q_{10}, $n_{Q_{10}} = 209$: "I take special measures (such as switching off some services etc.) to protect my privacy when using smart objects."). The mentioned measures are switching off the objects to avoid the data collection (35%), disuse of cloud connection, using local servers (6%), and checking all the privacy settings and disabling smart objects or features, which are not necessary (57%). One participant mentioned that s/he actually does not know any measures that really help to protect privacy (2%).

4.4 Information and Control Willingness

In the next step, respondents had to rate the following statements on a 5 point Likert scale. Based on the results regarding the statements, we investigate to which degree participants are willing to exercise control over the data collected and shared by smart objects (Q_{16}). The statements and the distribution of the values regarding those statements are presented in Table 1 and in Fig. 2, respectively.

The outliers present the divergent answers from the frequently mentioned answers of the participants and each number presents the data set of the correspondent anonymous respondent.

About 94% of the participants indicated that they want to have more information about the data collected by smart objects about themselves in a smart home environment ($Q_{16.1}$). 96% also precised that they want to have an overview of all the information collected by used smart objects ($Q_{16.2}$). About 94% of the participants would like to see a summary of the collected data over a given period, such as daily, weekly, monthly ($Q_{16.3}$). Additionally, about 84% of the consumers want to have more information about collected data in their own smart home environments in real-time ($Q_{16.4}$).

In addition, for more transparency approx. 92% of our sample want to have more information about the associated risks to their privacy resulting from sharing the collected data ($Q_{16.5}$). About 87% of the participants also want to have more information about the associated personal and social advantages by sharing the collected data of their own smart home environment ($Q_{16.6}$). Further analysis shows that there is a positive correlation between statements $Q_{16.5}$ and $Q_{16.6}$ (r = 0.608, significant at the 0.01 level - 2-tailed). This confirms that the users, who want to have information about associated risks to their privacy by data sharing, at the same time also want to have information about the associated personal and social advantages resulting from data sharing.

Almost all participants (97%) indicated that they would like to have control about the data collected and shared by smart objects ($Q_{16.7}$ and $Q_{16.8}$). Note that there are no statistically significant differences between the answers given by participants belonging to different user profiles (Q_5, Mann-Whitney test, $p - values > 0.05$) and users applying special measures to protect their privacy (Q_{10}), as shown by a multiple linear regression test ($p - values > 0.05$).

Additionally, a large majority would like to determine which third parties are able to access their collected data (95% for $Q_{16.9}$) and for which purpose (95% for $Q_{16.10}$). To exercise this control, only 86% of the participants are willing to spend time on auditing the collected data ($Q_{16.11}$). An automated system taking privacy decisions would be supported by 74% of the participants ($Q_{16.12}$). About 96% of the participants also mentioned that they are willing to have clear policies

Table 1. Submitted statements in the Q_{16}: "Please enter your answer regarding the following statements."

Q#	Statements
$Q_{16.1}$	I would like to have more information about the data collected by smart objects about me in a smart home environment. ($n_{Q16.1} = 206$)
$Q_{16.2}$	I would like to have an overview of all the information collected by my smart objects. ($n_{Q16.2} = 206$)
$Q_{16.3}$	I would like to have a summary of the collected data over a given period, e.g. daily, weekly, monthly, etc. ($n_{Q16.3} = 206$)
$Q_{16.4}$	I would like to know in real-time about the data collected in my smart home environment. ($n_{Q16.4} = 205$)
$Q_{16.5}$	I would like to have more information about the associated risks to my privacy by sharing the collected data. ($n_{Q16.5} = 205$)
$Q_{16.6}$	I would like to have more information about the associated personal and social advantages by sharing the collected data from my smart home environment. ($n_{Q16.6} = 206$)
$Q_{16.7}$	I would like to be able to control which information is collected about myself in a smart home environment. ($n_{Q16.7} = 206$)
$Q_{16.8}$	I would like to be able to control the data shared by my smart objects. ($n_{Q16.8} = 206$)
$Q_{16.9}$	I would like to be able to determine who is able to access my data. ($n_{Q16.9} = 206$)
$Q_{16.10}$	I would like to be able to determine which information is used for which purpose. ($n_{Q16.10} = 206$)
$Q_{16.11}$	I would be willing to spend time to audit the data collected about myself in a smart home environment. ($n_{Q16.11} = 205$)
$Q_{16.12}$	I would prefer having an automated system taking privacy decisions for me after learning about my privacy risk awareness. ($n_{Q16.12} = 206$)
$Q_{16.13}$	I would like to have clear policies with the provider regarding the collected data from my own smart home environment. ($n_{Q16.13} = 205$)

with the data provider regarding the data collection in smart home environments ($Q_{16.13}$).

We finally asked the participants to indicate their motivating factors to use smart home objects while having the control on the data collected and shared (Q_{17}, $n_{Q_{17}} = 209$) and most cited reasons were: "Having control about the usage of collected data about me" (32%) followed by "feeling myself secured and protected" (29%) as well as "avoiding to draw a digital biography" (22%) and "having information about the data consumer of my data" (15%). Two participants wrote in the free text box that nothing motivates them to use smart home objects while having the control on the data collected. Two further participants also mentioned that they are not going to use any smart objects because they believe that there is no security while using those objects. One participant

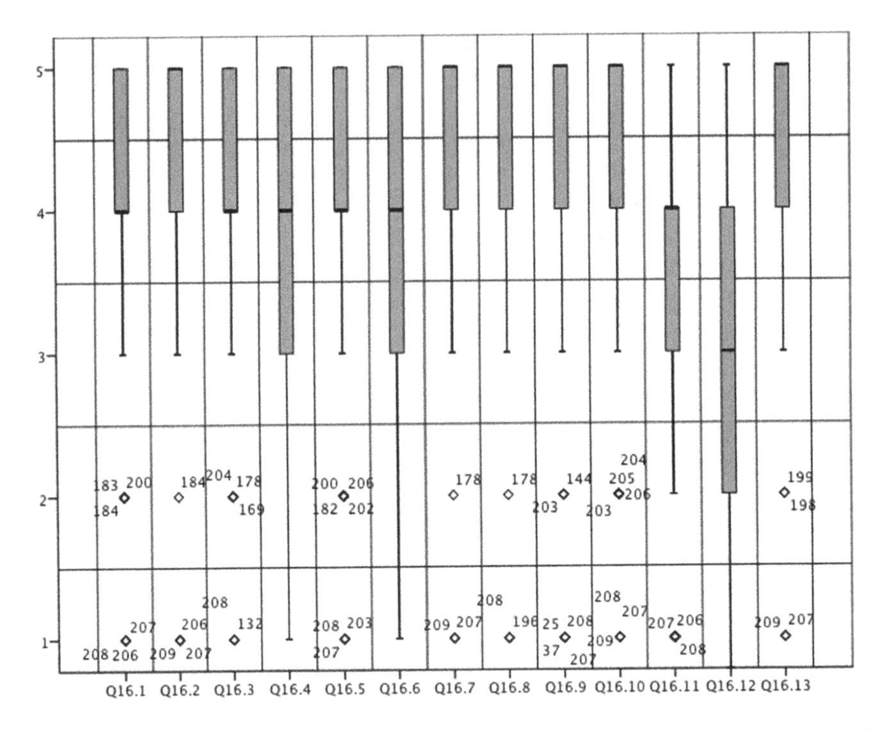

Fig. 2. Boxplots for submitted statements in the Q_{16} (1 = strong disagreement, while 5 = strong agreement)

explicitly indicated that s/he never wants to waste any time on validating or examining the collected data.

Derived Hypotheses: The analysis of participants' answers regarding Q_{16} make it obvious, that the majority of the participants want to have more transparency and associated data sharing information. It is still to be verified whether they want to have these information in order to consider this input while controlling the data sharing process. For this purpose, we derived and tested five hypotheses to verify, whether there is a significant dependency regarding the fact that the users want to have more transparency and associated data sharing information in order to consider this input while controlling the data sharing process. Fisher's Exact tests confirm the all five hypotheses with $p = 0.00 < 0.05$. This means that users want to control the information collected by smart objects (H_1) and are willing to have an overview of those information (H_3). The results also confirm that the users are willing to get information regarding the privacy risks arising from the publication of data (H_2). Similarly, the test results also confirm that the users want to determine who is able to access the shared data (H_4) and for which purpose the data are used (H_5), while controlling the data shared with third parties.

5 Derived Requirements for User-Centric-Privacy-Preserving Solutions

We leverage the results of the survey to derive requirements in form of control mechanisms for user-centric-privacy-preserving solutions in what follows. We define the identified control mechanisms as *User-Centric-Control-Points (UCCP).*

Data Object Tagging: Considering the results of the whole survey, we can derive that it will be useful to allow the user to tag his/her different smart objects as sensitive or non-sensitive object depending on the data the objects collect. For example, in one case a smart table mat could be non-sensitive, because it just collects information whether something is on the mat or not, while in another case smart fridge or calorie scanner could be tagged as sensitive, because those objects collect data regarding the users' eating and living habits. The users can consider these tagging when they make their decision whether they want to share the data while considering the associated privacy risks and advantages arising from sharing the collected data. These considerations allow us to derive the UCCP 1: Allowing the user to tag the smart object as a sensitive or non-sensitive object.

Data Minimization: Our results in Sects. 4.3 and 4.4 show that the participants do not have transparency and vast experience regarding the data collection and disclosure process in smart home environments. Participants' answers also outline that they want to have more information regarding the data collection process in smart home environments. Furthermore, the results underline that the participants want to control which information are collected in their smart home environment. These results help us to derive the following UCCP 2: Allowing the user to select which information are collected by the used smart objects in his environment.

Data Granularity: The results in Sect. 4.4 let us conclude that (1) the participants want to have an overview of all collected information and (2) that they want to review the collected data over a preferred period, such as weekly, monthly, and thus to determine the granularity of data collection. Based on these results, we can derive our next UCCP 3: Allowing the user to set in which granularity the data are collected for users' review.

Data Sharing: The participants' answers in Sect. 4.4 also outline that they want to have the opportunity to get more information regarding associated privacy risks, personal and social associated advantages resulting from sharing the collected data. This leads us to derive our next UCCP 4: Allowing the user to view the associated risks and social or personal advantages arising by sharing the collected context-data.

Data Disclosure Limitations: The results in Sect. 4.4 show that the users want to control the data shared. In this context, it is to be mentioned that the GDPR also demands to obtain consent for the processing of the personal data in understandable and simply accessible form from the users [1]. These results help us to derive another UCCP 5: Allowing the user to control the data sharing. This UCCP must include at least the following two options: Share the data or delete the data without any third parties getting access to the data.

Data Access Limitations: The results in Sect. 4.4 show that the participants want to have the control on who is able to access their data and for which purpose in case of data sharing. This leads us to derive the next UCCP 6: Allowing the users to determine who is able to access the data and for which purpose the data are used. This UCCP should also allow the users to set the settings, how the data are disclosed, anonymized or non-anonymized.

As listed above, the results of our survey allowed us to derive six UCCPs as requirements for end user-centric-privacy-preserving solutions for smart home environments. Furthermore, the derived UCCPs can also be categorized into three categories. These categories are transparency of data collection, data impli-

Table 2. Derived UCCPs for user-centric-privacy-preserving solutions based on the results

Category	UCCPs	Short description
Transparency of data collection	UCCP 1: Data object tagging	Allowing the user to tag the smart object as a sensitive or non-sensitive object
	UCCP 2: Data minimization	Allowing the user to set which information are collected by the used smart objects in his environment (Q_7, Q_8, Q_9, $Q_{16.1}$, $Q_{16.7}$)
	UCCP 3: Data granularity	Allowing the user to set in which granularity the data are collected and saved for users' review ($Q_{16.2}$, $Q_{16.3}$)
Data implication	UCCP 4: Data sharing	Allowing the user to view the associated risks and social or personal advantages arising by sharing the collected context-data ($Q_{16.5}$, $Q_{16.6}$)
Data access	UCCP 5: Data disclosure limitations	Allowing the user to control the data sharing: Share the data or delete data without any third party getting access to the data ($Q_{16.8}$)
	UCCP 6: Data access limitations	Allowing the users to determine who is able to access the data and for which purpose the data are used and how the data are disclosed, as anonymised or non-anonymised data ($Q_{16.9}$, $Q_{16.10}$)

cations and data access. In the first category, UCCPs are summarized, which allow users to gain more transparency regarding data collection. The second category includes UCCPs, which provide the data sharing information for users. The third category comprises UCCPs, which allow users to control the data sharing process (Table 2).

In future work, the privacy-preserving solutions with integrated UCCPs must be investigated in terms of their usability and applicability in everyday life.

6 Discussion

The answers to the questions on information collection and disclosure ($Q_{6,7,9}$) represent an interesting aspect. Regarding the results of Q_6 a majority (93%) of the participants are aware about the data collection in smart environments, but only 58% agreed in Q_7 that they know the information collected by smart objects. The comparison of the user profiles shows that the frequent and average users are more aware of the data collection than the non users. Furthermore, only 24% of the participants indicated in Q_9 that they know the third parties, who receive the data collected by smart objects. By considering these answers, it becomes obvious that the users have a lack of knowledge regarding the sensitive data collected by the smart objects and the third parties receiving those sensitive data without users' consent. This might have two reasons: (1) users have less transparency about the collected data and/or (2) users put less effort in finding out which information are collected, because they do not receive such information in an understandable way. Although approx. 93% indicated in Q_{15} that they "...believe that smart objects can endanger my privacy", only 55% in Q_{14} mentioned potential privacy risks in IoT-context and only 38% mentioned in Q_{10} that they are taking special measures to protect their own privacy. Regardless of the derived user profiles and users applying special privacy preserving measures, later in Q_{16} a majority indicated that they want to have control over data collection and disclosure in their smart home environment. This might mean that the missing transparency about collected as well as disclosed data and missing opportunities for the users to control the data collection and disclosure process give only limited permission for the users to be responsible for their own privacy protection. Additionally, it is not clear whether the 38% of the respondents (Q_{10}) apply those measures regularly or just once in a while. If the measures are applied regularly, then it is clear that those participants actively protect their privacy. Furthermore, there were also few participants, who mentioned that they do not have any motivation to deal with user-centric-privacy-preserving solutions, because they believe that there is no privacy in today's data-driven world.

Additionally, testing the five hypotheses (presented at the end of Sect. 4.4) helps us to conclude that users want to have more transparency and information regarding data collection and disclosure process in their smart home environment in order to consciously control the disclosure of the collected data. The results of $Q_{16.11}$ also provide the insight that the users want to be involved in their privacy protection while living in smart home environments. These results are not surprising and further emphasize that efficient user-centric-privacy-preserving solutions for data control are necessary. Consequently, the majority of our participants mentioned in Q_{17} that they are motivated to live in smart home environments while having control over the data collected and shared. In contrast, few participants pointed out that they will not use such devices due to privacy issues. Furthermore, based on the results regarding Q_7, Q_8, Q_9, Q_{16} we were able to derive six UCCPs as requirements for user-centric-privacy-preserving solutions for smart home environments, explained in Sect. 5. The derived UCCPs underline the aspects of GDPR [1] and can only provide added value if they are considered in the whole lifecycle of the personal data processing in smart home environments. Furthermore, the presented technical approaches in the second category in Sect. 2 can be considered in the implementation of the derived UCCPs, for instance RBAC mechanisms in the implementation of UCCP 6. The derived UCCPs based on users' answers represent their stated opinion and must be evaluated in a real smart home environment scenario. This will help us to find out whether the users accept to spend their time with such solutions in their everyday lives in order to protect their own privacy, as mentioned in $Q_{16.11}$.

Finally, our questionnaire-based survey has few limitations: As already preluded, the answers of the participants represent their opinion, but not necessarily their actual behavior. The participants may also be biased and not representative of the whole population. Indeed, the fact that they voluntarily answered the questionnaire, which was published on several Internet platforms and invitations sent by emails, may indicate that they may be more altruistic or that they are strongly willing to live or to deal with smart objects and environments than those who have not answered it. Ultimately, our findings mainly reflect the views of participants, who have access to the Internet.

7 Conclusions and Outlook

Within the scope of this paper, we have investigated based on the questionnaire-based survey (1) requirements for end user-centric privacy-preserving solutions and (2) users' readiness to be involved in their own privacy protection. Overall, our participants have indicated that they would like to have more transparency regarding data collection and more control over data collection and disclosure in smart home environments. Based on their answers, we have developed a set of requirements called UCCPs for privacy-preserving solutions in smart home environments that would allow users to exercise a control over their personal data. Our findings also underline that the participants want to be involved in their own privacy protection.

In future work, we plan to conduct user studies to investigate possible discrepancies between users' real behavior and stated opinion regarding the utilization of privacy-preserving solution with integrated UCCPs in smart home environments. We further plan to investigate the usability aspects of such solutions. Finally, further research work is needed to develop clear policy frameworks regarding the personal data processing in smart home environments, which have to be taken into account by the smart objects' providers.

Acknowledgement. We would like to thank Michael Friedewald for his helpful comments and the survey participants. Furthermore, we would like to thank Daniel Franke for providing us feedback on early versions of our questionnaire as well as Birgit Schuhbauer for proofreadings.

A Appendix - Survey questions

Q_1: Have you heard about the Internet of Things (IoT)?

Possible answers: *Yes/No/I prefer not to answer this question.*

Q_2: If yes, in which context have you heard about IoT?

Possible answers: *smart home/Industry 4.0/smart factory/smart city/smart things/Others: free text box for participants/I prefer not to answer this question.*

Q_3: Indicate the smart objects or applications (apps) that you know or use in your everyday life? (multiple choice possible)

Possible answers: *smart fridge/controlling home technology apps/smart grid apps/smart bulbs/smart alarm clock/smart toothbrush/smart washing machine/ smart voice control objects, such as Amazon echo/augmented/virtual reality glasses/smart scale/smart health devices/smart door/window locks/smartphone/ Others: free text box for participants/I do not utilize smart objects/I prefer not to answer this question.*

Q_4: If you already use smart objects, how do you get access to the collected data from your smart objects, through an app or web interface? (multiple choice possible and please click on respective smart object to choose the option between app and web interface)

Possible answers: *smart fridge/controlling home technology apps/smart grid apps/smart bulbs/smart alarm clock/smart toothbrush/smart washing machine/ smart voice control objects, such as Amazon echo/augmented/virtual reality glasses/smart scale/smart health devices/smart door/window locks/smartphone/ I prefer not to answer this question.*

Q_5: How frequently do you use a device connected to the Internet, such as smart scale, fridge, wearables, watch, etc.? (smartphone, computer, smart TV does not count as smart devices in this question).

Possible answers: *more than 20 times per day/less than 20 times per day/once per day/very rare/other options: free text box for participants/I do not use any smart objects/I prefer not to answer this question.*

Q_6: When using smart objects, I believe that those objects collect information about myself and my environment.

5-level Likert scale: *1 strongly disagree/2 disagree/3 partly agree/4 agree/5 strongly agree/Don't know/Prefer not to answer.*

Q_7: I believe that I know the information collected by smart objects. 5-level Likert scale: *1 strongly disagree/2 disagree/3 partly agree/4 agree/5 strongly agree/Don't know/Prefer not to answer.*

Q_8: Please indicate which kind of information you believe the smart objects collect about you. (multiple choice possible)

Possible answers: *location data/browsing data/data about your health, such as weight, movements, food purchase/personal data, e.g. bank details, relationships from voice control, finger prints from door locks etc./other information: free text box for participants/I prefer not to answer this question.*

Q_9: I believe that I know the parties who have access to collected data and receive the collected data from my smart objects. (Parties can be: hospital, doctors, insurance companies, institutes using data for statistics, etc.)

Possible answers: *I know/If you know, please mention in short key points the parties/I do not know/I prefer not to answer this question.*

Q_{10}: I take special measures (such as switching off some services etc.) to protect my privacy when using smart objects.

Possible answers: *Yes/If yes, please indicate the measures you usually take and the conditions/No/I prefer not to answer this question.*

Q_{11}: In a few years, I believe that it will be difficult to live without using smart objects.

5-level Likert scale: *1 strongly disagree/2 disagree/3 partly agree/4 agree/5 strongly agree/Don't know/Prefer not to answer.*

Q_{12}: What do you think about the advantages you have by using smart home objects? (Participants had the possibility to rate on each statement by using the following Likert Scale.)

5-level Likert scale: *1 strongly disagree/2 disagree/3 partly agree/4 agree/5 strongly agree/Don't know/Prefer not to answer.*

- facilitating to fulfill the everyday and routine tasks.
- high comfort and convenience
- low error rates (humans make mistakes more easily and frequently than machines)
- automatic adjustments of settings regarding my current lifestyle
- the smart objects record interesting information about myself and my surroundings.
- smart objects outline the optimization potentials regarding my everyday work or my health plan etc.
- specific things are automatically done by smart objects and releasing you from these tasks so you can spend time for more important things.
- no advantages
- Others: free text box for participants

Q_{13}: Please enter the answer for the following question to make sure, that you are not a robot: $150 + (2 \times 2) =$

Q_{14}: What do you know about privacy issues in the context of Internet of Things, specifically, to what extent do you understand potential privacy risks? (such as third parties get access to your data/to your house or to your bank account etc.).

Possible answers: *Please enter your answer here/I prefer not to answer this question.*

Q_{15}: I believe that smart objects can endanger my privacy.

5-level Likert scale: *1 strongly disagree/2 disagree/3 partly agree/4 agree/5 strongly agree/Don't know/Prefer not to answer.*

Q_{16}: Please enter your answer regarding the following statements. (Participants had the possibility to rate on each statement by using the following Likert Scale.)

5-level Likert scale: *1 strongly disagree/2 disagree/3 partly agree/4 agree/5 strongly agree/Don't know/Prefer not to answer.*

- $Q_{16.1}$: I would like to have more information about the data collected by smart objects about me in a smart home environment.
- $Q_{16.2}$: I would like to have an overview of all the information collected by my smart objects.
- $Q_{16.3}$: I would like to be able to control which information is collected about myself in a smart home environment.
- $Q_{16.4}$: I would like to know in real-time about the data collected in my smart home environment.
- $Q_{16.5}$: I would like to have a summary of the collected data over a given period, e.g. daily, weekly, monthly, etc.
- $Q_{16.6}$: I would like to have more information about the associated risks to my privacy by sharing the collected data.
- $Q_{16.7}$: I would like to have more information about the associated personal and social advantages by sharing the collected data from my smart home environment.
- $Q_{16.8}$: I would like to be able to control the data shared by my smart objects.
- $Q_{16.9}$: I would like to be able to determine which information is used for which purpose.
- $Q_{16.10}$: I would like to be able to determine who is able to access my data.
- $Q_{16.11}$: I would be willing to spend time to audit the data collected about myself in a smart home environment.
- $Q_{16.12}$: I would prefer having an automated system taking privacy decisions for me after learning about my privacy risk awareness.
- $Q_{16.13}$: I would like to have clear policies with the provider regarding the collected data from my own smart home environment.

Q_{17}: Indicate the factors that motivate you to use smart home objects while having the control about the data collected.

Possible answers: *Feeling myself secured and protected/It is not possible for third party data consumers to draw a digital biography from my daily routine/Having control about the usage of collected data about me/Others: free text box for participants/I prefer not to answer this question.*

Q_{18}: How old are you?

Q_{19}: What is your gender?

Possible answers: *Male/female/I prefer not to answer this question.*

Q_{20}: What is your nationality?

Q_{21}: What is your annual income range? (Euro values or equivalent in local currency)?

Possible answers: *Less than 25.000 Euro/25.000 Euro–40.000 Euro/40.000 Euro–75.000 Euro/75.000 Euro–100.000 Euro/More than 100.000 Euro/I prefer not to answer this question.*

References

1. Regulation (EU) 2016/679 of the European Parliament and of the Council of 27 April 2016 on the protection of natural persons with regard to the processing of personal data and on the free movement of such data, and repealing Directive 95/46/EC (General Data Protection Regulation). OJ L119/1, pp. 1–88 (2016)
2. Alcaide, A., Palomar, E., Montero-Castillo, J., Ribagorda, A.: Anonymous authentication for privacy-preserving iot target-triven applications. Comput. Secur. **37**, 111–123 (2013)
3. Apthorpe, N., Shvartzshnaider, Y., Mathur, A., Reisman, D., Feamster, N.: Discovering smart home Internet of Things privacy norms using contextual integrity. Proc. ACM Interact. Mob. Wearable Ubiquit. Technol. **2**(2), 1–23 (2018). Article 59
4. Barhamgi, M., et al.: Enabling end-users to protect their privacy. In: Proceedings of the 2017 ACM Asia Conference on Computer and Communications Security, pp. 905–907 (2017)
5. Cao, H., Liu, S., Guan, Z., Wu, L., Deng, H., Du, X.: An efficient privacy-preserving algorithm based on randomized response in IoT-based smart grid. In: 2018 IEEE SmartWorld, Ubiquitous Intelligence & Computing, Advanced & Trusted Computing, Scalable Computing & Communications, Cloud & Big Data Computing, Internet of People and Smart City Innovation, pp. 881–886 (2018)
6. Cao, J., Carminati, B., Ferrari, E., Tan, K.L.: CASTLE: continuously anonymizing data streams. IEEE Trans. Depend. Secur. Comput. **8**(3), 337–352 (2010)
7. Carretero, J., García, J.D.: The Internet of Things: connecting the world. Pers. Ubiquit. Comp. **18**(2), 445–447 (2014)
8. Chakravorty, A., Wlodarczyk, T., Rong, C.: Privacy preserving data analytics for smart homes. In: 2013 IEEE Security and Privacy Workshops, pp. 23–27 (2013)
9. Chan, E.M., Lam, P.E., Mitchell, J.C.: Understanding the challenges with medical data segmentation for privacy. In: Usenix Conference on Safety, Security, Privacy and Interoperability of Health Information Technologies, pp. 1–10 (2013)

10. Coopamootoo, K., Gross, T.: Why privacy is all but forgotten. Proc. Priv. Enhanc. Technol. **4**, 97–118 (2017)
11. Day, M., Turner, G., Drozdiak, N.: Amazon workers are listening to what you tell Alexa. https://www.bloomberg.com/news/articles/2019-04-10/is-anyone-listening-to-you-on-alexa-a-global-team-reviews-audio
12. Friedewald, M., Da Costa, O., Punie, Y., Alahuhta, P., Heinonen, S.: Perspectives of ambient intelligence in home environment. Telemat. Inform. **22**, 221–238 (2005)
13. Guo, L., et al.: A secure mechanism for big data collection in large scale internet of vehicle. IEEE Internet Things J. **4**(2), 601–610 (2017)
14. Huang, X., Craig, P., Lin, H., Yan, Z.: SecIoT: a security framework for the Internet of Things. Secur. Commun. Netw. **9**(16), 3083–3094 (2016)
15. Huang, X., Fu, R., Chen, B., Zhang, T., Roscoe, A.: User interactive Internet of Things privacy preserved access control. In: 2012 International Conference for Internet Technology And Secured Transactions, pp. 597–602 (2012)
16. Hussain, S.H., Geetha, S., Prabhakar, M.A.: Design and implementation of an adaptive model for sustainable home automation using Internet of Things (IoT). Int. J. Adv. Eng. Tech. **VII**(1), 827–829 (2016)
17. Jia, Y.J., et al.: ContexloT: towards providing contextual integrity to appified IoT platforms. In: Network and Distributed System Security Symposium (NDSS), pp. 1–15 (2017)
18. Karaboga, M., et al.: Das versteckte Internet: Zu Hause - im Auto - am Körper. White paper, Forum Privatheit und selbstbestimmtes Leben in der digitalen Welt (2015)
19. Khan, M.S.N., Marchal, S., Buchegger, S., Asokan, N.: chownIoT: enhancing IoT privacy by automated handling of ownership change. In: Kosta, E., Pierson, J., Slamanig, D., Fischer-Hübner, S., Krenn, S. (eds.) Privacy and Identity 2018. IAICT, vol. 547, pp. 205–221. Springer, Cham (2019). https://doi.org/10.1007/978-3-030-16744-8_14
20. Kokolakis, S.: Privacy attitudes and privacy behaviour: a review of current research on the privacy paradox phenomenon. Comput. Secur. **64**, 122–134 (2017)
21. Lee, H., Kobsa, A.: Understanding user privacy in Internet of Things environments. In: 2016 IEEE 3rd World Forum on Internet of Things (WF-IoT), pp. 407–412 (2016)
22. Li, X., Niu, J., Bhuiyan, M.Z.A., Wu, F., Karuppiah, M., Kumari, S.: A robust ECC-based provable secure authentication protocol with privacy preserving for industrial Internet of Things. IEEE Trans. Ind. Inform. **14**(8), 3599–3609 (2017)
23. Martin, K., Nissenbaum, H.: Measuring privacy: an empirical test using context to expose confounding variables. Columbia Sci. Technol. Law Rev. **18**, 176–218 (2016)
24. McReynolds, E., Hubbard, S., Lau, T., Saraf, A., Cakmak, M., Roesner, F.: Toys that listen: a study of parents, children, and internet-connected toys. In: Proceedings of the 2017 CHI Conference on Human Factors in Computing Systems, pp. 5197–5207 (2017)
25. Naeini, P.E., et al.: Privacy expectations and preferences in an IoT world. In: Thirteenth Symposium on Usable Privacy and Security (SOUPS 2017), pp. 399–412 (2017)
26. Ouaddah, A., Abou Elkalam, A., Ait Ouahman, A.: FairAccess: a new blockchain-based access control framework for the Internet of Things. Secur. Commun. Netw. **9**(18), 5943–5964 (2016)
27. Pasquale, F.: The Black Box Society: the Secret Algorithms that Control Money and Information. Harvard University Press, Cambridge (2015)

28. Perera, C., McCormick, C., Bandara, A.K., Price, B.A., Nuseibeh, B.: Privacy-by-design framework for assessing Internet of Things applications and platforms. In: Proceedings of the 6th International Conference on the Internet of Things (ACM), pp. 83–92 (2016)
29. Su, J., Cao, D., Zhao, B., Wang, X., You, I.: ePASS: an expressive attribute-based signature scheme with privacy and an unforgeability guarantee for the Internet of Things. Future Gener. Comp. Sys. **33**, 11–18 (2014)
30. Udoh, E.S., Alkharashi, A.: Privacy risk awareness and the behavior of smartwatch users: a case study of Indiana University Students. In: 2016 Future Technologies Conference (FTC), pp. 926–931 (2016)
31. Wang, X., Zhang, J., Schooler, E.M., Ion, M.: Performance evaluation of attribute-based encryption: toward data privacy in the IoT. In: 2014 IEEE International Conference on Communications (ICC), pp. 725–730 (2014)
32. Yang, J.C., Fang, B.X.: Security model and key technologies for the Internet of Things. J. China Univ. Posts Telecommun. **18**, 109–112 (2011)
33. Yang, L., Humayed, A., Li, F.: A multi-cloud based privacy-preserving data publishing scheme for the Internet of Things. In: Proceedings of the 32nd Annual Conference on Computer Security Applications (ACM), pp. 30–39 (2016)
34. Yang, W., Li, N., Qi, Y., Qardaji, W., McLaughlin, S., McDaniel, P.: Minimizing private data disclosures in the smart grid. In: Proceedings of the 2012 ACM Conference on Computer and Communications Security, pp. 415–427 (2012)
35. Yu, T., Sekar, V., Seshan, S., Agarwal, Y., Xu, C.: Handling a trillion (unfixable) flaws on a billion devices: rethinking network security for the Internet-of-Things. In: Proceedings of the 14th ACM Workshop on Hot Topics in Networks. Article no. 5 (2015)
36. Zeng, E., Mare, S., Roesner, F.: End user security and privacy concerns with smart homes. In: Proceedings of the Thirteenth USENIX Conference on Usable Privacy and Security (SOUPS 2017), pp. 65–80 (2017)
37. Zheng, S., Apthorpe, N., Chetty, M., Feamster, N.: User perceptions of smart home IoT privacy. In: Proceedings of the ACM on Human-Computer Interaction, vol. 2, Article no. (CSCW 200), pp. 1–20 (2018)
38. Zhou, W., Jia, Y., Peng, A., Zhang, Y., Liu, P.: The effect of IoT new features on security and privacy: new threats, existing solutions, and challenges yet to be solved. IEEE Internet Things J. **6**(2), 1606–1616 (2019)

Self-Sovereign Identity Systems
Evaluation Framework

Abylay Satybaldy(✉), Mariusz Nowostawski(✉), and Jørgen Ellingsen(✉)

Computer Science Department, NTNU, Gjøvik, Norway
{abylay.satybaldy,mariusz.nowostawski}@ntnu.no

Abstract. Digital identity systems have been around for almost as long as computers and have evolved with the increased usage of online services. Digital identities have traditionally been used as a way of authenticating to the computer systems at work, or a personal online service, such as an email. Today, our physical existence has a digital counterpart that became an integral part of everyday life. Self-Sovereign Identity (SSI) is the next step in the evolution of the digital identity management systems. The blockchain technology and distributed ledgers have provided necessary building blocks and facilities, that bring us closer to the realisation of an ideal Self-Sovereign Identity. But what exactly is an ideal Self-Sovereign Identity? What are the characteristics? Trade-offs? Here, we propose the framework and methodology that can be used to evaluate, describe, and compare SSI systems. Based on our comparison criteria and the evaluation framework, we present a systematic analytical study of existing SSI systems: uPort, Sovrin, ShoCard, Civic, and Blockstack.

Keywords: Identity management systems · Digital signatures · Peer-to-peer computing · Cryptographic protocols · Computer security

1 Introduction

Our world becomes increasingly digital and our lives heavily rely on digital systems. Data and information is valuable for governments and industry alike. Digital industry leaders have built their business on targeted marketing and big data for years and the public slowly realises the scope of the impact it has on the social structures, as well as on individuals. The legal systems in different countries take closer look into digital identities and the consequences of people not owing their own information. In Europe, the General Data Protection Regulation (GDPR) and other similar initiatives are put in place to ensure that governments and industry is managing personal information correctly and that the individuals' data is not misused.

As more and more businesses and governmental entities understand the value of data, personally-identifiable information, and digital identities there emerge new challenges related to information security and individual freedoms. There are inherent trade-offs that need to be addressed and explored. On the one hand,

Published by Springer Nature Switzerland AG 2020
M. Friedewald et al. (Eds.): Privacy and Identity 2019, IFIP AICT 576, pp. 447–461, 2020.
https://doi.org/10.1007/978-3-030-42504-3_28

we give up information about us to online services at a rapid rate, for the industry to store, analyse and process it all in new and creative ways. Sometimes to gain financial value, as with targeted marketing and content. Sometimes to achieve political leverage. On the other hand, governments want us to reveal the data and not use privacy-preserving techniques as they can be misused for criminal or terrorism purposes. The data sharing creates opportunities for businesses to reach their target audience with high accuracy. It also offers value to users, through sharing interesting content around topics that they care about.

As a society, we are trading privacy for convenience and the negative effects have just started to show up. In 2018 the scandal surrounding Facebook and Cambridge Analytica has been revealed and demonstrated how much value and impact social data has. Cambridge Analytica is suspected to have been able to influence the United States 2016 presidential election and the Brexit vote on the British referendum to leave the European Union [7,8].

While information sharing on social media is a choice, society today expects us to have an online presence and identity. We need to have an account on Google or Apple to use our smart phones and an e-mail address to register additional online accounts for various services. We need identity to access our bank account, to purchase travel tickets and to board an airplane. It is nearly impossible to function without having some form of online identity.

The concept of Self-Sovereign Identity (SSI) is that each user fully controls their own information. Users can add, remove and share attributes at their own discretion. They can share their email to a service provider and then subsequently revoke the rights to use this email. Federated identities made some of these options available by allowing users to register with one provider, and then use that identity to access other services that accepted the same standard. One of the major problems of this approach is that the federated provider(s) users choose to register with has all the user information and control over it. Under self-sovereign identity model, identities must not be held by a singular third-party entity.

When Bitcoin first launched in 2009 it introduced the notion of a decentralised ledger [23]. The blockchain technology and decentralised consensus mechanisms offer technological solution to the 3rd party trust problem. Despite the fact that the majority of industry and academic efforts focus on currencies and transferring ownership of value, there is a growing interest in the use of blockchain and related decentralised technologies for managing identity.

While distributed ledgers have taken identity systems a step closer to an ideal Self-Sovereign Identity, they continue to struggle with some fundamental challenges. Most of the proposed and implemented identity systems are built on the infrastructure of digital currencies, and interactions with the network require some monetary value to be transferred. Those that are not, are partly centralized to manage consensus in the network. An ideal Self-Sovereign Identity System should be free and decentralized and the solutions proposed and implemented today have made compromises [4].

2 Self-Sovereign Identity

Even though, the Self-Sovereign Identity as a term is now well-established, both in academia and in the industry, there is no agreed consensus upon the actual formal definition. Using Peter de Marneffe's principles for Self-Sovereignty [18] and Weik's definition of Identity [30], we can describe Self-Sovereign Identity in its simplest form as *a digital representation of the individuals characteristics, description and identifiers where no government, or organization, can violate our right to chose our level privacy or celebrity with our identity attributes.*

Cameron wrote *The Laws of Identity* in 2005 while working as Identity and Access Architect at Microsoft Corporation [9]. *The Laws of Identity* [9] precedes the first distributed ledger [22] and the first mention of the concept Self-Sovereign Identity [4]. While unaware of the technological advance of distributed ledgers in the years to come, Kim Cameron elaborate on Microsoft Passport and how privacy concerns and reliability on a single organization in part lead to the failure of it's mission to become the identity system for the internet. He defines the need for user control, minimal disclosure, and a portable and inter-operable system.

While *The Laws of Identity* is a good foundation for *identity*, one of the first references to identity sovereignty occurred in February 2012 by Moxie Marlinspike in his post about *Sovereign Source Authority* [17]. Subsequently, in 2016 Christopher Allen introduces the term Self-Sovereign Identity and defines it using 10 principles [4]. He expands on *the Laws of Identity* by defining how the identity should exist, why the system and its algorithms must be transparent, and how it must be persistent while still being portable and inter-operable.

Abraham in a whitepaper on Self-Sovereign Identity [1] details the requirements of a Self-Sovereign Identity Concept. The definitions of Abraham align well with Christopher Allen, although Christopher Allen's principles are noticeably more comprehensive. Abraham expands on the definition of control and adds *"all access of identity data of a user should be logged for later verification"*. This is trade-off between security and privacy, and should at least be optional for the user.

The concept of Self-Sovereign Identity (SSI) could become the next stage in the evolution in identity management. SSI can be defined as a permanent identity owned and controlled by the person or entity to whom it belongs to without the need to rely on any external administrative authority and without the possibility that this identity can be taken away. That requires not just the inter-operability of a user's identity across multiple locations, with the user's consent, but also, true user control of that digital identity, and full user autonomy. To accomplish this, a self-sovereign identity must be transportable; it cannot be locked down to one site, provider or locale. This can be enabled by an ecosystem that facilitates the acquisition and recording of attributes, and the propagation of trust among entities leveraging such identities.

3 Evaluation Framework

There is no definitive criteria of how to evaluate SSI systems, but the framework proposed by Allen represents a comprehensive spectrum of SSI requirements, encompassing security, data integrity and privacy. In addition to Allen's work, we have used Cameron's *"The Laws of Identity"* [9] - another well-established evaluation framework for digital identity systems. We decided to add the *"Usability"* law to our evaluation model as the role of user experience is essential in building successful digital identity system. We used these guiding principles as a reference to evaluate the current state of Self-Sovereignty in the published and proposed Self-Sovereign Identity Systems.

The requirements are as follows:

1. *User control and consent*
   Users must control their identities. Users should always be able to refer to it, update it and access their own data. All the claims and personal identity information must be easily retrieved by user when needed. Sharing of personal data must only occur with the consent of the identity owner.
2. *Privacy and protection*
   The rights of users must be protected on the protocol level. The users must be able to choose their privacy model. In order to support users' privacy, disclosure of claims must be minimized. When personal data is disclosed, that disclosure should involve the minimum amount of information necessary to accomplish the task at hand. Given the long-living ambition of SSI implementations, long-term (e.g., post-quantum, information theoretical, etc.) security guarantees must be taken into account.
3. *No trust in central authority*
   Identities must not be held by a single third-party entity, even if it is a trusted entity that is expected to work in the best interest of the user. The necessary guarantees and checks must be part of the protocol layer.
4. *Portability and persistence*
   Identities must be long-lived and preferably last for as long as the identity owner wishes; a user must have a *right to be forgotten*, which means, ability to remove some of the operational data from the SSI system. Personal information and services about identity must be transportable. Transportable identities ensure that the user remains in control of his identity and can also improve an identity's persistence over time. Identity owners should be able to recover their private keys and credentials in case of loss or theft of their primary access device.
5. *Transparency*
   Systems and protocols must be transparent. The systems used to administer and operate a network of identities must be open, both in how they function and in how they are managed and how they are updated. The algorithms should be free, well-known and architecture independent;

6. *Interoperability*
   Identities should be as widely usable as possible. The SSI system should enable global identities which could cross international boundaries and various system implementations. Transportable identity is sometimes mentioned as a requirement to be fulfilled.
7. *Scalability*
   As the user demands are increasing all the time, the identity systems must be highly scalable. SSI system should be able to maintain its effectiveness throughout even if there are additions or expansions in aspects such as resources or the number of end users without disrupting its functionality.
8. *Usability*
   The user experience must be consistent with user needs and expectations. Identity owners must be able to count on a consistent user experience across various technology platforms and services.

4 State-of-the-Art Developments in Practice

There is a number of start-ups and companies that directly tackle the problem of digital identity management. Examples include Sovrin, uPort, OLYMPUS [24], SelfKey [14], Blockstack, Civic, ShoCard, lifeID [15] and MultiChain [20]. Many of those systems utilize blockchain technology to solve current identity management challenges.

In this review, we evaluate five representative proposals: Sovrin, uPort, ShoCard, Civic and Blockstack. We selected these systems because they provide technical documentation, reports and white papers with the most technical details of their designs, have sizable online communities, and serve similar purpose to the broader landscape of self-sovereign identity management schemes.

4.1 Sovrin

The Sovrin Foundation have set out on a mission to standardize and create an infrastructure for Self-Sovereign identities, using blockchain as storage for Distributed Identities. In theory, anyone can issue or verify an identity [27]. The Sovrin blockchain has been designed only for identity, and is takes the digital trust away from centralized Certificate Authorities (CAs) to a web of trust model. The Sovrin SSI model is not dependent on any particular distributed ledger, but can work with any blockchain that meets the required properties. With Sovrin, trust is established using verifiable claims. As stated by [21] a verifiable claim is a claim shared by any person, organization, or thing that can be instantly verified by the receiving party. Verifiable claims, along with all private data, are stored off-ledger by each self-sovereign identity owner, wherever the owner decides. Sovrin utilizes a permissioned blockchain using nodes called *Stewards* to achieve global consensus. The *Stewards* are approved by the Sovrin Foundation, a non-profit foundation with a board of twelve trustees plus a Technical Governance Board. The open source code base was transferred to the Linux Foundation to become the Hyperledger Indy project [13], and was officially launched with the first 10 *Stewards* in 2017 [28].

Analysis. The main goal of Sovrin is to provide users with full control over all aspects of their digital identity. Each user can choose which attribute credentials are revealed and who can access them (1).

The selective disclosure uses an advanced privacy-enhancing technique known as a zero-knowledge proof (ZKP). Moreover, Sovrin provides pairwise-pseudonymous *Decentralized Identifiers* (DIDs) [25] and public keys for every relationship to protect the privacy of users without sacrificing functionality. By design, each DID is linked to pseudonymous network address provided by private agent, thus, user can securely exchange verifiable claims and any other data with another user over an encrypted private channel. These private agents can operate in the cloud layer or on edge devices (mobile phones, laptops, tablets, etc.). If encrypted data was stored on a public blockchain, the encryption could have been broken in the future (for instance, with quantum computing). Therefore, no private data is stored on the Sovrin ledger which makes the system satisfy the security aspects (2).

Although there is no central authority in the Sovrin Network, users must rely on agencies and on the Stewards. The trust and transparency are addressed through the web of trust and the reputation and non-collusion of the Stewards. Private data is stored on the users device or a chosen agent and does not exist in any service provider's system or database (3).

Sovrin expects to create a market for agencies who act on behalf of users and support portability of personal data. Identity owners can recover their private keys and credentials in case of loss as Sovrin provides a decentralized means of revocation using cryptographic accumulators. Data should use system-independent semantic graph formats such as JSON-LD to ensure portability across providers (4).

The Sovrin protocol is based entirely on open standards and software developed with open source licenses. The infrastructure support and core software is built on top of the Hyperledger Indy Project [13]. The Sovrin Network is governed by nonprofit Sovrin Trust Framework composed of stewards (volunteer experts) in digital identity, privacy, and policy from around the world (5).

The Sovrin Network consists of stewards from all over the world. The first 24 stewards span 11 countries and include different financial institutions, startups, non-governmental organizations and personal data authorities. As the SSI identity ecosystem expands, new agencies, and new stewards, will join. The Sovrin Foundation expects to collaborate and support interfaces with other existing digital identity systems (6).

To achieve high scalability, the Sovrin Network uses two rings of nodes: a ring of validator nodes to accept write transactions, and a much larger ring of observer nodes running read-only copies of the blockchain to process read requests (7).

User integration is not well-defined. It is unclear how it will happen and it remains an open question. The smart cryptographic tools deployed by the identity system should be transformed and delivered to end-users in a user-friendly way. Sovrin is still in the early development phase, and the user experience should be thoroughly addressed by developers and services joining the identity ecosystem (8).

4.2 uPort

uPort is an open identity system that allows users to register their own identity on Ethereum, send and request credentials, sign transactions, and securely manage keys and data. The uPort mobile app generates a key pair and deploys three smart contracts for each identity. A *Proxy Contract* is deployed as the user's unique identifier, a *Controller Contract* to provide identity access, and a *Recovery Quorum Contract* to help with recovery of a user's identity should they lose access to it [29]. For key recovery, identity owners must nominate trustees, who can activate a vote to create a new public key via the *Controller Contract*; once a quorum is reached, the controller replaces the lost public key with a new nominated key by invoking a dedicated function of the proxy. The uPort Registry cryptographically links profile data or attributes to a uPort identifier and stores the data as a plain JSON structure [16].

Analysis. uPort provides a framework for identity owners to gather attribute credentials from an ecosystem of identity providers and does not perform any identity proofing. User controls creation of uPortIDs and can share personal information with 3rd parties at their own discretion. The personal information is stored on-device and off-chain with IPFS and is always accessible by the user. uPort provides more control and responsibility over uPortIDs to the hands of its users (1).

For low-value accounts uPort identifiers can be created without disclosure of personal data. Moreover, the lack of inherent link between uPortIDs makes the identity system robust. The JSON profile of user in the registry is visible to public, which could leak information about specific attributes and compromise privacy of users (2).

Users can prove ownership of uPortID without relying on a central authority and the authentication of a user can be done on mobile device. As only identity owner alone has write access, a user can selectively discard their negative attributes such as a criminal conviction, a low credit score, and others that the user does not want to be associated with her or his account. Moreover, uPort has some centralized elements, such as the messaging server to transfer attributes, a push notification center and an application manager. Those can potentially represent a source of censorship or enforcement in the system (3).

uPort provides users with Self-Sovereign Wallet to manage keys, credentials and identity data. The private key is stored on the user's mobile device. Support of key recovery protocol helps users to maintain a persistent digital identity even after the loss or theft of mobile device. The key recovery protocol is based on the act of nominated trustees who can raise a vote to set a new public key via the controller smart contract (4).

uPort is an open identity system built on public, permissionless blockchain, Ethereum, and consists of open-source protocols and developer tools (5).

uPort provides tools for building user-centric Ethereum apps. Developers can freely create uPort compatible applications. Moreover, the platform supports simple authentication, single-sign-on, and easy integration for Dapps or other applications. uPort joined the Decentralized Identity Foundation [12], and aims to develop a standard for everyone (6).

uPort is building identity infrastructure on top of the Ethereum, the public blockchain has significant scalability problems. By having identity rooted on-chain uPort gets benefits of the verification and security of the Ethereum. The majority of interactions in uPort including the transactions and the data storage happen off chain which make the system more scalable. However, with the increase of user base and the expansion of system enabling faster and cheaper transactions could become a major hurdle for uPort (7).

Identity owner can access the service by the mobile application which provides a consistent user experience. Also, QR code-scanning feature makes it easy to initiate interactions with a relying party. However, users could find uPort's key recovery protocol and personal data storage schemes too cumbersome or difficult to understand (8).

4.3 ShoCard

ShoCard is a digital identity and authentication platform built on the public Bitcoin blockchain. User's identity information is stored in the form of signed cryptographic hashes in the blockchain. The blockchain is used to validate that information and confirm third parties that have certified the identity of the user. There is no store or central location that holds user's private information and pieces of a user's identification do not need to be spread in other services in order to authenticate or prove ownership of an account. On the blockchain, the user initiates an identity verification handshake with the third party. The information is fully encrypted and placed in a secure data envelope that only the recipient can decrypt. Once both identities are confirmed the transaction can proceed. The system can write five million user records on a publicly verifiable blockchain in 30 min. ShoCard was founded in 2015 [11,26].

Analysis. ShoCard allows users and entities to establish their identities with one another in a secure, verified way. Storage of personal information and sharing with 3rd parties is controlled by the end user (1).

The data is not available in any readable form to any third party or ShoCard without the user sharing information first. ShoCard only uses the blockchain to verify and does not store any personal data on it. ShoCardIDs are bootstrapped with an existing trusted identity document (for example, passport or driver's license). This may make ShoCard less attractive for low-value online accounts (2).

Although there is no central database of logins to become a target for hacking, ShoCard central servers act as an intermediary between users and relying parties. Attribute validation protocol relies on ShoCard servers that write the encrypted, signed credentials onto the blockchain (3).

ShoCard is partly centralized and it creates uncertainty about the longitudinal existence of a ShoCardID. If the ShoCard servers eventually stop working, identity owners would be unable to use their digital identities and credentials. Users are also not supported with cryptographic key management (4).

ShoCard identities are stored in the Bitcoin blockchain which is inherently public and transparent. Users keep their private keys safe on their own smartphones or computers, and they have a public key that can be used by services to verify their ID using ShoCard (5).

Companies can incorporate the ShoCard technology into their existing app or website through a software development kit. ShoCard facilitates for a multitude of different authentication and verification purposes, including KYC, authentication, auditable authorization, and attestation of credentials. Moreover, ShoCard provides enterprise-level identity authentication through mobile device (6).

Shocard relies on public blockchain, but the architecture of ShoCard is designed to be highly-scalable. The system can write five million user records on a publicly verifiable blockchain in 30 min. Moreover, ShoCard is designed to be blockchain-independent to position themselves to take advantage of future advances in technology (7).

The authentication process is simple to follow. It begins when a user downloads the app to create their ShoCard ID. They take a picture of a valid, government-issued piece of identification from which ShoCard extracts the personal information. The user confirms the data, self-certifies, and either creates a passcode or opts for their phone's fingerprint scan (8).

4.4 Civic

Civic is another system that creates an ecosystem for identity verification services based on a blockchain. Key pairs are generated by a third party wallet, and the identity information is stored on the user's device [31]. Civic and the blockchain only receives hashes of the data, and stores these as a ERC20 tokens on the Ethereum network. Civic's network accommodates three different but interdependent entities: users, validators, and service providers. The users are anyone who wishes to use the protocol to register an identity. Validators are responsible for verifying an identity's authenticity on the blockchain's distributed ledger. They can then sell this information to service providers who need to verify their customer's identities, exchanging the data for a Civic token (CVC). CVC will be used as a form of settlement between participants to an identity-related transaction within the ecosystem [10,19]. Civic is built on the Ethereum blockchain and uses smart contracts to oversee data attestation and payout for this work.

Analysis. Civic's identity platform uses a verified identity for multi-factor authentication on web and mobile apps without the need for usernames or passwords. Users are in control of their secured data and they only have to provide the information they are comfortable sharing (1).

Identity data is encrypted and stored only in the Civic app on user mobile devices. With third-party authenticated identity data, Civic cannot be compelled by a foreign government or criminal organization to invalidate identity data. Personal identity information that was attested are stored in the form of verified hash into a Merkle tree and recorded in the blockchain. The portions of the Merkle tree can selectively be revealed which enhances user control by allowing the identity owner to selectively reveal pieces of personal information in various circumstances (2).

The Civic ecosystem will incentivize participation by trusted identity verification providers known as *Validators* who run the nodes on the public blockchain and sign transactions. Civic reshapes the role of centralization and embraces an open ecosystem of validators. Thus, proposed identity system does not have a single point of failure, but it is not fully decentralized and has similar consensus mechanism as Sovrin (3).

Moreover, identity data is revocable by the authenticating authority. For example, if a user changes their last name, then the former/invalid last name data is revoked on the blockchain by the authenticating authority. Thus, to maintain a persistent digital identity, Civic users should rely on authentication authorities (4).

Civic uses the public blockchain and has no proprietary software or infrastructure which makes the system more transparent (5).

One of the advantages of Civic identity ecosystem is a healthy partner network which includes financial institutions, government entities, and utility companies. Civic wants to create identity verification market where banks, utility companies, local, state, and federal agencies, etc. will be able to verify the attributes of the identity of an individual or business on a blockchain and through the use of smart contracts, validators will be able to price their identity verification and offer them to other participants (6).

Although Civic is built on the Ethereum blockchain, the system retains its effective performance as it has a central role in their ecosystem and uses validators that verify identity information (7).

Users can download mobile app to gain access to Civic's identity platform. Moreover, Civic is planning to launch the *Civic Wallet*. By coupling identity with other features, this wallet will allow users to transact using traditional and cryptocurrencies more securely and easily than with other wallets. However, the project's development is in early stages and wider user adoption has not yet been achieved (8).

4.5 Blockstack

Blockstack is a decentralized computing network and app ecosystem that puts users in control of their identity and data. Instead of relying on servers operated by applications, users are able to provide their computation and storage resources. The Stacks blockchain provides the global consensus and coordination layer for the network and implements the native token of the Blockstack network called the Stacks token. A Blockstack ID is a decentralized identity which provides a user with a single identity to log into decentralized applications (DApps). Blockstack PBC, a Public Benefit Corp, along with open-source contributors develop the core protocols and developer libraries for Blockstack [2,3,6].

Analysis. Applications built on Blockstack enable users to own and control their data directly. Blockstack applications store data with the user (using their private data lockers) and don't need to store any user data or access credentials at the server side (1).

Sharing of content is achieved through a secure and encrypted medium. However, the set of profiles is globally visible and discoverable via the blockchain, which could leak information about specific attributes and compromise privacy of users (2).

Blockstack protocol removes central points of control and failure. Compared to traditional internet applications, the business logic and data processing runs on the client, instead of on centralized servers hosted by application providers (3).

A decentralized storage system, called Gaia [5], enables user-controlled private data lockers. Data on Gaia is encrypted and signed by user-controlled cryptographic keys. Users can host these data lockers on a cloud-provider or other data storage options like private hosting. Importantly, the user controls which provider to use. However, Blockstack does not have the key recovery protocol and users cannot reset their keys in case of loss or theft (4).

Blockstack's first-generation blockchain (Stacks) operates logically on top of the Bitcoin network. The Blockstack open-source repositories contain developer libraries for a number of different platforms. The open-source community behind the project maintains tutorials, API documentation, and system design documents which are available on Github (5).

Blockstack is modular, and developers can easily customize it and integrate alternative technologies. Blockstack takes a "full-stack" approach and provides default options for all the layers required to develop decentralized applications. As of early 2019, there are more than 100 applications built on Blockstack (6).

To achieve scalability, Blockstack minimizes application logic and data at blockchain layer. The use of off-chain name registrars enables over a hundred users to register in a single blockchain transaction, which could support hundreds of thousands of user registrations per day. The decentralized storage system also scales well because it does not index individual user files or file-chunks but indexes pointers to user's storage backends (7).

Blockstack provides users with a universal username that works across all applications without the need for any passwords. However, the Blockstack ecosystem is still in its early days and currently provides only the desktop version of *Blockstack Browser* that allows users to create and manage Blockstack IDs and explore decentralized apps (8).

Table 1. A summary of analysis based on SSI principles

SSI requirements	Sovrin	uPort	ShoCard	Civic	Blockstack
1. User control and consent	✔	✔	✔	✔	✔
2. Privacy and protection	✔	✖	✖	✔	✖
3. No trust in a central authority	✔	✔	✖	✔	✔
4. Portability and persistence	✔	✔	✖	✖	✖
5. Transparency	✔	✔	✔	✔	✔
6. Interoperability	✔	✔	✔	✔	✔
7. Scalability	✔	✖	✔	✔	✔
8. Usability	✖	✔	✔	✔	✖

*A table cell with ✔ indicates that we found evidence that a system complied with a specific requirement, and a cell with ✖ indicates that a system does not fully comply with a specific requirement.

5 Discussion

The definition presented in Sect. 2 represents requirements for an idealised Self-Sovereign Identity System. Distributed Ledger Technology has provided us with tools to decentralize applications that previously required a trusted third-party. Those new solutions and technologies present an opportunity to rethink how we manage identities and personal information digitally.

The academic landscape on the topic is sparse. Most of the information is published in whitepapers and through industry implementations. A true Self-Sovereign Identity system might have an unappealing non-profit requirement that limit the business validity of SSI as a Service, or SSI for profit.

All the compared Self-Sovereign Identity Systems in Sect. 4 provide the *identity owner* with full control over their identity and the ability to selectively disclose claims and attributes. They all also embrace the need for trust and transparency by providing source code available for review.

The evaluation in Sect. 4 and its accompanying Table 1 provide an overview of the current state of the Self-Sovereign Identity landscape. This comparison reveals four major shortcomings that are present in all of the discussed systems.

Centralization. The systems are all based on blockchains and inherit the security of the network making it resistant to third party influence. However, a system is not decentralized just by incorporating partial data storage in a blockchain. The collusion of a large mining pools in Bitcoin network and of validators in permissioned blockchains could potentially introduce censorship problems when maintaining an identity ecosystem. In order to create a truly decentralized system every aspect of the system must be outside any one organizations control. This would reduce the economic viability for organizations to pursue Self-sovereign Identity as a Service and the incentive to do research and development to create the underlying system.

Identity Revocation. Identity revocation represents one of the most challenging issues within SSI systems as there is no central server which can easily revoke users associated cryptographic keys. The systems presented here do not store the anonymous credentials and secret keys. The systems rely on a user to keep the data in a secure storage in his smartphone or PC. On-device storage increases security by being inaccessible for adversaries even in its encrypted state. However, this new approach that relies on nontechnical users to keep credentials safe comes with undeniable risks. The current on-device solution that is used in some of these systems are not persistent through failure or loss of device. For example, Blockstack and ShoCard do not support any end user key management. While Sovrin and uPort have proposed the promising concept of key recovery, their work are still in progress. Creating a secure, cost-efficient and usable management of identities is not a simple task. Self-sovereign identity requires innovative, effective and well-analyzed solutions to support it.

Human Integration. Self-sovereign identity systems should be designed to solve the challenges faced by end users. So far, we have seen that the evaluated implementations mainly focus on the underlying technology, not the user interaction. Usable interface and key management and privacy implications for users are not addressed yet in sufficient depth. The future SSI schemes with a novel technological underpinning but developed with impractical end user interaction are unlikely to create widespread uptake.

Economic Barriers. Traditional decentralized blockchains like Bitcoin and Ethereum require miners to reach consensus in the network. These hashing-operations are keeping the network safe by so called *proof of work* mechanism. This relies on having large computational power that any one adversary never will be able to outperform the legitimate network nodes. This computational race is power and hardware expensive and as long as this is the fundamental technology behind a Self-Sovereign Identity System there must be cost associated with usage. One alternative would be to shift the cost over to the service providers, but its hard to find other than economic incentives for them to participate in the system. Running a permissioned ledger like *Sovrin* does create a solution for the cost challenge but is at the same time it shifts the system towards a more centralized model.

6 Conclusion

In this article we have studied and provided our vision of the concept of Self-Sovereign Identity. We have analyzed the current state-of-the-art and investigated existing working implementations: Sovrin, uPort, ShoCard, Civic, and Blockstack. We investigated how these early experiments with SSI address the identity management challenges and how they map to the ideal, proposed SSI model. The current strengths and limitations of SSI systems were discussed by applying a new evaluation framework. The framework has been based on the literature and it represents a synthesised model based on frameworks proposed earlier by other authors. We can see that identity platforms presented in this paper have different level of decentralization and incorporate blockchain to achieve their goals – in an attempt at creating a true self-sovereign identity management system.

Technology innovations in the area of cryptographic protocols, blockchain and distributed ledger as well as decentralised consensus systems might provide us with the practical building blocks to implement and realise an SSI. The distributed and decentralised ledger creates a jurisdictional space that makes it harder to be manipulated by powerful actors and could provide necessary censorship resistance. A system for truly self-sovereign identities has not yet been achieved in the current state of the field, however the discussed systems represent various attempts that address the core challenges. The recommendation based on the research in this paper would be to re-evaluate how we formally approach the issue. Corporations and for-profit organizations might never benefit economically from a true self-sovereign identity system, and therefore, it is paramount that a non-profit organizations and academia take the lead in the effort and innovate new ways of managing digital identities.

References

1. Abraham, A.: Self-sovereign identity. https://www.egiz.gv.at/files/download/Self-Sovereign-Identity-Whitepaper.pdf. Accessed 1 Mar 2019
2. Ali, M., Nelson, J., Shea, R., Freedman, M.J.: Blockstack: a global naming and storage system secured by blockchains. In: 2016 USENIX Annual Technical Conference (USENIX ATC 2016), pp. 181–194 (2016)
3. Ali, M., Shea, R., Nelson, J., Freedman, M.J.: Blockstack: a new internet for decentralized applications. Technical whitepaper version 1 (2017)
4. Allen, C.: The path to self-sovereign identity (2016). http://www.lifewithalacrity.com/2016/04/the-path-to-self-soverereign-identity.html
5. Blockstack: A decentralized storage architecture. https://docs.blockstack.org/storage/overview.html. Accessed 10 Dec 2019
6. Blockstack: Blockstack technical whitepaper v 2.0, May 2019. https://blockstack.org/whitepaper.pdf. Accessed 21 Oct 2019
7. Cadwalladr, C.: The great British Brexit robbery: how our democracy was hijacked. Guardian **7** (2017)
8. Cadwalladr, C., Graham-Harrison, E.: The Cambridge analytica files. Guardian **21**, 6–7 (2018)
9. Cameron, K.: The laws of identity. Microsoft Corp. **5**, 8–11 (2005)
10. Capilnean, T.: Evolving trust with applied game theory: recent white paper update describes trust creation through smart contracts (2018). https://www.civic.com/blog/evolving-trust-with-applied-game-theory-recent-white-paper-update-describes-trust-creation-through-smart-contracts/
11. Dunphy, P., Petitcolas, F.A.: A first look at identity management schemes on the blockchain. IEEE Secur. Priv. **16**(4), 20–29 (2018)
12. Decentralized identity foundation. https://identity.foundation. Accessed 10 Dec 2019

13. Linux Foundation: Hyperledger indy project. https://www.hyperledger.org/projects/hyperledger-indy. Accessed 10 Dec 2019
14. The SelfKey Foundation: Selfkey technical whitepaper, September 2017. https://selfkey.org/wp-content/uploads/2019/03/selfkey-whitepaper-en.pdf. Accessed 21 Oct 2019
15. lifeID: An open-source, blockchain-based platform for self-sovereign identity, September 2017. https://lifeid.io/whitepaper.pdf. Accessed 21 Oct 2019
16. Lundkvist, C., Heck, R., Torstensson, J., Mitton, Z., Sena, M.: uPort: a platform for self-sovereign identity (2017). https://whitepaper.uport.me/uPort_whitepaper_DRAFT20170221.pdf
17. Marlinspike, M.: Sovereign source authority (2012). http://www.moxytongue.com/2012/02/what-is-sovereign-source-authority.html. Accessed 23 May 2019
18. de Marneffe, P.: Vice laws and self-sovereignty. Crim. Law Philos. **7**(1), 29–41 (2013). https://doi.org/10.1007/s11572-012-9157-x
19. Peer Mountain: Working together for better self-sovereign identity: civic, selfkey, and peer mountain (2018). https://medium.com/peermountain/working-together-for-better-self-sovereign-identity-civic-selfkey-and-peer-mountain-282bca9a8e4a
20. MultiChain: Multichain Open source blockchain platform, October 2019. https://www.multichain.com. Accessed 21 Oct 2019
21. Nabi, A.G.: Comparative study on identity management methods using blockchain. Master's thesis, University of Zurich (2017)
22. Nakamoto, S.: Bitcoin: a peer-to-peer electronic cash system. Bitcoin.org (2017). https://bitcoin.org/bitcoin.pdf. Accessed 19 Nov 2019
23. Nakamoto, S., et al.: Bitcoin: a peer-to-peer electronic cash system. Working Paper (2008). https://bitcoin.org/bitcoin.pdf
24. OLYMPUS: Oblivious identity management for private and user-friendly services, October 2019. https://olympus-project.eu. Accessed 21 Oct 2019
25. Reed, D., Sporny, M., Longley, D., Allen, C., Grant, R., Sabadello, M.: Decentralized identifiers (DIDs). W3C Credentials Community Group (2017)
26. ShoCard: Identity for a mobile world (2019). https://shocard.com
27. Sovrin: Control your digital identity (2019). https://sovrin.org
28. Tobin, A., Reed, D.: The inevitable rise of self-sovereign identity. The Sovrin Foundation, vol. 29 (2016)
29. uPort: Open identity system for the decentralized web (2019). https://www.uport.me
30. Weik, M.H.: Computer Science and Communications Dictionary. Springer, Boston (2001). https://doi.org/10.1007/1-4020-0613-6_8580
31. Bitcoin Wiki: Civic (2019). https://en.bitcoinwiki.org/wiki/Civic

Privacy in Location-Based Services and Their Criticality Based on Usage Context

Tom Lorenz(✉) and Ina Schiering(✉)

Ostfalia University of Applied Sciences, Wolfenbüttel, Germany
{tom.lorenz1,i.schiering}@ostfalia.de

Abstract. Location based services are an important trend for smart city services, mobility and navigation services, fitness apps and augmented reality applications. Because of the growing significance of location-based services, location privacy is an important aspect. Typical use cases are identified and investigated based on user perceptions of usefulness and intrusiveness. In addition criticality of services is evaluated taking the typical technical realization into account. In the context of this analysis the implication of privacy patterns is investigated. An overall criticality rating based on applied location privacy patterns is proposed and thoroughly discussed, while taking the decrease of usability into consideration.

Keywords: Location-based services · Smart city · Location privacy · Tracking · Augmented reality · Privacy risks

1 Introduction

The consideration of location and movement information is used by various location based services and smart city services. It was investigated especially in the context of ubiquitous computing as described by Bellavista et al. [2]. There are applications as navigation, tracking of children and pets, location-based mobile gaming, dating and fitness apps. Recent examples are games as Pokémon Go [31], the upcoming augmented reality-game "Harry Potter Wizards Unite"[1] and location based dating apps [14]. In the context of smart cities location data is used by public institutions for infrastructure and commerce planning [35]. Applications using location based information are frequently described as *Location Tracking* resp. *Location Awareness*. Also the notions *Participatory Sensing* and *Pervasive Location Awareness* are used [11].

Privacy risks concerning the usage of location based data and the concept of location privacy were investigated by Beresford et al. [3]. There location privacy is defined as "the ability to prevent other parties from learning one's current

[1] https://www.harrypotterwizardsunite.com/.

Published by Springer Nature Switzerland AG 2020
M. Friedewald et al. (Eds.): Privacy and Identity 2019, IFIP AICT 576, pp. 462–478, 2020.
https://doi.org/10.1007/978-3-030-42504-3_29

or past location". Finn et al. [13] define privacy of location and space as that "individuals have the right to move about in public or semi-public space without being identified, tracked or monitored". Location data is specified in the General Data Protection Regulation (GDPR) Article 4(1) as one means of identification of natural persons. In Recital 75 the analysis or prediction of location or movement is considered as a privacy risk.

The aim of this paper is to investigate location privacy in specific usage contexts. To this aim use cases for typical location based services are described. Based on the analysis of privacy risks and privacy concerns in the context of location based services, a categorization for privacy risks is proposed considering criticality of use cases based on typical realizations and user perceptions of intrusiveness and usefulness. Afterwards the applicability of privacy patterns and the impact on the analysis is investigated.

2 Background

2.1 Indoor and Outdoor Localization

Smart devices offer various possibilities for generating location information. For outdoor localization and navigation the use of *Global Positioning System* (GPS) sensors is the approach mainly used. The position can be efficiently calculated based on trilateration (see Fig. 1(a)[2]) on a low-cost basis and necessary hardware is easily obtainable due to large-scale production.

(a) Trilateration achieved by satellites in the case of GPS (b) relevant feature lines used in Visual Landmark Detection [6]

Fig. 1. Localization methods

For indoor localization other approaches exist since satellite signals are hardly receivable due to modern building style. Even for low signal strengths, positioning is not reasonable since due to indoor environment even a couple of meters

[2] Source: https://gisgeography.com/trilateration-triangulation-gps/.

could mean a difference in offices or floors respecting vertical accuracy. Therefore some approaches rely on additional infrastructure in the environment. Beside *Near-Field Communication* (NFC) and beacons also approaches based on 3D object detection as used for *Augmented Reality* (AR) devices as the Microsoft HoloLens [16] or behavior based approaches as proposed in [37] are used. Table 1 summarizes localization approaches.

Positioning algorithms using beacons or NFC work mainly similar to GPS based algorithms via trilateration (see Fig. 1(a)). For NFC based algorithms user devices may also include NFC reading technology and can therefore calculate their own position based on incoming signals [25]. Alternatively, users may have a fixed NFC tag used for identifying themselves on specific NFC Readers (e.g. used in public transport which could potentially also be used in location based games). Beacons or NFC Tags that are able to cover the needed building or area are required.

Other approaches use a device carried by the user itself. These behavior based approaches use multiple sensors like accelerometer, gyroscope or magnetometer to notice positional changes and obtain a position by mapping these changes and building hypotheses concerning the actual position. Often a combination of sensors is integrated in an *inertial measurement unit* (IMU). Due to the increasing number of smartphones and depending on the built in sensors of the device itself, this method is very flexible and cheap since a new area do not have to be equipped with beacons beforehand. Fitness trackers calculate the covered distance by a derived step size based on the body height. That this step size might change based on hurry or company is not taken into consideration in these distance estimations. In general, the results of these approaches are still not very robust.

Table 1. Analysis of localization approaches - yes (✓), depends/medium (❍), no (✗), high (↑), low (↓)

	GPS	Beacons	NFC	Visual	IMU	Hololens
Indoor localization	✗	✓	❍	❍	❍	✓
Outdoor localization	✓	❍	✗	❍	❍	❍
HW costs	↓	↑	❍	↓	↓	↑
SW complexity	↓	↓	↓	↑	❍	↑
Cloud computation needed	✗	✗	✗	✓	❍	❍
Computation on device sufficient	✓	✓	✓	✗	❍	✗

In addition to these sensor based approaches other localization methods use visual input for localization. One approach is to compute a positional change or motion based on the exact alteration of distinctive points in a video stream. This approach is called visual odometry [27,28]. Based on the use case visual odometry can come in quite handy since it is also viable with a single camera and can be calculated on the device if the video stream quality is reduced. It is also used by the NASA since it only relies on camera input [33].

Another viable approach are visual landmarks. Photos or outlines of distinctive environmental objects, e.g. buildings, sights or places can be used for a rough localization by comparing the current photo to a subset of existing similar photos of given landmarks (see Fig. 1(b)). Due to the high number of comparisons and the corresponding high amount of computational power needed, this technique is mostly realized via cloud services [6,30]. Although perspective transformations are possible, a big dataset of distinctive objects is needed. These transformations allow for an image correction of small angles but information of buildings backsides or other points of views have to be covered in the data set as well. A 3D Model of the designated building might be viable as well. Mapping the current 3D area to a previously scanned three dimensional space is possible as well [18], but as soon as the environment exceeds an office, additional enhancements are needed. Further the HoloLens uses a combination of sensors, e.g. distance sensors and 3D mapping to distinguish its position. To speed up certain processes these areas are recognized and distinguished by a WiFi signal [20].

2.2 Privacy Patterns for Localization Information

Since specific locations e.g. homes are directly bound to a person, anonymized location data can often be successfully deanonymized. By composing several distinct data sets, also called linking, tracking of individuals is possible therefore resulting in deanonymization of the given location data as described in [5,8,24]. To reduce the risk of deanonymization, countermeasures in the form of privacy enhancing techniques (privacy patterns) are investigated in the context of the use cases described in Sect. 5.1. In the following privacy patterns for localization information are explained and summarized:

- **Discretized Points on Grid:** is a pattern that aligns current GPS positions on a predetermined grid. These calculations are done locally on the device before sending locational information to the service provider. Hence privacy is increased by decreasing accuracy [22].
- **Delays:** are introduced to add noise to the point in time of a position. Both small (couple of seconds) and big delays (couple of hours) are considered. While adding always a fixed amount is insufficient, a random amount of time is added to the original time stamp of the positional info or waited on the device before sending a request to the service provider. A higher limit allows for bigger time gaps and therefore increased privacy in contrast to small delays which, on the other hand side might not influence the usability as much.
- **Fixed Time Slots:** are similar to the described delays, but instead of adding a delay to the point in time of a given position, the positioning itself is only done every x minutes or hours. Mostly fixed time slots are more user-friendly since even if the waiting time of a delay might be shown to the user before each transmission, it is still randomized. Using a fixed time slot allows the user or service provider some kind of predictability when to expect the next update.

- **Dummy Users:** can be applied by adding Gaussian noise to the positional data. Therefore instead of sending just one position to the server, multiple positions are sent. From the information received by the service the correct result can be filtered based on the current position of the user [22]. This method leads to very accurate results for the user itself, but also results in higher data transfer and more processing on the device and also for the service provider, since many of these requested information are not used after receiving them on the device. Also this approach might not be suitable for use cases, where exact positions of the users might needed to provide information for a whole community. The process of generating realistic noise is not that trivial in certain cases as well.

Mix Zones are also a common method for achieving location privacy. While k-anonymity cannot be achieved by a single device alone, it furthermore relies on a user base being in the same area with permanent changing pseudonyms to ensure ongoing privacy [4]. Furthermore linkability is feasible if the distance between mix zones varies and users need different a amount of times to reach them. Anonymity or at least pseudonymity can be achieved easily by adding a specific layer between user and service provider [29]. But these approaches hold a similar problem. If not enough user are in a close area, a person is easily traceable, even if multiple, anonymous GPS Positions are lining up for a potential route. Furthermore these approaches do not take into consideration, that leaving a mix zones on a specific place (e.g. office or home) are strong indicators to a specific individual and leads to personalized information. Since we can not improve these method on the device itself and traceability is still possible, mix zones are not considered in the list of privacy patterns investigated here.

3 Related Work

Concerning location privacy, attacks and adversary models, an overview is given in [36]. In the context of privacy and provenance the risk of location information is investigated as one aspect [32]. Also in [7,35] location information is taken into account for assessing privacy risks.

User perceptions about privacy risks including risks concerning location privacy are the focus of a user study in [12]. The willingness of users to share location information depending on the context are investigated in [11,14,17,21]. Disclosure in dating apps is discussed in detail in [14]. [21] is describing the intrusiveness and consequences of location based ads. [17] introduces a mobile recommendation system for tourism considering privacy. [11] shows how users tend to think about their own location data to create a transparent bus delaying information service.

User ratings regarding location based service usefulness and mostly intrusiveness can be found in an early use case study [1]. Additional ethical questions about tracking mobile phone users were portrayed [34].

There are several technical approaches which address privacy enhancing technologies in the area of location information as e.g. mix zones [4] or the use of pseudolocations to achieve k-anonymity [29]. Also there are approaches measuring the effectiveness of measurements to remove location information and the risk of reconstruction of this information [19].

Privacy enhancing patterns to improve location privacy are discussed in detail in [22]. Furthermore added fake requests to create dummy users for a different car route are explained [26]. Additionally the trade-off between location obfuscation and quality of service is described in [15].

4 Methodology

Based on the identified approaches in Sect. 3, a classification of utilized sensors resp. localization approaches, usage contexts and privacy perceptions concerning location privacy is developed. The most common approaches distinguish applications that persistently request and/or send location information of users, and those that allow users to send their location only upon request. On mobile phones several applications running in the background are steadily aware of users positions. We focus on application contexts and consider both types of applications in the classification. On request location aware services make use of certain privacy patterns as e.g. fixed time slots already obsolete.

For the consideration of typical contexts, several use cases are investigated in the following analysis. For each use case a standard solution is the basis of the investigation. There typical intervals of position requests and accuracy of the localization information are stated.

As a baseline of the analysis, the use cases are rated according to usefulness and intrusiveness. Usefulness and intrusiveness are criteria that are often investigated in the context of surveys about user acceptance of location tracking [1,11,12,21]. Usefulness or being 'useful' is described as 'can be used to advantage; serviceable; helpful; beneficial; often, having practical utility' by [10]. In the context of user studies usefulness was rated by the user and sometimes combined with the average number of uses per day of the given application. The notion 'intrusive' is defined as 'something that invades personal space, that becomes too involved or that comes too close without being invited' by [9]. Also concerning intrusiveness user perceptions are considered.

Whereas in most cases these user perceptions are mostly subjective, sometimes also general aspects as e.g. detecting the delay of buses [11] are taken into consideration. The rating of use cases is derived from these surveys. In addition the criterion criticality is considered. The original criticality is based on the analysis of use cases from a mainly technical perspective. There factors as times of localization and accuracy are taken into account to derive a criticality rating.

In a second step of the analysis, the impact of privacy-enhancing patterns on the evaluation of use cases is considered. To this aim, increase of privacy and also decrease of usability are investigated and the criticality in the context is derived.

5 Analysis

5.1 Use Cases of Location-Based Services

The following use cases give a broad overview of location-based services in daily life.

Social Events in City: Information about events in a certain area, e.g. a city or a region, is important for tourists and inhabitants. People organizing such events or organizations as city marketing or ticket sales organizations are interested in gaining attention to attract participants or increase ticket sales. Many implementations use exact locations and refresh rates are not transparent or even adjustable for users. A forced refresh of the information in the app by the user triggers an obvious update of the position, but in addition frequent updates in the background are triggered which is not transparent for the user.

ATM or Store Close by (On Request): This describes a use case, where users can get position information of nearby stores or points of interest specified according to their current position. In this particular case users send their position and receive results on demand and are therefore not tracked most of the time.

ATM or Store Close by (On Push): Instead of informing users only on request, in this use case they are informed periodically. Users could either be notified if they enter a certain area and the system knows and recognizes the device. Otherwise devices could send requests based on the current location all the time.

Close Loop Navigation (Turn by Turn): Besides classical offline navigation systems, where maps are updated via file transfer and accidents are transmitted via very high frequency (same as vhf radio stations), meanwhile systems with a permanent internet connection are often used, which collect position information of users for navigation and to detect traffic jams. Therefore positional information is updated permanently and as precise as possible.

Nearby Friends: By this type of application information is provided, if a particular person of the friends list of the user is in a certain distance to the user. Positional updates are usually done every couple of seconds. Functions like these have become available in social networks, messengers and numerous dating apps, even more specific in the following use case. Sometimes there is information provided, how to decrease the localization accuracy from high to medium on mobile phones, but mainly due to less battery usage than to increase privacy.

Locate People on a Map: In addition to the information that friends are nearby, maps are provided with position information of these friends. That does not only provide the distance of the persons to a user, but the specific location of them. Regarding privacy most of these applications provide possibilities to restrict the visibility of position and e.g. images. Irrespective of these visibility

Fig. 2. Snapchats introduced Snap Map

settings, providers store all of the positional information. Sharing location information happens in real time and is updated frequently with a high accuracy (see Fig. 2[3]).

Notification of Traffic Jam: In addition to turn by turn navigation the applications are notifying users about speed cameras ahead on the route, density of traffic, information about accidents and construction sites. This information is also taken into account in addition to the position of users to enhance route planning and the estimated time of arrival.

Location Based Ads: Location based advertisement is a very import part of targeted advertisement. Whereas targeted advertisement might happen while browsing through the web, it is also used in big shopping centers to track customer locations [23]. Beside tracking when customers enter or leave the shopping center, also their routes through the building might be tracked. User profiles could be derived by common fingerprinting approaches.

Location Based Games: Location based games are a recent trend. The general aim is to reach a specific position, earn points and collect achievements. Well-known examples are games as "Pokémon GO" (see Fig. 3(b)[4]) or "Harry Potter Wizards Unite" (see Fig. 3(a)[5]). Games are based on collectibles that are randomly placed on the map. These items are necessary for a good gaming experience and a virtual progress. They can be found while walking around the environment as the users characters in the game and adapt their positioning based on your real position and movement. Therefore positional information about the user is collected quite frequently and with high accuracy. The usefulness of that type of applications is quite subjective since it is a game and merely fun is the main purpose.

---

[3] Source: https://www.turn-on.de/tech/ratgeber/so-nutzt-du-die-neue-snap-map-in-snapchat-eine-anleitung-280917.

[4] Source: https://nianticlabs.com/de/support/pokemongo/.

[5] Source: https://www.imore.com/harry-potter-wizards-unite-what-are-inns-and-how-do-they-work.

Location During Emergency Call: Transmitting the location during an emergency is a special use case taking location information into consideration. Emergency call (eCalls) is a technology transmitting important vehicular data e.g. airbag deployment, impact sensors, but also location data to the authorities in case of an emergency. This technique is mandatory for cars that are sold in the European Union since April 2018. Transmission of data happens only once in case of an accident and is highly accurate.

(a) Harry Potter Wizards Unite (b) Pokémon GO

Fig. 3. Location based games

5.2 Analysis of Typical Realizations of Use Cases

The evaluated use cases represent a broad variety and controversial necessity of stated applications. While differentiation between the type of localization were not made especially in these use cases, the types of localization and the place (e.g. indoor or outdoor) would lead to no difference since accuracy and the times of localization would be similarly realized in both settings.

The localization method and therefore the place of calculation implies additional threats besides location leakage. Based on [12] location leakage (and therefore probably other data e.g. from an IMU) is less crucial than image leakage which could happen if visual localization is obtained by vulnerable cloud services. In Table 2 these ratings are summarized.

As a first step of the analysis, the use cases presented in Sect. 5.1 are evaluated based on surveys about user acceptance of location tracking [1,11,12,21] and typical technical realizations.

5.3 Privacy Patterns in Location Based Services

In the second step of the analysis the impact of applying privacy patterns for localization information is investigated. Therefore changes for intervals of localization and accuracy are considered and the implication on privacy is evaluated. In addition it is also important to consider changes to the usability resp. usefulness of the services, since these approaches minimize or obfuscate data to increase privacy.

Table 2. Use cases and their usefulness (from +1 to +4), intrusiveness (from -1 to -4) and criticality (from -1 to -4).

Criteria	Use cases				
	Social events in city	ATM close by (request)	ATM close by (push)	Close loop navigation	Nearby friends
Usefulness	+1	+3	+2	+3	+2
Accuracy	10 m	10 m	10 m	10 m	10 m
Times of loc.	30 s	Once	10 s	1 s	5 s
Intrusiveness	-3	-2	-3	-2	-3
Criticality	-4	-2	-4	-3	-4
Criteria	Use cases				
	Locate people on map	Notification traffic jam	Location based ads	Location based games	Location during emergency
Usefulness	+2	+3	+1	+2	+4
Accuracy	10 m	10 m	10 m	10 m	10 m
Times of loc.	5 s	1 s	5 s	1 s	Once
Intrusiveness	-4	-2	-4	-3	-1
Criticality	-4	-3	-4	-4	-1

For each use case and privacy pattern, the resulting criticality of the use case is rated as follows based on the other criteria for the use case:

$$Criticality = UF + I + Crit_{orig} + IoP + DoU \tag{1}$$

The criticality is described as the sum of the previous weighted elements as seen in (1). There UF denotes UseFulness, I is Intrusiveness, IoP describes the Increase of Privacy, DoU the Decrease of Usability. All elements were weighted corresponding to the weights stated in Table 3. The evaluation of all parameters and the overall criticality is described below and summarized in Table 4.

The weights are chosen based on the corresponding considerations. To minimize personal preferences of users the initial usefulness is not considered by

Table 3. Applied weights to the elements in Eq. 1

Element	Weight	Normalized
Usefulness	0	0
Intrusiveness	1	0.14
Original criticality	3	0.43
Increase of privacy	1	0.14
Decrease of usability	2	0.29

choosing the weight for usefulness as 0, since the application is already used and hence considered as useful. Since the overall intrusiveness of the application itself still has an impact even if privacy patterns are applied, the initial intrusiveness has the weight of 1. It was tried to enhance given intrusiveness, by a technical view e.g. location aware or location tracking, on demand or continuing location data resulting in the original criticality of the use case. This original criticality is one of the most important factors resulting in a weighting of 3. We chose the Increase of Privacy weight as 1 since it is important that privacy patterns lead to an increase of privacy. Additionally we decided to choose a higher weight of 2 for Decrease of Usability taking the users point of view into consideration. Since privacy patterns will not be applied, resp. users will not consider using an application that is barely usable anymore.

Location based services for identifying **social events in a city** is a good example for improvements concerning privacy. Standard localization (e.g. via GPS) is pretty accurate (around 10 m) and might be done every 30 s (depending on settings in the app or chosen by the developer). Decreasing the accuracy to around 20 km could be coped by providing more information resulting in more events. This results in more flexibility for users but also in the need of filtering the results more thoroughly manually and thus decreasing the usability. Overall this is quite a viable strategy to increase privacy.

Adding a slight variable delay to location data in a non time-critical application has no substantial impact and hence leads to mainly no decrease in usability but also only to a slight increase in privacy. A larger delay would improve privacy farther but also lead to a further decrease of usability. For applications based on information that is merely static, fixed time slots, e.g. every 3 h or even only once a day, seems to be a viable option. Especially events which need a thorough preparation and sometimes additional approvals are typically known far longer than a day before. Therefore larger time slots would be manageable. Short term updates of certain events could be achieved by subscribing to events which users intend to visit. Even if this might lead to different private issues, really tight localization timers or position request become obsolete. Furthermore planning for a trip might be coped with a search function for cities or areas. Moreover adding noise to the location data, like requesting data for multiple cities across the country, might be a good approach to increase privacy as well, since the service is not directly disrupted by requests based on different locations.

The drawbacks are that more data has to be filtered on the device and also the increased work load for sending these data and handling of requests on the server side. Also dummy users could be inserted. From the service providers point of view the load only increases if the number of users is substantially increased by dummy users.

Regarding **ATMs or stores close by (on request)** one can see that the original criticality in contrast to the **(on push)** application is lower. This is due to the fact that updates and positional information are only sent upon manual request of users, which leads to a high grade of control. On the other hand frequently requesting data based on users movements might be more useful for some but results in a higher criticality.

In both cases delaying and decreasing the times of localization helps in general but results only in a small increase of privacy. Using big time gaps increases privacy further, but results also in a substantial decrease of usability. While aligning the positional information to an underlying grid might be a good idea, localization frequency would be still high. Further adding dummy users has the same effect as in the previously described use case.

Table 4. Analysis of criticality, privacy and usability - increase of privacy from +1 (worst) to +4 (best), decrease of usability -1 (minimal) to -4 (disrupted) and criticality from -1 (minimal) to -4 (bad) - or not applicable (x).

Criteria	Use case						
	Normal	Points on grid	Delay		Time slots - every		Dummy users
			slight (x s)	big (x h)	(x min)	(x h)	
Social event in city							
Times of localization	30 s	30 s	+10 s	+1 h	180 min	24 h	30 s
Accuracy	10 m	20 km	10 m	10 m	10 m	10 m	10 m
Increase of privacy	None	+3	+1	+2	+2	+3	+4
Decrease of usability	None	-1	-1	-2	-2	-2	-1
Criticality	-4	-2	-2	-3	-3	-2	-2
ATM or store close by (on request)							
Times of localization	Once	Once	+10 s	+1 h	Once	Once	Once
Accuracy	5 m	5 km	5 m	5 m	5 m	5 m	5 m
Increase of privacy	None	+2	+1	+3	x	x	+4
Decrease of usability	None	-3	-2	-4	x	x	-2
Criticality	-2	-2	-2	-2	x	x	-1
ATM or store close by (on push)							
Times of localization	10 s	10 s	+10 s	+1 h	5–10 min	1 h	10 s
Accuracy	5 m	5 km	5 m	5 m	5 m	5 m	5 m
Increase of privacy	None	+2	+1	+3	+2	+3	+4
Decrease of usability	None	-3	-1	-3	-2	-3	-2
Criticality	-4	-3	-2	-3	-2	-3	-2

(*continued*)

Table 4. (*continued*)

Criteria	Normal	Points on grid	Delay: slight (x s)	Delay: big (x h)	Time slots - every (x min)	Time slots - every (x h)	Dummy users
Close loop navigation (turn by turn)							
Times of localization	1 s	1 s	+10 s	+1 h	1 min	1 h	5 s
Accuracy	5 m	250 m	5 m	5 m	5 m	5 m	5 m
Increase of privacy	None	+2	+1	+2	+2	+3	+3
Decrease of usability	None	-3	-3	-4	-3	-4	-1
Criticality	-3	-2	-2	-3	-2	-2	-1
Nearby friends							
Times of localization	5 s	5 s	+10 s	+1 h	5–10 min	1 h	5 s
Accuracy	100 m	10 km	100 m	100 m	100 m	100 m	100 m
Increase of privacy	None	+3	+1	+2	+2	+3	+3
Decrease of usability	None	-2	-1	-4	-2	-4	-3
Criticality	-4	-2	-2	-3	-2	-3	-3
Locate people on a map							
Times of localization	5 s	5 s	+10 s	+1 h	5–10 min	1 h	5 s
Accuracy	100 m	10 km	100 m	100 m	100 m	100 m	100 m
Increase of privacy	None	+3	+1	+2	+2	+3	+4
Decrease of usability	None	-3	-1	-3	-2	-3	-4
Criticality	-4	-2	-2	-3	-3	-3	-3
Notification of traffic jam							
Times of localization	5 s	5 s	+10 s	+1 h	5 min	1 h	5 s
Accuracy	100 m	10 km	100 m	100 m	100 m	100 m	100 m
Increase of privacy	None	+3	+1	+3	+3	+4	+4
Decrease of usability	None	-2	-1	-3	-2	-3	-4
Criticality	-3	-2	-2	-2	-2	-2	-2
Location based ads							
Times of localization	5 s	5 s	+10 s	+1 h	5–10 min	1 h	5 s
Accuracy	5 m	5 km	5 m	5 m	5 m	5 m	5 m
Increase of privacy	None	+2	+1	+3	+3	+3	+4
Decrease of usability	None	-3	-1	-3	-2	-3	-1
Criticality	-4	-3	-3	-3	-2	-3	-1
Location based games							
Times of localization	1 s	1 s	+10 s	+1 h	5 min	1 h	1 s
Accuracy	10 m	1 km	10 m	10 m	10 m	10 m	10 m
Increase of privacy	None	+2	+2	+2	+3	+4	+4
Decrease of usability	None	-3	-2	-3	-2	-3	-4
Criticality	-4	-3	-2	-3	-2	-2	-3
Location during emergency call							
Times of localization	Once	Once	+10 s	+1 h	Once	Once	Once
Accuracy	1 m	50m	1 m	1 m	1 m	1 m	1 m
Increase of privacy	None	+2	+1	+1	x	x	+4
Decrease of usability	None	-4	-3	-4	x	x	-4
Criticality	-1	-2	-3	-3	x	x	-2

A reasonable approach could be to combine lower times of localization with a lower accurate positioning and in addition adding dummy users if possible. Also since positions do not change regularly, information might be saved locally on the device beforehand. An update of these stored positions could still be triggered manually beforehand and distance calculations etc., could be done locally.

Turn by Turn navigation relies on close loop positional data to guide users around the streets. While probably not all positional information is directly sent to the service provider during navigation, some information would often be collected and sent to the service provider at least afterwards to improve quality of service. Aligning the estimated positions on a grid could make navigation much more difficult. The current position might jump between streets, or blocks, if the granularity of the grid is not adequately chosen. Therefore this is problematic regarding usability since the service might e.g. confuse intersections. Also the current direction and velocity would be impossible to calculate due to inaccurate positions if no additional sensor data is available.

Adding delays complicates navigation for the user as well since the whole process starts lagging. Fixed time slots for orientation might work, but the user would have to remember turns in between updates which makes it more complicated to navigate in cities. Adding Dummy Users depends on how the additional users are aligned. If there is just a random cloud of additional data around a certain position all the time, the original route is still traceable. Furthermore realistic routes have to be simulated.

Nearby Friends, **locate people on a map** and the **notification of traffic jams** have similar problems. Decreasing the accuracy of positional information substantially will render the service nearly unusable. Most of these applications update their positional information too often (e.g. people on a map are usually updated every 2 s). These update cycles could be lowered and still offer full functionality. Adding dummy users could be a possibility, but incorporates the risk of distorting these services.

Location based games might lead to another problem. While delaying is annoying in some games, adding dummy users could lead to a disruptive or even not playable game anymore since some of the games measure the distance and time between positions. Some games restrict, penalize or punish for fast movement. Sending multiple positions might be colluding with the game itself, since most actions of the user have to be confirmed by the server, before a certain progress in the game is possible. Furthermore it could be considered as a way of cheating. Adding a big delay might decrease the usability substantially. Hence decreasing the accuracy might be a more viable option.

The last use case investigated is the **emergency call**. Decreasing accuracy as well as only sending positions in fixed, predetermined time slots, delaying the data transmission or adding additional dummy positions might lead to delays or confusion for emergency forces. Applying privacy patterns in that use case might be questionable for such a crucial task due to a low original criticality rating.

6 Discussion and Conclusion

Location-based services are already used ubiquitously and constitute an important innovation in the area of smart services. Especially the success of smartphones and the development of AR devices as the Microsoft Hololens foster this trend. Therefore it is important to develop frameworks for location privacy and to reduce the privacy risk based on context and user perceptions.

Location requests triggered by users ensure the lowest intrusiveness, since the user knows about sending current location information and can control these types of requests. Certainly retrieving data on request is preferable, persistent requests could be pruned. For example fixed time slots for providing the same kind of information can be useful. Neglecting computational power adding noise and therefore additional requests to services is a quite reliable way as long as the service is still usable. Otherwise decreasing accuracy should be kept in mind as well since applications often update positions every couple of seconds close enough to tenth of meters, even if it is not necessary.

In general it is preferable to compute positions and routes on the local device instead of employing cloud services to minimize data. But as summarized in Table 1, a variety of localization methods are not suitable to be calculated on a smart phone only. But even if cloud computation is needed, data minimization should be a top priority. Positional data minimization might be accomplished as well, if only edges of a building are send at a certain location, triggered by the user.

The development and investigation of privacy patterns for location privacy is an important topic for future research. A central problem is that often the usefulness decreases substantially such that some services are nearly rendered unusable. Based on this general analysis, it would be important to investigate privacy perceptions of users in different usage contexts in more detail.

Acknowledgment. This work was supported by the Federal Ministry of Education and Research (BMBF) as part of SmarteInklusion (01PE18011C).

References

1. Barkhuus, L., Dey, A.K.: Location-based services for mobile telephony: a study of users' privacy concerns. In: INTERACT 2003, vol. 3, pp. 702–712. Citeseer (2003)
2. Bellavista, P., Küpper, A., Helal, S.: Location-based services: back to the future. IEEE Pervasive Comput. **7**(2), 85–89 (2008)
3. Beresford, A.R., Stajano, F.: Location privacy in pervasive computing. IEEE Pervasive Comput. **2**(1), 46–55 (2003)
4. Beresford, A.R., Stajano, F.: Mix zones: user privacy in location-aware services. In: Proceedings of the Second IEEE Annual Conference on Pervasive Computing and Communications Workshops, pp. 127–131. IEEE (2004)
5. Boutet, A., Mokhtar, S.B., Primault, V.: Uniqueness assessment of human mobility on multi-sensor datasets (2016)
6. Cipolla, R., Robertson, D., Tordoff, B.: Image-based localization. In: Proceedings International Conference Virtual Systems and Multimedia, vol. 2004 (2004)

7. De, S.J., Le Métayer, D.: PRIAM: a privacy risk analysis methodology. In: Livraga, G., Torra, V., Aldini, A., Martinelli, F., Suri, N. (eds.) DPM/QASA -2016. LNCS, vol. 9963, pp. 221–229. Springer, Cham (2016). https://doi.org/10.1007/978-3-319-47072-6_15
8. De Montjoye, Y.A., Hidalgo, C.A., Verleysen, M., Blondel, V.D.: Unique in the crowd: the privacy bounds of human mobility. Sci. Rep. **3**, 1376 (2013)
9. YourDictionary: Definition of "intrusive" (2015). https://www.yourdictionary.com/intrusive
10. YourDictionary: Definition of "useful" (2015). https://www.yourdictionary.com/useful
11. Dou, E., Eklund, P.W., Gretzel, U.: Location privacy acceptance: attitudes to transport-based location-aware mobile applications on university campus (2016)
12. Felt, A.P., Egelman, S., Wagner, D.: I've got 99 problems, but vibration ain't one: a survey of smartphone users' concerns. In: Proceedings of the Second ACM Workshop on Security and Privacy in Smartphones and Mobile Devices, pp. 33–44. ACM (2012)
13. Finn, R.L., Wright, D., Friedewald, M.: Seven types of privacy. In: Gutwirth, S., Leenes, R., de Hert, P., Poullet, Y. (eds.) European Data Protection: Coming of Age, pp. 3–32. Springer, Dordrecht (2013). https://doi.org/10.1007/978-94-007-5170-5_1
14. Fitzpatrick, C., Birnholtz, J., Brubaker, J.R.: Social and personal disclosure in a location-based real time dating app. In: 2015 48th Hawaii International Conference on System Sciences, pp. 1983–1992. IEEE (2015). http://ieeexplore.ieee.org/document/7070049/
15. Gardner, Z., Leibovici, D., Basiri, A., Foody, G.: Trading-off location accuracy and service quality: privacy concerns and user profiles. In: 2017 International Conference on Localization and GNSS (ICL-GNSS), pp. 1–5. IEEE (2017)
16. Garon, M., Boulet, P.O., Doironz, J.P., Beaulieu, L., Lalonde, J.F.: Real-time high resolution 3d data on the HoloLens. In: 2016 IEEE International Symposium on Mixed and Augmented Reality (ISMAR-Adjunct), pp. 189–191. IEEE (2016)
17. Gavalas, D., Konstantopoulos, C., Mastakas, K., Pantziou, G.: Mobile recommender systems in tourism. J. Netw. Comput. Appl. **39**, 319–333 (2014)
18. Hito, G.: Overlaying virtual scale models on real environments without the use of peripherals (2018)
19. Hossain, A., Quattrone, A., Tanin, E., Kulik, L.: On the effectiveness of removing location information from trajectory data for preserving location privacy. In: Proceedings of the 9th ACM SIGSPATIAL International Workshop on Computational Transportation Science, pp. 49–54. ACM (2016)
20. Hübner, P., Weinmann, M., Wursthorn, S.: Marker-based localization of the Microsoft HoloLens in building models. Int. Arch. Photogram. Remote Sens. Spat. Inf. Sci. **42**(1), 195–202 (2018)
21. Ketelaar, P.E., et al.: "Opening" location-based mobile ads: how openness and location congruency of location-based ads weaken negative effects of intrusiveness on brand choice. J. Bus. Res. **91**, 277–285 (2018)
22. Krumm, J.: A survey of computational location privacy. Pers. Ubiquit. Comput. **13**(6), 391–399 (2009)
23. Li, K., Du, T.C.: Building a targeted mobile advertising system for location-based services. Decis. Support Syst. **54**(1), 1–8 (2012)
24. Ma, C.Y., Yau, D.K., Yip, N.K., Rao, N.S.: Privacy vulnerability of published anonymous mobility traces. IEEE/ACM Trans. Netw. (TON) **21**(3), 720–733 (2013)

25. Meng, P., Fehre, K., Rappelsberger, A., Adlassnig, K.P.: Framework for near-field-communication-based geo-localization and personalization for android-based smartphones - application in hospital environments. eHealth **198**, 9–16 (2014)
26. Meyerowitz, J., Roy Choudhury, R.: Hiding stars with fireworks: location privacy through camouflage. In: Proceedings of the 15th Annual International Conference on Mobile Computing and Networking, pp. 345–356. ACM (2009)
27. Nister, D., Naroditsky, O., Bergen, J.: Visual odometry. In: Proceedings of the 2004 IEEE Computer Society Conference on Computer Vision and Pattern Recognition, CVPR 2004, vol. 1, pp. 652–659. IEEE (2004). http://ieeexplore.ieee.org/document/1315094/
28. Parra, I., Sotelo, M.A., Llorca, D.F., Ocaña, M.: Robust visual odometry for vehicle localization in urban environments. Robotica **28**(3), 441–452 (2010)
29. Peng, T., Liu, Q., Wang, G.: Enhanced location privacy preserving scheme in location-based services. IEEE Syst. J. **11**(1), 219–230 (2014)
30. Qu, X.: Landmark based localization: detection and update of landmarks with uncertainty analysis, p. 191, October 2016
31. Rauschnabel, P.A., Rossmann, A., tom Dieck, M.C.: An adoption framework for mobile augmented reality games: the case of Pokémon Go. Comput. Hum. Behav. **76**, 276–286 (2017)
32. Reuben, J., Martucci, L.A., Fischer-Hübner, S., Packer, H.S., Hedbom, H., Moreau, L.: Privacy impact assessment template for provenance. In: 2016 11th International Conference on Availability, Reliability and Security (ARES), pp. 653–660, August 2016
33. Swank, A.J.: Localization Using Visual Odometry and a Single Downward-Pointing Camera. NASA Glenn Research Center, Cleveland (2012)
34. Taylor, L.: No place to hide? The ethics and analytics of tracking mobility using mobile phone data. Env. Plan. D Soc. Space **34**(2), 319–336 (2016)
35. Van Zoonen, L.: Privacy concerns in smart cities. Gov. Inf. Q. **33**(3), 472–480 (2016)
36. Xue, M., Liu, Y., Ross, K.W., Qian, H.: I know where you are: thwarting privacy protection in location-based social discovery services. In: 2015 IEEE Conference on Computer Communications Workshops (INFOCOM WKSHPS), pp. 179–184. IEEE (2015)
37. Zhou, B., Li, Q., Mao, Q., Tu, W., Zhang, X.: Activity sequence-based indoor pedestrian localization using smartphones. IEEE Trans. Hum. Mach. Syst. **45**(5), 562–574 (2014)

Author Index

Printed in the United States
by Baker & Taylor Publisher Services

Printed in the United States
by Baker & Taylor Publisher Services